Chemical and Process Thermodynamics

Third Edition

PRENTICE HALL INTERNATIONAL SERIES
IN THE PHYSICAL AND CHEMICAL ENGINEERING SCIENCES

NEAL R. AMUNDSON, SERIES EDITOR, *University of Houston*

ADVISORY EDITORS
ANDREAS ACRIVOS, *Stanford University*
JOHN DAHLER, *University of Minnesota*
THOMAS J. HANRATTY, *University of Illinois*
JOHN M. PRAUSNITZ, *University of California*
L. E. SCRIVEN, *University of Minnesota*

Chemical and Process Thermodynamics

Third Edition

B. G. Kyle

Emeritus Professor of Chemical Engineering
Kansas State University

Prentice Hall PTR
Upper Saddle River, New Jersey 07458

Library of Congress Cataloging-in-Publication Data

Kyle, B. G. (Benjamin Gayle)
    Chemical and Process Thermodynamics / B. G. Kyle. — 3rd ed.
      p. cm. — (Prentice Hall international series in the physical
  and chemical engineering sciences)
    Includes index.
    ISBN 0-13-087411-6
    1. Thermodynamics. I. Title. II. Series.
  QD504.K94   1999
  660.2'969—dc20   91–30970

Acquisitions editor: Bernard Goodwin
Editorial/production supervisor: Mary Sudul
Cover design director: Jerry Votta
Cover design: Anthony Gemmellaro
Manufacturing manager: Alan Fischer

© 1999, 1992, 1984 by Prentice Hall PTR
Prentice-Hall, Inc.
A Simon & Schuster Company
Upper Saddle River, New Jersey 07458

Prentice Hall books are widely used by corporations and government agencies for training, marketing, and resale. The publisher offers discounts on this book when ordered in bulk quantities. For more information, contact Corporate Sales Department, Phone: 800-382-3419; FAX: 201-236-7141; E-mail: corpsales@prenhall.com
Prentice Hall PTR, One Lake Street, Upper Saddle River, NJ 07458.

Printed in the United States of America
10 9 8 7 6 5 4 3 2

ISBN 0-13-087411-6

Prentice-Hall International (UK) Limited, London
Prentice-Hall of Australia Pty. Limited, Sydney
Prentice-Hall Canada, Inc., Toronto
Prentice-Hall Hispanoamericana, S.A., Mexico
Prentice-Hall of India Private Limited, New Delhi
Prentice-Hall of Japan, Inc., Tokyo
Simon & Schuster Asia Pte. Ltd., Singapore
Editora Prentice-Hall do Brasil, Ltda., Rio de Janeiro

To Trisha
even now and more than ever

CONTENTS

5 The Thermodynamic Network

6 Heat Effects

7 Equilibrium and Stability

Appendix C   **734**

Appendix D   **742**

Appendix E   **746**

Appendix F   **751**

Index   **753**

This edition represents an attempt to produce a thermodynamics text suitable for the age of the personal computer. Steps toward this goal include providing a variety of computing resources on the companion CD-ROM, rewriting several sections, supplying many worked examples and student problems requiring the use of these resources, and adding Chapter 17 dealing with modeling of thermodynamic systems. The CD-ROM contains (a) spreadsheets and executable programs for specific applications, (b) POLYMATH, a general purpose numerical analysis program, (c) EQUATIONS OF STATE, a graphics program with tutorials illustrating various concepts with three-dimensional *PVT* plots, and (d) spreadsheet solutions to several worked examples in the text. A complete listing and brief description of these resources can be found in Appendix E.

Chapter 9, Principles of Phase Equilibrium, has been reorganized to include calculation of phase equilibrium via an equation of state as this topic is now an integral part of the thermodynamics of phase equilibrium. This change carries over into Chapter 10, Applied Phase Equilibrium, where the determination of the Peng-Robinson interaction parameter from VLE data parallels the determination of Wilson parameters from VLE data and both processes are viewed from the perspective of information processing. Executable programs are provided for these tasks.

Chapter 11, Additional Topics in Phase Equilibrium, has been augmented by the inclusion of executable programs for the determination the Peng-Robinson interaction parameter from solid-gas equilibrium data and for the calculation of solid-gas equilibrium from the Peng-Robinson equation. There is also an executable program using the UNIFAC method for estimating activity coefficients.

Chapter 13, Complex Chemical Equilibrium, has been rewritten in an attempt to clarify the phase rule as applied to chemically reacting systems and to accommodate the use of POLYMATH for solving the non-linear equations arising from free energy minimization.

Chapter 17, Thermodynamics and Models, presents the philosophy of modeling engineering systems and illustrates this with several realistic examples. Also, the technique of combining the UNIFAC model with the Peng-Robinson model for the prediction of phase equilibria is presented and illustrated with an example.

While there is an increased emphasis on computing, the major objective of this edition, as well as that of the two previous editions, is to expose and explain the rationale of thermodynamics. My philosophical stance toward computing in the curriculum is that it should never be an end in itself but should free students of burdensome and tedious calculations so that they can use this savings in time and attention to focus on fundamentals and be able to tackle larger and more complex problems. I also believe that students should understand what a particular program does and not regard it simply as a magic black box. Accordingly, while I have attempted to make the executable programs user-friendly, they are not written for a passive user but require the student's input and decisions.

The companion CD-ROM also contains auxiliary text: a multichaptered essay on entropy which presents the microscopic view of entropy, examines the connection between entropy and information, and discusses some of the paradoxes associated with entropy.

When applied to a thermodynamics course, the term *advanced* usually has the connotation of increased breadth or complexity of application and greater depth of understanding rather than a higher level of required mathematical skills. Consistent with this view, the book should also be suitable for a graduate level course. Many of the chapters contain topics that could be omitted in an undergraduate course but included in a graduate course. These topics are placed toward the end of the chapters and are marked with an asterisk in the Contents. It is intended that the omission of these topics not affect the orderly development of topics in subsequent chapters. Actually, few if any of these topics are beyond the reach of an undergraduate, and perhaps it is more appropriate to regard them as optional material, allowing flexibility in course design.

The underlying theme that permits the unification of what otherwise would be disparate topics resides in the following definition. *Thermodynamics is an approach to processing, evaluating, and extending experimentally gained information about systems capable of existing in equilibrium states.* While acknowledging the inadequacy of this definition, I believe that it has pedagogical value because it suggests that the subject should be presented and viewed from an experimental perspective. Indeed, this is especially true when it is recognized that the application of thermodynamics often involves the use of variables which are inherently abstract and therefore difficult to conceptualize. Students can hardly be expected to understand these applications if they lack the basic understanding of the phenomenon or system under study. This understanding should include how, and with what difficulty, the experimentally measured variables are determined. Equally, one can be comfortable with abstract thermodynamic variables only after their evaluation has been tied to experimentally determined quantities. A distinguishing feature of this book is the development of this experimental perspective.

In keeping with the experimental perspective, thermodynamic functions are not introduced until the need for them has been established or at least suggested. Hopefully, this allows the functions to be viewed as tools needed to accomplish a specific objective and not simply meaningless, abstract, and arbitrarily defined functions. This approach has led to an unorthodox development of phase equilibrium wherein partial molar properties are not introduced until the fourth of a four-chapter sequence. Thus the treatment of many common topics and a delineation of the rationale for the thermodynamic approach can occur before the student is encumbered with yet another type of function.

S.I. units are used extensively but not exclusively. For some time to come, engineers will find pressure gauges calibrated in psi, thermometers in °F, and physical property data reported in various units. They therefore should be able to work in any system of units. Additionally, in the major applications of thermodynamics—phase equilibrium and chemical equilibrium—S.I. units offer little computational advantage. In fact, because the standard state remains one atmosphere, their use in the treatment of chemical equilibrium can often cause confusion or error. For this reason, fugacities are always expressed in atmospheres in Chapters 12 and 13.

I thank my colleague, Shaoyi Jiang, who read parts of the manuscript and made valuable suggestions. I also thank my colleagues Richard G. Akins and Larry A. Glasgow, who rendered invaluable help and advice with the computing phase of the work. I am grateful to Dr. Y. L. Huang for his work in the development phase of the software on the companion CD-ROM. I also thank Professor Michael Cutlip who has been very helpful regarding the use of POLYMATH. Mr. Harvey Wilson provided a special version of his Shareware steam tables, WASP, for inclusion on the CD-ROM, but unfortunately was unable to accept the terms of the software-use agreement. I am grateful for his efforts and regret that his work could not be included. I am deeply indebted to Professor Kenneth Jolls for the special version of EQUATIONS OF STATE and for his heroic effort in perfecting some rather crude tutorials that I devised. Finally, I especially appreciate the efforts of Nancy Vesta, my copy editor, who skillfully surmounted some rather primitive technology and whose glad grace and good cheer made an otherwise tiresome task bearable.

B. G. Kyle
Manhattan, Kansas

Notation

Latin Symbols

A	Area	E	Total energy, extensive
A	Helmholtz free energy, extensive	e	Total energy, intensive
		$\mathbf{E}$	Work equivalent, extensive
A, B, C	Empirical constants	$\mathbf{e}$	Work equivalent, intensive
$\mathbf{A}$	Surface Area	E_K	Kinetic energy, extensive
a, b, c	Empirical constants	E_P	Potential energy, extensive
A_i	A chemical specie	F	Feed rate
a_i	Activity of component i	$\mathbf{F}$	Force
a_{12}, a_{21}	Parameters in Wilson equation		Generalized force
B	Second virial coefficient		Degrees of freedom
B	Availability, extensive	f	Friction factor
b	Availability, intensive	f	Fugacity
b_l	Number of atomic weights of element l in the system		Fraction liquefied
		f_i	Fugacity of component i
C	Number of components	G	Gibbs free energy, extensive
$\mathbf{c}$	Sonic velocity	$\mathbf{G}$	Mass flux
C_P	Constant-pressure heat capacity, extensive	g	Gibbs free energy, intensive
		$\mathbf{g}$	Acceleration of gravity
c_P	Constant-pressure heat capacity, intensive	G_{12}, G_{21}	Parameters in Wilson equation
		ΔG^f	Free energy of formation
C_V	Constant-volume heat capacity, extensive	H	Enthalpy, extensive
		h	Enthalpy, intensive
		$\mathbf{H}$	Magnetic field strength
C_v	Constant-volume heat capacity, intensive	ΔH^f	Heat of formation
		Δh^*	Enthalpy of departure
		Δh_s	Integral heat of solution
$\bar{c}_P$	Mean heat capacity	I	Magnetization intensity
D	Diameter	I	Ionic strength

K	Equilibrium constant	V	Vapor rate
K_i	Equilibrium ratio	V	Volume, extensive
k_i	Henry's law constant	v	Volume, intensive
k_i'	Henry's law constant based on molality	$\mathbf{v}$	Velocity
L	Liquid rate	W	Work
L	Length	$\mathbf{W}$	Total work (work + equivalent work of heat exchanges) in an open system
l	Displacement or distance		
L_m	Latent heat of melting		
M	Any extensive property	$\mathbf{w}$	Total work in a closed system
m	Any intensive property	W_e	Electrical work
m	Mass	W_f	Fluid work
m	Molality	W_s	Shaft work
$\mathbf{M}$	Mach number	w_s	Shaft work per unit of flowing fluid
m	Mass flow rate		
m_i	Molality of component i	W'	Net, or useful, work
N	Number of chemical species	W^E	Equivalent work of heat
n	Number of mols	X^l	The lth element
P	Pressure	X_i	Solid mol fraction of component i
$P°$	Vapor pressure	x_i	Liquid mol fraction of component i
p_i	Partial pressure of component i in a mixture		
Q	Heat flow	y_i	Vapor mol fraction of component i
$\mathbf{Q}$	Thermodynamic function for a mixture	Z	Compressibility factor
q	Heat flow per unit of flowing fluid	z	Elevation
		z	Electrical charge
R	Gas law constant	z_i	Mol fraction of compound i in feed stream
R	Number of independent chemical reactions		

Greek Symbols

α	Relative volatility
α	Extent of reaction
β	Formula coefficient matrix
β_{ij}	Specie formula coefficient
γ	Heat capacity ratio
γ_i	Activity coefficient of component i
Γ_i	Activity coefficient of component i in a solid solution
δ	Solubility parameter
η	Efficiency, general, but always based on second law

$\mathbf{R}$	Reaction matrix
r	Generalized displacement
S	Number of stoichiometric constraints
S	Entropy, extensive
s	Entropy, intensive
$\Delta s*$	Entropy departure
T	Absolute temperature
t	Empirical temperature
T_m	Melting temperature
T_0	Temperature of the medium
U	Internal energy, extensive
u	Internal energy, intensive

θ	Tentative absolute temperature
θ	Ideal-gas temperature
θ	Freezing point depression
λ	Lagrangian multiplier
μ_i	Chemical potential of component i
v_i	Stoichiometric coefficient
π	Number of phases
π	Lagrangian multiplier
ρ	Density
ρ	Rank of formula coefficient matrix
ρ	Lagrangian multiplier
σ	Created entropy
σ	Surface tension
ϕ_i	Fugacity coefficient of component i
φ_i	Volume fraction
ω	Acentric factor

Operators

δ	Denotes a virtual variation
Δ	Finite change in a state property
d	Infinitesimal change or total differential operator
∂	Partial differential operator
$\int$	Integral operator
$\oint$	Integral operator for closed path
$\ln$	Natural logarithm operator (base e)
$\log$	Common logarithm operator (base 10)
$\prod$	Cumulative product operator
Σ	Cumulative summation operator

Special Notation

$\wedge$ (as in $\hat{f}_i$)	Denotes property of a component in a mixture
$^{-}$(as in $\overline{H}_i$)	Denotes partial molar property

Subscripts

$A, B, C, \dots,$	Denotes system states or species in a mixture
c	Critical property
P	Constant pressure
r	Reduced property
s	Saturated phase
V	Constant volume
0	Reference state, initial state or dead state
$1, 2, 3, \dots,$	Denotes system states or species in a mixture
$\pm$	Denotes a mean ion property

Superscripts

e	With operator Δ denotes excess property change upon mixing
i	With operator Δ denotes property change on forming an ideal solution
L, S, V, α, β	Phase identification
$'$	The prime designates the ideal-gas state or a liquid phase
$''$	The double prime designates a liquid phase
$\circ$	Standard state. With operator Δ denotes a standard property change
$*$ or $\square$	Hypothetical pure component state extrapolated from infinite dilution behavior

Chemical and Process Thermodynamics

Third Edition

Introduction

1-1 THE ANATOMY OF THERMODYNAMICS

Before committing a great deal of time and effort to the study of a subject, it is reasonable to ask the following two questions: What is it? What is it good for? Regarding thermodynamics, the second question is more easily answered, but an answer to the first is essential to an understanding of the subject. Although it is doubtful that many experts or scholars would agree on a simple and precise definition of thermodynamics, necessity demands that a definition be attempted. However, this is best accomplished after the applications of thermodynamics have been discussed.

Applications of Thermodynamics.   There are two major applications of thermodynamics, both of which are important to chemical engineers:

1. The calculation of heat and work effects associated with processes, as well as the calculation of the maximum work obtainable from a process or the minimum work required to drive a process

2. The establishment of relationships among the variables describing systems at equilibrium

The first application is suggested by the name *thermodynamics*, which implies heat in motion. Most of these calculations can be made by the direct implementation of the first and second laws. Examples are calculating the work of compressing a gas, performing an energy balance on an entire process or a process unit, determining the minimum work of separating a mixture of ethanol and water, or evaluating the efficiency of an ammonia synthesis plant.

The application of thermodynamics to a particular system results in the definition of useful properties and the establishment of a network of relationships among the properties and other variables such as pressure, temperature, volume, and mol fraction. Actually, application 1 would not be possible unless a means existed for evaluating the necessary thermodynamic property changes required in implementing the first and second laws. These property changes are calculated from experimentally determined data via the established network of relationships. Additionally, the network of relationships among the variables of a system allows the calculation of values of variables that are either unknown or difficult to determine experimentally from variables that are either available or easier to measure. For example, the heat of vaporizing a liquid can be calculated from measurements of the vapor pressure at several temperatures and the densities of the liquid and vapor phases, the heat of mixing two liquids can be determined by measuring the equilibrium pressure and compositions of coexisting liquid and vapor phases at several temperatures, and the maximum conversion obtainable in a chemical reaction at any temperature can be calculated from calorimetric measurements performed on the individual substances participating in the reaction.

The Nature of Thermodynamics.   The laws of thermodynamics have an empirical or experimental basis, and in the delineation of its applications the reliance upon experimental measurement stands out. Thus, thermodynamics might be broadly defined as a means of extending our experimentally gained knowledge of a system or as a framework for viewing and correlating the behavior of the system. To understand thermodynamics, it is essential to keep an experimental perspective, for if we do not have a physical appreciation for the system or phenomenon studied, the methods of thermodynamics will have little meaning. We should always ask the following questions: How is this particular variable measured? How, and from what type of data, is a particular property calculated?

Often it is easy to miss the intimate experimental dependence of thermodynamics if we are concerned only with the direct applications of the first and second laws. Here the required thermodynamic property changes are usually obtained from a convenient tabulation (e.g., steam tables), or we calculate property changes for a simple substance such as an ideal, monatomic gas with $c_p = 5/2R$. In any event we seldom question the origin of our data and thereby remove thermodynamics from its experimental context, rendering it a lifeless and meaningless set of equations.

Because of its experimental foundation, thermodynamics deals with macroscopic properties, or properties of matter in bulk, as opposed to microscopic properties which are assigned to the atoms and molecules constituting matter. Macroscopic properties are either directly measurable or calculable from directly measurable properties without recourse to a specific theory. Conversely, while microscopic properties are ultimately determined from experimental measurements, their authenticity depends on the validity of the particular theory applied to their calculation. Herein lies the power and authority of thermodynamics: Its results are independent of theories of matter and are thus respected and confidently accepted.

In addition to the certitude accorded its results, thermodynamics enjoys a broad range of applicability. Thus, it forms an integral part of the education of engineers and scientists in many disciplines. Nevertheless, this panoramic scope is often unappreciated because each discipline focuses only on the few applications specific to it. Actually, any system that can exist in observable and reproducible equilibrium states is amenable to the methods of thermodynamics. In addition to fluids, chemically reacting systems, and systems in phase equilibrium—which are of major interest to chemical engineers—thermodynamics has also been successfully applied to systems with surface effects, stressed solids, and substances subjected to gravitational, centrifugal, magnetic, and electric fields.

Through thermodynamics the potentials that define and determine equilibrium are identified and quantified. These potentials also determine the direction in which a system will move as well as the final state it will reach but offer no information concerning the time required to attain the final state. Thus, time is not a thermodynamic variable, and the study of rates is outside the bounds of thermodynamics except in the limit as the system nears equilibrium. Here rate expressions should be thermodynamically consistent.[1]

The experiments and observations on which the laws of thermodynamics are based are neither grand nor sophisticated. Also, the laws themselves are stated in rather pedestrian language. Yet, from this apparently unimpressive beginning a grand structure has evolved which is a tribute to the inductive powers of the human mind. This never fails to inspire awe in the thoughtful and serious student and has led Lewis and Randall[2] to refer to thermodynamics as a cathedral of science. The metaphor is well chosen, for in addition to technical accomplishment and structural integrity, one sees beauty and grandeur. It is no small wonder that the study of thermodynamics can be technically rewarding, intellectually stimulating, and for some, a pleasurable experience.

---

[1] See, for example, K. G. Denbigh, *The Principles of Chemical Equilibrium*, 3rd ed., Cambridge University Press, New York, 1971, Chap. 15.

[2] G. N. Lewis and M. Randall, *Thermodynamics and the Free Energy of Chemical Substances*, McGraw-Hill, New York, 1923.

1-2 THE TERMINOLOGY OF THERMODYNAMICS

Some words that are a part of our everyday vocabulary have a special or more restricted meaning in a thermodynamic context. These words and their thermodynamic meaning are presented here.

System. A *system* is any part of the universe we choose to study or analyze. The system may be enclosed within real boundaries, or the boundaries may be imaginary. Also, the boundaries may be either rigid or movable. Often the analysis of a given problem can be greatly simplified by our choice of the system and its boundaries. Therefore, considerable care should be exercised in defining the system.

State. The term *state* refers to the condition in which the system exists. In thermodynamics this always means a state in which equilibrium is obtained, and the terms *state* and *equilibrium state* are used interchangeably. In Sec. 1-4 more attention will be given to the definition and discussion of the concept of equilibrium; however, here it is sufficient to define it as a condition that is both time-invariant and reproducible.

Surroundings. Having chosen the system and its boundaries, the remainder of the universe becomes the *surroundings*. More realistically, the surroundings may be visualized as only that part of the universe which is affected by processes occurring within the system.

Closed System. A *closed system* does not exchange mass with the surroundings. It may exchange heat and work with the surroundings and thereby undergo changes in energy and volume or experience other property changes, but its mass remains constant.

Open System. An *open system* does exchange mass with its surroundings. Sometimes the boundaries of an open system are defined as a specific volume of space through which mass may enter or leave. An example would be a chemical reactor receiving reactants and discharging products while exchanging heat with the surroundings. To further illustrate the distinction between open and closed systems, consider a liquid and vapor mixture of a pure substance confined to a cylinder fitted with a piston. Movement of the piston causes changes in the quantities of each phase with concomitant heat exchange with the surroundings. Selection of either individual phase results in an open system which can increase or decrease in mass, while the two phases together (or simply the contents of the cylinder) constitute a closed system.

Isothermal System. If any process occurring in a system takes place at constant temperature, the system, as well as the process, is said to be *isothermal*. This applies to both open and closed systems. To maintain isothermal conditions, it is usually necessary for the system to exchange heat with the surroundings.

Adiabatic System. When a process occurs in either an open or closed system without the exchange of heat with the surroundings, both the process and the system are termed *adiabatic*. An adiabatic system is insulated perfectly against the flow of heat. Because in the strict sense this is impossible, the concept of adiabaticity is, of course, an idealization. Nevertheless, it is a condition that can be closely approached in many situations and is therefore a very useful concept.

Isolated System. A system is *isolated* if it exchanges neither mass, heat, nor work with the surroundings. The surroundings are unaffected by processes occurring in isolated systems. A chemical reaction occurring in a well-insulated vessel of constant volume might well represent a process occurring in an isolated system.

1-3 THE VARIABLES AND QUANTITIES OF THERMODYNAMICS

Here we will discuss some of the commonly encountered variables of thermodynamics and how they are measured or determined. These variables can be classified as either *state variables* (*state properties*) or *path variables*; here *path variables* will never be referred to as *properties*. *Properties* are further classified as either *extensive* or *intensive*.

State Variables. The terms *state variable, state property, property, point property,* and *point function* are used interchangeably and refer to a variable whose value depends on the state in which the system exists. Between two states the change in a state variable is always the same regardless of which path the system travels. Temperature, pressure, and volume are examples of state variables.

Path Variables. Changes in *path variables* depend on the path traveled by the system. These variables have meaning only when applied to a process in which the path between two states is specified. Work and heat are path variables; they exist only when a system has experienced a change in state. Therefore, while we may say that a system in a given state has a certain volume (*state property*), it is incorrect to state that it possesses a certain quantity of heat or work.

Extensive Properties. An *extensive property* depends on the size of the system. Volume is an example of an extensive property. Extensive properties are also additive. If the system consists of several identifiable parts, the volume of the system is the sum of the volumes of its parts. Besides the volume, many thermodynamic functions, which will be defined later, are extensive properties. In this text extensive properties will be designated by capital letters.

Intensive Properties. Properties which do not depend on the size of the system are called *intensive properties*. Examples are temperature, pressure, and specific volume (or density). When the system is homogeneous, an extensive property can be converted into an intensive property by dividing by the total quantity of the system.[3] In this text intensive properties, with the exception of temperature and pressure, will be designated by lowercase letters.

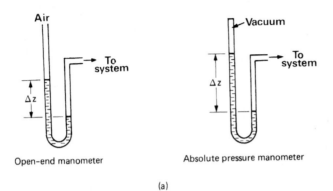

Open-end manometer   Absolute pressure manometer

(a)

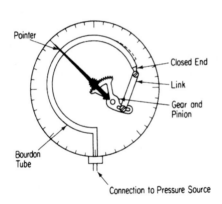

(b) Bourdon gauge

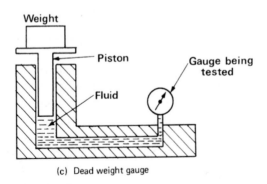

(c) Dead weight gauge

Figure 1-1
Pressure measuring devices. [Part b from David M. Himmelblau, *Basic Principles and Calculations in Chemical Engineering*, 4th ed., 1982, p. 40. Reprinted by permission of Prentice Hall, Inc., Englewood Cliffs, NJ.]

---

[3] Intensive properties may be placed on a mol basis (molar properties) or a mass basis (specific properties).

Pressure. Pressure is perhaps one of the few thermodynamic variables for which you will already have a good physical appreciation. It is simply force per unit area, and it can be measured with a manometer or a Bourdon gauge. Simplified drawings of these instruments are shown in Fig. 1-1. A pressure difference between the inside and outside of the Bourdon tube causes it to straighten out. This movement is translated into pointer motion through a mechanical linkage. For precise work, gauges are usually calibrated by a dead weight gauge, also shown in Fig. 1-1, where the pressure is determined directly from the mass of the weight and the cross-sectional area of the piston.

In the S.I. system, the unit of pressure is a N/m^2, which is also called a pascal, Pa. A more convenient unit is the bar: 10^5 Pa. A standard atmosphere is 1.01325 bar.

Temperature. Many would agree that temperature is a numerical measure of the *degree of hotness*, but few could probably supply a satisfactory scientific definition of it. Here we will not be concerned with a precise definition but will attempt to attach physical significance to the concept of temperature. Even though the term "degree of hotness" implies a physiological, and hence subjective, sensation, it is experienced by all, and all would agree, at least qualitatively, on the definition of *hotness*. Therefore, if the concept could be quantified, an acceptable definition of temperature should result.

We may define a thermometer as any instrument that provides a numerical measure of degree of hotness. The thermometer should possess some readily measurable property in which value depends on the degree of hotness. By fixing the numerical value of the temperature at two easily reproducible reference points, a temperature scale could be established based on the following relationship:

$$\frac{t-0}{100-0} = \frac{X_t - X_0}{X_{100} - X_0} \tag{1-1}$$

In this relationship we have set $t = 0$ and $t = 100$ at the two reference points (the ice point and the normal boiling point of water) where the property has the values X_0 and X_{100}, respectively. The temperature is defined to be linear with respect to the measured property. Linearity is desirable for a temperature scale, for if the properties chosen for different thermometers do not exhibit the same type of behavior, the thermometers will not give identical readings except at the reference points. Fortunately, a thermometer based on the *PVT* behavior of an ideal gas has been shown to possess linearity.

The ideal-gas thermometer is based on measuring either the pressure of a given mass of gas confined to a constant volume or the volume of a mass of gas confined at constant pressure. The concept of a gas thermometer is simple, but in practice complex apparatus and meticulously detailed technique are required; here we will be concerned only with the concept. Figure 1-2 is an idealized representation of the behavior of several ideal-gas thermometers each containing a different working fluid. Gases A, B, and C might be different gases or the same gas at different densities. Each thermometer is calibrated at the reference points of 0 and 100°C where the values of P_0 and P_{100} required in Eq. (1-1) are obtained.

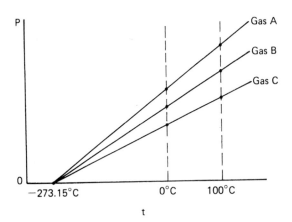

Figure 1-2   Conceptual behavior of ideal-gas thermometers.

Each thermometer gives the same temperature readings based on Eq. (1-1) including $t = -273.15°C$ at zero pressure. Thus, we are assured of linearity.

When the preceding zero-pressure point is used to evaluate a parameter, Eq. (1-1) becomes

$$t + 273.15 = \frac{100\,P_t}{P_{100} - P_0} \tag{1-2}$$

where the pressure P has replaced the general property X. Equation (1-2) suggests that an *ideal-gas temperature* θ, defined as

$$\theta = t + 273.15 \tag{1-3}$$

would be directly proportional to the pressure and thus more convenient to work with. There is some degree of universality associated with the ideal-gas temperature scale because it is independent of the identity of the gas employed as the working fluid.

In Chap. 4 it will be shown that a temperature scale based on the second law can be established that is totally independent of the nature of the working substance. Such a scale is absolute and is referred to as the *absolute temperature scale*. It is easily shown that the ideal-gas temperature scale is identical to the absolute temperature scale. In thermodynamics this temperature scale is used exclusively and is designated T.

Because of the experimental complexity associated with using an ideal-gas thermometer, this instrument is not used routinely. Rather it has been used to determine a number of reproducible reference points (e.g., the freezing point of gold) which are then used to calibrate other instruments of greater practicality.

Temperatures defined by Eq. (1-2) and represented by t are on the Celsius scale, while those represented by Eq. (1-3) are on the Kelvin scale. The Fahrenheit scale is established by setting the reference temperatures at the ice point and water normal boiling point at 32 and

212, respectively. In *absolute temperature* the Rankine scale is the counterpart to the Fahrenheit scale. They are related by °R = 459.67 + °F. In the S.I. system the Kelvin scale is used with the unit designated a Kelvin (1 K = 1°C = 1.8°F = 1.8°R).

Work. Defined as the product of a force **F** and the displacement it produces, the work dW associated with an infinitesimal displacement dl is

$$dW = \mathbf{F}\,dl \tag{1-4}$$

For a finite displacement this becomes

$$W = \int \mathbf{F}\,dl \tag{1-5}$$

In engineering thermodynamics the predominant type of work is that associated with the expansion or contraction of a fluid. Consider a gas in a cylinder fitted with a piston. The force required to produce a volume change dV is the pressure P times the area A of the piston. Also recognizing that dV is the product $A\,dl$, where dl is the linear movement of the piston, we have

$$dW = \mathbf{F}\,dl = PA\left(\frac{dV}{A}\right) = P\,dV \tag{1-6}$$

Also,

$$W = \int P\,dV \tag{1-7}$$

From Eq. (1-7) we see that for a finite process, work may be visualized as the area under a curve of P vs. V. Figure 1-3 shows two alternate paths a system comprised of a gas in a cylinder might take between two terminal states designated 1 and 2. These two paths represent the actual succession of states, expressed in terms of P and V, through which the gas passes in moving from state 1 to state 2. The areas under curves 1-2 and 1-3-2 are different, and hence the work, a path variable, is indeed seen to depend on the path.

Electrical work is often involved in processes we will wish to analyze. For a battery or electrochemical cell the electrical work is $W = nZ\mathfrak{F}E$ where E is the voltage and $nZ\mathfrak{F}$ is the quantity of charge (electrons) flowing expressed as the product of the number of ions reacted n, the number of charges per ion Z, and Faraday's constant, 96,479 C/g equivalent.

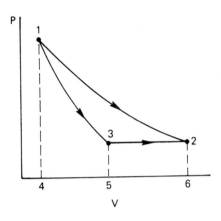

Figure 1-3 Path 1-2: work = area 1264.
Path 1-3-2: work = area 1354 + area 3265.

When considering equipment either driven by or generating electricity, the product of voltage E and current i, a flow of charge, gives the rate of performing work, or power. The electrical work performed over a given time interval τ is

$$W = \tau i E$$

Because it is theoretically possible to achieve conversion between mechanical and electrical work, the term *work* can be used to refer to either quantity.

Work has meaning only when exchanged. One body does work, and another receives it. The expanding gas does work on the piston face. In turn, this energy can be translated into other types of work through appropriate arrangements of rods, crankshafts, gears, etc.

Because quantities such as force, distance, pressure, and volume are measurable, it is possible by application of Eq. (1-5) or (1-7) to calculate directly the work exchanged by a system under study. The unit of work in the S.I. system is the joule, J.

Heat. Like temperature, heat is familiar but difficult to define precisely. Like work, it is a path quantity and has meaning only when exchanged.

When two bodies of different temperature are placed in contact, we always observe that they eventually reach the same temperature, and we postulate that something, called "heat," flowed from the hotter to the colder body. Suppose that one body is a kilogram of water and that by placing a hotter body, say a copper block, in the water a temperature rise is observed. We say that heat flowed from the block to the water. However, we also know that the same temperature rise can be achieved by adding a quantity of work, say electrical work in the form of resistance heating, to the water. Thus, there is an equivalence between heat and work. Further elucidation of the nature of heat and work and their equivalence must await the development of the first law of thermodynamics. However, it should be noted that the quantity of heat cannot be directly measured except as the equivalent amount of work. Through application of the first law, the quantity of heat can be calculated. Like work, heat is expressed in joules, J.

1-4 EQUILIBRIUM AND THE EQUILIBRIUM STATE

Because thermodynamics is based on experimentation, a definition of equilibrium should be given in that context. Therefore, a *system will be said to be in equilibrium when its measurable properties are not observed to change with time.* It is implied that the system is not interacting with the surroundings, and we should be careful to exclude from this definition steady-state open systems which, while exhibiting time invariance, exchange mass, heat, or work with the surroundings. Because of their interactions with the surroundings, steady states should be easily discernible from equilibrium states, and no confusion should occur.

An equilibrium state is characterized by the properties of the system in this time-invariant condition. In addition, an equilibrium state must be reproducible. The question of reproducibility entails the specification of a certain number of variables describing the system and is dealt with under the subject *phase rule* (Sec. 1-5).

With the special attention given to the definition of equilibrium states and to the need for assurance that systems under study exist in such states, one often tends to ascribe to them a mystical quality and obtains the impression that they are not routinely encountered. Such notions are unfounded, for our apparent preoccupation with this subject stems from a desire to define precisely the limits of applicability of thermodynamics. Actually, these limits are rather broad, as systems amenable to thermodynamic treatment are regularly studied in the laboratory and routinely encountered in the practice of science and engineering. All spontaneous changes move a system toward an equilibrium state; therefore, all closed systems eventually reach equilibrium.

Later, as we apply the methods of thermodynamics, we will find that certain intensive properties, called "thermodynamic potentials," must be uniform throughout systems in equilibrium.[4] Among the potentials are temperature, pressure, and chemical potential; their uniformity implies no transfer of heat, mechanical work, or mass—a condition we have already noted.

As pertains to a nonreacting fluid, specification of two intensive variables[5] will serve to define an equilibrium state. The requirement that the thermodynamic potentials be uniform throughout the fluid can also be seen as necessary in order that a specified temperature, pressure, or composition (or any other intensive property) have meaning.

Finally, while our emphasis here has been placed on defining and identifying equilibrium states, it should be noted that in a great many applications—particularly those in phase equilibrium and chemical systems at equilibrium—we know that the system is in equilibrium, or we wish to determine the relationships among thermodynamic variables for systems at equilibrium. Whether these results can be applied to a particular system under study is a determination to be made by the observer. However, even if our system is not, or will not be, at equilibrium, beneficial results may still be obtained. For example, it is useful to know the

---

[4] While exceptions exist, e.g., systems in gravitational or centrifugal fields, they are encountered only rarely and must be accorded special treatment.

[5] Because they are variables we can most easily control, temperature and pressure are usually chosen.

equilibrium conversion for a chemical reaction even though our system is not expected to reach equilibrium. This represents the best which could be expected and thus places an upper limit on the performance of our system.

1-5 THE PHASE RULE

We have already encountered the problem of determining when a system is in a reproducible equilibrium state and how many variables need be specified to ensure this condition. The phase rule provides the answer to this problem. It can be deduced through thermodynamic reasoning,[6] but here it is simply stated without proof:

$$\mathscr{F} = C + 2 - \pi \tag{1-8}$$

This expression applies to systems at equilibrium and relates the number of components in the system C and the number of phases π to the degrees of freedom $\mathscr{F}$. The degrees of freedom are the number of *intensive* variables that must be fixed before the equilibrium state is properly defined and thus the system is reproducible.

A phase is any identifiable state of matter which is homogeneous throughout and will be either gas, liquid, or solid. Except under extremely high pressures, all substances mix completely in the gas phase, and thus only a single gas phase will exist in any system considered here. Complete mixing is not always found among liquids, and it is not uncommon to encounter systems with more than one liquid phase present at equilibrium.[7] More variety is found among solid phases. Here it is possible to find pure solids, solid solutions, and stoichiometric compounds (e.g., hydrates) with the possibility that each can exist in more than one crystal habit. For example, the two crystalline forms of sulfur, rhombic and monoclinic, are identifiable and are treated as separate phases. While a phase must be homogeneous, it need not be continuous. Consider a system of a saturated solution in equilibrium with salt crystals. We observe many individual crystals, but because each crystal is identical in composition and crystal habit, it is considered to be a part of a single solid phase.

It should be emphasized that the phase rule deals only with intensive variables and thus says nothing about the size of a system or the relative proportions of its phases. For example, the phase rule states that in preparing a vapor-liquid mixture ($\pi = 2$) of a pure substance ($C = 1$) we may control only one intensive variable ($\mathscr{F} = 1$). From the standpoint of the phase rule, a system containing 99% vapor and 1% liquid is equivalent to one containing 1% vapor and 99% liquid because both possess identical intensive properties. Similarly, the system comprised of salt crystals and saturated solution would have unchanged intensive properties if all but one salt crystal were removed.

---

[6] This will be done in Chap. 9 for nonreacting systems and in Chap. 13 for chemically reacting systems.
[7] A system containing 10 liquid phases has been described: J. H. Hildebrand and R. L. Scott, *Regular Solutions*, Prentice–Hall, Englewood Cliffs, NJ. 1962.

1-6 THE REVERSIBLE PROCESS

A process that moves a system from state A to state B is said to be reversible if the work and heat effects realized from the process are sufficient to restore the system to its original state (state A). Actually, we are considering a cycle $A \rightarrow B \rightarrow A$ consisting of two reversible steps which when completed leaves no changes in either the system or surroundings.

To better visualize the concept of reversibility and its implications, let us examine a specific cyclic process: the adiabatic expansion and subsequent adiabatic compression of gas in a cylinder fitted with a piston. Because there is no heat exchange, we need only focus on the work effects. In an actual process there will be a certain amount of friction encountered in moving the piston against the cylinder walls. Improved design may reduce this, but it can never be completely eliminated. To overcome the force of friction, it will be necessary during an expansion for the fluid pressure P_f to exceed the pressure P_e corresponding to the force applied to the external face of the piston. Thus, the work done by the fluid is $\int P_f dV$, and the work delivered to an external agency is $\int P_e dV$. Because $P_f > P_e$, we know that

$$\int_A^B P_f \, dV > \int_A^B P_e \, dV$$

Thus, the work done by the gas in expanding exceeds that delivered to the surroundings by an amount necessary to overcome the effect of friction. This work is wasted and appears as frictional heat. In the compression step the existence of friction requires $P_e > P_f$, hence,

$$\int_B^A P_e \, dV > \int_B^A P_f \, dV$$

Therefore, the work required from the surroundings is greater than that of compressing the gas by the amount lost to friction. If the gas travels the same path from A to B as from B to A, then

$$\int_A^B P_f \, dV = -\int_B^A P_f \, dV$$

or the work required to compress the gas is numerically equal to the work done by the gas in expanding. However, the surroundings receive less and must supply more work than this and therefore incur a deficit. Clearly, the presence of friction precludes the existence of a reversible process.

Friction is not the only factor which determines the reversibility of a process. Energy may be dissipated within the expanding or contracting gas because of turbulence caused by gradients in temperature and pressure. The energy lost to turbulence cannot be delivered to the surroundings as work; hence a process with no friction would still not be reversible unless turbulence was eliminated. Turbulence can be reduced by slowing the rate at which changes occur. The process should be carried out slowly so that temperature and pressure are uniform throughout the gas. The uniformity of these thermodynamic potentials is, as we have already seen, a condition of equilibrium, and we are thereby specifying that at all times the gas exists in an equilibrium state. To realize this would require that, in the absence of friction, the difference between the fluid pressure and the external pressure be infinitesimal.

From this specific example we may—by inference—generalize the necessary conditions for reversibility: the absence of dissipative processes such as friction, the existence of the system in equilibrium states at all times, and the maintenance of only infinitesimal differences in thermodynamic potential between the system and its surroundings. When heat is exchanged, this would require that the uniform temperature of the system differ only infinitesimally from that of the surroundings.

Clearly, the reversible process can never be carried out. However, it represents a limiting case which can be approached by an actual process, and it is the only type of process for which work can be calculated. To understand this restriction, we should realize that paths such as those shown in Fig. 1-3 are actually a succession of equilibrium states through which the system passes. Because only equilibrium states can be represented on diagrams such as this, the path is necessarily the trajectory of a reversible process.

As we have seen, a reversible process produces the maximum or requires the minimum amount of work. If we wish to know the actual work, we first calculate the reversible work and correct it by means of an efficiency. Defined as a ratio involving actual work and reversible work, the efficiency is determined empirically for the system of interest or for a class of similar systems.

CHAPTER 2

The First Law of Thermodynamics

2-1 THE FIRST LAW AND INTERNAL ENERGY

In Chap. 1, heat and work were seen to be quantities which have meaning only when transferred between bodies. The commodity transferred is called *energy*—mechanical energy in the case of work and thermal energy in the case of heat.

In addition to being exchanged, energy can also be stored. For example, mechanical energy (work) obtained from the expansion of a gas in a cylinder could be used to raise a weight. The weight is said to possess potential energy which can be stored indefinitely or used to transfer mechanical energy to another body. Similarly, thermal energy can also be stored. The addition of thermal energy (heat) to a body is often manifested as a temperature rise. If the body is then perfectly insulated, it will maintain its temperature indefinitely or could at some later time transfer thermal energy to another body.

Another characteristic of energy is its convertibility among its various forms. Consider, for example, a hydroelectric generating plant where the potential energy of water, which it possesses because of its elevation, is converted to kinetic energy, which is in turn imparted to the blades of a turbine and results in mechanical energy, the turning of a shaft. The turbine shaft delivers mechanical energy to the generator where it is converted into electrical energy. Electrical energy is easily transported and may be used elsewhere to run a motor which delivers mechanical work, or it could be used to heat homes through electrical resistance heating. Except for the last process where heat is produced, it is theoretically[1] possible to continue the energy transformations indefinitely. We can always convert mechanical energy completely into heat, and we can freely convert one type of mechanical energy into another, but we cannot completely convert heat into work or mechanical energy.[2]

The concept of energy can be clarified and its various manifestations reconciled through the first law of thermodynamics. This law requires the conservation of energy, where

change in energy of system = net exchange of energy with surroundings

For closed systems work and heat are the only modes of energy exchange,[3] and we write

net exchange of energy with surroundings = $Q + W$

where Q and W represent heat and work, respectively. Here the following convention is observed with respect to the sign of Q and W.[4] They are each taken positive when the exchange is from the surroundings to the system. When more than one exchange of heat or work occurs, the terms Q and W are algebraic sums of individual exchanges and therefore represent net exchanges.

We know that a system can possess kinetic energy by virtue of its velocity and potential energy by virtue of its elevation and should therefore include these forms of energy in our accounting. We also know that in exchanging heat or work with the surroundings a system experiences a change of state which could be manifested by a temperature change or a phase change. This suggests that the system possesses another form of energy, called *internal energy*, which is associated with the state of the system and will be designated by the symbol U. For the energy change of the system we therefore write

$$\text{change in energy of system} = \Delta U + \Delta E_K + \Delta E_P = \Delta E$$

---

[1] All processes are assumed reversible.

[2] This limitation is dealt with by the second law of thermodynamics.

[3] In an open system, energy flows may also be associated with mass flows.

[4] The convention regarding work is not universally accepted. Often the reverse is employed. Although these conventions result in different algebraic statements, when consistency is observed, the results of the applications of the first law are unaffected.

where ΔU, ΔE_K, and ΔE_P are, respectively, the changes in internal energy, kinetic energy, and potential energy and ΔE represents the total energy change of the system. For a mass m the kinetic and potential energy changes are

$$\Delta E_K = \frac{m\Delta \mathbf{v}^2}{2}$$

$$\Delta E_P = mg\Delta z$$

where $\mathbf{v}$ and z are velocity and elevation, respectively, and $\mathbf{g}$ is the gravitational acceleration. Because all types of energy changes are expressed as extensive property changes, energy changes for a composite system can be expressed as the sum of the contributions of its component parts. In the S.I. system the unit of all these energy changes is the joule, J. In addition to the previously listed forms of energy other forms exist and on rare occasions should be included in ΔE. Some of these are surface energy and energy due to external fields such as electric or magnetic.

For the closed system the first law can now be written as

$$\Delta E = Q + W \tag{2-1}$$

For many processes the system is stationary and does not experience changes in kinetic or potential energy. Even when the system possesses these forms of energy, it is found quite often that the change in internal energy for a process overwhelmingly predominates,[5] and it is appropriate to state the first law as

$$\Delta U = Q + W \tag{2-2}$$

For infinitesimal changes we write

$$dU = dQ + dW \tag{2-2a}$$

In ensuring the conservation of energy, the first law defines the internal energy as a state property. The absolute value of this property can never be known; only changes between states can be determined. However, this in no way limits the application of the first law.

Our conception of heat can now be quantified. For any process occurring in a closed system the quantity of heat exchanged Q can be calculated from W and ΔU via Eq. (2-2). As we have seen in Chap. 1, work can be measured, and, as we shall soon see, ΔU may be determined for any prescribed change of state of the system. As a specific example, the quantity of heat required to raise the temperature of 1 kg of water from 298 to 299 K could be determined through a measurement carried out adiabatically ($Q = 0$). The water contained in a well-insulated vessel of constant volume could be heated by electrical energy flowing through a resistor immersed in the water, and the quantity of electrical work required to pro-

---

[5] See Ex. 2-5.

duce the temperature change could be accurately measured. From this measured quantity of work, application of Eq. (2-2) determines ΔU for the particular change in state. Because U is a state property, ΔU for this change in state will always be the same regardless of the path followed. We may now consider a process where this same change in state is accomplished solely by the addition of heat ($W = 0$). This could be accomplished by placing the water in contact with a warmer body. Application of Eq. (2-2) results in $Q = \Delta U$. In this experiment we have actually measured the quantity of work which would accomplish the same change in state as an equal quantity of heat.

As seen by the preceding example the first law allows us to store experimentally determined information about the system as values of a state function. Information deposited in this manner is readily available to calculate heat and work effects for other processes the system may undergo. This convenience, of course, would not be possible but for the fact that the internal energy is a state property.

Internal energy is defined by the first law. It is a property that matter must possess in order that the behavior of a system exchanging heat and work can be properly explained. This law is based on considerable experimental evidence and has been subjected to and survived many tests since its inception; therefore, its authenticity is unquestioned. This empirical basis endows the results of thermodynamics with authority and certitude; however, its detachment from theories of matter denies us insights that would derive from theoretical models. Such is the case with internal energy. Thermodynamics requires the existence of a function we choose to call internal energy and provides a means of determining its changes. While this is sufficient for any application of thermodynamics, we are uncomfortable with abstractions and prefer to attach physical significance to the quantities we deal with. Fortunately, our familiarity with molecular theory allows us to visualize internal energy in terms of the various types and levels of energy available to the atoms and molecules of matter.

EXAMPLE 2-1

A system consists of a gas in a cylinder confined by a piston. On the *PV* diagram of Fig. 2-1, the state of the gas is represented by a point, and the path of a process is represented by a line. The system is taken from state *A* to state *B* along the path *ABC*. The process from *A* to *C* is constant pressure, and the system receives 50 J of work and gives up 25 J of heat to the surroundings. The process from *C* to *B* is constant volume, and the system receives 75 J of heat.

The return path from *B* to *A* is adiabatic. How much work is exchanged with the surroundings?

Solution 2-1

For the return path with $Q = 0$ the first law becomes

$$\Delta U_{BA} = U_A - U_B = W$$

EXAMPLE 2-1 CON'T

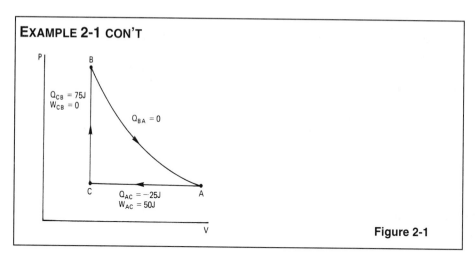

Figure 2-1

Therefore, we need only ΔU_{BA} for the process B to A. Because U is a state property, we write

$$\Delta U_{BA} = -\Delta U_{AB}$$

and can obtain ΔU_{AB} from the information supplied for the process ABC. Because the process CB is at constant volume, there is no work involved, and we write

$$\Delta U_{AB} = Q_{AC} + Q_{CB} + W_{AC} + 0$$

$$\Delta U_{AB} = -25 + 75 + 50 = 100\text{J}$$

Therefore,

$$\Delta U_{BA} = -100\text{ J}$$

and

$$W_{BA} = -100\text{ J}$$

Work in the amount of 100 J is done by the system for the adiabatic return path.

EXAMPLE 2-2

A rigid tank of 0.1 m³ contains a mixture of saturated steam and saturated water at a pressure of 2 bar. By volume the mixture is 10% liquid. How much heat must be added so that the tank contain only saturated steam? What will be the pressure in the tank?

Solution 2-2

From Table C-1 in the Appendix the properties of saturated liquid and vapor at 2 bar are

$$v^G = 0.8857 \text{ m}^3 / \text{kg} \qquad u^G = 2529.5 \text{ kJ} / \text{kg}$$
$$v^L = 0.001061 \text{m}^3 / \text{kg} \qquad u^L = 504.5 \text{ kJ} / \text{kg}$$

The amounts of liquid and vapor are

$$0.1(0.1) = 0.01 \text{ m}^3 = 0.001061 m^L$$

$$m^L = 9.425 \text{ kg}$$

$$0.1(0.9) = 0.09 \text{ m}^3 = 0.8857 m^G$$

$$m^G = 0.102 \text{ kg}$$

$$m = m^L + m^G = 9.425 + 0.102 = 9.527 \text{ kg}$$

Our system is the 9.527 kg of water in the tank. Application of the first law for this change, where $W = 0$ (no volume change), yields

$$\Delta U = Q$$

The initial state of our system is known, but the final state is not. However, we know that there will be 9.527 kg of saturated steam occupying the volume of 0.1 m³ and therefore know an intensive variable, the specific volume:

$$v^G = \frac{0.1}{9.527} = 0.01049 \text{ m}^3 / \text{kg}$$

Through interpolation we find from Table C-1 that this specific saturated vapor volume occurs at a pressure of 148.7 bar where it is found that u^G = 2462.8 kJ/kg. We may now evaluate ΔU :

$$\Delta U = 2462.8(9.527) - (504.5)9.425 - (2529.5)0.102$$

$$= 18,437 \text{ kJ} = Q$$

This is the amount of heat that must be added to accomplish the desired internal energy change.

2-2 THE ENTHALPY

In making calculations involving the first law, we often encounter the internal energy U grouped with the pressure-volume product PV and therefore find it convenient to define a new thermodynamic variable, the *enthalpy H*:

$$H = U + PV \qquad (2\text{-}3)$$

As U, P, and V are state variables, the enthalpy is also a state variable. As with internal energy, the joule is the unit of the enthalpy in the S.I. system.

For a constant-pressure process it is seen from Eq. (2-3) that the enthalpy change is

$$\Delta H = \Delta U + P\Delta V \qquad (2\text{-}4)$$

Now consider a process where heat is exchanged with a system that is always maintained at a constant pressure. From the first law, Eq. (2-2), we have

$$Q_P = \Delta U - W_P \qquad (2\text{-}2)$$

If the system is capable of performing only PV work and a change of volume against a constant pressure occurs reversibly, we may write[6]

$$W_P = -\int P\, dV = -P\Delta V \qquad (1\text{-}7)$$

$$Q_P = \Delta U + P\Delta V \qquad (2\text{-}5)$$

Comparison of Eqs. (2-4) and (2-5) shows that

$$Q_P = \Delta H \qquad (2\text{-}6)$$

Under conditions of constant pressure and when the system is capable of only PV work , the enthalpy change equals the heat effect of a reversible process. In applications the restriction to reversible processes is routinely relaxed and with some justification.[7]

For an ideal gas it can be shown that the enthalpy and internal energy depend only on temperature.[8] For real gases the effect of pressure on h or u is quite small; liquids and solids

---

[6] The negative sign is consistent with our convention. A decrease in volume ($\Delta V < 0$) represents work done on the system and results in $w > 0$.

[7] It is usually the case that constant-pressure conditions are imposed by a surrounding fluid and that while changes within the system might occur so rapidly that its states could not be properly defined, the work may still be evaluated from a volume change observed in the surrounding fluid.

[8] This will be proved in Chap. 5.

behave similarly. In most applications a system will be assumed to undergo no change in internal energy or enthalpy unless it experiences a change in temperature, composition, or phase.

EXAMPLE 2-3

A table of property values is desired for a substance which can be regarded as an ideal gas. Experimentally, the gas confined at constant volume will be heated in an adiabatic calorimeter by the addition of a measured quantity of current passing through a resistor across which a known voltage is imposed. In a particular run, 0.00100 m³ of gas—initially at 298 K and 1 atm—is heated to 308 K by the addition of 11.93 J of electrical energy. Using a reference state of 298 K and 1 atm, determine numerical values of the enthalpy and internal energy of the gas at 308 K and 1 atm.

Solution 2-3

For this adiabatic process ($Q = 0$) involving electrical work W_e but no PV work, the first law becomes

$$\Delta U = U_{308} - U_{298} = W_e$$

For convenience we will choose $U = 0$ in the reference state of 298 K and obtain

$$U_{308} = 11.93 \, J$$

The 11.93 J refers to the internal energy of 0.00100 m³ of gas, and to be more useful, the values in our table should be intensive properties based on a kmol. From the ideal-gas law,

$$n = \frac{PV}{RT} \qquad R = 8314.3 \frac{N \cdot m}{kmol \cdot K}$$

$$n = \frac{1.01325(10^5)N \cdot m^{-2}(0.00100 m^3)}{8314.3 N \cdot m \cdot kmol^{-1} K^{-1}(298 \, K)}$$

$$= 4.090(10^{-5}) \, kmol$$

The molar internal energy is

$$u_{308} = \frac{11.93}{4.090(10^{-5})} = 2.917(10^5) \, J / kmol$$

Because of the defining equation [Eq. (2-3)], we cannot arbitrarily set both h and u equal to zero in the reference state. We have set u equal to zero, and therefore we find h in the reference state from

$$h_{298} = u_{298} + Pv = 0 + RT$$

$$= 8314.3\, \text{J} \cdot \text{kmol}^{-1} \cdot \text{K}^{-1}(298\ \text{K}) = 24.78(10^5)\ \text{J} / \text{kmol}$$

Similarly, we find h at 308 K as

$$h_{308} = u_{308} + RT = 2.917(10^5) + 8314.3(308)$$

$$= 28.53(10^5)\ \text{J} / \text{kmol}$$

We now have the following two entries in our table:

Temperature	Internal Energy	Enthalpy
(K)	(10^5 J kmol^{-1})	(10^5 J kmol^{-1})
298	0	24.78
308	2.917	28.53

Because the internal energy and enthalpy of an ideal gas depend only on temperature, it is necessary to specify only this independent variable in our table.

EXAMPLE 2-4

For 0.100 kmol of the gas of Ex. 2-3, calculate the heat required to increase the temperature from 298 to 308 K under conditions of constant pressure.

Solution 2-4

For a constant-pressure heating Eq. (2-6) applies:

$$Q_P = \Delta H = n(h_{308} - h_{298})$$

Using enthalpies tabulated in Ex. 2-3, we have

$$Q_P = 0.100(28.53 - 24.78)(10^5) = 2.75(10^4)\ \text{J}$$

In Ex. 2-3 experimental observation of the gas undergoing a constant-volume heating was used to calculate and tabulate values of u and h. These tabulated property data were subsequently used in Ex. 2-4 to calculate the behavior of the gas when undergoing a different type of process. Thus, it could be said that the first law enables us to obtain the maximum use of experimentally gained information.

EXAMPLE 2-5

Demonstrate that kinetic and potential energy effects are small in comparison with thermal energy effects.

Solution 2-5

To increase the velocity of a kilogram of matter from 0 to 50 m/s (112 mph) requires the addition of

$$\Delta E_K = \frac{m\mathbf{v}^2}{2} = \frac{1(50)^2}{2} = 1250 \text{ J}$$

For an increase in elevation of 100 m, a kilogram of matter requires the addition of

$$\Delta E_P = m\mathbf{g}\Delta z = 1(9.807)(100) = 980.7 \text{ J}$$

The changes in kinetic and potential energy are independent of the nature of the substance; however, internal energy or enthalpy changes corresponding to a definite temperature change are not. For example, enthalpy and internal energy changes corresponding to a change from 100 to 101°C for liquid water and for steam are[9]

Liquid Water	Steam
$\Delta u = 4050 \text{ J / kg}$	$\Delta u = 1000 \text{ J / kg}$
$\Delta h = 4200 \text{ J / kg}$	$\Delta h = 1500 \text{ J / kg}$

The kinetic and potential energy changes considered are much larger than those normally encountered, yet at best they are barely equivalent to the thermal energy corresponding to a temperature change of a single degree.

---

[9] Based on data from Table C-1 in the Appendixes.

2-3 THE HEAT CAPACITY

It is customary to define a heat capacity C in terms of the heat required to produce a temperature change when the system is constrained to a specified path:

$$C \equiv \left(\frac{dQ}{dT} \right)_{path} \tag{2-7}$$

Paths of constant volume and constant pressure are of most interest. During a constant-volume heating we have $dW = 0$, and from Eq. (2-2a) we state that

$$dU = dQ_V$$

The constant-volume heat capacity C_V is now expressed as

$$C_V = \left(\frac{\partial U}{\partial T} \right)_V \tag{2-8}$$

For a constant-pressure heating Eq. (2-6) stated for an infinitesimal change is

$$dH = dQ_P$$

and the constant-pressure heat capacity C_P becomes

$$C_P = \left(\frac{\partial H}{\partial T} \right)_P \tag{2-9}$$

Because U and H are state properties, so are their temperature derivatives C_V and C_P. While it is possible to use heat capacities based on the extensive properties H and U, it is more convenient to work with the intensive properties c_V and c_P defined as

$$c_V = \left(\frac{\partial u}{\partial T} \right)_V \tag{2-8a}$$

$$c_P = \left(\frac{\partial h}{\partial T} \right)_P \tag{2-9a}$$

Experimentally, the constant-pressure heat capacity is determined from enthalpy change measurements. As shown in Ex. 2-3, adiabatic calorimetry can be used. While the concept of adiabatic calorimetry is simple, sophisticated equipment and painstaking technique are required to minimize heat leaks and thereby obtain good quality data.[10] Constant-volume heat capacities are usually calculated from constant-pressure heat capacities and other physical property data via the network of thermodynamic relationships.[11]

From Eqs. (2-8a) and (2-9a) we see that internal energy and enthalpy changes can be expressed as

$$\left[\Delta u = \int c_V \, dT \right]_V \qquad \qquad (2\text{-}10)$$

$$\left[\Delta h = \int c_P \, dT \right]_P \qquad \qquad (2\text{-}11)$$

Thus, information about a system may be stored either in the form of enthalpy and internal energy or heat capacity. While both presentations are encountered, heat capacity data are more common, probably because an absolute value of the heat capacity is obtained, whereas the presentation of enthalpy data requires the specification of a reference state.

2-4 THE FIRST LAW FOR OPEN SYSTEMS

To derive the first law as applied to open systems, we begin with our previous conservation statement:

change in energy of system = net exchange of energy with surroundings

However, we recognize that in addition to the normal modes of energy transfer, heat and work, the right-hand side should also include energy exchanges due to mass entering and leaving the system. Each unit of mass carries with it the amount of energy

$$e = u + \mathbf{g} z + \frac{\mathbf{v}^2}{2} \qquad \qquad (2\text{-}12)$$

Our system is that volume of space containing an apparatus, a process, or any part of a process. Heat and work are exchanged with the surroundings, and any number of streams transport matter into and out of the system.

We will first consider the most general case, an unsteady-state system—one in which mass or properties are changing. Over an interval where the total energy of the system changes by ΔE, our conservation statement becomes

---

[10] For a good overview of practical calorimetry, see M. L. McGlashan, *Chemical Thermodynamics*, Academic Press, New York, 1979, Chap. 4.

[11] See Ex. 5-3.

$$\Delta E = Q + W + \sum_{\substack{\text{all} \\ \text{influent} \\ \text{streams}}} \int e_i \, dm_i - \sum_{\substack{\text{all} \\ \text{effluent} \\ \text{streams}}} \int e_j \, dm_j \qquad (2\text{-}13)$$

As before, Q and W may be algebraic sums, and ΔE may refer to a composite system. Because the conditions of a stream may change over the interval, the energy flow associated with each stream (indexed i for influent and j for effluent) is written as an integral. The integration is performed over the mass that flowed during the interval. The sum of these integrals for the influent streams represents an energy flow into the system, while the summation over effluent streams represents flow of energy out of the system.

There are two types of work that should be identified: shaft work W_s and fluid work W_f, where

$$W = W_s + W_f \qquad (2\text{-}14)$$

Fluid work is associated with moving fluid into and out of the system. A unit of incoming fluid of volume v and at the pressure P has the amount of work Pv done on it by the upstream fluid as it enters the system. Similarly, a unit of fluid leaving the system does work Pv on the downstream fluid. Thus, the total fluid work W_f is

$$W_f = \sum_{\substack{\text{all} \\ \text{influent} \\ \text{streams}}} \int P_i v_i \, dm_i - \sum_{\substack{\text{all} \\ \text{effluent} \\ \text{streams}}} \int P_j v_j \, dm_j \qquad (2\text{-}15)$$

Again, integrals are written because conditions may change over the interval in which a quantity of mass m_i enters or a quantity of mass m_j leaves the system. The shaft work, being defined as the total work less the fluid work, is seen to be the net work required to drive the process or obtainable from the process. This will usually be in the form of a rotating or reciprocating shaft but could also be electrical work as in the case of an electrochemical device.

Substitution of Eqs. (2-15) and (2-14) into Eq. (2-13) yields

$$\Delta E = Q + W_s + \sum_{\substack{\text{all} \\ \text{influent} \\ \text{streams}}} \int \left(e_i + P_i v_i \right) dm_i - \sum_{\substack{\text{all} \\ \text{effluent} \\ \text{streams}}} \int \left(e_j + P_j v_j \right) dm_j \qquad (2\text{-}16)$$

The term $e_i + P_i v_i$ can be written as

$$u_i + P_i v_i + g z_i + \frac{\mathbf{v}_i^2}{2} = h_i + g z_i + \frac{\mathbf{v}_i^2}{2}$$

and our most general first law statement becomes

$$\Delta E = Q + W_s + \sum_{\substack{\text{all} \\ \text{influent} \\ \text{streams}}} \int \left(h_i + \mathbf{g} z_i + \frac{\mathbf{v}_i^2}{2} \right) dm_i$$

$$- \sum_{\substack{\text{all} \\ \text{effluent} \\ \text{streams}}} \int \left(h_j + \mathbf{g} z_j + \frac{\mathbf{v}_j^2}{2} \right) dm_j \tag{2-17}$$

For steady-state conditions, Eq. (2-17) can be simplified. At steady state, $\Delta E = 0$ and all quantities under integrals are constant, which leads to

$$0 = Q + W_s - \Delta H - \Delta E_P - \Delta E_K \tag{2-18}$$

where

$$\Delta H = \sum_{\substack{\text{all} \\ \text{effluent} \\ \text{streams}}} h_j m_j - \sum_{\substack{\text{all} \\ \text{influent} \\ \text{streams}}} h_i m_i \tag{2-19}$$

$$\Delta E_P = \sum_{\substack{\text{all} \\ \text{effluent} \\ \text{streams}}} m_j \mathbf{g} z_j - \sum_{\substack{\text{all} \\ \text{influent} \\ \text{streams}}} m_i \mathbf{g} z_i \tag{2-20}$$

$$\Delta E_K = \sum_{\substack{\text{all} \\ \text{effluent} \\ \text{streams}}} \frac{m_j \mathbf{v}_j^2}{2} - \sum_{\substack{\text{all} \\ \text{influent} \\ \text{streams}}} \frac{m_i \mathbf{v}_i^2}{2} \tag{2-21}$$

and m_i and m_j are the flows of matter into and out of the system, respectively, during the interval under consideration.

The special case of one entering and one leaving stream dictates

$$m = m_i = m_j$$

which results in

$$\Delta H = m\left(h_j - h_i\right) = m\Delta h \tag{2-22}$$

$$\Delta E_P = m\mathbf{g}\left(z_j - z_i\right) = m\mathbf{g}\Delta z \tag{2-23}$$

$$\Delta E_K = \frac{m}{2}\left(\mathbf{v}_j^2 - \mathbf{v}_i^2\right) = \frac{m\Delta\mathbf{v}^2}{2} \tag{2-24}$$

Where thermal effects are present they commonly predominate, and the most frequently used form of the first law for steady-state open systems is

$$\Delta H = Q + W_s \tag{2-25}$$

While this equation bears a close resemblance to the first law for a closed system [Eq. (2-2)], it must be remembered that W_s is not the total work but the shaft, or net, work exchanged. For a steady state with one inlet and outlet stream the relationship between these quantities is easily visualized. We focus on a unit of fluid as it passes through the system entering with P_1, v_1 and leaving with P_2, v_2. For the unit of fluid the total work is[12]

$$W = -\int P\, dv$$

and the fluid work is

$$W_f = P_1 v_1 - P_2 v_2$$

From Eq. (2-14) the shaft work is

$$W_s = -\int P\, dv - \left(P_1 v_1 - P_2 v_2\right)$$
$$= \Delta(Pv) - \int P\, dv$$
$$= \int P\, dv + \int v\, dP - \int P\, dv$$
$$= \int v\, dP$$

To appreciate the significance of this result, it is instructive to consider a specific example, the compression of a gas, with the process represented on the PV plot of Fig. 2-2. The compressor is considered to be a cylinder fitted with inlet and outlet valves and a piston. The compression process consists of a cycle of three steps, intake, compression, and discharge. These steps are shown in Fig. 2-2 as 0 to 1, 1 to 2, 2 to 3, respectively. Also, in Fig. 2-3 the position of the valves and piston, as well as the direction of travel of the piston, are shown at a point on each of the steps marked a, b, and c. We take as our system the quantity of gas acted on by one cycle, and therefore for each step, assumed reversible, the work is $-\int P\, dV$.

For the intake step the entering gas moves the piston back against a constant force equal to $P_1 A$ (where A is the area of the piston face). The work done by the gas is

---

[12] We are now dealing with a closed system—a unit of fluid.

$-P_1(V_1 - 0)$, where $V_1 = mv_1$. This work is represented in Fig. 2-2 by the area of the rectangle $P_1 V_1$.

During the compression step the work done on the gas is $-\int_{v1}^{v2} P\, dV$ and is represented in Fig. 2-2 as the area enclosed by $1\text{-}2\text{-}V_2\text{-}V_1$.

In the discharge step which occurs at the constant downstream pressure P_2, the work done on the gas is $-P_2(0-V_2)$, where $V_2 = mv_2$. This work is represented by the rectangle $P_2 V_2$ in Fig. 2-2.

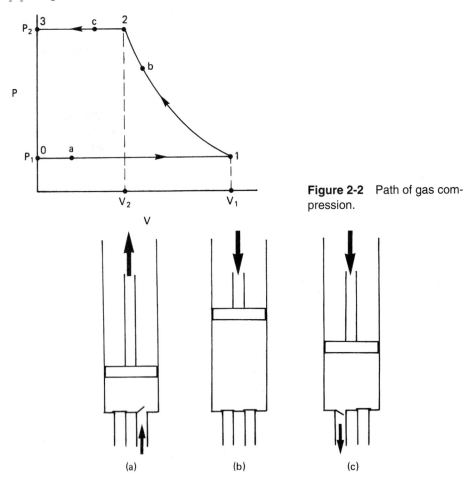

Figure 2-2   Path of gas compression.

Figure 2-3   Gas compression steps. (a) Intake step: intake valve open, discharge valve closed, (b) Compression step: intake and discharge valves closed, (c) Discharge step: intake valve closed, discharge valve open.

The net work for this flow process is obtained by algebraic addition of work for the individual steps or the areas displayed in Fig. 2-2. By noting that we have designated the sign of the rectangle P_1V_1 negative and the signs of the area 1-2-V_2-V_1 and the rectangle P_2V_2 positive, the net area is seen to be 0-1-2-3, or the area between the curve 1-2 and the P axis. This, of course, is $\int V\,dP$ and is in agreement with Eq. (2-26).

Equations (2-17) and (2-18) are the most general first law statements for unsteady-state and steady-state open systems, respectively. In the analysis of any given problem we first determine which equation applies and then begin to examine our system in regard to each term in the equation. This systematic approach assures us that nothing has been forgotten, and our treatment of each term allows us to readily identify assumptions or approximations made. Application of these equations will be demonstrated in the following examples.

EXAMPLE 2-6

Many industrial plants generate their own steam. The boiler cannot be operated so as to respond immediately to demand fluctuations, and it is convenient to use an accumulator in the steam supply system. This is simply a large insulated tank in which steam can be stored when demand is low and withdrawn when demand is high. In this particular plant the accumulator has a volume of 10 m³ and is initially two-thirds full of water at 80°C. How much steam saturated at 15 bar can be stored? After the accumulator is fully loaded with the 15-bar steam, how much steam at 5 bar can be withdrawn?

Solution 2-6

Initially, the accumulator contains liquid water and saturated vapor at 80°C. Compared to the liquid, the mass of vapor is negligible. Steam from a steady supply enters the accumulator and continues to flow until the accumulator pressure reaches 15 bar. This is an unsteady-state process and requires the use of Eq. (2-17), which we write as

$$\Delta U = \int_0^{m_i} h_i\,dm_i = h_i m_i$$

Because the influent steam is always saturated at 15 bar, the integration is easily carried out where m_i is the mass of steam which has entered the accumulator when the pressure reaches 15 bar. The following data are read from Table C-1 of the Appendix:

at 80° $\quad u^L = 334.8$ kJ / kg $\quad\quad v^L = 0.001029$ m³ / kg

at 15 bar $\quad u^L = 843.16$ kJ / kg $\quad\quad v^L = 0.001154$ m³ / kg

$$u^G = 2594.5 \text{kJ} / \text{kg} \qquad\qquad v^G = 0.13177 \text{m}^3 / \text{kg}$$
$$h^G = 2792.2 \text{kJ} / \text{kg}$$

The first law statement is now

$$(m_o + m_i)\left[2594.5x + 843.16\,(1-x)\right] - 334.8m_o = 2792.2m_i$$

where x is the fraction of the accumulator contents by weight which is vapor and m_o is the original mass of water in the accumulator. A volume statement for the original condition yields

$$10(0.667) = 0.001029m_o$$

$$m_o = 6482 \text{ kg}$$

Another volume statement provides the additional equation needed for the determination of the two unknowns, m_i and x:

$$10 = (6482 + m_i)\left[0.13177x + 0.001154(1-x)\right]$$

When these two equations are solved simultaneously, we obtain

$$m_i = 1695 \text{kg} \quad x = 0.000533$$

With the addition of 1695 kg of steam the accumulator is almost completely filled with saturated water at 15 bar (198.3°C).

For the withdrawal of steam from the accumulator we again employ Eq. (2-17), which reduces to

$$\Delta U = -\int_0^{m_j} h_j \, dm_j$$

The steam leaving the accumulator will always be saturated, but it is initially at 15 bar and finally at 5 bar. The enthalpies for these conditions are 2792.2 and 2748.7 kJ/kg, respectively. As this is a small percentage change, we are justified in employing an average enthalpy of 2770.5 kJ/kg so that the integral becomes $2770.5m_j$, and the first law statement becomes

$$(m_o - m_j)\left[u^G x + u^L(1-x)\right] - m_o\left[2594.5x_o + 843.16(1-x_o)\right] + 2770.5m_j = 0$$

Our volume statement is now

$$10 = (m_o - m_j)\left[v^G x + v^L(1-x)\right]$$

From the first part of the problem we have

$$m_o = 6482 + 1695 = 8177 \text{ kg}$$

$$x_o = 0.000533$$

From Table C-1 we obtain the following saturation properties at 5 bar:

$$u^L = 639.68 \text{ kJ / kg} \qquad v^L = 0.001093 \text{ m}^3 / \text{kg}$$
$$u^G = 2561.2 \text{ kJ / kg} \qquad v^G = 0.3749 \text{ m}^3 / \text{kg}$$

Substitution of these values into the preceding two equations and subsequent simultaneous solution yields

$$m_j = 781 \text{ kg}$$

EXAMPLE 2-7

Wet steam containing 2% by weight of entrained liquid (98% quality) at 5 bar is available at the rate of 1 kg/s. It is desired to mix this wet steam with steam at 5 bar and 200°C to obtain dry saturated steam at 5 bar. The mixing will be considered adiabatic. At what rate should the superheated steam be added?

Solution 2-7

For this steady-state flow process Eq. (2-18) simplifies to

$$\Delta H = 0$$

From Table C-1 of the Appendix the following enthalpy values are obtained:

sat. liquid	640.23 kJ / kg
sat. vapor	2748.7 kJ / kg
superheated vapor	2855.4 kJ / kg

Based on Eq. (2-19), for each kilogram of wet steam and x kg of superheated steam we write

$$\Delta H = 0 = 2748.7(1 + x) - [0.98(2748.7) + 0.02(640.23)] - 2855.4x$$

$$x = 0.392 \text{ kg}$$

One kg/s of wet steam and 0.392 kg/s of the superheated steam when mixed adiabatically produce 1.392 kg/s of *dry* saturated steam.

EXAMPLE 2-8

Compare the work of pumping a liquid to that of compressing a vapor between the same two pressures. Use water as the fluid.

Solution 2-8

For steady-state flow the work is

$$W_s = \int v \, dP$$

For the vapor we will choose an isothermal path and evaluate the integral graphically from Pv data taken from Table C-2. We choose $t = 200°C$, $P_1 = 0.1$ MPa, and $P_2 = 0.5$ MPa and read the following values:

P(MPa)	0.1	0.2	0.3	0.4	0.5
v(m³/kg)	2.172	1.0803	0.7163	0.5342	0.4249

These data are plotted in Fig. 2-4 as v vs. P,

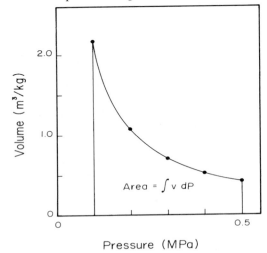

Figure 2-4 Graphical evaluation of $\int v \, dP$.

and the area under the curve from $P_1 = 0.1$ MPa to $P_2 = 0.5$ MPa is found by counting squares to be

$$\text{area} = \int v \, dP = 0.339 \frac{\text{m}^3 \cdot \text{MPa}}{\text{kg}}$$

The work is

$$W_s = \int v \, dP = 0.339 \frac{\text{m}^3}{\text{kg}} \times 10^6 \frac{\text{N}}{\text{m}^2} = 339 \text{ kJ / kg}$$

It is customary to regard the liquid as incompressible and write

$$W_s = \int v \, dP = v \Delta P$$

At 200°C the specific volume of liquid water is 0.00116 m³/kg, and for the same pressure change of 0.4 MPa we have

$$W_s = 0.00116(0.4)(10^6) = 0.46 \text{ kJ / kg}$$

It is seen that considerably less work is required to pump a liquid than to compress a vapor.

PROBLEMS

2-1. Use the first law to show why leaving the refrigerator door open will not cool the kitchen.

2-2. One kmol of an ideal gas originally at 300 K and 1 bar is heated at constant pressure to a temperature of 406 K and then compressed isothermally to a volume equal to its initial volume. The gas has a constant-pressure heat capacity (assumed constant) of 30 kJ/kmol · K. For this two-step process, provide as much information as possible regarding (a) ΔU, (b) ΔH, (c) Q, (d) W.

2-3. A mol of ideal gas undergoes a process such that $V_1 = V_2$, although the volume does not remain constant during the process. Indicate whether the following statements are true or false and justify your answer:

(a)　$\Delta U = \int C_V \, dT$ for reversible or irreversible process.
(b)　$W = 0$ only for a reversible process.
(c)　$Q = \int C_V \, dT$ only for a reversible process.

2-4. The heating of one kmol of an ideal gas at constant volume from 300 K to 400 K requires 2500 kJ. Use this information and assume constant heat capacities to calculate the heat required to heat the gas at constant pressure between these two temperatures.

2-5. A closed rigid vessel having a volume of 1 m³ is filled with steam at 10 bar and 573 K. Heat is removed until the temperature reaches 473 K. Determine the quantity of heat removed.

2-6. What can you say about ΔU, Q, and W for the following changes?

    (a)   A chemical reaction occurs irreversibly in a well-insulated bomb. There is an increase in the number of mols such that pressure increases considerably.

    (b)   One kmol of an ideal gas (c_p = 30 kJ/kmol · K, c_V = 21.7 kJ/kmol · K) undergoes a process where its temperature changes from 290 to 340 K. The gas volume is the same at the beginning and end of the process, although during the process the volume may change.

    (c)   One kmol of the preceding gas expands irreversibly and adiabatically between P = 10 bar, T = 400 K and P = 1 bar, T = 200 K.

2-7. A bomb having a volume of 0.028 m³ contains 0.45 kg of water. Initially, the bomb and contents are at 278 K.

    (a)   The bomb is heated to 422 K. What is the pressure inside the bomb? How much heat has been transferred to the bomb?

    (b)   At what temperature will the bomb contain dry saturated steam?

2-8. A tank of 1 m³ volume contains 1.651 kg of water and the pressure is 0.15 MPa.

    (a)   How much heat must be added in order that the tank contain saturated steam of 100% quality?

    (b)   What will be the pressure in the tank?

2-9. A rigid tank contains 0.28 m³ of steam at 7 bar and 478 K and 0.21 kg of saturated liquid water at 373 K. The steam and water are initially separated by a partition. If the partition is broken and the steam and water allowed to mix, how much heat will have to be added or removed in order that the tank contain dry saturated steam? What will be the pressure in the tank?

2-10. Two cylinders, each with a volume of 0.01 m³, are connected and are immersed in a water bath so that their contents are always at 300 K. Initially, one cylinder contains an ideal gas at a pressure of 5 bar and the other is evacuated. A valve is opened and the pressures are allowed to equalize. For this change calculate

    (a)   the total heat exchange between bath and cylinders resulting from this change.

    (b)   the heat exchange between the high-pressure cylinder and the bath.

2-11. A chemical reaction involving only gaseous species occurs in a constant-volume bomb of 0.01-m³ capacity. During the reaction the bomb is insulated and loses no heat. The reaction goes to completion, and the temperature is observed to rise from 298 to 415 K while the pressure rises from 1 to 3.22 bar. The insulation is then removed, and the bomb and contents are cooled to 298 K by the removal of 25.52 kJ of heat. No reaction occurs during this cooling step, and the final pressure is 2.31 bar.

    (a)   Calculate ΔU and ΔH for the two-step process that begins and ends at 298 K.

    (b)   If, instead of in a bomb, the reaction were carried to completion isothermally at 298 K and at a constant pressure of 1 bar, calculate the heat effect .

2-12. An ideal gas (c_p = 30 kJ/kmol · K) flows steadily through a long capillary tube that is well insulated. The gas enters the capillary tube at 2 bar and 300 K and leaves at 1 bar. What is its exit temperature?

2-13. Steam at 7 bar is flowing through a pipe. The packing around a valve is defective so that steam leaks slowly to the atmosphere. A thermometer placed in the dry steam issuing from the leak reads 394 K. What is the quality (the fraction that is vapor) of the steam in the pipe?

2-14. An ideal gas (c_p = 30 kJ/kmol · K) is compressed at the rate of 1 kmol/min from 1 to 10 bar. The gas enters the compressor at 294.4 K and leaves at 405.6 K. Cooling water circulates through the compressor at a rate of 1 kg/s and undergoes a temperature rise of 15.6 K. How much horsepower is used in this compression? Does your answer depend on the process being reversible? Explain.

2-15. A turbine operating adiabatically is fed with steam at 400°C and 8.0 MPa at the rate of 1000 kg/h. Process steam saturated at 0.5 MPa is withdrawn from an intermediate location in the turbine at the rate of 300 kg/h and the remaining steam leaves the turbine saturated at 0.1 MPa. What is the power output of the turbine?

2-16. Dry saturated steam at 0.8 MPa flows through a heater and emerges at 0.8 MPa and 400°C. The steam then flows through a turbine operating adiabatically and exits saturated at 0.1 MPa. How much

  (a) heat is added to each kg of steam flowing through the heater?
  (b) work is done by each kg of steam flowing through the turbine?

2-17. The hydraulic ram is a device that requires no work but uses the kinetic energy of a moving column of water to raise a portion of the water to a higher elevation.[13] The ram is located below a source and is connected by a "fall pipe" in which the water gains kinetic energy.

A farmer has a stream running through her property, which she has dammed to form a pond. Just below the pond the stream falls into a ravine, where the elevation is 30 ft below the surface of the pond. The farmer is considering locating a hydraulic ram on this spot to lift water for irrigation purposes to an elevation 10 ft above the level of the pond. If friction and other sources of inefficiency were absent, what fraction of the water entering the ram could be raised to the higher elevation?

2-18. A hydroelectric power station takes water from a dam to operate a turbine. The water level in the reservoir is 300 feet above the turbine, and water leaving the turbine passes through a diffuser which spreads the horizontally traveling exit water stream with the same type of pattern produced by the nozzle on a garden hose. Some of the water leaving the diffuser reaches a height of 20 feet above the turbine. A 3-foot diameter pipe connects the turbine and diffuser. From this limited information estimate the power generated by the station. Identify sources of error and their effect upon your estimate.

---

[13] See G. G. Brown et al., *Unit Operations*, Wiley, New York, 1950, Chap. 14.

2-19. Figure 15-21 shows a Rankine cycle. For the following conditions at locations A, B, E, and F, calculate Q_1, Q_2, W_p, and W_T based on 1 kg of water flowing through the cycle. The turbine operates adiabatically.

POSITION

	A	B	E	F
t (°C)	54	54	350	54
P (MPa)	0.015	1.0	1.0	0.015
State	Liquid	Liquid	S.h. vapor	Sat. vapor

Also, apply the first law to the closed system of the power plant. What fraction of the heat supplied, Q_2, is converted into work?

2-20. Low-pressure steam, saturated at 0.200 MPa, is upgraded by an adiabatic compression to 0.600 MPa, where its temperature is 250°C. Based on 1 kg of flowing steam, find the

    (a)   work required for the compression.
    (b)   heat available if the stream is condensed to saturated liquid at 0.600 Mpa.
    (c)   fraction of this heat represented by latent heat and the temperature at which it is delivered.

2-21. Steam at 2 MPa and 300°C enters a long underground pipeline and emerges at 1.8 MPa and 225°C. Calculate the heat loss per kilogram of flowing steam. If the pipeline could be insulated so that the heat loss were only 15 kJ per kg of flowing steam, what would be the exit steam temperature if the exit pressure stayed at 1.8 MPa?

2-22. A water chiller, fed with a continuous stream of water at 84°F, produces a continuous stream of chilled water at 55°F by adiabatic evaporation. An ejector continuously removes the water vapor, which is saturated at 55°F. What fraction of the entering water must be evaporated?

2-23. Your hometown plans to install a water fountain in the city park. The proposed design calls for a pool 20 ft in diameter and 2 ft deep with a pump mounted above the pool surface and having a short, vertical discharge pipe. The pump takes water from the pool and directs it vertically to form the fountain. It is desired that the fountain rise 20 ft above the pool surface. According to the plans, the pump's discharge pipe is 2 in. (I.D. = 2.07 in.) and terminates at a distance of 2 ft above the pool surface. Estimate the horsepower requirement, ignoring frictional losses.

2-24. An ejector is a device in which the kinetic energy of a high-velocity fluid stream is used to entrain and compress a second fluid stream. There are no moving parts and adiabatic operation is approximated. A particular ejector uses saturated steam at 0.8 MPa to remove saturated water vapor at 0.025 MPa from an evaporator. One kilogram of high-pressure steam entrains 0.75 kg of vapor from the evaporator, and the mixed stream leaves the ejector at a pressure of 0.1 MPa. What is the state of the exit stream?

2-25. A rigid, well-insulated tank contains 100 kg of liquid water at 21°C. The tank is connected through a valve to a steam main where the pressure is 1 MPa. The valve is opened and steam enters the tank until the water temperature rises to 65°C, at which time the valve is closed. It is determined that there are now 107.96 kg of water in the tank. Estimate the quality of steam in the main.

2-26. A well-insulated flask, initially containing liquid water at 24°C, is attached to a vacuum pump that removes water vapor from the flask. As the system is adiabatic, the heat of vaporization is supplied by the water with a concomitant cooling of the water. If the vacuum pump runs until the flask contains ice at 0°C, what fraction of the original water remains as ice? The heat of fusion of water is 143.4 Btu/lb.

2-27. A constant-volume bomb of capacity 0.028 m³ contains an ideal gas at a pressure of 10 bar and a temperature of 311 K. Connected to the bomb is a capillary tube through which the gas may slowly leak out into the atmosphere. Surrounding the bomb and capillary is a water bath, which keeps the bomb and its contents at 311 K. Find the quantity of heat exchanged between the bomb and the bath when gas no longer escapes from the bomb. The gas has a constant c_p of 30 kJ/kmol · K.

2-28. A constant-volume bomb of 0.05 m³ capacity initially contains an ideal gas at a pressure of 1 bar. Gas from a reservoir held at 10 bar and 300 K flows into the bomb until the pressure in the bomb reaches 5 bar. The bomb is immersed in a constant-temperature water bath which keeps the bomb and its contents at 300 K. During the interval when the bomb pressure changes from 1 to 5 bar, how much heat is added or removed by the water bath? Specify if heat is added or removed. The gas has a constant c_p of 30 kJ/kmol · K.

2-29. A well-insulated, evacuated tank of 0.1 m³ volume is attached to a supply system containing an ideal gas at 10 bar and 300 K. A valve is opened and the tank is filled to a pressure of 10 bar. What is the temperature of gas in the tank? The gas has a constant c_p of 30 kJ/kmol · K.

2-30. A bomb contains 5 kg of water (liquid and vapor) and is kept at a temperature of 162 ± 0.1° C by means of a surrounding temperature bath. Attached to the bomb is a long capillary tube through which steam leaks slowly to the atmosphere. The capillary is also immersed in the bath so that the leaving steam is at the bath temperature. How much heat must be supplied in order to vaporize 1 kg of the water in the bomb?

2-31. An insulated, evacuated tank having a volume of 1.42 m³ is attached to a steam line containing steam at 3.5 bar and 422 K. The steam is allowed to flow into the tank until the pressure rises to 3.5 bar. Assuming the tank to have negligible heat capacity, how many kilograms of steam enter the tank?

2-32. Cylinders of compressed gas are being filled from a supply system where the pressure and temperature are 13.6 bar and 294 K. The cylinders are originally evacuated and then connected to the supply system and filled rapidly until the cylinder pressure is equal to the supply pressure. The cylinder is then disconnected and stored in a room where the temperature is 294 K. Assuming that filling occurs adiabatically, what will be the cylinder pressure when the cylinder and contents have reached 294 K? Assume ideal-gas behavior and a constant c_p of 30 kJ/kmol · K.

CHAPTER 3

The Behavior of Fluids

A great many applications of thermodynamics are concerned with the behavior of fluids, mainly gases. Providing the data needed to evaluate thermodynamic property changes are available, the first and second laws permit us to calculate heat and work effects associated with various types of processes involving fluids. As will be seen in Chap. 5, this requires *PVT* data as well as thermal data such as heat capacities. Because of the importance of *PVT* data, this chapter begins with a general description of fluid behavior and its experimental determination. This is followed by a discussion of equations of state. Several specific equations which enjoy wide usage in chemical engineering are presented. Next, heat and work effects for an ideal gas undergoing specific processes are calculated. These results are useful because they very often approximate real-gas behavior. In addition, the use of a substance so simply described allows the illustration of principles without the encumbrance of lengthy computations. Finally, two approaches to generalized equations of state are presented: one utilizing the compressibility factor and the other based on a modified van der Waals equation. These approaches allow the *PVT* behavior of a gas to be estimated from only a knowledge of its critical temperature and pressure.

3-1 THE *PVT* BEHAVIOR OF FLUIDS

A pure fluid is in an equilibrium state when two intensive variables are fixed and all properties are uniform throughout.[1] Because three variables characterize the behavior of a fluid, a complete representation would require a three-dimensional *PVT* plot. Such a plot can be viewed by means of the tutorial program PVT3D that runs in the software EQUATIONS OF STATE on the CD-ROM accompanying this text.[2] Lacking a three-dimensional plot, the *PVT* behavior of a fluid is usually displayed on a *Pv* diagram on which constant temperature curves, isotherms, are drawn. Figure 3-1 shows the general behavior of a pure fluid on these coordinates.

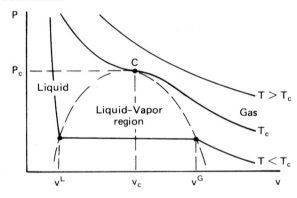

Figure 3-1   *Pv* diagram for a pure substance

Three isotherms are shown: the critical temperature, one higher temperature, and one lower temperature. Above the critical temperature where no liquid phase can exist, isotherms simply show the volume decreasing with increasing pressure in a monotonic fashion. Below the critical temperature isotherms show discontinuity at the two-phase envelope: the dashed curve. To the left of this envelope is the liquid region where it is seen that the isotherm is represented by an almost vertical line showing that the volume of a liquid changes only slightly with large changes in pressure. As the temperature is increased to approach the critical temperature, where liquid and vapor become indistinguishable, the length of the hori-

---

[1] This also applies to nonreacting mixtures as long as the composition is specified. For example, application of the phase rule to air (21% oxygen, 79% nitrogen) results in $C = 2$, $\pi = 1$, and $\mathscr{F} = 3$; however, specification of composition reduces the degrees of freedom to 2.

[2] See App. E for a description of this software.

zontal segment, $v^G - v^L$, approaches zero. This means that at the critical point the critical isotherm exhibits a point of inflection where

$$\left(\frac{\partial P}{\partial v}\right)_{T_c} = 0 \qquad (3\text{-}1)$$

$$\left(\frac{\partial^2 P}{\partial v^2}\right)_{T_c} = 0 \qquad (3\text{-}2)$$

Sometimes *PVT* data are plotted as isochores—lines of constant density—on a *PT* diagram. Figure 3-2 shows the general behavior of a pure fluid on such a plot. The heavy curve terminating at the critical point C is the vapor pressure curve which separates the vapor and liquid regions. Isochores are shown as dashed curves emanating from the vapor pressure curve. If one seals in a tube a liquid-vapor mixture of a pure substance, represented by point A, and raises the temperature, the point representing the state of the system moves along the vapor pressure curve until one of three things occurs. If the overall density is greater than that at the critical point, the meniscus separating the phases will rise in the vertically held tube until at point B the tube contains all liquid. On further heating the system traces a path along a liquid isochore, BE. If the overall density is less than the critical density, the meniscus falls on heating until point D is reached, where the tube contains all vapor. Further heating moves the system along a vapor isochore, DG. If the tube is charged with exactly the critical density, the meniscus remains near the center of the tube until the critical point C is reached, where it then disappears. On further heating the system follows the critical isochore, CF. This is found to have the same slope as the vapor pressure curve at the critical point. The sealed-tube, disappearing-meniscus technique has been widely used to determine the critical temperature of substances.

In Figure 3-2, a third fluid-phase region is shown: the supercritical fluid region lying beyond the critical point where T exceeds T_c and P exceeds P_c. The supercritical fluid has a

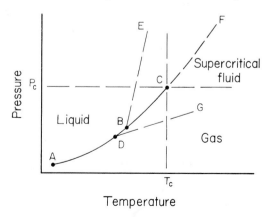

Figure 3-2 *PT* diagram for a pure substance.

density close to that of a normal liquid, but possesses a much lower viscosity and a higher diffusivity. This facilitates mass transport and is one of the reasons for the recent interest in supercritical fluids as extraction and leaching solvents.[3] They are especially attractive in treating food products because they readily evaporate and leave no undesirable solvent residue.[4]

The experimental determination of *PVT* data usually follows one of two courses: measurement of pressure as a function of volume at a series of fixed temperatures or measurement of pressure as a function of temperature at a series of fixed volumes. While these measurements are easy to conceptualize, the task of obtaining data of sufficient accuracy for thermodynamic calculations is quite demanding. These measurements, detailed in the literature,[5] are characterized by a fairly complex apparatus and painstaking technique.

Figure 3-3 shows the apparatus used by Beattie,[6] which consisted of four major parts: a bomb containing the gas, a mercury injector, a deadweight gauge, and a pressure-balancing device. A known quantity of gas was weighed into the bomb, which had an accurately known volume. The introduction or removal of known volumes of mercury from the calibrated mercury injector determined the gas volume. Corrections to the bomb volume were made for the effects of temperature and pressure. The pressure was measured with an oil deadweight gauge. An electrically actuated device allowed the balancing of the oil and mercury pressures. The accuracy of the results was estimated by Beattie to be 0.004% for pressure, 0.01–0.02°C for temperature, 0.05–0.1% for volume, and 0.2 mg for the mass.

3-2 EQUATIONS OF STATE

Often, *PVT* data are fitted to algebraic equations called "equations of state." Many such equations of varying degrees of complexity have been proposed, and while a few have some basis in molecular theory, all are used empirically. The advantages offered by equations of state are mainly data reduction and ease of use in subsequent calculations. A short tutorial program, EQSTATE, that runs in the software EQUATIONS OF STATE provides general information regarding equations of state.[7]

---

[3] For a summary of industrial applications, see E. Stahl, K. W. Quirin, and D. Gerard, *Dense Gases for Extraction and Refining*, Springer-Verlag, Berlin, 1988 and M.A. McHugh and V.J. Krukonis, *Supercritical Fluid Extraction,* 2nd ed., Butterworth-Heinemann, Boston, 1994.

[4] For example, caffeine is removed from coffee by extraction with supercritical carbon dioxide.

[5] See J. S. Rowlinson, "The Properties of Real Gases," in *Handbuch der Physik*, Vol. 12 (S. Flügge, ed.), Springer, Berlin, 1958, Chap. 1, or B. LeNeindre and B. Vodar, eds., *Experimental Thermodynamics*, Vol. 2, Butterworth's, London, 1971.

[6] J. A. Beattie, *Proc. Am. Acad. Arts Sci., 69,* 389 (1934).

[7] See App. E for a description of this software.

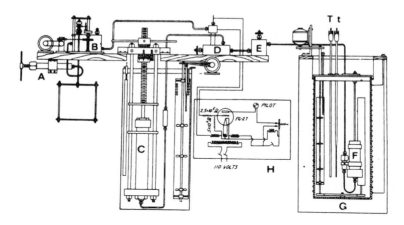

A and B	Dead–weight pressure gauge
C	Calibrated mercury injector
D, E, and H	Electrically–actuated pressure balance
F	Bomb
G	Thermostat
T	Measured temperature
t	Thermostat control temperature

Figure 3-3 Beattie's *PVT* apparatus. [Reprinted with permission from J. A. Beattie, *Proc. Am. Acad. Arts Sci.*, *69*, 389 (1934).]

Virial Equation. The virial equation expressing the compressibility factor Z as a power series in density, or reciprocal volume, can be derived from statistical mechanics:

$$Z = \frac{Pv}{RT} = 1 + \frac{B}{v} + \frac{C}{v^2} + \ldots \tag{3-3}$$

The parameters B and C are called second and third virial coefficients, respectively, and are functions only of temperature. This equation finds its greatest use at low to moderate pressures where it can be safely truncated after the second term. Its great popularity is undoubtedly due to its theoretical derivation, which also prescribes how the coefficients for a mixture depend on composition. This equation forms the basis for a good number of vapor-liquid equilibrium correlation and prediction methods.[8]

Because it is pressure-explicit, Eq. (3-3) is often more difficult to use than a corresponding volume-explicit series

$$Z = \frac{Pv}{RT} = 1 + B'P + C'P^2 + \ldots \tag{3-3a}$$

---

[8]For example, see J. M. Prausnitz, *Molecular Thermodynamics of Fluid-Phase Equilibria,* Prentice-Hall, Englewood Cliffs, NJ, 1969, Chap. 5.

Like B and C of Eq. (3-3), B' and C' are functions only of temperature. The two sets of parameters are related by

$$B' = \frac{B}{RT}$$

$$C' = \frac{C - B^2}{(RT)^2}$$

The van der Waals Equation.   Undoubtedly, the best known equation of state is the van der Waals equation:

$$P = \frac{RT}{v - b} - \frac{a}{v^2} \tag{3-4}$$

According to van der Waals's derivation, the term a/v^2 accounts for attractive forces between molecules, and b is a correction for the volume occupied by the molecules themselves.

     This equation is cubic in v. At temperatures below the critical, there are three real roots, as indicated in Fig. 3-4. It can be shown that the segment of the isotherm lying between A and B represents unstable states, and therefore in the pressure range P_A to P_B there are only two values of v that correspond to stable states. While this suggests

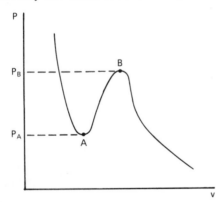

Figure 3-4   Van der Waals isotherm below T_c.

that the application of the van der Waals equation is appropriate for to calculations representing the two-phase region, the form of the equation is far too simple to be capable of closely representing the isotherm in this region.

     When an equation of state contains only two parameters, it is always possible to express these parameters in terms of the critical temperature and pressure by applying Eqs. (3-1) and (3-2). Evaluating these derivatives and the equation of state at the critical point results in three equations involving a, b, T_c, P_c, and v_c. Because v_c is not always available, it can be eliminated and the remaining equations solved for a and b in terms of T_c and P_c. For the van der Waals equation, this results in

$$a = \frac{27R^2T_c^2}{64P_c} \qquad b = \frac{RT}{8P_c}$$

Because T_c and P_c are known for a great many substances, it is convenient to evaluate parameters in this manner. However, it should be recognized that the minimum amount of data has been used and from a region distant from where the equation will usually be applied. Parameters so evaluated would certainly be inferior to those evaluated from sufficient data in the region of intended use, but they can be used to provide rough estimates of *PVT* behavior in the absence of other experimental information.

Modified van der Waals Equations. We have just seen that van der Waals isotherms roughly approximate fluid behavior in the two-phase region. The equation is therefore a logical starting point in the development of a fairly simple equation of state capable of reasonably representing *PVT* data in this region. Many modifications have been proposed; three that appear regularly in the chemical engineering literature[9] are the Redlich-Kwong equation[10]

$$P = \frac{RT}{v-b} - \frac{a}{T^{1/2}v(v+b)} \tag{3-5a}$$

the Soave equation[11]

$$P = \frac{RT}{v-b} - \frac{a(T)}{v(v+b)} \tag{3-5b}$$

the Peng-Robinson equation[12]

$$P = \frac{RT}{v-b} - \frac{a(T)}{v^2 + 2bv - b^2} \tag{3-5c}$$

All three equations retain the repulsive term and introduce temperature dependence into the attractive term of the original van der Waals equation. All three are capable of representing vapor and liquid phases reasonably well.

Benedict-Webb-Rubin Equation. This equation of state has played a prominent role in the calculation of thermodynamic properties and phase equilibria of light hydrocarbons and their mixtures. The original eight-parameter version is[13]

[9]See R. C. Reid, J. M. Prausnitz, and B. E. Poling, *The Properties of Gases and Liquids,* 4th ed., McGraw-Hill, New York, 1987, Chap. 3.
[10]O. Redlich and J. N. S. Kwong, *Chem. Rev., 44,* 233 (1949).
[11]G. Soave, *Chem. Eng. Sci., 27,* 1197 (1972).
[12]D. Y. Peng and D. B. Robinson, *Ind. Eng. Chem. Fund., 15,* 59 (1976).
[13]M. Benedict, G. B. Webb, and L. C. Rubin, *J. Chem. Phys., 8,* 334 (1940).

$$P = RT\rho + \left(B_0 RT - A_0 - \frac{C_0}{T^2}\right)\rho^2 + (bRT - a)\rho^3 + a\alpha\rho^6 + \frac{c\rho^3}{T^2}\left(1 + \gamma\rho^2\right)e^{-\gamma\rho^2}$$

Several modified versions contain additional parameters to improve its ability to accurately fit *PVT* data. This equation is capable of closely representing both liquid and vapor phases. A discussion of the application of this equation and a tabulation of its parameters for 33 substances is given by Reid et al.[14]

3-3 THE IDEAL GAS

As the pressure is decreased and the temperature increased, all gases approach ideal-gas behavior, and for many calculations the assumption of ideal-gas behavior is reasonable. It is therefore worthwhile to examine the behavior of an ideal gas in some detail, recognizing that the ideal gas is a simplified model of real-gas behavior. An ideal gas obeys the equation of state[15]

$$PV = nRT \qquad \text{or} \qquad Pv = RT \qquad R = 8314.3 \text{ N} \cdot \text{m} / \text{kmol} \cdot \text{K} \qquad (3\text{-}6)$$

and its enthalpy or internal energy depends only on temperature. When applied to Eqs. (2-10) and (2-11), the ideal gas definition results in

$$\Delta u = \int c_V \, dT \qquad\qquad (3\text{-}7)$$

$$\Delta h = \int c_P \, dT \qquad\qquad (3\text{-}8)$$

or

$$\frac{du}{dT} = c_V \qquad\qquad (3\text{-}9)$$

$$\frac{dh}{dT} = c_P \qquad\qquad (3\text{-}10)$$

Based on Eq. (2-3), the enthalpy and internal energy of an ideal gas are related by

$$h = u + RT \qquad\qquad (3\text{-}11)$$

Differentiation with respect to *T* yields

---

[14] R. C. Reid, *op. cit.*
[15] See inside, front cover for a listing of *R* in various units.

$$\frac{dh}{dT} = \frac{du}{dT} + R$$

and from Eqs. (3-9) and (3-10) we obtain

$$c_P = c_V + R \tag{3-12}$$

Note that because u and h depend only on T the same is true for c_p and c_v.

The Isothermal Process.   Here we consider the reversible isothermal compression or expansion of an ideal gas. From Eq. (3-7) the condition of isothermality results in $\Delta u = 0$, and for a closed system the first law becomes

$$Q = -W \tag{3-13}$$

The work per mol of gas is calculated from

$$W = -\int_{v_1}^{v_2} P \, dv$$

$$= -RT \int_{v_1}^{v_2} \frac{dv}{v} = -RT \ln \frac{v_2}{v_1} \tag{3-14}$$

For a compression $v_2 < v_1$ with $W > 0$, and for an expansion $v_2 > v_1$ with $W < 0$, which is consistent with our convention for the sign of W. From Eq. (3-13) it is seen that heat must be added during an expansion and removed during a compression to maintain a constant temperature. Because of isothermality and Eq. (3-6), the volume ratio v_2/v_1 in Eq. (3-14) may be replaced by P_1/P_2:

$$Q = -W = RT \ln \frac{P_1}{P_2} \tag{3-15}$$

For an open system the shaft work per mol of gas is

$$W_s = \int_{P_1}^{P_2} v \, dP = RT \int_{P_1}^{P_2} \frac{dP}{P} \tag{2-26}$$

$$= RT \ln \frac{P_2}{P_1} \tag{3-16}$$

This is seen to be identical with the isothermal work for a closed system.

The Adiabatic Process.   With $dQ = 0$, in a closed system we have

$$du = dW = -P \, dv = -RT \frac{dv}{v}$$

Because u depends only on T, we write

$$c_V \, dT = -RT \frac{dv}{v}$$

For constant c_V this integrates to

$$\frac{T_2}{T_1} = \left(\frac{v_1}{v_2} \right)^{R/c_V} \tag{3-17}$$

We now define $\gamma = c_P / c_V$, and from Eq. (3-12) write

$$\frac{R}{c_V} = \frac{c_P - c_V}{c_V} = \gamma - 1$$

so that Eq. (3-17) can now be written as

$$\frac{T_2}{T_1} = \left(\frac{v_1}{v_2} \right)^{\gamma - 1} \tag{3-18}$$

From the ideal gas law,

$$\frac{v_1}{v_2} = \frac{T_1 P_2}{T_2 P_1}$$

and substitution in Eq. (3-18) results in

$$\frac{T_2}{T_1} = \left(\frac{P_2}{P_1} \right)^{(\gamma - 1)/\gamma} \tag{3-19}$$

Elimination of T_2 / T_1 between Eqs. (3-18) and (3-19) yields

$$\left(\frac{v_1}{v_2} \right)^{\gamma - 1} = \left(\frac{P_2}{P_1} \right)^{(\gamma - 1)/\gamma} \tag{3-20}$$

or

$$P_1 v_1^\gamma = P_2 v_2^\gamma = \text{constant} \tag{3-21}$$

Equations (3-18), (3-19), and (3-21) each describe the path followed by an ideal gas undergoing a reversible adiabatic expansion or compression in a closed system.

For an open system a derivation paralleling that of Eq. (3-18) begins with

$$dh = dW_s = v \, dP$$

where

$$dh = c_P \, dT = RT \frac{dP}{P}$$

Integration of this equation leads to Eq. (3-19). Equations (3-18) and (3-21) also apply, and thus it is seen that the path followed by an ideal gas undergoing a reversible adiabatic process is the same for both closed and open systems. However, the reversible work is different. For a closed system,

$$W = \Delta u = c_V(T_2 - T_1) \qquad \longleftarrow \quad \text{closed} \qquad (3\text{-}22)$$

while for an open system,

$$W_s = \Delta h = c_P(T_2 - T_1) \qquad \longleftarrow \quad \text{open} \qquad (3\text{-}23)$$

To have the work expressed in terms of pressures, we write

$$W = -\int P \, dv$$

$$W_s = \int v \, dP$$

Substitute Eq. (3-21) and integrate to obtain

$$-W = \frac{RT_1}{\gamma - 1}\left[1 - \left(\frac{P_2}{P_1} \right)^{(\gamma-1)/\gamma} \right] \qquad (3\text{-}24)$$

and

$$-W_s = \frac{\gamma \, RT_1}{\gamma - 1}\left[1 - \left(\frac{P_2}{P_1} \right)^{(\gamma-1)/\gamma} \right] \qquad (3\text{-}25)$$

Comparison of Eqs. (3-24) and (3-25) as well as Eqs. (3-22) and (3-23) shows that W_s is larger than W by the factor γ.

EXAMPLE 3-1

One kmol of an ideal gas is taken through a four-step cyclic process as displayed on the Pv diagram of Fig. 3-5. The gas is subjected successively to an isothermal expansion at 600 K from 5 to 4 bar (A to B), an adiabatic expansion to 3 bar (B to C), a constant-pressure cooling (C to D), and constant-volume heating (D to A). All processes are assumed reversible. For these processes it is reasonable to assume c_p is constant and equal to 30 kJ/kmol · K. Calculate Q, W, Δu, and Δh for each step and for the entire process.

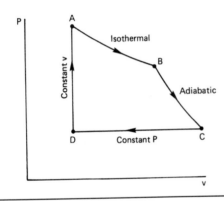

Figure 3-5

Solution 3-1

The following summarizes the state of the gas corresponding to points A, B, C, and D in Fig. 3-5:

$T_A = 600$ K	$T_C = 554$ K
$P_A = 5$ bar	$P_C = 3$ bar
$V_A = 9.98$ m^3	$V_C = 15.4$ m^3
$T_B = 600$ K	$T_D = 360$ K
$P_B = 4$ bar	$P_D = 3$ bar
$V_B = 12.5$ m^3	$V_D = 9.98$ m^3

The underlined values are fixed from the problem statement; the remaining values are calculated as follows:

V_A from T_A and P_A via Eq. (3-6).
V_B from T_B and P_B via Eq. (3-6).
T_C from T_B, P_B, and P_C via Eq. (3-19).
V_C from T_C and P_C via Eq. (3-6).
V_D must equal V_A.
T_D from V_D and P_D via Eq. (3-6).

We now calculate Δu, Δh, Q, and W for each step.

A to B, isothermal expansion

$$Q = -W = RT \ln \frac{P_1}{P_2}$$

$$Q = -W = 8.3143(600) \ln \frac{5}{4} = 1113 \text{ kJ}$$

$$\Delta u = 0$$

$$\Delta h = 0$$

B to C, adiabatic expansion

$$Q = 0$$

$$T_C = T_B \left(\frac{P_C}{P_B} \right)^{(\gamma - 1)/\gamma}$$

$$c_V = 30 - 8.3143 = 21.7 \text{ kJ / kmol} \cdot \text{K}$$

$$\gamma = \frac{30}{21.7} = 1.38$$

$$T_C = 600 \left(\frac{3}{4} \right)^{(1.38 - 1)/1.38} = 554 \text{ K}$$

$$\Delta u = c_V (T_C - T_B) = 21.7(554 - 600) = -998 \text{ kJ}$$

$$\Delta h = c_P (T_C - T_B) = 30(554 - 600) = -1380 \text{ kJ}$$

$$W = \Delta u = -998 \text{ kJ}$$

C to D, constant-pressure cooling

$$Q_P = \Delta h = c_P(T_D - T_C) = 30(360 - 554) = -5820 \text{ kJ}$$

$$\Delta u = c_V(T_D - T_C) = 21.7(360 - 554) = -4210 \text{ kJ}$$

$$W = \Delta u - Q = -4210 - (-5820) = 1610 \text{ kJ}$$

D to A, constant-volume heating

$$W = 0$$

$$\Delta u = c_V(T_A - T_D) = 21.7(600 - 360) = 5208 \text{ kJ}$$

$$\Delta h = c_P(T_A - T_D) = 30(600 - 360) = 7200 \text{ kJ}$$

$$Q = \Delta u = 5208 \text{ kJ}$$

These values are compiled in the following table:

Process		W	Q	Δu	Δh
A to B		-1113	1113	0	0
B to C		-998	0	-998	-1380
C to D		1610	-5820	-4210	-5820
D to A		0	5208	5208	7200
	Σ	-501	501	0	0

For the four-step cyclic process the enthalpy and internal energy changes are zero, as required for a state property. For $\Delta u = 0$ the first law requires $Q + W = 0$; it is seen that this condition is satisfied.

An extensive exercise of the first law is available on the tutorial program 1STLAW that runs in the software EQUATIONS OF STATE.

3-4 THE COMPRESSIBILITY FACTOR

Defined as Pv/RT, the compressibility factor Z is seen to be a dimensionless or reduced quantity. Many such reduced properties have been successfully correlated by the principle of corresponding states. This principle has some basis in molecular theory, but its justification is overwhelmingly empirical. It specifies that systems which are in the same reduced state exhibit the same reduced properties. The reduced state is defined in terms of reduced temperature T_r and reduced pressure P_r.

$$T_r = \frac{T}{T_c}$$

$$P_r = \frac{P}{P_c}$$

where T_c and P_c are the critical temperature and pressure. It is expected that at a specified T_r and P_r every gas will have the same compressibility factor, and thus a correlation of Z in terms of T_r and P_r would apply to all gases. This is found to be a very good approximation for substances of similar chemical structure, but no single universal correlation appears possible. As expected, the situation improves considerably when a parameter characterizing the chemical nature of substances is incorporated into the correlation. Two such three-parameter correlations have been developed. One, developed at the University of California, uses an acentric factor which is determined from the vapor pressure curve.[16] The other, developed at the University of Wisconsin, uses the critical compressibility factor Z_c as the characterizing parameter.[17] Based on the study of 82 compounds, the Wisconsin group found the critical compressibility factor $P_c v_c / RT_c$ to range from 0.23 to 0.30; however, 60% of the compounds, including most hydrocarbons, fell within the range 0.26–0.28. They prepared a compressibility factor correlation based on the average Z_c of 0.27 and determined the corrections to be made for compounds of differing Z_c. Figure 3-6 shows the compressibility factor as a function of P_r and T_r based on compounds for which $0.26 \le Z_c \le 0.28$. This figure will be sufficient for the problems found in this text; however, when good estimates are needed for compounds outside this range of Z_c, the original reference should be consulted.

Note that the principle of corresponding states does not specify the functionality of the relationship between the reduced property and the reduced temperature and pressure; it simply states that a relationship should exist. The actual relationship must be determined from experimental observation of many compounds. Thus, the reduced isotherms in Fig. 3-6 are

---

[16] See G. N. Lewis and M. Randall, *Thermodynamics*, 2nd ed., revised by K. S. Pitzer and L. Brewer, McGraw-Hill, New York, 1961, App. 1.

[17] See O. A. Hougen, K. M. Watson, and R. A. Ragatz, *Checmial Process Principles*, Part II, 2nd ed., Wiley, New York, 1959, Chap. 14.

the best curves drawn through many data points. The average deviation of experimental values from the correlation is stated to be about 2.5%.

The equation

$$Pv = ZRT \qquad (3\text{-}26)$$

together with Fig. 3-6, which represents the functionality

$$Z = Z(T_r, P_r; Z_c) \qquad (3\text{-}27)$$

constitute a generalized equation of state in which functionality may be stated as

$$f(P, v, T; T_c, P_c, Z_c) = 0 \qquad (3\text{-}28)$$

Once the three characterizing parameters—T_c, P_c, and Z_c—are known, the *PVT* behavior of any substance can be estimated.

3-5 GENERALIZED EQUATIONS OF STATE

We have seen that the application of Eqs. (3-1) and (3-2) at the critical point allows two parameters in an equation of state to be expressed in terms of T_c and P_c. In functional notation a two-parameter equation so generalized (e.g., van der Waals) could then be stated as

$$f(P, v, T; T_c, P_c) = 0 \qquad (3\text{-}29)$$

The similarity of this equation to Eq. (3-28) suggests that this is another application of the principle of corresponding states; however, there are two major differences. First, Eq. (3-29) contains one less parameter and, second, its form is dictated by the equation of state instead of being determined from experimental data.

These difficulties have largely been surmounted with the development of modified van der Waals equations, which contain an additional parameter, the acentric factor, and have been algebraically modified to better fit experimental *PVT* data. Thus, the application of Eqs. (3-1) and (3-2) to the Peng-Robinson equation [Eq. (3-5c)] results in

$$b = 0.07780 \, \frac{RT_c}{P_c} \qquad (3\text{-}5d)$$

$$a(T_c) = 0.45724 \frac{R^2 T_c^2}{P_c} \qquad (3\text{-}5e)$$

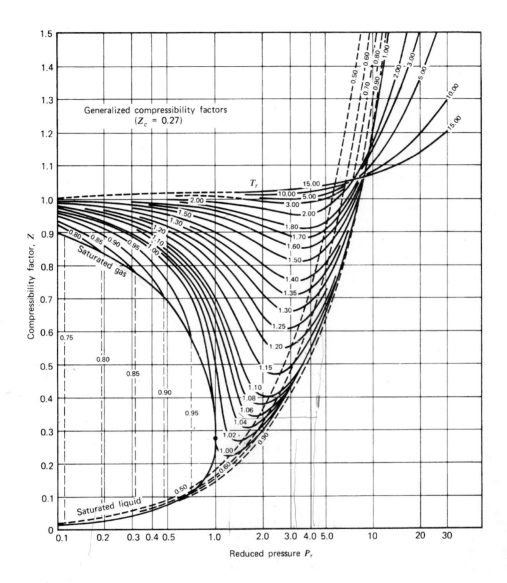

Figure 3-6 [From O. A. Hougen, K. M. Watson, and R. A. Ragatz, *Chemical Process Principles Charts*, 2nd ed., Wiley, New York, 1960, as adapted in *Chemical and Engineering Thermodynamics* by S. I . Sandler, Wiley, New York, 1977, with permission of John Wiley & Sons, Inc.]

where b is a constant and a is made a function of temperature and acentric factor, ω [18], according to

$$a(T) = a(T_c) \cdot \alpha(T, \omega) \tag{3-5f}$$

The form of the function α was determined by fitting vapor pressure data for many nonpolar substances:[19]

$$\sqrt{\alpha} = 1 + \kappa\left[1 - \sqrt{\frac{T}{T_c}}\right] \tag{3-5g}$$

where

$$\kappa = 0.37464 + 1.54226\omega - 0.26992\omega^2 \tag{3-5h}$$

As required by Eq. (3-5f), we observe that Eq. (3-5g) yields $\alpha = 1$ at the critical temperature.

When combined, Eqs. (3-5c)–(3-5h) constitute the eneralized Peng-Robinson equation, which can be stated functionally as

$$f(P,v,T;T_c,P_c,\omega) = 0 \tag{3-30}$$

This is equivalent to the three-parameter, corresponding-states's correlation of compressibility factor and provides estimates of comparable accuracy. The generalized equation of state has the advantage of being in algebraic rather than in graphical form and is therefore more useful for computerized calculations.

For computational convenience, the Peng-Robinson equation can be transformed to

$$Z^3 + (B - 1)Z^2 + (A - 3B^2 - 2B)Z + (B^3 + B^2 - AB) = 0 \tag{3.5i}$$

where

$$A = \frac{a(T)P}{(RT)^2} \tag{3-5j}$$

---

[18] The acentric factor is defined in terms of the reduced vapor pressure at a reduced temperature of 0.7:

$$\omega = -1.000 - \log\left[\frac{P^\circ}{P_c}\right]_{T_r=0.7}$$

Acentric factors for various substances are listed in Table A-1.

[19] For the determination of vapor pressure via an equation of state, see Sec. 9-12.

$$B = \frac{Pb}{RT} \qquad (3\text{-}5\text{k})$$

The parameters A and B can be cast into generalized form by employing Eqs. (3-5d)–(3-5g):

$$A = 0.45724 \frac{P_r}{T_r^2} \left[1 + \kappa \left(1 - T_r^{\frac{1}{2}} \right) \right]^2 \qquad (3\text{-}5\text{l})$$

$$B = 0.07780 \frac{P_r}{T_r} \qquad (3\text{-}5\text{m})$$

Equation (3-5i), together with Eqs. (3-5h), (3-5l), and (3-5m), provides the value of Z for specified values of T and P when T_c, P_c, and ω are known. As with all of the modified van der Waals equations, there are three real roots to Eqs. (3-5c) and (3-5i) for temperatures below the critical temperature. The middle root is unstable and the two stable roots of Eq. (3-5i) can be found with a simple numerical procedure—such as Newton's method—using starting points of $Z = 0$ and $Z = 6$ for liquid and vapor respectively. The program PREOS.EXE on the accompanying CD-ROM uses this method.

EXAMPLE 3-2

Find the density of chlorine gas, Cl_2, at 155 bar and 521 K using the
(a) compressibility factor chart.
(b) generalized Peng-Robinson equation of state.

Solution 3-2

From Table A-1 we find

$$\text{M.W.} = 70.91 \qquad \omega = 0.090$$
$$T_c = 417 \text{ K} \qquad P_c = 76.1 \text{ atm} = 77.1 \text{ bar}$$

The reduced conditions are

$$T_r = 521/417 = 1.25$$

$$P_r = 155/77.1 = 2.01$$

(a) From Fig. 3-6, we read $Z = 0.645$ and determine v:

$$v = \frac{ZRT}{P} = \frac{0.645(8314)(521)}{155(10^5)} = 0.177 \text{ m}^3 / \text{kmol}$$

$$\rho = 70.91 \frac{\text{kg}}{\text{kmol}} \times \frac{1}{0.177} \frac{\text{kmol}}{\text{m}^3} = 401 \text{ kg} / \text{m}^3$$

(b) We first determine κ from Eq. (3-5h)

$$\kappa = 0.37464 + 1.54226(0.090) - 0.26992(0.090)^2 = 0.511$$

and determine A and B to be

$$A = \frac{0.45724(2.01)}{(1.25)^2} \left[1 + 0.511\left(1 - \sqrt{1.25}\right)\right]^2 = 0.51937$$

$$B = \frac{0.07780(2.01)}{1.25} = 0.1251$$

The resulting equation,

$$Z^3 - 0.8749 Z^2 + 0.2222 Z - 0.04737 = 0$$

will have only one real root for $T_r > 1.0$. By trial and error, the root is found to be

$$Z = 0.644$$

From this we find

$$v = \frac{0.644(8314)(521)}{155(10^5)} = 0.180 \text{ m}^3 / \text{kmol}$$

and

$$\rho = \frac{70.91}{0.180} = 394 \text{ kg} / \text{m}^3$$

The agreement between the two methods is seen to be quite good.

For an occasional calculation, Eq. (3-5i) can be solved graphically or by trial and error; however, for repeated calculations a numerical technique better suited to computerized computation is desirable. The accompanying CD-ROM contains the program PREOS.EXE, which solves Eq. (3-5i) and provides v and Z for liquid or vapor phases for a specified set of temperatures and pressures. See App. E.

EXAMPLE 3-3

Use the *Simultaneous Algebraic Equation Solver* of the POLYMATH software (on the accompanying CD-ROM) to plot the left-hand side of Eq. (3-5i), $f(Z)$, vs Z for conditions where three real roots exist.

Solution 3-3

When there is only one nonlinear equation, POLYMATH displays the equation graphically in addition to determining its roots. For chlorine at $T_r = 0.7$ and $P_r = 0.1$ the following is a facsimile of the POLYMATH worksheet. This problem is in the POLYMATH library as EX3-3.

Equations:
$f(Z) = Z^3 + (B - 1)*Z^2 + (A - 3*B^2 - 2*B)*Z + B^3 + B^2 - A*B$
$A = (0.45724*Pr/Tr^2)*(1 + k*(1 - sqrt(Tr)))^2$
$B = 0.0778*Pr/Tr$
$k = 0.37464 + 1.54226*w - 0.26992*w^2$
$Tr = 0.7$
$Pr = 0.1$
$w = 0.090$
$Z_{min} = 0, Z_{max} = 1.0$

The values of Z which make $f(Z) = 0$, the roots, were found to be 0.0151, 0.0809, and 0.8929 and a plot of $f(Z)$ vs Z over the range $Z = 0$ to $Z = 1$ is shown in Fig. 3-7a. As this figure does not show clearly the

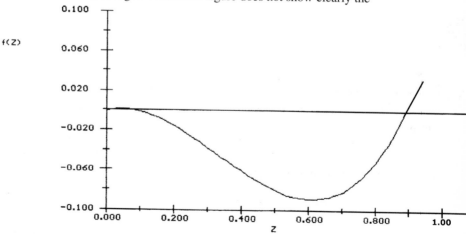

Figure 3-7a

location of the two smallest roots, the problem was re-solved changing the range to $Z_{min} = 0$ and $Z_{max} = 0.12$. This plot, shown in Fig. 3-7b, clearly reveals the location of the two smallest roots. The middle root, 0 .0809, corresponds to the middle root of the van der Waals equation and is unstable.

Thus, at this reduced temperature and pressure the compressibility factor is 0.0151 for the liquid phase and 0.8929 for the vapor phase. It should be emphasized that while we have obtained Z's for the liquid and vapor phases at a specified temperature and pressure, these phases are not necessarily saturated (they may not be in equilibrium). This can be seen from the fact that for a $T_r = 0.7$ there will be a range of pressures that will yield results resembling those of Fig. 3-7. Also, in terms of Fig. 3-4 one can see that for the isotherm shown, any pressure between P_A and P_B will yield values of the liquid and vapor volumes and subsequently Z's. However, for that temperature there can be only pressure that corresponds to saturation.

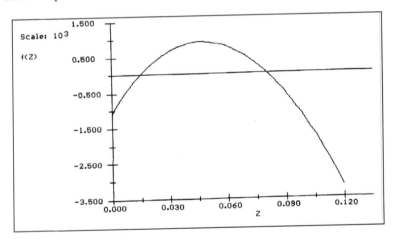

Figure 3-7b

EXAMPLE 3-4

The POLYMATH *Simultaneous Differential Equation Solver* (on the accompanying CD-ROM) is capable of constructing plots showing the numerical solutions of ordinary differential equations. Use this facility to prepare a plot of compressibility factor, Z, vs reduced pressure, P_r, when carbon dioxide is represented by the van der Waals and Peng-Robinson equations at a reduced temperature of 1.1.

Solution 3-4

At constant temperature the van der Waals and Peng-Robinson equations [Eqs. (3-4) and (3-5c)] are differentiated to obtain respectively

$$\frac{dP}{dv} = -\frac{RT}{(v-b)^2} + \frac{2a}{v^3}$$

$$\frac{dP}{dv} = -\frac{RT}{(v-b)^2} + \frac{2a(T)(v+b)}{(v^2 + 2vb - b^2)^2}$$

these derivatives can be rearranged to

$$\frac{dv}{dP_r} = P_c \left[-\frac{RT}{(v-b)^2} + \frac{2a}{v^3} \right]^{-1}$$

for the van der Waals equation and

$$\frac{dv}{dP_r} = P_c \left[-\frac{RT}{(v-b)^2} + \frac{2a(T)(v+b)}{(v^2 + 2vb - b^2)^2} \right]^{-1}$$

for the Peng-Robinson equation. The independent variable is P_r and before solving the problem it will be necessary to distinguish the Peng-Robinson volume from the van der Waals volume. Henceforth, v_{pr} will be used for the Peng-Robinson volume and POLYMATH will numerically determine v and v_{pr} as functions of P_r over the specified range of P_r. Auxiliary functions such as $Z_{vdw} = Pv/RT$ and $Z_{pr} = Pv_{pr}/RT$ can be added to the problem statement and plotted as functions of the independent variable P_r. This problem is stored in the POLYMATH library as EX3-4; a facsimile of the POLYMATH worksheet is shown below.

Equations: Initial value
d(V)/d(Pr)=Pc/(-R*T/(V-b)^2+2*a/V^3)   3.67648
d(Vpr)/d(Pr)=Pc/(-R*T/(Vpr-B)^2+2*A*(Vpr+B)/(Vpr^2+2*B*Vpr-   3.65996
B^2)^2)
a=27*R^2*Tc^2/(64*Pc)
b=R*Tc/(8*Pc)
A=0.45724*R^2*Tc^2*(1+k*(1-sqrt(Tr)))^2/Pc
B=0.0778*R*Tc/Pc
k=0.37464+1.54226*w-0.26992*w^2
Zvdw=P*V/(R*T)
Zpr=P*Vpr/(R*T)
P=Pr*Pc
Tr=T/Tc

Equations: <u>Initial value</u>
R=0.08206
Tc=304.2
Pc=72.9
w=0.239
T=334.6
$Pr_0 = 0.1$ $Pr_f = 5$

Note that the Peng-Robinson parameters are in capital letters and the van der Waals parameters are lower case. Before these equations can be solved, it is necessary to have the initial values of V and V_{pr}. These are obtained from POLYMATH's *Simultaneous Algebraic Equation Solver*. These problems are stored in POLYMATH's library as PREOS and VDW.

The POLYMATH plot of Z_{vdw} and Z_{pr} vs. P_r is shown in Fig. 3-8 where it is seen that the results from the two equations do not diverge appreciably until P_r is about 1.0. Beyond this point, the divergence increases as P_r increases and is sizeable.

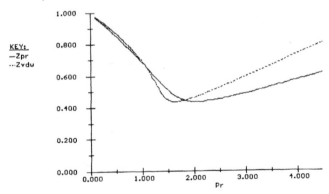

Figure 3-8

PROBLEMS

3-1.   According to the catalog of a compressed gas supplier, a full cylinder of ethylene (C_2H_4) contains 30 lb of gas with a cylinder pressure of 1200 psig. Estimate the amount of ethylene in the cylinder when the pressure is 500 psig.

3-2. To what pressure would a 3-ft³ cylinder have to be filled at 70°F if one wished to store 50 lb of ethylene in it?

3-3. A CO_2 cylinder of 1-ft³ volume is fitted with a pressure gauge that is calibrated in pounds of CO_2 in the cylinder (at room temperature). Instead of CO_2, the cylinder contains nitrogen. When the gauge reads 8.4 lb of CO_2, estimate the pounds of nitrogen in the cylinder.

3-4. Show that it is necessary to use at least the third virial coefficient in order that the virial equation predict the correct trend for the Pv curve near the critical point.

3-5. Show that on the conventional compressibility factor chart (Z vs. P_r with T_r as a parameter), the critical isotherm has an infinite slope at the critical point. Also show that on a plot of Z vs. $1/V_r$ (where V_r is reduced volume) the slope of the critical isotherm at the critical point is finite.

3-6. Find the compressibility factor at the critical point for a gas obeying the van der Waals equation of state.

3-7. The Boyle point is often used to characterize gases. When isotherms of Pv are plotted vs. P, it is found that the slopes at $P \rightarrow 0$ are greater than zero at high temperatures and less than zero at lower temperatures. The temperature at which the slope passes through zero is called the Boyle point or the Boyle temperature. The virial coefficients are functions only of temperature. What type of behavior does the second virial coefficient exhibit at the Boyle temperature?

3-8. A polytropic process can be defined as one that occurs along a path such that the heat capacity along the path c is constant. Show that a reversible polytropic expansion or compression of an ideal gas can be represented by the equation

$$Pv^n = \text{constant}$$

where n is a constant.

3-9. For a reversible adiabatic process involving an ideal gas with constant c_p we obtained the expression

$$\frac{T_2}{T_1} = \left[\frac{P_2}{P_1} \right]^{\frac{\gamma-1}{\gamma}}$$

Determine the equivalent expression for an ideal gas with

$$c_P = a + bT + cT^2$$

3-10. An ideal gas (c_p = 30 kJ/kmol · K) is contained in a vessel of 0.1-m³ volume. The gas, originally at 1 bar and 298 K, is heated at constant volume to 400 K. For this process, find (a) ΔU, (b) ΔH, (c) Q, (d) W.

3-11. An ideal gas with c_p = 29.3 kJ/kmol · K undergoes the following three-step process:

Step 1:   The gas is heated at constant volume from 300 K and 0.1 MPa until the pressure reaches 0.2 MPa.
Step 2:   The gas is expanded adiabatically and reversibly to a pressure of 0.1 MPa.
Step 3:   At a constant pressure of 0.1 MPa, the gas is cooled to 300 K.

Determine the heat and work effects for each step and an efficiency defined as the ratio of work done to heat supplied.

3-12. It is reported that a gas (assumed ideal) is taken through a series of five operations, finally arriving at the same pressure, temperature, and volume it possessed originally:

Step 1:   The system absorbs 220 kJ of heat and performs 300 kJ of work on the surroundings.
Step 2:   The system loses 110 kJ of heat while the volume remains constant.
Step 3:   110 kJ of work is done on the system, and it absorbs 60 kJ of heat.
Step 4:   The system loses 160 kJ of heat while the volume remains constant.
Step 5:   The system absorbs 290 kJ of heat and does 220 kJ of work on the surroundings.

(a)   Can you say whether this report is correct or not? Explain.
(b)   Would your answer be different if you were told that an isomerization reaction ($A = B$) was possible? Explain.
(c)   What if a reaction of the type $2A = A_2$ were possible? Explain.

3-13. A gas cylinder of 1-ft³ volume containing an ideal gas (c_p = 7 Btu/lb mol °F), initially at a pressure of 10 atm and a temperature of 70°F, is connected to another cylinder of 1-ft³ volume that is evacuated. A valve between the two cylinders is opened until the pressures in both cylinders equalize. Find the final temperature and pressure in each cylinder if each cylinder is insulated from the surroundings and from the other cylinder. List and discuss any assumptions required to solve this problem.

3-14. Nitrogen gas is stored in a tank at a pressure of 15 bar and 294 K. The tank has a capacity of 0.057 m³. Gas is allowed to flow from the tank through a partially opened valve to a gas holder where the pressure is constant at 1.15 bar. When the pressure in the tank has dropped to 5 bar, calculate the mass of nitrogen removed from the tank under the following conditions:

(a)   if the process occurred slowly enough so that the temperature remained constant and
(b)   if the process occurred rapidly enough that heat transfer was negligible.

Assume ideal-gas behavior.

3-15. An ideal gas (c_p = 7 Btu/lb mol °F) at a pressure of 100 psia and 200°F flows through a turbine that operates adiabatically. The gas leaves the turbine at atmospheric pressure and 0°F.

(a)   How many Btus of work were obtained per lb-mol of gas?

(b) What is the maximum amount of work in Btus that could be obtained from the gas if it is expanded adiabatically to atmospheric pressure?

3-16. An ideal gas (c_p = 30 kJ/kmol K) passes through a centrifugal blower at a rate of 1 kmol/s where it is compressed adiabatically from 1 to 2.0 bar. The gas enters the blower at 298 K and leaves at 404 K. Find

(a) the power supplied to the blower.
(b) the efficiency of the blower based on an adiabatic reversible compression between the same initial and final pressures.

3-17. In the production of ammonia it is necessary to purge part of a recycle stream in order to prevent the buildup of inerts. The purge gas at 2 atm and 400°F passes through an adiabatically operated turbine and exits at atmospheric pressure. The plant is being surveyed to determine its energy utilization efficiency, and you have been given the job of assessing the efficiency of this turbine. It is impractical to measure the power output of the turbine, but the exit gas temperature is known to be 306°F. Assume that the gas is ideal, with c_p = 7 Btu/lb mol °F, and compare the actual performance of the turbine with the maximum work obtainable from an adiabatic expansion of the gas to atmospheric pressure.

3-18. An ideal gas at 2 atm and 400°F flows into a turbine and leaves at 1 atm and 306°F. The turbine operates adiabatically and the gas has a c_p of 7.0 Btu/lb mol °F. Compare the work obtained from this turbine with that which could be obtained from a reversible adiabatic expansion of the gas to a pressure of 1 atm.

3-19. The force of an explosion is often measured against an equivalent amount of TNT. In military parlance, nuclear warheads are rated in kilotons or megatons of TNT, but on a less apocalyptic level, the TNT equivalence of 2000 Btu/lb is useful for assessing plant safety hazards. Determine the TNT equivalence for an exploding tank of compressed gas, assuming that the work, or energy available for destruction, can be approximated by a reversible adiabatic expansion to atmospheric pressure. A tank of 100-ft³ capacity containing air at ambient temperature fails at a pressure of 20 atm. Assume ideal-gas behavior.

3-20. A portable compressed air tank is frequently used to inflate flat tires. The tank is initially filled with compressed air at 150 psia and 70°F, and the air in the tire may be assumed to be at atmospheric pressure. After connecting the air hose to the tire, a valve is opened and air rushes into the tire. In the specific circumstances under study, the valve is closed after the pressure in the tank has fallen to 75 psia. The tank volume is 0.5 ft³ and that of the tire is 1 ft³. Assume that the tire is rigid so that its walls do not move during the filling operation and that the filling occurs rapidly enough that the process may be considered adiabatic. Assume air is an ideal gas with constant heat capacity of 7 Btu/lb mol °F.

(a) What are the temperatures in the tank and tire immediately after filling?
(b) What will be the eventual pressures in the tank and tire after thermal equilibrium with the surroundings at 70°F has occurred?
(c) If the tire is not rigid during filling, will your answer in part (a) be too high or too low? Explain.

3-21. Estimate the altitude at which cumulus clouds will form. Because pressure decreases with increasing altitude, air on rising will expand and cool until condensation occurs thus forming cumulus clouds. At ground level the pressure is 740 mmHg and the air has a temperature of 85°F and a dew point of 75°F. The following relation between altitude, Z, and pressure, P, can be derived by means of a simple differential force balance

$$P = P_0 e^{-\frac{Mg\,Z}{RT}}$$

where P_0 is the ground-level pressure, M is molecular weight, and **g** is the gravitational acceleration. Assume that the expansion follows a reversible adiabatic path.

3-22. Using the technique of Ex. 3-4, compare compressibility factor isotherms at $T_r = 1.1$ for carbon dioxide as given by the Redlich-Kwong, Soave, and Peng-Robinson equations [Eqs. (3-5a), (3-5b), and (3-5c) respectively]. Use the range $0.1 \le P_r \le 5$.

3-23. Use the technique of Ex. 3-3 to prepare plots of $f(Z)$ vs Z for chlorine at $T_r = 0.7$ and for P_r values of 0.01, 0.2, and 1.0.

The Second Law of Thermodynamics

The first law requires the conservation of energy and gives equal weight to all types of energy changes. While no process is immune to its authority, this law does not recognize the quality of energy nor does it explain why spontaneously occurring processes are never observed to spontaneously reverse themselves. The repeatedly confirmed observation that work may be completely converted into heat but that the reverse transformation never occurs quantitatively leads to the recognition that heat is a lower quality of energy. The second law, with its origins deeply rooted in the study of the efficiency of heat engines, recognizes the quality of energy. Through this law the existence of a heretofore unrecognized property, the entropy, is revealed, and it is shown that this property determines the direction of spontaneous change. The second law in no way diminishes the authority of the first law; rather it extends and reinforces the jurisdiction of thermodynamics.

4-1 HEAT ENGINES AND THE CARNOT CYCLE

A heat engine is a device, considered a closed system operating in a cyclic manner, that is supplied with heat and performs work. Because heat cannot be completely converted into work, a portion of the input heat must be rejected to a low-temperature reservoir (e.g., ambient air or cooling water). Hence, there is at least one input and one outflow of heat associated with the device. Any cyclic process can form the basis of a heat engine, although the working substance is usually a fluid that is taken through a succession of steps such as vaporization, condensation, compression, expansion, and heating or cooling. The tutorial RANKINE that runs in the software EQUATIONS OF STATE on the accompanying CD-ROM[1] describes a modern-day power cycle (heat engine).

During the Industrial Revolution, a great deal of activity was devoted to constructing engines to power machinery using coal as a fuel to supply input heat. After much experimentation it was found that while improvements in design and construction resulted in increased efficiency, there appeared to be an upper limit, considerably less than 100%, to the efficiency with which heat could be converted into work. Thus, it appeared that something other than mechanical irreversibilities governed the efficiency of heat engines. Reflection on years of accumulated experience with heat engine design and operation led Carnot to make the following statement, which has come to be known as Carnot's principle:

> *The maximum efficiency of a heat engine depends only on the temperatures between which it operates and is independent of the nature of the cyclic process.*

Carnot's principle forms the basis for the second law of thermodynamics, which is usually stated as follows:

> *It is impossible for a device operating in a cyclic manner to completely convert heat into work.*

From Carnot's principle two corollaries are easily proved. Because these statements are intuitively acceptable, they are presented without proof.[2]

Corollary 1. For given operating temperatures the maximum efficiency is attained by a reversible heat engine.

Corollary 2. For given operating temperatures all reversible heat engines have the same efficiency regardless of the nature of the working substance or cycle.

---

[1] See App. E for a description of this software.

[2] For a proof, see, for example, K. Denbigh, *The Principles of Chemical Equilibrium*, Cambridge University Press, New York, 1971, Chap. 1 or J. A. Beattie and I. Oppenheim, *Principles of Thermodynamics*, Elsevier, Amsterdam, 1979, Chap. 8.

The emphasis here is on reversible heat engines because we wish to remove from our consideration any type of mechanical irreversibility so that the principle that describes the limits of the conversion of heat into work can be discerned. Because of corollary 2, we will base all derivations in this chapter exclusively on the simplest of all heat engines—the Carnot engine. Any result obtained through the use of a Carnot engine will then be applicable to all reversible heat engines.

The Carnot engine undergoes a four-step cycle, known as the Carnot cycle, which consists of two isothermal and two adiabatic steps carried out reversibly. There is no restriction on the working substance; however, calculations are more easily executed if an ideal gas is chosen. From this description it appears that the Carnot engine is more a concept than a reality. This is true; however, we know in principle that a well-designed and carefully crafted heat engine could closely approach reversible operation, and it is quite conceivable that operating conditions could be chosen so that a gaseous working fluid would behave ideally. The Carnot engine is a very useful tool for thermodynamic derivations and for thermodynamic analysis of processes, and we can be comfortable using it only if we are convinced that it is not a nebulous abstraction but a reasonable and approachable standard.

4-2 THE IDEAL-GAS CARNOT CYCLE

A Carnot engine operates between two constant-temperature reservoirs as sketched in Fig. 4-1. With a gaseous working fluid the Carnot cycle can be represented on a Pv plot as

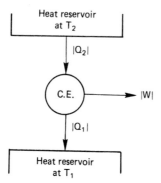

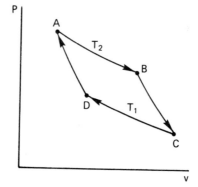

Figure 4-1 Carnot engine.

Figure 4-2 Representation of a Carnot cycle.

shown in Fig. 4-2. Heat is taken in during the isothermal expansion at T_2 while the gas performs work, the gas does additional work in expanding adiabatically to T_1, heat is rejected during the isothermal compression at T_1, and an adiabatic compression restores the gas to its initial state. Work is done on the gas in the last two steps, and the net work performed during the cycle is the algebraic sum for all four steps. Because work in a closed system is calculated from $\int P\,dv$, it is easily shown that the net work is the area enclosed by the cycle $ABCD$ in Fig. 4-2.

For an isothermal expansion or compression of an ideal gas in a closed system we have seen that $\Delta u = 0$ and that

$$Q = -W = RT \ln \frac{P_1}{P_2} \qquad (3\text{-}15)$$

Applying this to the Carnot cycle sketched in Fig. 4-2 gives

$$Q_2 = -W_{AB} = RT_2 \ln \frac{P_A}{P_B} \qquad (4\text{-}1)$$

$$Q_1 = -W_{CD} = RT_1 \ln \frac{P_C}{P_D} \qquad (4\text{-}2)$$

For the adiabatic steps, from the first law we write

$$W_{BC} = \Delta u_{BC} = \int_{T_2}^{T_1} c_V\,dT$$

$$W_{DA} = \Delta u_{DA} = \int_{T_1}^{T_2} c_V\,dT$$

and observe that

$$W_{BC} = -W_{DA} \qquad (4\text{-}3)$$

The paths of reversible adiabatic processes are described by

$$\frac{T_1}{T_2} = \left(\frac{P_C}{P_B} \right)^{(\gamma-1)/\gamma}$$

$$\frac{T_2}{T_1} = \left(\frac{P_A}{P_D} \right)^{(\gamma-1)/\gamma} \qquad (3\text{-}19)$$

Manipulation of these two expressions results in

$$\frac{P_A}{P_D} = \frac{P_B}{P_C}$$

or

$$\frac{P_A}{P_B} = \frac{P_D}{P_C} \tag{4-4}$$

Substitution of Eq. (4-4) in Eq. (4-2) yields

$$Q_1 = -W_{CD} = -RT_1 \ln \frac{P_A}{P_B} \tag{4-5}$$

The net work produced by the cycle is

$$-W = -(W_{AB} + W_{BC} + W_{CD} + W_{DA})$$

which on application of Eq. (4-3) and substitution of Eqs. (4-1) and (4-5) reduces to

$$-W = RT_2 \ln \frac{P_A}{P_B} - RT_1 \ln \frac{P_A}{P_B}$$

$$-W = R(T_2 - T_1) \ln \frac{P_A}{P_B} \tag{4-6}$$

The efficiency, defined as the ratio of net work produced $-W$, to heat supplied Q_2 is

$$\eta = \frac{-W}{Q_2} = \frac{R(T_2 - T_1) \ln \left(\dfrac{P_A}{P_B} \right)}{RT_2 \ln \left(\dfrac{P_A}{P_B} \right)}$$

$$\eta = \frac{T_2 - T_1}{T_2} \tag{4-7}$$

Carnot's principle has now been quantified. This result, obtained for the special case of a Carnot cycle with ideal-gas working fluid, is applicable by virtue of corollary 2 to all reversible heat engines. It specifies the best we could hope to achieve with any heat engine we might construct to operate between two fixed temperatures. The term $(T_2 - T_1) / T_2$, often called "the Carnot factor," determines the quality of heat—which we see is measured in terms of its convertibility into work. The tutorial CARNOT that runs in the software EQUATIONS OF STATE on the accompanying CD-ROM deals with the Carnot cycle.

4-3 THE ABSOLUTE TEMPERATURE SCALE

Up to this point the symbol T has stood for a temperature measured on the ideal-gas temperature scale discussed in Sec. 1-3. This scale possesses some degree of universality because it is independent of the nature of the working gas; however, it would be desirable to go a step further and establish a temperature scale totally independent of the nature of the working substance—a scale that would be truly absolute. Such a scale was established by Thomson (Lord Kelvin) based on Carnot's principle.

The derivation of the absolute temperature scale is based on an unspecified Carnot engine. Any substance that can exchange work can be the working substance. In addition to ideal gases, Carnot cycles are possible based on, for example, the stretching and relaxing of a strip of rubber, the magnetization and demagnetization of a paramagnetic solid, or the extension and contraction of a surface film. In general, any cycle can be represented as two isotherms and two adiabats on a plot of generalized force $\mathscr{F}$ vs. generalized displacement r. For a fluid the coordinates are P vs. v; for the rubber strip, force vs. length; for the paramagnetic solid, magnetic field strength vs. intensity of magnetization; and for the surface film, surface tension vs. area. Figure 4-3 shows two adiabats on a generalized $\mathscr{F}r$ plane upon which Carnot cycles can be represented. Several isotherms of the temperature scale we wish to establish are drawn in this figure. The upper isotherm corresponds to the ideal-gas temperature T, and the lower isotherm corresponds to an ideal-gas temperature of zero. Between these two ideal-gas temperatures will be n isotherms, where n is the number of units on the absolute scale corresponding to the difference between these two fixed ideal-gas temperatures. Two adjacent isotherms along with the adiabats serve to define a Carnot cycle, and between T and zero there are n such Carnot engines. Each engine in this cascade receives the heat rejected by the one above it at a higher temperature, and its rejected heat flows into the one below it at a lower temperature. Because of Eq. (4-7), the last engine with a lower temperature of zero rejects no heat.

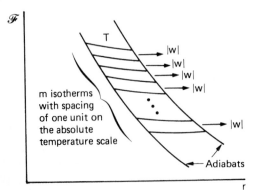

Figure 4-3 Carnot cycles defining the absolute temperature scale.

The absolute temperature scale will be set by specifying that all engines perform the same amount of work $|w|$.[3] Because no heat is rejected from the entire cascade and the engines have undergone a complete cycle ($\Delta U = 0$), the first law requires the heat input to the first engine to be

$$|Q| = n|w|$$

We now consider the first m high-temperature engines as a composite engine which receives heat $n|w|$ and produces work $m|w|$. The efficiency of this composite engine is

$$\eta = \frac{m|w|}{n|w|} = \frac{m}{n}$$

Now n is the number of absolute temperature units between zero and T on the ideal-gas scale, and if we match the zeros of the two scales, then n is the absolute temperature θ corresponding to T. Likewise, m is the number of absolute temperature units between which the composite engine operates and can be designated $\Delta\theta$. The efficiency of the composite engine is now seen to be

$$\eta = \frac{\Delta\theta}{\theta}$$

In terms of the ideal-gas temperature, from Eq. (4-7) the efficiency of the composite engine is

$$\eta = \frac{\Delta T}{T}$$

and we see that the absolute and ideal-gas temperatures are related by

$$\theta = bT$$

where b is a proportionality constant. If we require θ and T be equal at a reference temperature such as the water triple point, we set b equal to unity and have

$$\theta = T$$

Thus, it has been shown that the temperature scale originally based on the extrapolated properties of gases coincides with one based on Carnot's principle. It is completely general and independent of the working substance. For this reason it is termed the "absolute" temperature.

It should be noted that what we have called the absolute temperature scale is not the only scale that could be established from Carnot's principle. Indeed, Kelvin originally proposed an absolute temperature scale based on each Carnot engine of the cascade of Fig. 4-3 having equal efficiency. This results in an entirely different absolute temperature scale. This

---

[3] In this section it is convenient to disregard the signs on w and Q and use absolute values.

scale would be valid and would allow the development of the thermodynamic network of variables and connecting relationships, but there is no compelling reason to accept it over the much simpler ideal-gas/absolute temperature scale. From here on the symbol T designates absolute temperature measured on a scale coinciding with the ideal-gas temperature scale.

4-4 THE ENTROPY FUNCTION

For a reversible heat engine operating between T_2 and T_1 the efficiency was shown to be

$$\eta = \frac{|W|}{Q_2} = \frac{T_2 - T_1}{T_2} \tag{4-7}$$

When the engine has executed an integral number of cycles, we have $\Delta U = 0$, and the first law requires that

$$|W| = |Q_2| - |Q_1|$$

Substitution of this into Eq. (4-7) yields

$$\frac{|Q_2| - |Q_1|}{|Q_2|} = 1 - \frac{|Q_1|}{|Q_2|} = 1 - \frac{T_1}{T_2}$$

or

$$\frac{|Q_1|}{T_1} = \frac{|Q_2|}{T_2} \tag{4-8}$$

Dropping the absolute value signs and recognizing that Q_1 and Q_2 are opposite in sign, we write

$$\frac{Q_1}{T_1} + \frac{Q_2}{T_2} = 0 \tag{4-9}$$

This result applies to a closed system (the heat engine) undergoing a cyclic process with heat added at T_2 and removed at T_1. Because the sum of property changes for a cyclic process is zero, Eq. (4-9) suggests that Q_1/T_1 and Q_2/T_2 are property changes for the isothermal steps of the cycle where heat is received or rejected. This implies the existence of a property, called entropy and designated S, that is defined by

$$dS = \frac{dQ_{rev}}{T} \tag{4-10}$$

The developments which lead us to so define the entropy were based on the performance of a Carnot engine where all operations are reversible. For this reason the entropy is defined in terms of the heat effect in a reversible process as designated by the notation dQ_{rev}.

To prove that the entropy is a property, it is necessary to show that the integral of dS around any arbitrary closed path is zero, or in mathematical notation

$$\oint dS = 0 \tag{4-11}$$

Figure 4-4 depicts an arbitrary closed path on generalized $\mathscr{F}r$ coordinates around which a body is taken in a reversible manner. Imposed on this path are two adiabats $\overline{BDFH}$ and $\overline{CAGE}$. Also, an isotherm, such as $\overline{CD}$ is drawn so that the area under the path $ACDB$ is equal to the area under AB. This area, $\int \mathscr{F} dr$, represents the work between states A and B. Between these two states the internal energy change is fixed, and by specifying the work by the two paths to be equal, we may thus equate the heat effects. The same procedure can be followed in locating the isotherm GH. We now have a Carnot cycle $CDFHGA$ that receives and rejects the same quantities of heat as that portion of the arbitrary cycle lying between the two adiabats.

As the distance between adiabats is decreased, the zigzag path $ACDB$ more closely approaches the actual path AB, and an isothermal approximation for this path becomes more reasonable. With the heat effects already shown to be equal, the isothermality approximation leads to equal Q/T for that portion of the arbitrary cycle and its Carnot approximation.

We now visualize the closed path intersected by numerous, narrowly spaced adiabats with isothermal segments that closely approximate the actual path and define numerous minute Carnot cycles. Each of the cycles exchanges minute quantities of heat Q_{2i} and Q_{1i} at temperatures and T_{2i} and T_{1i}, for the totality of these cycles Eq. (4-9) requires

$$\sum \frac{Q_{2i}}{T_{2i}} + \sum \frac{Q_{1i}}{T_{1i}} = 0 \tag{4-12}$$

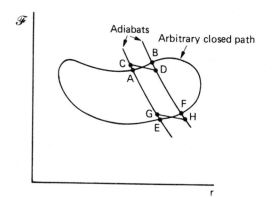

Figure 4-4 Approximation of an arbitrary closed path with Carnot cycles.

Because of the arbitrariness of the cycle being approximated, there is no need to retain the subscripts 1 and 2, and we write

$$\sum \frac{Q_i}{T_i} = 0 \tag{4-13}$$

In the limit as the spacing between adiabats approaches zero, Eq. (4-13) becomes

$$\oint \frac{dQ_{\text{rev}}}{T} = \oint dS = 0 \tag{4-14}$$

In this limit, Eq. (4-14)—which applies to the unbounded number of infinitesimally small Carnot cycles—applies equally to the arbitrary closed cycle. Thus, for the substance undergoing the cycle, the existence of another thermodynamic property has been proved. Because the nature of the substance is unspecified, this result is general, and we may state that the entropy is a property of all matter.

4-5 ENTROPY AND THE SPONTANEITY OF NATURAL PROCESSES

To examine the role played by the entropy in determining the direction of change, we consider a spontaneous process that occurs in a closed adiabatic system. No restrictions are placed on the nature of this process which takes the system from state A to state B and may have a quantity of work W_{AB} associated with it.

   We now return the system to its original state A by means of a reversible process that is not necessarily adiabatic. For the return path $B \rightarrow A$ the quantities of heat and work are Q_{BA} and W_{BA}, respectively. The sum of the two steps $A \rightarrow B$ and $B \rightarrow A$ is a cyclic process for which we may state $\Delta U = 0$ and employ the first law to obtain the result

$$Q_{BA} = -\left(W_{AB} + W_{BA}\right)$$

The second law requires that Q_{BA} be zero or negative, for if positive, the system would have received heat and converted it completely into work. Because the process $B \rightarrow A$ is reversible, we may employ Eq. (4-10) to calculate the entropy change

$$S_A - S_B = \int \frac{dQ_{BA}}{T}$$

and because $Q_{BA} \leq 0$, it is seen that $S_A - S_B \leq 0$. For this two-step, cyclic process all property changes must equal zero, and for the spontaneous process $A \rightarrow B$ we may state that

$$S_B - S_A \geq 0 \tag{4-15}$$

The equality sign in Eq. (4-15) refers to a reversible adiabatic process, a result which could be obtained directly from application of Eq. (4-10). For the spontaneous—or irreversible—process $A \to B$, the result is

$$S_B - S_A > 0$$

Thus, a spontaneous process occurring in a closed adiabatic system is accompanied by an increase in entropy. For such systems it can be said that the entropy is that property of the system that determines the direction of change—processes for which $\Delta S > 0$ are possible and processes for which $\Delta S < 0$ are impossible. It is this characteristic of the entropy which allows the specification of conditions of equilibrium and thereby permits the application of thermodynamics to topics such as phase equilibrium and chemical equilibrium, so important to chemical engineers.

Equation (4-15) was derived for a process occurring in a closed adiabatic system and therefore may seem to have limited application. However, the desired generality of this result may be achieved by considering the adiabatic system to consist of several subsystems. These subsystems would be the primary system, or the system of interest, plus any number of heat reservoirs that exchange heat with the primary system. For the composite system, which is adiabatic, Eq. (4-15) applies. When our focus is on the primary system, this equation may be restated as

$$\Delta S + \Delta S_{\text{surr}} \geq 0 \qquad (4\text{-}16)$$

This statement is valid because the entropy of the system, as with any extensive property, can be obtained by summing the entropy of its various parts. Here ΔS signifies the entropy change of the primary system and ΔS_{surr} signifies the entropy change of all other parts of the composite system (these other parts become the surroundings when the primary system is designated the system). Thus, for a process occurring in a nonadiabatic closed system, the entropy criterion still obtains; however, its implementation requires that changes in the surroundings must also be considered.

4-6 CALCULATION OF ENTROPY CHANGES

It has been shown that matter possesses a property called "entropy" and that this property is useful in determining the direction of change in a system. We now concern ourselves with the procedure for calculating entropy changes. Because the entropy change requires the reversible heat effect for its evaluation and this is a quantity which can never be measured, it is natural to inquire as to how this evaluation is to occur. The answer is really quite simple. The entropy is a property, and therefore changes in this property depend only on the initial and final states of the system and are independent of path. If we wish to calculate the entropy change experienced by a system undergoing an irreversible process between states A and B, we merely need to devise a reversible path between these two states and evaluate the entropy change via this path. The entropy change for this devised reversible path must be identical to the entropy change for the actual irreversible path. This calculation technique will be illustrated for some commonly encountered situations.

Heat Exchange with a Reservoir.   Suppose a reservoir exchanges a quantity of heat in an irreversible manner[4] and that this exchange produces a definite change in state of the reservoir. If the reservoir temperature were maintained constant by virtue of a two-phase mixture (e.g., ice and water), the exchange of the heat would result in a definite change in the portions of the phases. If the reservoir were simply a large body (e.g., a lake or the ambient air), we could, in principle, still visualize a change in temperature, although it might be imperceptibly small, thus justifying the assumption of a constant-temperature reservoir. In any event a given quantity of heat will bring about the same change in state of the reservoir regardless of whether it is transferred reversibly or irreversibly. Thus,

$$Q = Q_{rev}$$

and if the reservoir temperature remains constant, from Eq. (4-10) we have

$$\Delta S = \frac{1}{T} \int dQ_{rev} = \frac{Q_{rev}}{T} = \frac{Q}{T} \qquad (4\text{-}17)$$

Change in Temperature.   When a body changes temperature as a result of exchanging heat, the heat effect is expressible in terms of the heat capacity.[5] For a constant-pressure process,

$$dQ_P = dH = C_P \, dT$$

and for a constant-volume process,

$$dQ_V = dU = C_V \, dT$$

As we have already seen, a given quantity of heat produces the same state change in our system whether exchanged reversibly or irreversibly, and for the constant-pressure process we write

$$\Delta S_P = \int \frac{dQ_P}{T} = \int \frac{C_P \, dT}{T} \qquad (4\text{-}18)$$

Likewise, for a constant-volume heat exchange,

---

[4] Reversible heat exchange occurs with an infinitesimal temperature difference. Irreversible heat exchange occurs across a finite temperature difference.

[5] C_P and C_V are used for the system containing n mol of the substance. $C_P = nc_p$, and $C_V = nc_V$.

$$\Delta S_V = \int \frac{dQ_V}{T} = \int \frac{C_V \, dT}{T} \tag{4-19}$$

When the heat capacity may be regarded constant over the temperature range, Eqs. (4-18) and (4-19) integrate to

$$\Delta S_P = C_P \ln \frac{T_2}{T_1} \tag{4-20}$$

$$\Delta S_V = C_V \ln \frac{T_2}{T_1} \tag{4-21}$$

Phase Change.   A reversible phase change can occur only under conditions where the two phases are in equilibrium. For the determination of these conditions the phase rule provides guidance. For example, when applied to a pure substance $(C = 1)$, the result is one degree of freedom when two phases are in equilibrium $(\pi = 2)$ Therefore, if we wish to reversibly vaporize a pure substance at a fixed pressure, the temperature must be the saturation temperature and cannot be specified otherwise. For a pure substance, the phase change occurs at constant temperature and pressure, and for a mol or unit mass we have

$$Q_P = \Delta h$$

and

$$\Delta s = \frac{\Delta h}{T} \tag{4-22}$$

Changes in Pressure.   As with the enthalpy and internal energy, the entropy of condensed phases changes little with changes in pressure, and therefore such changes are often neglected. This is not true of gases. Here we will consider only an ideal gas; entropy changes of real gases will be considered in Chap. 5.

For a mol of ideal gas undergoing a reversible isothermal change in pressure from Eq. (3-15) we have

$$\Delta s_T' = \frac{Q}{T} = R \ln \frac{P_1}{P_2} \tag{4-23}$$

The superscript on Δs indicates that this equation is applicable only to ideal gases, and the subscript denotes conditions of isothermality.

Computational Path.   In calculating the entropy change of a substance corresponding to a specified change of state, a reversible path between the initial and final state must be devised. This will be referred to as "the computational path." To illustrate the delineation of a computational path and the attendant calculations, consider the process where a mol of pure liquid at T_1 and P_1 is changed into a gas at T_2 and P_2. It will be assumed that the final conditions are such that the gas behaves ideally. The computational path is illustrated by the sequence of dashed arrows in Fig. 4-5, where the actual change is shown by the heavy arrow. From the figure it is seen that the desired entropy change Δs is the sum of entropy changes for four separate steps. The computational path may also be represented on a phase diagram as shown in Fig. 4-6.

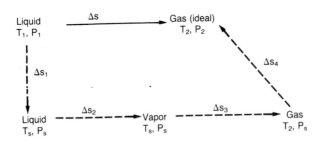

Figure 4-5 Computational path for entropy calculations.

Δs_1.   This is a constant-pressure heating, and Eq. (4-18) applies:

$$\Delta s_1 = \int_{T_1}^{T_s} \frac{c_P(l)}{T} \, dT$$

where T_s is the saturation temperature corresponding to P_1 and $c_p(l)$ is the liquid heat capacity. This process is shown in Fig. 4-6 as a horizontal line connecting point 1 to point a.

Δs_2.   The liquid is vaporized with T_s and P_1 remaining constant, and

$$\Delta s_2 = \frac{\Delta h}{T_s}$$

The system remains at point a in Fig. 4-6 because the vaporization process occurs at constant T and P.

Δs_3.   The vapor is heated at constant pressure from the saturation temperature to T_2, and

$$\Delta s_3 = \int_{T_s}^{T_2} \frac{c_P'}{T} \, dT$$

where c_P' is the ideal-gas heat capacity. This step is shown as a horizontal line between points a and b in Fig. 4-6.

Δs_4.   At the temperature T_2 the pressure is changed to P_2 and from Eq. (4-23)

$$\Delta s_4 = R \ln \frac{P_1}{P_2}$$

This step is represented by the vertical line connecting point b to point 2 in Fig. 4-6.

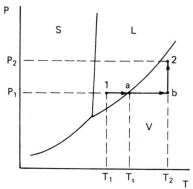

Figure 4-6 Representation of a computational path on a phase diagram.

The total entropy change is

$$\Delta s = \int_{T_1}^{T_s} \frac{c_P(l)\,dT}{T} + \frac{\Delta h}{T_s} + \int_{T_s}^{T_2} \frac{c_P'\,dT}{T} + R \ln \frac{P_1}{P_2}$$

This is the molar entropy change between these specified terminal states regardless of the actual path taken by the system. The entropy change must be calculated for a reversible path; however, once calculated, this property change applies to irreversible processes as well. The molar entropy or specific entropy s has the units kJ/kmol · K or kJ/kg · K, whereas the entropy of a given quantity of a substance S has the units kJ/K.

EXAMPLE 4-1

It is possible to cool liquid water below its freezing point of 273.15 K without the formation of ice if care is taken to prevent nucleation. A kilogram of subcooled liquid water at 263.15 K is contained in a well-insulated vessel. Nucleation is induced by the introduction of a speck of dust, and a spontaneous crystallization process ensues. Find the final state of the water, and calculate the entropy change of the water, the surroundings, and the total entropy change.

Solution 4-1

Our system will be the 1 kg of water. This is visualized as a constant-pressure process for which we have shown that

$$Q_P = \Delta H \qquad (2\text{-}6)$$

For the adiabatic process we have $\Delta H = 0$ and may determine the final state from the two-step computational path shown in Fig. 4-7. Because the enthalpy is a property, its change between any two states is independent of path, and the path we choose for the computation is simply one of convenience. First, the liquid is heated from 263.15 to 273.15 K, where

$$\Delta H_1 = 1\int_{263.15}^{273.15} c_P \, dT$$

With $c_P = 4.185$ kJ / kg $\cdot$ K and assumed constant over this small temperature range we have

$$\Delta H_1 = 1(4.185)(10) = 41.85 \, \text{kJ}$$

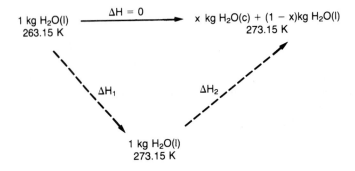

Figure 4-7

In the second step x kg of ice are formed at 273.15 K, where the heat of fusion is 334 kJ/kg. The enthalpy change for this step is

$$\Delta H_2 = x\left(-\Delta h_{\text{fusion}}\right) = -334x$$

From the condition

$$0 = \Delta H_1 + \Delta H_2$$

we find that $x = 0.125$ kg, and hence the final state is an ice-water mixture at 273.15 K in which 12.5% of the mixture is ice.

The same computational path is used to calculate the entropy change of the water. For the heating step,

$$\Delta S_1 = 1 \int_{263.15}^{273.15} \frac{c_P \, dT}{T} = 1(4.185) \ln \frac{273.15}{263.15} = 0.156 \, \text{kJ} \, / \, \text{K}$$

For the crystallization step,

$$\Delta S_2 = \frac{x(-\Delta h_{\text{fusion}})}{T} = \frac{0.125(-334)}{273.15} = -0.153 \, \text{kJ} \, / \, \text{K}$$

For the process $\Delta S = \Delta S_1 + \Delta S_2 = 0.003$ kJ / K, and the entropy change experienced by the water during this spontaneous process is seen to be positive in accordance with the entropy criterion as stated by Eq. (4-15). Because the process is adiabatic, the surroundings undergo no change as a result of this process, and hence $\Delta S_{\text{surr}} = 0$.

EXAMPLE 4-2

A copper block having a mass of 10 kg and at a temperature of 800 K is placed in a well-insulated vessel containing 100 kg of water initially at 290 K. Calculate the entropy change for the block, the water, and the total process. The heat capacities are 4.185 kJ/kg · K for water and 0.398 kJ/kg · K for copper.

Solution 4-2

The system will be the block, and the water and the process will be visualized as constant pressure. With $Q_P = \Delta H$ and $Q_P = 0$ we have $\Delta H = 0$ and can write

$$10(0.398)(T - 800) + 100(4.185)(T - 290) = 0$$

$$T = 294.8 \, \text{K}$$

In the final state both water and copper block are at 294.8 K. The copper block has experienced a change of state in which its temperature has decreased from 800 to 294.8 K. By Eq. (4-20) the entropy change for this state change is

$$\Delta S_{\text{Cu}} = 10(0.398) \ln \frac{294.8}{800} = -3.973 \, \text{kJ} \, / \, \text{K}$$

Similarly, for the water,

$$\Delta S_{\text{H}_2\text{O}} = 100(4.185) \ln \frac{294.8}{290} = 6.870 \, \text{kJ} \, / \, \text{K}$$

For this process nothing in the surroundings is affected, and hence the total entropy change is $6.870 - 3.973 = 2.897$ kJ / K. The total entropy change is positive, as required by the criterion of Eq. (4-15).

EXAMPLE 4-3

Two perfectly insulated tanks each having a volume of 1 m³ are connected by means of a small pipeline containing a valve. Initially, one tank contains an ideal gas at 2 bar and 290 K, and the other is completely evacuated. The valve is opened, and the pressures and the temperatures are allowed to equalize.
(a) What is the final temperature and pressure in the tanks?
(b) What is the entropy change of the gas?

Solution 4-3

(a) The system is the gas. No heat or work is exchanged between the system and the surroundings, and hence $\Delta U = 0$. Because the internal energy of an ideal gas depends only on temperature, $\Delta T = 0$. The final temperature is equal to the original temperature, and because the volume has doubled, the ideal-gas law requires

$$P_2 = \frac{1}{2} P_1 = 1 \, \text{bar}$$

(b) To evaluate the entropy change, we must use a reversible path between $T = 290$ K, $P = 2$ bar and $T = 290$ K, $P = 1$ bar. A reversible isothermal expansion is such a path, and Eq. (4-23) gives the molar entropy change:

$$\Delta s = R \ln \frac{P_1}{P_2} = 8.314 \ln \frac{2}{1} = 5.763 \text{kJ / kmol} \cdot \text{K}$$

The quantity of gas in the system is

$$n = \frac{PV}{RT} = \frac{2(10^5)(1)}{8.314(10^3)(290)} = 0.0830 \, \text{kmol}$$

and the entropy change for the gas is

$$\Delta S = n \Delta s = 0.0830(5.763) = 0.478 \text{ kJ / K}$$

As the process is adiabatic, the surroundings undergo no change resulting from heat exchange and hence experience no entropy change. This obviously irreversible process exhibits a positive entropy change in accordance with our entropy criterion.

It should be noted that in each of these three examples the system was adiabatic. However, all processes were irreversible, and therefore the actual heat exchange $Q = 0$ was not a reversible heat effect and could not be used to compute the entropy change. Such a calculation would have produced the result $\Delta S = 0$ and would have implied that the processes were reversible. This we know is clearly not the case. Yet it may seem contradictory that the actual heat effect $Q = 0$ was used to evaluate the entropy change of the surroundings but cannot be used to evaluate the entropy change of the system. The difference is that in each example the system experienced a change in state, whereas the surroundings were unaffected by the process occurring in the system and therefore experienced no change in state and hence no entropy change.

4-7 OPEN SYSTEMS

Up to this point in the chapter our consideration has been restricted to closed systems to the apparent neglect of open systems. Because of the separate treatment given the application of the first law to open systems, we may have been led to expect a similar approach with the second law. This is unnecessary. While all derivations have been based on closed systems, the essence of the chapter is the establishment of the entropy as a property in which value depends only on the state of a substance. This state is independent of whether the substance is static and constitutes a closed system or is flowing through an open system. We saw this when determining enthalpy changes for application of the first law to open systems. Similarly, reservoirs that exchange heat experience an entropy change which depends only on the quantity and not the mode of the heat exchanged.

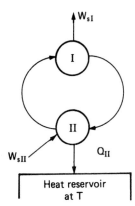

Figure 4-8 Closed system consisting of two devices each operating as open systems.

It now remains to be determined whether the entropy criterion, Eq. (4-15) or (4-16), derived for closed systems is also valid for open systems. For this determination, visualize two devices in a closed loop through which fluid is circulated, as sketched in Fig. 4-8.

Device I operates adiabatically and irreversibly, performing work W_{sI}, while device II operates reversibly, exchanging heat Q_{II} and work W_{sII} with the surroundings. The circulating fluid undergoes the state change A to B in device I and the reverse change B to A in device II. The entire loop is a closed system even though each device is an open system. When steady-state conditions are obtained, there is no time-dependent change in the closed system, and we may state that $\Delta U = 0$ with respect to any time interval. With this result the first law requires that

$$Q_{II} = -\left(W_{sI} + W_{sII}\right)$$

where these are the quantities of heat and work exchanged over the interval. The second law requires $Q_{II} \leq 0$; otherwise heat has been completely converted into work.

Device II remains in a steady-state condition, and the removal of heat Q_{II} through the device affects only the flowing fluid, which must then undergo a negative entropy change. In flowing through device I the fluid experiences the reverse, or a positive, entropy change. As with a closed system, the criterion stated by Eq. (4-15) applies equally to adiabatic and irreversible flow processes. In the case of an open system at steady state it is the flowing fluid that experiences the entropy change. By considering device I to be a composite system which includes heat reservoirs, we can, with the same argument employed in Sec. 4-5, show that Eq. (4-16) also applies to open systems.

4-8 APPLICATIONS OF THE SECOND LAW

The first law allows the calculation of heat and work effects for any change in state we may specify, but the second law allows us to determine the direction and extent of change which could occur. Therefore, the two laws are complementary and taken together form a powerful set of tools for analyzing processes. Certain types of problems where the change in state is known may not require the use of the second law for calculation of heat and work effects. These we have seen in Chap. 2. However, if we wish to know the maximum work available from a process, or the minimum amount of work required to drive a process, or the efficiency of a process, or how far a process will go from a given initial state, or whether a specified process is possible, we obtain the answer by using the second law—usually in combination with the first law.

For most problems we consider the process to be reversible and set the sum of all entropy changes in the system and surroundings equal to zero. This second law statement along with the first law provides two equations. If in formulating the problem more than two unknowns appear, other equations will be needed. The equations will arise out of the problem statement or result from reasonable approximations or assumptions. Those equations arising from the problem statement could consist of material balances or statements of constancy of volume or pressure. Equations arising from approximations and assumptions could result from specifying an equation of state (e.g., ideal gas) or conditions of operation (e.g., adiabatic or isothermal). The following examples will illustrate the application of these principles.

EXAMPLE 4-4

How much work could have been obtained from the copper block and water described in Ex. 4-2?

Solution 4-4

Here it is desired to find the maximum work obtainable from a process which requires us to specify reversibility. Because of the temperature difference, heat will flow from the block into the water, and it is possible, at least in principle, to harness this process by employing a reversible heat engine (Carnot engine). This engine would receive heat from the block and reject heat to the water. Operation of the engine would thus lower the block temperature and raise the water temperature and would continue with decreasing work output until the temperature difference approached zero. Our system will be chosen to be the copper block and water plus the Carnot engine and is therefore adiabatic. The Carnot engine will be sized so that it has undergone an integral number of cycles when all the possible work has been extracted and therefore undergoes no property changes. We may then write

$$\Delta S_{Cu} + \Delta S_{H_2O} = 0$$

$$10(0.398)\ln\frac{T}{800} + 100(4.185)\ln\frac{T}{290} = 0$$

$$T = 292.79\,\text{K}$$

This temperature is lower than that found in Ex. 4-2 for the irreversible process because the adiabatic system has done work. The quantity of work may be calculated by applying the first law to the system. For $Q = 0$,

$$\Delta U = W$$

For condensed phases we have seen that ΔU differs only negligibly from ΔH. For this property change we have

$$\Delta U = \Delta H = 10(0.398)(292.79 - 800) + 100(4.185)(292.79 - 290)$$

$$W = \Delta U = -851\,\text{kJ}$$

Thus, 851 kJ of work is possible. When the process was carried out irreversibly as in Ex. 4-2, this ability to do work was lost.

EXAMPLE 4-5

Determine whether the following process violates the laws of thermodynamics. An ideal gas of constant heat capacity (c_p = 30 kJ/kmol · K) at 10 bar and 295 K enters a device which is thermally and mechanically insulated from the surroundings. One-half of the gas leaves the device at 355 K and 1 bar, while the other half leaves at 235 K and 1 bar.

Solution 4-5

We have a steady-state open system with $Q = 0$ and $W_S = 0$ for which the first law dictates $\Delta H = 0$. This will now be tested:

$$\Delta H = \frac{1}{2}(30)(355 - 295) + \frac{1}{2}(30)(235 - 295) = 0$$

There is no violation of the first law. Second law conformance will be tested by applying the criterion of Eq. (4-15). The entropy changes are as follows:

Hot gas

$$\Delta s_{\text{H}} = R\ln\frac{10}{1} + 30\ln\frac{355}{295} = 24.70\,\text{kJ} / \text{kmol} \cdot \text{K}$$

Cold gas

$$\Delta s_{\text{C}} = \ln\frac{10}{1} + 30\ln\frac{235}{295} = 12.32\,\text{kJ} / \text{kmol} \cdot \text{K}$$

For a kmol of gas flowing through the device

$$\Delta S = \frac{1}{2}\Delta s_H + \frac{1}{2}\Delta s_C = \frac{1}{2}(24.70 + 12.32) = 18.51\,\text{kJ} / \text{K}$$

As the total entropy change for this adiabatic process is definitely positive, there is no violation of the second law. We may say that such a device is possible because neither law is violated. Actually, such a device exists and is called a Ranque-Hilsch vortex tube. It contains no moving parts; compressed gas enters tangentially with a hot stream leaving one end of the tube and a cold stream leaving the other. Much effort has been expended in understanding how the device works, but no simple explanation is available.

EXAMPLE 4-6

Determine the maximum temperature spread obtainable from the Ranque-Hilsch vortex tube when fed with the same gas under the same operating conditions as in Ex. 4-5.

Solution 4-6

The best performance will be obtained for reversible operation where the total entropy change is zero:

$$\Delta S = 0 = \frac{1}{2}\left(R \ln \frac{10}{1} + 30 \ln \frac{T_H}{295} + R \ln \frac{10}{1} + 30 \ln \frac{T_C}{295} \right)$$

The first law provides the additional equation needed to solve for T_H and T_C:

$$\Delta H = 0 = \frac{1}{2}(30)(T_H - 295) + \frac{1}{2}(30)(T_C - 295)$$

Solving simultaneously gives

$$T_H = 545.5 \, \text{K} \qquad\qquad T_C = 44.5 \, \text{K}$$

Some error may have been introduced with the assumption of constant heat capacity, but we see that an extremely large temperature spread is possible.

The Carnot engine can be useful in determining the minimum required work or maximum available work associated with a process. When operated in reverse, it becomes a heat pump or refrigerator–air conditioner depending on whether the rejected high-temperature heat is used for heating or the low-temperature input heat is removed from the system. We have seen that the efficiency of a Carnot engine is $(T_2 - T_1)/T_2$ and can easily determine the coefficient of performance (COP) for a heat pump, defined as Q_2/W, or a refrigerator–air conditioner, defined as Q_1/W. Figure 4-9 shows a Carnot engine operating in the reverse direction removing heat $|Q_1|$ from the reservoir at T_1 and delivering heat $|Q_2|$ to the reservoir at T_2. This requires that work be supplied. From the first law,

$$W = |Q_2| - |Q_1|$$

and from the second law,

$$\frac{|Q_2|}{T_2} = \frac{|Q_1|}{T_1}$$

These equations are combined to yield the coefficients of performance:

Heat pump

$$\text{COP} = \frac{|Q_2|}{W} = \frac{T_2}{T_2 - T_1} \qquad (4\text{-}24)$$

Refrigerator–air conditioner

$$\text{COP} = \frac{|Q_1|}{W} = \frac{T_1}{T_2 - T_1} \qquad (4\text{-}25)$$

These expressions are not restricted to Carnot cycles but apply to any reversible device operating between the two specified temperatures.

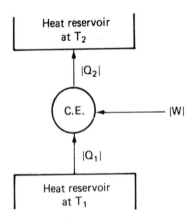

Figure 4-9   Carnot engine operating in reverse.

EXAMPLE 4-7

From a thermodynamic perspective, compare the effectiveness of the following methods of heating a house to be maintained at 295 K (71°F):
(a) gas furnace.
(b) electrical resistance heating.
(c) electrically driven heat pump.

     Through the power plant and transmission system 33% of the heat of combustion of the gas is converted to electricity delivered to the house. The heat pump takes heat from the outside air at the average temperature of 275 K (35°F). The heat pump has an overall efficiency

> **EXAMPLE 4-7 CON'T**
>
> of 25%, and the gas furnace delivers 75% of the heat of combustion for house heat (the other 25% goes up the flue). The heat of combustion of the gas (methane) is 890.9 MJ/kmol.

Solution 4-7

As a basis of comparison we take 10^6 kJ of heat delivered to the house.
(a) For the gas furnace,

$$\frac{10^6}{0.75(890,900)} = 1.50 \text{ kmol of gas}$$

(b) All the electrical energy entering the residence is effective for heating; however, we desire the amount of gas required to generate the electricity:

$$\frac{10^6}{0.33(890,900)} = 3.40 \text{ kmol of gas}$$

(c) We first calculate the work required for a reversible heat pump and then use the overall efficiency to obtain the actual work. The coefficient of performance of the heat pump is

$$\frac{Q_2}{W} = \frac{T_2}{T_2 - T_1} = \frac{295}{20} = 14.75$$

$$W = \frac{10^6}{14.75} = 6.78(10^4) \text{ kJ}$$

This is the electrical energy required to operate a reversible heat pump. To determine the gas consumption at the power plant, we correct for heat pump efficiency and for power plant conversion efficiency:

$$\frac{6.78(10^4)}{0.25(0.33)(890,900)} = 0.922 \text{ kmol of gas}$$

The heat pump is seen to be more efficient than the gas furnace, which in turn is more efficient than electrical resistance heating.

The combustion of the gas is an irreversible step and therefore all three heat delivery systems contain an inherent inefficiency. Much more efficient operation would be obtained if the oxidation of the gas could be carried out reversibly in a device[6] which would perform work and exchange heat with the surroundings at the temperature T. For such a device operating in a steady-state fashion, we write

$$\Delta H = Q + W_s$$

$$\Delta S + \Delta S_{surr} = 0$$

$$Q = -Q_{surr}$$

and obtain

$$Q = T\Delta S$$

and

$$\Delta H - T\Delta S = W_s$$

ΔH and ΔS are property changes for the oxidation of methane and are -890.9 MJ/kmol and -0.2426 MJ/kmol $\cdot$ K, respectively at 298 K. For each kmol of methane oxidized in the device at 298 K, there would be heat evolved and work done equal to

$$Q = 298(-242.6) = -72,330 \text{ kJ}$$

$$W_s = -890,900 - 298(-242.6) = -818,569 \text{ kJ}$$

When supplied to the heat pump, this work would deliver heat to the residence in the amount of

$$Q = 14.75(818,569) = 12.07(10^6) \text{ kJ / kmol of gas}$$

and to supply 10^6 kJ of heat would require

$$\frac{1}{12.07} = 0.0829 \text{ kmol of gas}$$

This is the amount of gas required for a completely reversible process and should serve as a basis for evaluating the efficiency of each mode of heating. Referred to this standard, the following efficiencies are obtained:[7]

Gas furnace:	5.5%
Electrical resistance heating:	2.4%
Electrically driven heat pump:	9.0%

---

[6] A reversible fuel cell.
[7] See also W. D. Metz, *Science, 188,* 820 (1975).

The second law with its concept of quality of energy judges the performance of these heating systems rather harshly. While the furnace could be said to be 75% efficient from a total energy recovery standpoint (a first law efficiency), the second law calls our attention to the fact that a high-grade energy source (chemical energy) has been used for a task more befitting a low-grade energy source (low-temperature heat). The same can be said of electrical resistance heating. This type of thermodynamic overkill has been likened to using a chain saw to cut butter. While it is discouraging to see this wasteful use of energy as judged by the second law, we must realize that this means that there is a great potential for energy savings in making space-heating systems more efficient.[8] Consistent with a reversible, and hence thermodynamically ideal, process being one where all potentials are balanced, it could be alternatively stated that such a process should have its energy source closely matched in quality to its energy needs.

In the preceding few details were given concerning the design or mode of operation of the devices employed in the execution of the processes studied. This may at first cause consternation because it seems that in addition to working in terms of abstract quantities like ΔH and ΔS, which yet may have little physical meaning, we also appear to be arbitrary and vague in describing our system. One might ask the following: How, in Ex. 4-6, can we determine the performance of a device that is nothing more than a *black box*? Or one might ask: How, in Ex. 4-7, we can determine the maximum work attainable from the oxidation of methane by specifying only something as nebulous as a *reversible device*?

The reason we need not detail the device is that the quantities we seek—for example, the maximum available work, the minimum required work, the maximum quantity of heat removed or delivered, or the final state of a system—are expressible in terms of changes in state properties. Therefore, we need only supply details of our system sufficient to fix the initial and final states because regardless of what happens inside the process, the property changes are fixed. Viewing the situation positively, we can say that our thermodynamic tools are quite flexible and produce results of maximum generality.

These results merely provide a limit to what is possible. If we desire to exploit this potential, we must then be concerned with the specifics of the process.

4-9 THE MICROSCOPIC VIEW OF ENTROPY

We have provided a mathematical definition of entropy, shown that it is a state property, developed means of calculating its changes, and shown its usefulness in determining maximum and minimum work and direction and extent of change. While this is sufficient for the application of thermodynamics, we would be more comfortable with entropy if, like other physical properties, it could be given an interpretation on a microscopic level. However, Before attempting such an explanation, it is necessary to reiterate that the results of thermo-

---

[8] Approximately 18% of America's energy consumption is due to space heating. See E. Hirst and J. C. Moyers, *Science, 179*, 1299 (1973).

dynamics are not dependent on the validity of any particular theory or model that describes matter in terms of ultimate particles.

From quantum theory[9] we know that the energy of a particle can assume only certain discrete values although the spacing between values may be quite small. Each of these energy values is referred to as a *particle quantum state*. It is also possible to define a quantum state for an entire assemblage of particles. This *total quantum state* depends on the positions of the individual particles as well as their energies. Macroscopic systems of thermodynamic interest contain an astronomical number of particles, and an even larger number of accessible total quantum states—each representing a particular spatial configuration of the particles and particle energy distribution corresponding to a fixed total energy. Through the methods of statistical mechanics it can be shown that in an isolated thermodynamic system the entropy is proportional to the logarithm of the number of accessible *total quantum states* Ω:

$$S = k \ln \Omega \tag{4-26}$$

Thus, for a spontaneous change occurring in an isolated system, we write

$$S_2 - S_1 = k \ln \frac{\Omega_2}{\Omega_1} \tag{4-26a}$$

and note that the required condition $S_2 > S_1$ dictates $\Omega_2 > \Omega_1$. This means that the more stable state is characterized by a larger number of accessible quantum states and an increase in Ω can be visualized as a spreading of the system over an increased number of accessible quantum states. Therefore, the system moves in the direction of increasing possibilities.

This exercise may not be perceived as wholly satisfactory as we seem to have traded one abstraction, the entropy, for another, Ω, the number of accessible total quantum states. Because Ω is not an easily visualized physical quantity, there is no simple satisfying microscopic interpretation for entropy. Nevertheless, there are two widely held alternative, but not mutually exclusive, interpretations of entropy:

1. It is a measure of order (or disorder) in the microscopic system.

2. It is a measure of information (or ignorance) on a microscopic level possessed by an observer.

From a molecular viewpoint, the association of a positive entropy change with an increase in disorder seems quite reasonable for ordinary phase changes and for the process of mixing. For other processes the association is less obvious, and for at least one process (e.g., the adiabatic crystallization of a subcooled liquid, Ex. 4-1) it fails completely. One can also object to this interpretation because order and disorder are not precise objective terms.

---

[9] A detailed account of the quantum-statistical-mechanical derivation of entropy can be found in Chap. 2 of the essay ENTROPY on the accompanying CD-ROM (See App. E); only a brief discussion of the results is given here.

With the information interpretation,[10] we reason that because $\Omega_2 > \Omega_1$ means the system has moved from fewer to more possibilities, we as observers have suffered a loss of certainty regarding its microscopic state. Thus, a decrease in information is said to accompany an increase in entropy. The major objection to this view is that entropy, a physical quantity, has been represented by information, an entirely subjective quantity.

We see that despite its importance for our understanding of physical phenomena, entropy has eluded our attempts to give it a simple microscopic interpretation. The only unambiguous interpretation is macroscopic and arises from considerations presented in Sec. 4-8: The entropy change measures the work lost when a process falls short of the reversible ideal.[11]

4-10 THE THIRD LAW OF THERMODYNAMICS

The first and second laws have resulted in the definition of state properties, or, more strictly, changes in state properties. These laws provide no information concerning the absolute value of internal energy or entropy. The third law defines no new property but allows absolute values of the entropy to be determined. Statements of the third law take several forms in the literature;[12] however, because our ultimate interest lies mainly in applications, the formulation most suited for that purpose is presented here:

The entropy of a perfect crystalline substance is zero at zero absolute temperature.

This statement allows an absolute value of the entropy of a substance to be calculated from the equations developed in Sec. 4-6. For a substance which is gaseous at the temperature T we obtain[13]

$$s_T = \int_0^{T_F} \frac{c_P(c)}{T} dT + \frac{\Delta h^{\text{fusion}}}{T_F} + \int_{T_F}^{T_V} \frac{c_P(l) dT}{T} + \frac{\Delta h^{\text{vap}}}{T_V} + \int_{T_V}^{T} \frac{c_P(g) dT}{T} \qquad (4\text{-}27)$$

If the substance were in the liquid state, we would drop the last two terms and carry out the second integration between T_F and T. If more than one solid phase existed, then appropriate terms involving heat capacity and latent heat would be added. The appropriate form of Eq. (4-27) can be used to calculate absolute entropies of substances for which the necessary calorimetric data are available. These values find their greatest use in calculations involving

---

[10] The information interpretation of entropy is fully explored in Chap. 4 of the essay ENTROPY on the accompanying CD-ROM.

[11] See Sec. 14-2.

[12] For a detailed discussion of the third law see Chap. 3 of the essay ENTROPY on the accompanying CD-ROM (see App. E).

[13] For details of evaluating the first integral, see K. G. Denbigh, *Principles of Chemical Equilibrium*, 3rd ed., Cambridge University Press, New York, 1971, Chap. 13.

chemical equilibrium, as will be shown in Chap. 12. The successful use of absolute entropies for this purpose provides a convincing proof of the validity of the third law.

The law differs from the first two laws of thermodynamics in that its implementation requires a knowledge of the microscopic nature of our system. The term *perfect crystal* implies such knowledge and is best understood from Eq. (4-26). At absolute zero all particles will be in the lowest energy state, and hence there is only a single particle quantum state. Therefore, if there is only one spatial configuration, as with a perfect crystal, then there is only one total quantum state, $\Omega = 1$, and the entropy is zero. The vast majority of substances form perfect crystals, although several substances are known to form crystals which are not perfect. Absolute entropies for these anomalous substances are found by calculating the entropy of the crystal at absolute zero through the methods of statistical and quantum mechanics.

PROBLEMS

4-1. A prospector's cabin is located beside a large Alaskan lake. In the winter the average air temperature is 239 K, but beneath the ice cover the water in the lake is close to 273 K. The prospector, an amateur thermodynamicist, believes that he could heat his cabin by making use of this temperature difference. Is this possible? Explain.

4-2. Determine the thermodynamic temperature scale originally proposed by Kelvin, in which degrees are defined so that each engine has the same efficiency. See Sec. 4-3. Discuss the pros and cons of such a scale.

4-3. Assuming that unlimited heat exchange with the environment at 298 K is possible,

    (a)   Can steam be used to produce ice?
    (b)   Can ice be used to produce steam?

Justify your answers with calculations where steam refers to saturated steam at 1 atm (212°F).

        Heat of fusion: 143 Btu/lb
        Heat of vaporization: 970 Btu/lb
        c_p ice: 0.44 Btu/lb °F
             water: 1.0 Btu/lb °F
             steam:0.50 Btu/lb °F

4-4. Calculate the minimum amount of work required to convert 1 g mol of oxygen gas at 298 K and 1 atm to the liquid state at 90.13 K (the atmospheric boiling point) in a reversible process in which heat is transferred to a heat reservoir at 298 K. The following data are given: c_p for oxygen gas: 6.96 cal/g mol °C, Δh of vaporization at 90.13 K: 1628.8 cal/g mol.

4-5. A heat reservoir consisting of 100 lb of liquid water initially at 212°F furnishes heat to a Carnot engine that discards heat to a sink consisting of 100 lb of liquid water initially at 32°F. Calculate the maximum work obtainable from the engine.

4-6. A process has been proposed whereby an ideal gas (c_p = 30 kJ/kmol · K) is taken from P = 10 bar and T = 300 K to P = 1 bar and T = 500 K in a closed system. During the process the system does 1000 kJ of work and receives 5430 kJ of heat from the surroundings at 300 K.

(a) Is this process possible?
(b) Would it be possible for the gas to undergo the same change in state in a spontaneous process occurring adiabatically? Explain.

4-7. A heat reservoir consisting of 100 lb of water initially at 212°F furnishes heat to a heat engine which discards heat to a sink which is initially 100 lb of ice at 32°F. Calculate the maximum heat obtainable from the engine.

4-8. Nitrogen gas flowing at the rate of 1000 kg/h enters a compressor at 1 bar and 300 K and leaves at 10 bar and 420 K. Cooling water flows through the compressor at the rate of 1500 kg/h, entering at 300 K and leaving at 320 K.

(a) How much power is supplied to the compressor?
(b) With the same cooling water flow rate and inlet temperature, calculate the minimum power required to attain the same change in state of the nitrogen.

4-9. Steam at 200 psia and 600°F flows through a turbine operating adiabatically and exits at atmospheric pressure. For every pound of steam flowing through the turbine there are 150 Btu of shaft work delivered.

(a) What is the final condition of the exit steam?
(b) What is the maximum amount of work that could be obtained per pound of steam for adiabatic operation between these two pressures?

4-10. It is proposed to air condition a house using an absorption refrigeration system with heat supplied from water heated in solar panels. The house is to be maintained at 75°F, and the average ambient air temperature is 95°F. If hot water is to be taken from the solar panels at 200°F and returned at 160°F, find the minimum hot water circulation rate if 12,000 Btu/h must be removed from the house. State all assumptions made.

4-11. It is proposed to use a limited supply of liquid nitrogen for the generation of power on a space vehicle. Compute the maximum work available from 1 lb of liquid nitrogen when a reversible heat engine of the required complexity is available.

The temperature and pressure of the nitrogen supply are −320° F and 1 atm, respectively. The latent heat of vaporization of nitrogen is 85.5 Btu/lb, and the specific heat of nitrogen vapor is 0.25 Btu/lb °F and may be assumed constant. A heat source of 300°F is available.

4-12. A steam turbine operates adiabatically and produces 4000 hp. Steam enters the turbine at 300 psia and 900°F. Exhaust steam from the turbine is saturated vapor at 5 psia. What is the steam rate (in pounds per hour) through the turbine? What is the efficiency of the turbine compared with isentropic operation?

4-13. Calculate the minimum amount of work required to produce 1 ton of ice in a flow process starting with water at 70°F. Assume the surroundings to be at 70°F. The heat of fusion of water is 143 Btu/lb.

4-14. In a plant, the overhead vapor from a distillation column is condensed with cooling water entering the condenser at 80°F and leaving at 100°F. The vapor is essentially water, and the condenser operates at atmospheric pressure. Vapor flows into the condenser at a rate of 8000 lb/h, and the condensate is subcooled to 160°F. How much work could be obtained and by what percentage could the cooling water rate be reduced if the condenser were replaced with a reversible heat engine? As before, cooling water will enter at 80°F and leave at 100°F.

4-15. As a result of energy conservation measures, a chemical manufacturing facility has an abundant supply of low-pressure steam. A reasonable supply of cooling water is also available, and it is desired to consider the possibility of using these resources to produce chilled water for air conditioning. The steam is available, saturated at 1 atm, and will leave the chiller as saturated liquid. Cooling water is available at 80°F and will leave the chiller at 100°F. The chilled water will enter the chiller at 75°F and leave at 55°F. Based on 1 lb of steam, determine the cooling water requirement and the maximum quantity of chilled water that can be obtained.

4-16. As you are well aware, this institution is trying desperately to trim its operating budget. As one of the largest expenditures is for winter heating, we want to consider every possible opportunity to save on this energy cost. It has been suggested that disconnecting the electricity to drinking fountains in campus buildings during the winter would save money; however, someone has pointed out that these devices function as heat pumps and may be beneficial for heating. Ignoring for the moment the minuscule savings potential of this proposed action, give us your informed opinion based on thermodynamic reasoning.

> Water comes into buildings at 55°F and is cooled to 40°F in the drinking fountain. Buildings are maintained at 70°F, and we will assume no heat exchange between incoming or outgoing water and the building. The refrigeration unit in the drinking fountain rejects heat to the inside building space and will be assumed to be 50% efficient.

> If a Btu of electricity costs four times as much as a Btu of heating from fuel, would the proposed disconnection policy result in savings?

4-17. A steady stream of air at 300°F and 5 atm is available. How much work can be obtained from each pound-mol of air if it flows through a reversible device and leaves at 77°F and 1 atm? The device may exchange heat with the surroundings at 77°F. Assume ideal-gas behavior and constant heat capacity ($c_p = 7.0$ Btu/lb mol °R).

4-18. Find the maximum amount of work that could be obtained from a kg of steam at 8 MPa and 500°C in a steady-state flow process where the exit stream is saturated liquid at 1 bar. Unlimited heat exchange with the surroundings at 289 K is possible.

4-19. We have an excess of saturated steam available at 3 MPa and plenty of exhaust steam available at 0.3 MPa, but have a need for steam at 1.0 MPa. It has been proposed that the low-pressure steam be compressed to 1.0 MPa using work available from the expansion of the high-pressure steam to 1.0 MPa. Assume that the compression and the expansion can be carried out adiabatically and reversibly. Determine the relative amounts of the two streams.

4-20. Superheated steam is available at 3.0 MPa and 500°C but saturated steam at 3.0 MPa would meet our process needs. It has been suggested that liquid water at 24°C be injected into the superheated steam to produce saturated steam at 3.0 MPa.

(a) How many kgs of saturated steam could be produced from a kg of superheated steam?
(b) The mixing process of part *a* would not appear to be an efficient operation. For a kg of superheated steam, calculate the maximum amount of heat that could be delivered to the process at the condensing temperature of 3.0 MPa saturated steam. Assume unlimited heat exchange with the surroundings at 24°C. Compare this with the amount of heat delivered by the saturated steam of part *a*.

4-21. Compare the performance of the ejector described in Prob. 2-24 with a reversible device operating adiabatically, fed with saturated steam at 0.8 MPa, and exhausting saturated vapor at 0.025 MPa to the surroundings at 1 bar. How many kilograms of low-pressure vapor can be removed from the evaporator per kilogram of high-pressure steam?

4-22. Process steam, saturated at 30 psig, is presently being generated in a boiler fired with methane. The efficiency (first law basis) of the boiler is 85%. Consideration is being given to cogeneration of electricity and steam. If we can expect an efficiency (Carnot) of 37% in producing electricity, how much electricity and 30-psig steam can be produced per lb-mol of methane burned? If the generated electricity is used to power a heat pump with 55% efficiency, taking heat from a reservoir at 70°F and delivering it to the process at the condensing temperature of 30-psig steam, what is the total amount of heat delivered to the process (cogenerated steam plus pumped heat) per lb-mol of methane? How does this compare with the simple steam generation process?

4-23. It has been noted by J. L. McNichols, W. S. Ginell, and J. S. Cory [*Science, 203*, 167 (1979)] that a temperature difference of up to 18°C exists in several hydroelectric reservoirs in the western United States. They have determined the potential energy equivalent of this thermal energy by

$$\mathbf{g}h = C\Delta T \tag{1}$$

where $\mathbf{g}$ is the acceleration due to gravity, C is the specific heat of water, ΔT is the maximum temperature difference within the reservoir, and h is the equivalent hydrostatic head. Recognizing the Carnot limitation, they also determined h', the maximum realizable equivalent hydrostatic head:

$$h' = h\frac{\Delta T}{T} = \frac{C\Delta T^2}{\mathbf{g}T} \tag{2}$$

where T is the warm water temperature. This result was challenged by Crouch [*Science, 208,* 1292 (1980)], who obtained, for the maximum power output **P**,

$$\mathbf{P} = mC[T + \alpha \, \Delta T - (T + \Delta T)^{\alpha} \mathrm{T}^{\,1-\alpha}] \tag{3}$$

which for $0 \le \Delta T / T \ll 1$ may be approximated by

$$\mathbf{P} \simeq \frac{mC\Delta T^2}{T} \left[\frac{\alpha(1-\alpha)}{2} \right] \tag{4}$$

In deriving the equation, Crouch considered a flow rate of m of which a fraction α is at a temperature $T + \Delta T$ and a fraction $1 - \alpha$ is at T. Derive all four equations, and attempt to resolve the differences, pointing out any flawed thermodynamic reasoning.

4-24. An air-driven turbine is attached to a well-insulated tank which is initially evacuated. The turbine is operated by air that enters the tank from the atmosphere and operates until the tank pressure reaches atmospheric pressure. For a tank of 1-m^3 volume, determine the reversible work obtainable when

    (a)   the turbine is operated isothermally at 298 K.
    (b)   the turbine is operated adiabatically.

Also determine the final temperature of the air in the tank for each mode of operation.

4-25. For a portable power source it is proposed to use a small expansion engine connected to a tank of 1-ft^3 volume. The tank will contain air (assumed to behave as an ideal gas with $c_p = 7$) at an initial pressure of 10 atm. Assuming that the operation of this system (tank and engine) can be made isothermal at 530°R, calculate the maximum work which could be obtained.

4-26. An air-driven turbine is attached to a tank of 1-m³ volume which is initially evacuated. The turbine is operated by air that enters the tank from the atmosphere and operates until the tank pressure reaches atmospheric pressure. Determine the maximum work obtainable from this system when the turbine operates isothermally at 298 K and the air in the tank is kept at 298 K by heat exchange with the surroundings at 298 K. Also determine the quantity of heat exchange between the tank and the surroundings.

4-27. A tank of 10-m³ volume contains an ideal gas ($c_p = 30$ kJ/kmol · K) at 298 K and 5 bar. If gas from the tank is allowed to flow through a reversible device and exhaust to the atmosphere, what is the maximum work that could be obtained if both the tank and reversible device could be kept isothermal at 298 K?

The Thermodynamic Network

We have seen how ordinary and simple statements summarizing the results of many years of experimentation with energy conversion, prompted by the need for engines to power the machines of the Industrial Revolution, have formed the basis for the first and second laws of thermodynamics. Through the application of mathematical logic and without any simplifying assumptions or approximations these statements were fashioned into the two basic laws. These laws are mathematical statements that define two hitherto unrecognized properties of matter: the internal energy U and the entropy S. Additionally, these laws have been shown to be quite useful in dealing with systems in which energy transformations occur.

Another major application of thermodynamics involves extending our knowledge of a system by calculating properties that are unknown or would be difficult to measure from known or easily measured properties. To accomplish this we must first establish relationships among these properties. The major thrust of this chapter centers on the enthalpy and the entropy: development of equations relating these properties to other measurable properties, the implementation of these equations with available experimental data for the determination of enthalpy and entropy changes, and the use of the corresponding states principle and generalized equations of state to estimate enthalpy and entropy changes when experimental PVT data are lacking. A general and systematic procedure for determining relationships among the various thermodynamic properties and their derivatives, the Jacobian method, is also presented.

5-1 THE FREE ENERGY FUNCTIONS

For calculations to follow, it is now convenient to introduce two new thermodynamic functions. The Helmholtz free energy A, defined as

$$A = U - TS \tag{5-1}$$

and the Gibbs free energy G, defined as

$$G = U + PV - TS = H - TS \tag{5-2}$$

As these functions are defined in terms of other state properties, they are also state properties.[1] The introduction of these functions will facilitate the establishment of the thermodynamic network and the treatment of equilibrium. In addition, they will be shown later to be useful measures of the maximum work obtainable in an isothermal process. For this reason they are often called "work functions."

5-2 THE CLAUSIUS INEQUALITY AND THE FUNDAMENTAL EQUATION

For any process occurring in a closed system we have seen that

$$\Delta S_{sys} + \Delta S_{surr} \geq 0 \tag{5-3}$$

where the equality refers to a reversible process and the inequality refers to an irreversible process. Consider a closed system surrounded by a heat reservoir always at the temperature T, and let the system absorb an amount of heat Q. The entropy change for the surroundings is $-Q/T$, and Eq. (5-3) becomes

$$\Delta S \geq \frac{Q}{T} \tag{5-4}$$

---

[1] Again we will follow the convention of representing extensive properties with capital and intensive properties with lowercase symbols.

As it is no longer necessary to distinguish between the system and the surroundings, the subscript has been dropped; unsubscripted properties will henceforth refer to the system.

Equation (5-4), the second law statement, will now be substituted into the first law statement ($\Delta U = Q + W$) to obtain

$$\Delta U - T\Delta S - W \leq 0 \tag{5-5}$$

At this point it will be convenient to identify two types of work: work exchanged with the environment at a constant pressure due to changes in the system volume[2] and all other forms of work W' (e.g., shaft work or electrical work). Thus, Eq. (5-5) can be written as

$$\Delta U - T\Delta S + P\Delta V - W' \leq 0 \tag{5-6}$$

Equations (5-5) and (5-6) are alternative forms of the Clausius inequality so useful in determining conditions of equilibrium and stability. These combined first and second law statements are quite general. Equation (5-5) is restricted only to the case of constant-temperature surroundings, while Eq. (5-6) is useful where the surroundings are at constant temperature and pressure. There are no restrictions on the type of process occurring within the system. For what follows it is convenient to consider an infinitesimal change and write Eq. (5-6) as

$$dU - T\,dS + P\,dV - dW' \leq 0 \tag{5-7}$$

The fundamental equation of thermodynamics

$$dU = T\,dS - P\,dV \tag{5-8}$$

which relates property changes, can be obtained from the Clausius inequality. Equation (5-6) applies to a process between two equilibrium states. An infinitesimal displacement from an equilibrium state meets the conditions of reversibility, and the equality in Eq. (5-7) must hold. Restriction of consideration to systems capable of only PV work eliminates dW'. Thus, in going from Eq. (5-6) to (5-8), the focus of our attention has been shifted from that of analyzing a system undergoing a process to that of relating property changes within the system. Specifically, Eq. (5-8) describes the relationship between changes in U, S, and V for an infinitesimal displacement from an equilibrium state characterized by given values of T and P.

The information embodied in Eq. (5-8) can also be expressed in terms of H, A, and G. For example,

$$H = U + PV$$

with

$$dH = dU + P\,dV + V\,dP$$

---

[2] Recall that in accordance with our convention PV work is $-P\Delta V$ and $W = W' - P\Delta V$.

and substitution of Eq. (5-8) for dU gives

$$dH = T\,dS + V\,dP \tag{5-9}$$

Similarly, using the defining equations for A and G, Eqs. (5-1) and (5-2), we obtain

$$dA = -P\,dV - S\,dT \tag{5-10}$$

$$dG = V\,dP - S\,dT \tag{5-11}$$

Equations (5-8)–(5-11) are equally valid statements of the fundamental equation. These equations play a major role in establishing the network of relationships among the thermodynamic variables.

5-3 THE THERMODYNAMIC NETWORK

The various statements of the fundamental equation, Eqs. (5-8)–(5-11), are written so that the left-hand side is the differential of a state function. They all have the form of an exact differential,

$$dz = \left(\frac{\partial z}{\partial x}\right)_y dx + \left(\frac{\partial z}{\partial y}\right)_x dy \tag{5-12}$$

where $z = z(x, y)$, which when written as

$$dz = M\,dx + N\,dy \tag{5-13}$$

is called an exact differential equation. A useful property of this type of equation is

$$\left(\frac{\partial M}{\partial y}\right)_x = \left(\frac{\partial N}{\partial x}\right)_y \tag{5-14}$$

Comparison of Eqs. (5-12) and (5-13) shows that this reciprocal relation is merely the statement that the order of differentiation is immaterial:

$$\frac{\partial^2 z}{\partial x \partial y} = \frac{\partial^2 z}{\partial y \partial x}$$

When Eqs. (5-8) through (5-11) are compared with Eq. (5-12), the following relations are apparent:

$$\left(\frac{\partial U}{\partial S}\right)_V = \left(\frac{\partial H}{\partial S}\right)_P = T \tag{5-15}$$

$$\left(\frac{\partial U}{\partial V}\right)_S = \left(\frac{\partial A}{\partial V}\right)_T = -P \qquad (5\text{-}16)$$

$$\left(\frac{\partial G}{\partial T}\right)_P = \left(\frac{\partial A}{\partial T}\right)_V = -S \qquad (5\text{-}17)$$

$$\left(\frac{\partial G}{\partial P}\right)_T = \left(\frac{\partial H}{\partial P}\right)_S = V \qquad (5\text{-}18)$$

Application of Eq. (5-14) to Eqs. (5-8)–(5-11) results in what are called Maxwell relations:

$$\left(\frac{\partial T}{\partial V}\right)_S = -\left(\frac{\partial P}{\partial S}\right)_V \qquad (5\text{-}19)$$

$$\left(\frac{\partial T}{\partial P}\right)_S = \left(\frac{\partial V}{\partial S}\right)_P \qquad (5\text{-}20)$$

$$\left(\frac{\partial P}{\partial T}\right)_V = \left(\frac{\partial S}{\partial V}\right)_T \qquad (5\text{-}21)$$

$$\left(\frac{\partial V}{\partial T}\right)_P = -\left(\frac{\partial S}{\partial P}\right)_T \qquad (5\text{-}22)$$

Equations (5-8)–(5-11) and (5-15)–(5-22), along with defining equations, constitute a network of equations which interrelate the various thermodynamic properties. In addition, other equations can be generated by performing various mathematical operations on these equations. For example, if we desire to know $(\partial H / \partial P)_T$, we may start with

$$dH = T\,dS + V\,dP \qquad (5\text{-}9)$$

and write

$$\frac{dH}{dP} = T\frac{dS}{dP} + V$$

This relation is general, and conditions of constant temperature can be imposed to give

$$\left(\frac{\partial H}{\partial P}\right)_T = T\left(\frac{\partial S}{\partial P}\right)_T + V \qquad (5\text{-}23)$$

Substitution of Eq. (5-22) yields

$$\left(\frac{\partial H}{\partial P}\right)_T = V - T\left(\frac{\partial V}{\partial T}\right)_P \qquad (5\text{-}24)$$

With the large number of equations and the varied operations that can be performed on them it should not be difficult to imagine the multitude of possible equations which can be derived. Often in the derivation of a desired relationship the starting point and the method of approach are not obvious, and much time can be spent generating unwanted expressions and following circular paths. A systematic procedure for determining relationships among the thermodynamic properties, the Jacobian method, will be treated in Sec. 5-8.

5-4 MEASURABLE QUANTITIES

Because the major advantage of the thermodynamic network is the determination of one quantity from information available for others, it is essential that the derived equations contain quantities that can be determined experimentally. For example, Eq. (5-23) expresses the isothermal pressure dependence of the enthalpy in terms of $(\partial S/\partial P)_T$, a quantity which is not directly measurable. However, Eq. (5-24) is useful because the desired quantity, $(\partial H/\partial P)_T$, is expressed in terms of V, T, and $(\partial V/\partial T)_P$ which are measurable.

As concerns pure substances or mixtures of unchanging composition the quantities most often measured are heat capacity and PVT data.

Heat Capacities. The constant-pressure heat capacity C_p and the constant-volume heat capacity C_V were defined in Chap. 2 where methods of experimental determination of these quantities were also indicated. From the defining equations it is observed that

$$C_P = \left(\frac{\partial H}{\partial T}\right)_P \qquad (5\text{-}25)$$

$$C_V = \left(\frac{\partial U}{\partial T}\right)_V \qquad (5\text{-}26)$$

The heat capacities may also be expressed in terms of the entropy. This may be done by dividing Eqs. (5-8) and (5-9) by dT, restricting Eq. (5-9) to a constant-pressure change, and imposing conditions of constant volume on Eq. (5-8). The results are

$$\left(\frac{\partial H}{\partial T}\right)_P = T\left(\frac{\partial S}{\partial T}\right)_P = C_P \qquad (5\text{-}27)$$

$$\left(\frac{\partial U}{\partial T}\right)_V = T\left(\frac{\partial S}{\partial T}\right)_V = C_V \qquad (5\text{-}28)$$

PVT **Data.** In Chap. 3 we saw how the relationship between volume of a fluid and temperature and pressure may be determined experimentally. While the general approach to this experimental endeavor is easy to visualize, many refinements in both apparatus and technique are necessary to minimize experimental errors. A high degree of accuracy is required because quite often the various derivatives involving P, V, and T are needed for calculation of other properties through the network of thermodynamic equations. To obtain a desired degree of accuracy in a derivative of a measured quantity, it is necessary for the accuracy associated with the quantity itself to be at a high level.

5-5 CALCULATION OF *H* AND *S* AS FUNCTIONS OF *P* AND *T*

The enthalpy and the entropy are the thermodynamic properties most often needed for engineering applications. Here we will be concerned with the determination of changes in these properties for any specified change of state over which the system remains a single phase. While it is possible to express the change of state in terms of any two state properties, the choice of pressure and temperature is most convenient. For purposes of representation it is convenient to arbitrarily assign a value of zero to the enthalpy and entropy in a datum, or reference, state[3] and thereby tabulate or plot values of ΔH and ΔS referred to this datum state. Although these properties may be labeled H and S, it should be remembered that they are not absolute values.[4] Fortunately, most applications require ΔH or ΔS, and the lack of absolute values presents no problem.

We desire to represent the enthalpy and entropy as functions of temperature and pressure, and from the calculus we may write

$$H = H(T, P) \qquad (5\text{-}29)$$

$$dH = \left(\frac{\partial H}{\partial T}\right)_P dT + \left(\frac{\partial H}{\partial P}\right)_T dP$$

$$S = S(T, P) \qquad (5\text{-}30)$$

---

[3] In the steam tables the datum state is saturated liquid at 32°F.
[4] It would be possible based on the third law to assign absolute values to the entropy; however, this is not usually done for fluid properties diagrams or tabulations.

$$dS = \left(\frac{\partial S}{\partial T}\right)_P dT + \left(\frac{\partial S}{\partial P}\right)_T dP$$

Substituting Eqs. (5-24) and (5-25) into Eq. (5-29) results in

$$dH = C_P \, dT + \left[V - T\left(\frac{\partial V}{\partial T}\right)_P\right] dP \tag{5-31}$$

and substitution of Eqs. (5-22) and (5-27) into Eq. (5-30) yields

$$dS = \frac{C_P}{T} \, dT - \left(\frac{\partial V}{\partial T}\right)_P dP \tag{5-32}$$

These two equations allow changes in H and S to be calculated for prescribed changes in P and T if heat capacity and PVT data are available. This calculation can be demonstrated for a change between the states defined by P_1, T_1 and P_2, T_2 as illustrated in Fig. 5-1. Because the change in a thermodynamic property is desired, any path between these states may be chosen for the calculation. Two possible paths are illustrated:

$$P_1, T_1 \rightarrow P_2, T_1 \rightarrow P_2, T_2$$

and

$$P_1, T_1 \rightarrow P_0, T_1 \rightarrow P_0, T_2 \rightarrow P_2, T_2$$

The path chosen for the calculation must be one for which the data are available to implement the calculation. For example, evaluation of ΔH and ΔS for the step P_2, $T_1 \rightarrow P_2$, T_2 requires C_p as a function of T at the pressure P_2, and because heat capacity data are usually available only at low pressure, the alternative path involving the step P_0, $T_1 \rightarrow P_0$, T_2 will be chosen. Following this path the property changes are calculated by a three-step process.

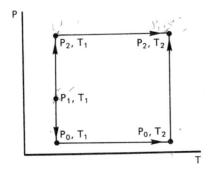

Figure 5-1   Computational path for enthalpy and entropy changes.

Step 1. $\quad P_1, T_1 \rightarrow P_0, T_1$

$$\Delta H_{\mathrm{I}} = H_{P_0,\ T_1} - H_{P_1,\ T_1} = \int_{P_1}^{P_0} \left[V - T\left(\frac{\partial V}{\partial T}\right)_P \right]_{T_1} dP$$

$$\Delta H_{\mathrm{I}} = \int_{P_1}^{P_0} [V]_{T_1}\, dP - T_1 \int_{P_1}^{P_0} \left[\left(\frac{\partial V}{\partial T}\right)_P \right]_{T_1} dP$$

$$\Delta S_{\mathrm{I}} = S_{P_0,\ T_1} - S_{P_1,\ T_1} = -\int_{P_1}^{P_0} \left[\left(\frac{\partial V}{\partial T}\right)_P \right]_{T_1} dP$$

To evaluate these integrals, we require V and $(\partial V/\partial T)_P$ as functions of P at the temperature T_1. If the available PVT data have been fitted to an equation of state, then analytical integration is possible; if not, graphical evaluation is necessary. The graphical method will be illustrated. In Fig. 5-2 a representation of V vs. T for constant values of P is shown. At the temperature T_1 we may read off values of V at the various pressures and could then evaluate $\int_{P_1}^{P_0} V\, dP$ by plotting V vs. P and determining the area under the curve between P_1 and P_0. Required values of $(\partial V/\partial T)_P$ can be obtained from the slope of the isobars at T_1. Thus, the other integral can be evaluated by plotting $(\partial V/\partial T)_P$ vs. P and determining the area under the curve between P_1 and P_0.

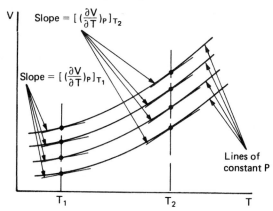

Figure 5-2 Graphical representation of PVT data for evaluation of Δh and Δs.

Step II. $\quad P_0, T_1 \rightarrow P_0, T_2$

$$\Delta H_{\mathrm{II}} = H_{P_0,\ T_2} - H_{P_0,\ T_1} = \int_{T_1}^{T_2} [C_P]_{P_0}\, dT$$

$$\Delta S_{\mathrm{II}} = S_{P_0,\ T_2} - S_{P_0,\ T_1} = \int_{T_1}^{T_2} \left[\frac{C_P}{T} \right]_{P_o} dT$$

To evaluate these changes, we need to know the constant-pressure heat capacity as a function of temperature at the pressure P_0. An analytic representation would allow direct integration, while a tabular relationship would require graphical integration.

Step III. $P_0,\ T_2 \rightarrow P_2,\ T_2$. For this step the calculations are similar to those of step I with the exception that V and $(\partial V / \partial T)_P$ must now be known as functions of P at the temperature T_2.

The sum of the ΔH's and the ΔS's for these three steps gives the desired property changes. It should be obvious that any change of state within the single phase region can be viewed as a series of isothermal pressure changes and isobaric temperature changes with the exact path being chosen on the basis of availability of the necessary C_p and PVT data.

Because many of the equations developed in this and the previous section are ultimately based on the fundamental thermodynamic equation (Eq. 5-8), it would behoove us to examine the generality of this relationship. The main restriction would appear to arise from setting dW' in Eq. (5-7) equal to zero because consideration was limited to systems capable of only PV work. This rules out work that could be accomplished by chemical reaction or mixing[5] and therefore limits applicability to systems of unchanging composition. Our application to single phases of unchanging composition is therefore valid.

The fact that the fundamental equation was derived from consideration of closed systems does not limit its generality because it simply relates property changes. For a single phase of unchanging composition all properties are defined by specifying two state properties (say T and P) and do not depend on whether we regard the system as open or closed. Thus, the enthalpy and entropy of steam at a given T and P are fixed and do not depend on whether the steam is confined to a vessel (closed system) or flowing through a pipeline (open system).

EXAMPLE 5-1

Execute the calculations of Δh and Δs for an ideal gas.

Solution 5-1

Equations (5-31) and (5-32), written in terms of extensive properties, may be restated in terms of intensive properties:

---

[5] The potentiality of obtaining work from a chemical reaction or mixing process is measured in terms of the chemical potentials. See Sec. 7-2.

$$dh = c_P \, dT + \left[v - T\left(\frac{\partial v}{\partial T} \right)_P \right] dP \qquad (5\text{-}31a)$$

$$ds = \frac{c_P}{T} \, dT - \left(\frac{\partial v}{\partial T} \right)_P dP \qquad (5\text{-}32a)$$

The equation of state of an ideal gas is

$$v' = \frac{RT}{P}$$

and it follows that

$$\left(\frac{\partial v'}{\partial T} \right)_P = \frac{R}{P}$$

Equation (5-31a) now becomes

$$dh' = c'_P \, dT + \left[v' - T\left(\frac{R}{P} \right) \right] dP$$

$$= c'_P \, dT$$

In the preceding we have designated ideal-gas properties by primes, and we see that, as previously stated, the enthalpy of an ideal gas does indeed depend only on temperature. Inclusion of this statement within the definition of an ideal gas is now seen to be superfluous as it is derivable from the equation of state. Because the heat capacity is expressible as the temperature derivative of the enthalpy, it too is seen to depend only on temperature.

Equation (5-32a) now becomes

$$ds' = \frac{c'_P}{T} \, dT - \frac{R}{P} \, dP$$

For a change between state 1 at P_1, T_1 and state 2 at P_2, T_2 we have

$$\Delta h' = \int_{T_1}^{T_2} c'_P \, dT \qquad (5\text{-}33)$$

$$\Delta s' = \int_{T_1}^{T_2} \frac{c'_P}{T} \, dT - R \ln \frac{P_2}{P_1} \qquad (5\text{-}34)$$

Because c_p' is independent of pressure and the pressure coefficient R/P is independent of temperature, the integrations may be performed directly without specification of the path.

In addition to providing the means of calculating enthalpy and entropy changes, Eqs. (5-31) and (5-32) can be written so as to describe the path taken by a fluid undergoing certain specific processes. For an adiabatic throttling process, the enthalpy remains constant, and in a reversible adiabatic process the entropy remains constant. For a throttling process we write

$$c_p\,dT + \left[v - T\left(\frac{\partial v}{\partial T}\right)_P \right]dP = 0 \tag{5-31b}$$

The behavior of a gas in such a process is usually characterized by the Joule-Thomson coefficient $(\partial T/\partial P)_H$.[6] From Eq. (5-31b) this is seen to be

$$\left(\frac{\partial T}{\partial P}\right)_H = \frac{1}{c_P}\left[T\left(\frac{\partial v}{\partial T}\right)_P - v \right] \tag{5-31c}$$

For an isentropic process Eq. (5-32) becomes

$$\frac{c_P\,dT}{T} - \left(\frac{\partial v}{\partial T}\right)_P dP = 0 \tag{5-32b}$$

The application of this equation to the special case of an ideal gas results in Eq. (3-19).

5-6 PROPERTY ESTIMATION FROM CORRESPONDING STATES

Often the PVT data required for the calculation of enthalpy and entropy changes are unavailable, or accurate values of these property changes are not needed, and estimates obtained at less computational expense will suffice. Such estimates for isothermal changes of enthalpy and entropy are based on the use of the compressibility factor correlation.It supplies the necessary relationship among the variables P, v, and T. Because isothermal changes of enthalpy and entropy with pressure are easily computed for ideal gases, it is convenient to work with

---

[6] The tutorial JT that runs in the software EQUATIONS OF STATE thoroughly examines Joule-Thomson phenomena. See Appendix E.

residual properties. The residual property (designated Δh^* and Δs^*) is obtained by subtracting the values of h and s at the specified T and P from the values of these properties the substance would exhibit were it an ideal gas in this state (designated h' or s'). The defining equations are

$$\Delta h^* = h' - h \tag{5-35}$$

$$\Delta s^* = s' - s \tag{5-36}$$

Equations (5-35) and (5-36) may be viewed as representing an impossible, or hypothetical, change in state because it is not always possible for a substance to be both a real gas and an ideal gas at the same temperature and pressure.[7] While Δh^* and Δs^* are hypothetical property changes,[8] their use in a computational path will be legitimate if consistency is observed in the calculation procedure. We have already seen that thermodynamic property changes may be calculated via any path, and here we are merely including a hypothetical state within the computational path. The use of this computational path for the change in state presented in Fig. 5-1 would result in the following steps:

$$
\begin{array}{ccccccccc}
 & \text{I} & & \text{II} & & \text{III} & & \text{IV} & \\
P_1, T_1 & \xrightarrow{\hspace{1cm}} & P_1, T_1 & \xrightarrow{\hspace{1cm}} & P_1, T_2 & \xrightarrow{\hspace{1cm}} & P_2, T_2 & \xrightarrow{\hspace{1cm}} & P_2, T_2 \\
\text{(real gas)} & & \text{(ideal gas)} & & \text{(ideal gas)} & & \text{(ideal gas)} & & \text{(real gas)}
\end{array}
$$

For the individual steps Δh and Δs are as follows:

Step I. In going from a real gas to an ideal gas at P_1, T_1 we write

$$\Delta h_{\mathrm{I}} = h'_{P_1, T_1} - h_{P_1, T_1} = \Delta h^*_{P_1, T_1}$$

$$\Delta s_{\mathrm{I}} = s'_{P_1, T_1} - s_{P_1, T_1} = \Delta s^*_{P_1, T_1}$$

Step II. For the isobaric heating of the ideal gas we employ Eqs. (5-33) and (5-34):

$$\Delta h_{\mathrm{II}} = \int_{T_1}^{T_2} c'_P \, dT$$

$$\Delta s_{\mathrm{II}} = \int_{T_1}^{T_2} \frac{c'_P}{T} \, dT$$

---

[7] Of course, real gases approach ideal-gas behavior as $P \to 0$ or $T \to \infty$ and hence Δh^* and Δs^* approach zero. However, this is not the case for a similarly defined residual volume.

[8] The residual properties Δh^* and Δs^* are often called "enthalpy departure "and "entropy departure."

Because we are dealing with an ideal gas, the heat capacity c'_P is independent of pressure, and it is not necessary to carry out this step at low pressure as we have done for a real gas.

Step III.   By using Eqs. (5-33) and (5-34), the change in pressure of the ideal gas results in

$$\Delta h_{III} = 0$$

$$\Delta s_{III} = -R \ln \frac{P_2}{P_1}$$

Step IV.   At the state P_2, T_2 we now convert the ideal gas to a real gas with

$$\Delta h_{IV} = h_{P_2, T_2} - h'_{P_2, T_2} = -\Delta h^*_{P_2, T_2}$$

$$\Delta s_{IV} = s_{P_2, T_2} - s'_{P_2, T_2} = -\Delta s^*_{P_2, T_2}$$

For the change of state between P_1, T_1 and P_2, T_2 on summing the four separate steps we have

$$\Delta h = \Delta h^*_{P_1, T_1} + \int_{T_1}^{T_2} c'_P \, dT - \Delta h^*_{P_2, T_2} \tag{5-37}$$

$$\Delta s = \Delta s^*_{P_1, T_1} + \int_{T_1}^{T_2} \frac{c'_P}{T} \, dT - R \ln \frac{P_2}{P_1} - \Delta s^*_{P_2, T_2} \tag{5-38}$$

Having demonstrated the utility of the concept of residual properties, we now turn our attention to the estimation of these properties.

By virtue of definition we expect the values Δh^* and Δs^* to approach zero as the pressure approaches zero. This will provide a convenient boundary condition for the determination of pressure dependency of these properties. We begin this by differentiating Eqs. (5-35) and (5-36) with respect to pressure at constant temperature:

$$\left(\frac{\partial \Delta h^*}{\partial P} \right)_T = \left(\frac{\partial h'}{\partial P} \right)_T - \left(\frac{\partial h}{\partial P} \right)_T \tag{5-39}$$

$$\left(\frac{\partial \Delta s^*}{\partial P} \right)_T = \left(\frac{\partial s'}{\partial P} \right)_T - \left(\frac{\partial s}{\partial P} \right)_T \tag{5-40}$$

The derivatives involving h' and s' were determined in Ex. 5-1 and are 0 and $-R/P$, respectively. The derivatives involving h and s will be evaluated from the following equation of state:

$$v = \frac{ZRT}{P}$$

where $Z = Z(P_r, T_r)$, from which we obtain

$$\left(\frac{\partial v}{\partial T}\right)_P = \frac{R}{P}\left[T\left(\frac{\partial Z}{\partial T}\right)_P + Z\right] \tag{5-41}$$

Employing this derivative in Eqs. (5-24) and (5-22) gives

$$\left(\frac{\partial h}{\partial P}\right)_T = v - \frac{ZRT}{P} - \frac{RT^2}{P}\left(\frac{\partial Z}{\partial T}\right)_P = -\frac{RT^2}{P}\left(\frac{\partial Z}{\partial T}\right)_P \tag{5-42}$$

$$\left(\frac{\partial s}{\partial P}\right)_T = -\frac{R}{P}\left[T\left(\frac{\partial Z}{\partial T}\right)_P + Z\right] \tag{5-43}$$

and substitution of these expressions into Eqs. (5-39) and (5-40) yields

$$\left(\frac{\partial \Delta h^*}{\partial P}\right)_T = \frac{RT^2}{P}\left(\frac{\partial Z}{\partial T}\right)_P \tag{5-44}$$

$$\left(\frac{\partial \Delta s^*}{\partial P}\right)_T = \frac{(Z-1)R}{P} + \frac{RT}{P}\left(\frac{\partial Z}{\partial T}\right)_P \tag{5-45}$$

These equations are now integrated for an isothermal change from $P = 0$, where Δh^* and Δs^* are zero, to the pressure P:

$$\Delta h^* = RT^2 \int_0^P \left(\frac{\partial Z}{\partial T}\right)_P \frac{dP}{P} \tag{5-46}$$

$$\Delta s^* = R \int_0^P \left[\frac{Z-1}{P} + \frac{T}{P}\left(\frac{\partial Z}{\partial T}\right)_P\right] dP \tag{5-47}$$

These equations can be put into reduced form with the following substitutions:

$$P = P_c P_r \qquad dP = P_c dP_r$$
$$T = T_c T_r \qquad dT = T_c dT_r$$

The result is

$$\frac{\Delta h^*}{RT_c} = T_r^2 \int_0^{P_r} \left(\frac{\partial Z}{\partial T_r}\right)_{P_r} \frac{dP_r}{P_r} \tag{5-48}$$

$$\frac{\Delta s^*}{R} = \int_0^{P_r} \left[Z - 1 + T_r\left(\frac{\partial Z}{\partial T_r}\right)_{P_r}\right] \frac{dP_r}{P_r} \tag{5-49}$$

Compressibility factor correlations such as shown in Fig. 3-6 can be used for Z and $(\partial Z/\partial T_r)_{P_r}$ needed to evaluate the integrals. It is seen that the quantities $\Delta h^*/RT_c$ and $\Delta s^*/R$ are functions of reduced properties, and therefore the theorem of corresponding states can also be applied to them. Correlations of these reduced residual properties are given in Figs. 5-3 and 5-4. When used in Eqs. (5-37) and (5-38), these quantities can be used to estimate enthalpy and entropy changes from the critical temperature and pressure and the ideal-gas, or low-pressure, heat capacity.[9]

The uncertainty in Z, stated as 2.5%, introduces a larger uncertainty in the reduced residual properties because the derivative of Z is required for their calculation. Therefore, if accurate values of Δh and Δs are needed, the method outlined in Sec. 5-5 should be used when the PVT data are available. When such data are unavailable and reliable estimates of reduced residual properties are desired, the original literature should be consulted.[10]

5-7 PROPERTY ESTIMATION VIA GENERALIZED EQUATIONS OF STATE

In Sec. 3-5 we noted that generalized equations of state produced estimates of PVT data comparable to the corresponding states compressibility factor but had the advantage of being computationally more useful. Here we will develop expressions for Δh^* and Δs^* in terms of the generalized Peng-Robinson equation of state. Because this equation is not explicit in v, the equations of the previous section will not provide a useful starting point. For the isothermal changes represented by Δh^* and Δs^*, it is convenient to choose v as the independent variable. For Δh^* we begin with

$$\Delta h^* = h' - h = (u' + Pv') - (u + Pv) \tag{5-50}$$

---

[9] Methods are available for the estimation of ideal-gas heat capacities from molecular parameters. For example, see R. C. Reid, J. M. Prausnitz, and B. E. Poling, *Properties of Gases and Liquids*, 4th ed., McGraw-Hill, New York, 1987 Chap. 6.

[10] Three-parameter correlations of the reduced residual properties can be found in G. N. Lewis and M. Randall, *Thermodynamics*, 2nd ed., revised by K. S. Pitzer and L. Brewer, McGraw-Hill, New York, 1961, Appendix 1; and O. A. Hougen, K. M. Watson, and R. A. Ragatz, *Chemical Process Principles*, Part II, 2nd ed., Wiley, New York, 1959, Chap. 14.

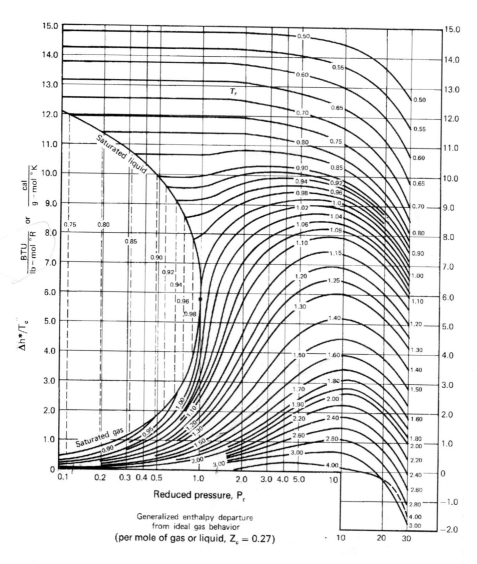

Figure 5-3 [From O. A. Hougen, K. M. Watson, and R. A. Ragatz, *Chemical Process Principles Charts*, 2nd ed., Wiley, 1960 as adapted in *Chemical and Engineering Thermodynamics* by S. I. Sandler, Wiley, 1977, with permission of John Wiley & Sons, Inc.]

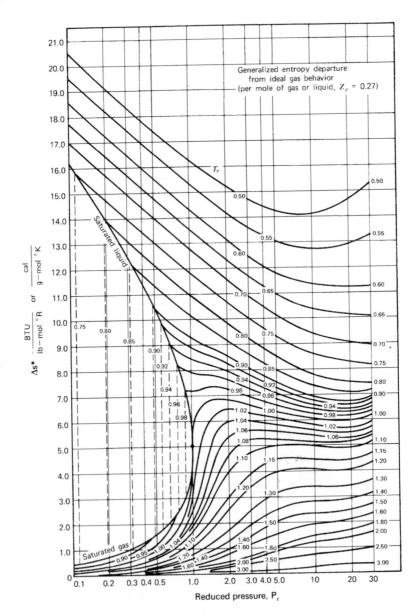

Figure 5-4 [From O. A. Hougen, K. M. Watson, and R. A. Ragatz, *Chemical Process Principles Charts*, 2nd ed., Wiley, 1960 as adapted in *Chemical and Engineering Thermodynamics* by S. I. Sandler, Wiley, 1977, with permission of John Wiley & Sons, Inc.]

$$\Delta h^* = RT - Pv + (u' - u) \tag{5-51}$$

and evaluate the last right-hand term by means of [11]

$$\left(\frac{\partial u}{\partial v}\right)_T = T\left(\frac{\partial P}{\partial T}\right)_v - P \tag{5-52}$$

which is integrated between v and $v = \infty$, where ideal-gas behavior is expected:

$$u' - u = \int_v^\infty \left[T\left(\frac{\partial P}{\partial T}\right)_v - P \right] dv \tag{5-53}$$

On combining Eqs.(5-51) and (5-53) we obtain

$$\Delta h^* = RT(1 - Z) + \int_\infty^v \left[P - T\left(\frac{\partial P}{\partial T}\right)_v \right] dv \tag{5-54}$$

To obtain Δs^* we begin with

$$\left(\frac{\partial s}{\partial v}\right)_T = \left(\frac{\partial P}{\partial T}\right)_v \tag{5-21}$$

Because s becomes infinite as v becomes infinite, it is not possible to integrate Eq. (5-21) directly. Instead we consider Δs^* to be comprised of two parts,

$$\Delta s^* = s' - s = (s' - s^\infty) + (s^\infty - s)$$

which will be integrated separately:

$$\Delta s^* = \int_\infty^{v'} \left(\frac{\partial P}{\partial T}\right)_v dv + \int_v^\infty \left(\frac{\partial P}{\partial T}\right)_v dv \tag{5-55}$$

Integration of the first integral is over the ideal-gas range where

$$\left(\frac{\partial P}{\partial T}\right)_v = \frac{R}{v}$$

The problem of limits is circumvented by adding the quantity

$$\int_\infty^v \frac{R}{v} dv - \int_\infty^v \frac{R}{v} dv$$

---

[11] This partial derivative can be obtained by applying conditions of constant temperature to Eq. (5-8) and substituting Eq. (5-21).

to the right-hand side of Eq. (5-55) to obtain

$$\Delta s^* = \int_{\infty}^{v} \left[\frac{R}{v} - \left(\frac{\partial P}{\partial T} \right)_v \right] dv \; - \; R \ln \frac{v}{v'} \qquad (5\text{-}56)$$

or

$$\Delta s^* = \int_{\infty}^{v} \left[\frac{R}{v} - \left(\frac{\partial P}{\partial T} \right)_v \right] dv \; - \; R \ln Z \qquad (5\text{-}57)$$

The Peng-Robinson equation (Eq. 3-5c) is explicit in P, and therefore $(\partial P / \partial T)_v$ is easily obtained. When Eqs. (3-5c) through (3-5h) are used in Eqs. (5-54) and (5-57), we obtain

$$\frac{\Delta h^*}{RT_c} = 2.078(1+\kappa)\left[1 + \kappa\left(1 - T_r^{1/2}\right)\right] \ln \left[\frac{Z + \left(1 + \sqrt{2}\right)B}{Z + \left(1 - \sqrt{2}\right)B} \right] - T_r(Z - 1) \qquad (5\text{-}58)$$

$$\frac{\Delta s^*}{R} = 2.078\kappa \left(\frac{1+\kappa}{T_r^{1/2}} - \kappa \right) \ln \left[\frac{Z + \left(1 + \sqrt{2}\right)B}{Z + \left(1 - \sqrt{2}\right)B} \right] - \ln \left(Z - B \right) \qquad (5\text{-}59)$$

where

$$Z^3 + (B - 1)Z^2 + \left(A - 3B^2 - 2B \right)Z + \left(B^3 + B^2 - AB \right) = 0 \qquad (3\text{-}5i)$$

$$A = 0.45724 \frac{P_r}{T_r^2} \left[1 + \kappa\left(1 - T_r^{1/2}\right) \right]^2 \qquad (3\text{-}5l)$$

$$B = 0.07780 \frac{P_r}{T_r} \qquad (3\text{-}5m)$$

$$\kappa = 0.37464 + 1.54226\omega - 0.26992\omega^2 \qquad (3\text{-}5h)$$

This set of equations allows Z, Δh^*, and Δs^* to be evaluated at specified sets of T and P from a knowledge of T_c, P_c, and the acentric factor, ω, and is easily adaptable to computerized calculations. The program PREOS.EXE on the accompanying CD-ROM executes these calculations. See App. E.

EXAMPLE 5-2

Estimate Δh^* and Δs^* for chlorine gas, Cl_2, at 155 bar and 521 K using
(a) figures 5-3 and 5-4.
(b) the generalized Peng-Robinson equation of state.

Solution 5-2

In Ex. 3-2 we found for these conditions

$$T_r = 1.25 \qquad P_r = 2.01$$

(a) At these reduced conditions we find from Figs. 5-3 and 5-4

$$\frac{\Delta h^*}{T_c} = 3.5 \frac{\text{cal}}{\text{g mol} \cdot \text{K}}$$

$$\Delta s^* = 2.1 \frac{\text{cal}}{\text{g mol} \cdot \text{K}}$$

(b) Also from Ex. 3-2 we obtain

$$\kappa = 0.511 \qquad B = 0.1251 \qquad Z = 0.644$$

On substitution of these values into Eqs. (5-58) and (5-59) we obtain

$$\frac{\Delta h^*}{RT_c} = 2.078(1.511)\left[1 + 0.511\left(1 - \sqrt{1.25}\right)\right]$$

$$\cdot \ln\left[\frac{0.644 + 2.414(0.1251)}{0.644 - 0.414(0.1251)}\right] - 1.25(0.644 - 1)$$

$$\frac{\Delta h^*}{T_c} = 3.63 \frac{\text{cal}}{\text{g mol} \cdot \text{K}}$$

$$\frac{\Delta s^*}{R} = 2.078(0.511)\left[\frac{1.511}{\sqrt{1.25}} - 0.511\right] \ln\left[\frac{0.644 + 2.414(0.1251)}{0.644 - 0.414(0.1251)}\right]$$

$$-\ln\,(0.644 - 0.1251)$$

$$\Delta s^* = 2.13\frac{\text{cal}}{\text{g mol} \cdot \text{K}}$$

The results of the two methods are seen to be in excellent agreement.

5-8 THE METHOD OF JACOBIANS

The Jacobian is a very useful tool in mathematics and was used by Shaw[12] to develop a systematic method of obtaining relationships among thermodynamic variables. A Jacobian is a determinant, the elements of which are partial derivatives. Here we are concerned only with second-order Jacobians and consider the variables x and y, each of which is a function of the independent variables α and β:

$$x = x(\alpha, \beta)$$

$$y = y(\alpha, \beta)$$

The Jacobian of x and y, denoted $J(x, y)$, is defined as

$$J(x, y) = \begin{vmatrix} \left(\dfrac{\partial x}{\partial \alpha}\right)_\beta & \left(\dfrac{\partial x}{\partial \beta}\right)_\alpha \\ \left(\dfrac{\partial y}{\partial \alpha}\right)_\beta & \left(\dfrac{\partial y}{\partial \beta}\right)_\alpha \end{vmatrix}$$

Therefore,

$$J(x, y) = \left(\frac{\partial x}{\partial \alpha}\right)_\beta\left(\frac{\partial y}{\partial \beta}\right)_\alpha - \left(\frac{\partial x}{\partial \beta}\right)_\alpha\left(\frac{\partial y}{\partial \alpha}\right)_\beta$$

Jacobians possess the following useful properties:

1. $J(x, y) = J(y, -x) = J(-y, x) = -J(y, x)$ (5-60)

---

[12] A. N. Shaw, *Philos. Trans. R. Soc. London, A234*, 299 (1935).

2. $J(x, x) = 0$ (5-61)

3. A partial derivative may be written as

$$\left(\frac{\partial y}{\partial x}\right)_z = \frac{J(y, z)}{J(x, z)}$$ (5-62)

where $z = z(\alpha, \beta)$

4. $J(k_1 x, k_2 y) = k_1 k_2 J(x, y)$, where k_1 and k_2 are constants. (5-63)

5. The exact differential equation

$$dx = M\, dy + N\, dz$$ (5-64)

may be expressed as

$$J(x, \alpha) = MJ(y, \alpha) + NJ(z, \alpha)$$ (5-65)

The first four properties derive directly from the defining determinant. The fifth may be verified by dividing Eq. (5-64) by $d\beta$ and subsequently holding α constant to obtain

$$\left(\frac{\partial x}{\partial \beta}\right)_\alpha = M\left(\frac{\partial y}{\partial \beta}\right)_\alpha + N\left(\frac{\partial z}{\partial \beta}\right)_\alpha$$

which on applying property 3 becomes

$$\frac{J(x, \alpha)}{J(\beta, \alpha)} = M\frac{J(y, \alpha)}{J(\beta, \alpha)} + N\frac{J(z, \alpha)}{J(\beta, \alpha)}$$

Multiplying by $J(\beta, \alpha)$ produces the desired result. As just shown, Jacobians can be manipulated in normal algebraic fashion.

Recognizing that in the exact differential equation

$$M = \left(\frac{\partial x}{\partial y}\right)_z = \frac{J(x, z)}{J(y, z)}$$

and

$$N = \left(\frac{\partial x}{\partial z}\right)_y = \frac{J(x, y)}{J(z, y)}$$

Eq. (5-65) may be written as

$$J(x, \alpha)J(y, z) + J(y, \alpha)J(z, x) + J(z, \alpha)J(x, y) = 0$$ (5-66)

This provides a relationship among the Jacobians that later will be shown to be quite useful.

In applying this method to thermodynamics, we recognize that property relationships are usually desired for single-phase systems of unchanging composition where the state of the system is defined by specification of two variables. Thus, any thermodynamic variable is a function of two independent variables, and the operations just developed are applicable. The variables P, V, T, and S appear most frequently,[13] and therefore the six possible Jacobians formed from these variables are given a letter designation for the sake of brevity:

$$J(V, T) = a$$
$$J(P, V) = b$$
$$J(P, S) = c$$
$$J(P, T) = l$$
$$J(V, S) = n$$
$$J(T, S) = J(P, V) = b$$

The last relationship follows from the Maxwell relation

$$\left(\frac{\partial T}{\partial V}\right)_S = -\left(\frac{\partial P}{\partial S}\right)_V \tag{5-19}$$

and application of property 3

$$\frac{J(T, S)}{J(V, S)} = -\frac{J(P, V)}{J(S,V)} = \frac{J(P, V)}{J(V, S)}$$

Actually, all four Maxwell relations can be expressed by the single Jacobian relationship[14]

$$J(T, S) = J(P, V) \tag{5-67}$$

---

[13] Here extensive variables are used. Equation (5-63) permits the transformation between extensive and intensive variables.

[14] Divide by $J(P, S)$ to obtain Eq. (5-20), by $J(T, V)$ to obtain Eq. (5-21), and by $J(T, P)$ to obtain Eq. (5-22).

TABLE 5–1

VALUES OF JACOBIANS, $J(x,y)$

x \ y	P	V	T	S	U	H	A
P	0	b	l	c	$Tc - Pb$	Tc	$-Sl - Pb$
V	$-b$	0	a	n	Tn	$Tn - Vb$	$-Sa$
T	$-l$	$-a$	0	b	$Tb + Pa$	$Tb - Vl$	Pa
S	$-c$	$-n$	$-b$	0	Pn	$-Vc$	$Sb + Pn$
U	$-Tc + Pb$	$-Tn$	$-Tb - Pa$	$-Pn$	0	$-TVc \\ -P(Tn - Vb)$	$T(Sb + Pn) \\ + PSa$
H	$-Tc$	$-Tn + Vb$	$-Tb + Vl$	Vc	$TVc \\ + P(Tn - Vb)$	0	$T(Sb + Pn) \\ - V(Sl + Pb)$
A	$Sl + Pb$	Sa	$-Pa$	$-Sb - Pn$	$-T(Sb + Pn) \\ - PSa$	$-T(Sb + Pn) \\ + V(Sl + Pb)$	0
G	Sl	$Sa + Vb$	Vl	$-Sb + Vc$	$-T(Sb - Vc) \\ - P(Sa + Vb)$	$-T(Sb - Vc) \\ + VSl$	$-SVl \\ - P(Sa + Vb)$

This is hardly surprising in view of the fact that the Maxwell relations derive from the four forms of the fundamental equation. As we have seen, these four equations convey the same information but in terms of different variables.

By setting x, y, z, and α equal to P, V, T, and S, respectively, Eq. (5-66) becomes

$$J(P, S)J(V, T) + J(V, S)J(T, P) + J(T, S)J(P, V) = 0$$

and in terms of the letter designation we have

$$b^2 - nl + ac = 0 \tag{5-66a}$$

This relationship is used to eliminate any particular Jacobian from an expression.

To facilitate use of the method a table of Jacobians of any pair of thermodynamic variables is prepared (Table 5-1). As an example of how the table is constructed, take the fundamental equation

$$dH = T\,dS + V\,dP \tag{5-9}$$

This can be written as

$$J(H, \alpha) = TJ(S, \alpha) + VJ(P, \alpha)$$

Setting $\alpha = T$, we obtain

$$J(H, T) = TJ(S, T) + VJ(P, T)$$

or

$$J(H, T) = -Tb + Vl$$

Setting $\alpha = S$ yields

$$J(H, S) = TJ(S, S) + VJ(P, S)$$

or

$$J(H, S) = 0 + Vc$$

The following examples illustrate the use of the method and the table of Jacobians.

EXAMPLE 5-3

Determine a relationship between C_p and C_v expressed in terms of measurable quantities.

Solution 5-3

Let us begin with the definition of C_P,

$$C_P = \left(\frac{\partial H}{\partial T} \right)_P$$

which we then write in Jacobian form as

$$C_P = \frac{J(H,P)}{J(T,P)} = \frac{Tc}{l}$$

Eliminating c with Eq. (5-66a) gives

$$C_P = T \left(\frac{nl - b^2}{al} \right)$$

$$= T \left[\frac{J(V,\,S)J(P,\,T) - J(P,\,V)^2}{J(V,\,T)J(P,\,T)} \right]$$

$$= T \frac{J(V,\,S)}{J(V,\,T)} - T \frac{J(P,\,V)}{J(V,\,T)} \frac{J(P,\,V)}{J(P,\,T)}$$

$$= T \left(\frac{\partial S}{\partial T} \right)_V + T \left(\frac{\partial P}{\partial T} \right)_V \left(\frac{\partial V}{\partial T} \right)_P$$

From Eq. (5-28), the first term is seen to be C_V, and we have the desired relationship:

$$C_P - C_V = T \left(\frac{\partial P}{\partial T} \right)_V \left(\frac{\partial V}{\partial T} \right)_P$$

EXAMPLE 5-4

Find a relation involving the adiabatic compressibility β_S in terms of other measurable quantities.

Solution 5-4

By definition,

$$\beta_S = -\frac{1}{V}\left(\frac{\partial V}{\partial P}\right)_S$$

and we will therefore seek to relate the partial derivative to other measurable quantities:

$$\left(\frac{\partial V}{\partial P}\right)_S = \frac{J(V,S)}{J(P,S)} = \frac{n}{c}$$

We can use Eq. (5-66a) to eliminate either n or c. By eliminating n,

$$\left(\frac{\partial V}{\partial P}\right)_S = \frac{b^2 + ac}{lc}$$

$$\left(\frac{\partial V}{\partial P}\right)_S = \frac{J(P,V)^2 + J(V,T)J(P,S)}{J(P,T)J(P,S)}$$

This expression may be divided by the product of any two Jacobians; here we may choose $J(P,T)^2$ and obtain

$$\left(\frac{\partial V}{\partial P}\right)_S = \frac{\left[J(P,V)^2/J(P,T)^2\right] + \left[J(V,T)J(P,S)/J(P,T)J(P,T)\right]}{J(P,T)J(P,S)/J(P,T)J(P,T)}$$

In terms of partial derivatives this is

$$\left(\frac{\partial V}{\partial P}\right)_S = \frac{(\partial V/\partial T)_P^2 + (\partial V/\partial P)_T(\partial S/\partial T)_P}{(\partial S/\partial T)_P}$$

Now, from Eq. (5-27) we have

$$\left(\frac{\partial V}{\partial P}\right)_S = \frac{(\partial V/\partial T)_P^2 + (C_P/T)(\partial V/\partial P)_T}{C_P/T}$$

and

$$-\beta_S = \frac{T}{VC_P}\left(\frac{\partial V}{\partial T}\right)_P^2 + \frac{1}{V}\left(\frac{\partial V}{\partial P}\right)_T$$

The second term is the negative of the isothermal compressibility β_T, and

$$\beta_S = \beta_T - \frac{T}{VC_P}\left(\frac{\partial V}{\partial T}\right)_P^2$$

EXAMPLE 5-5

Find $(\partial P / \partial V)_H$ in terms of measurable quantities.

Solution 5-5

$$\left(\frac{\partial P}{\partial V}\right)_H = \frac{J(P, H)}{J(V, H)} = \frac{Tc}{Tn - Vb}$$

Using Eq. (5-66a) to eliminate n, we obtain

$$\left(\frac{\partial P}{\partial V}\right)_H = \frac{Tc}{\left[(Tb^2 + Tac)/l\right] - Vb} = \frac{Tcl}{Tb^2 + Tac - Vbl}$$

Replacing the symbols with their corresponding Jacobians gives

$$\left(\frac{\partial P}{\partial V}\right)_H = \frac{TJ(P, S)J(P, T)}{TJ(P, V)^2 + TJ(V, T)J(P, S) - VJ(P, V)J(P, T)}$$

This expression may be divided by the product of any two Jacobians. Using $J(P, T)^2$, we obtain

$$\left(\frac{\partial P}{\partial V}\right)_H =$$

$$\frac{T\left[J(P, S)J(P, T)/J(P, T)J(P, T)\right]}{T\left[J(P, V)^2/J(P, T)^2\right] + T\left[J(V, T)J(P, S)/J(P, T)J(P, T)\right] - V\left[J(P, V)J(P, T)/J(P, T)J(P, T)\right]}$$

$$= \frac{T(\partial S/\partial T)_P}{T(\partial V/\partial T)_P^2 + T(\partial V/\partial P)_T(\partial S/\partial T)_P - V(\partial V/\partial T)_P}$$

By using Eq. (5-27), the final expression becomes

$$\left(\frac{\partial P}{\partial V}\right)_H = \frac{C_P}{T(\partial V/\partial T)_P^2 + C_P(\partial V/\partial P)_T - V(\partial V/\partial T)_P}$$

In these examples a general five-step procedure should be discerned: (1) The derivative is expressed in terms of the five basic Jacobians by using the table. (2) One of these Jacobians may be eliminated by the use of Eq. (5-66a). (3) In the resulting expression the letters are replaced by their corresponding Jacobians. (4) The resulting Jacobian expression is divided by a Jacobian expression of the proper order. (5) Jacobian notation is transformed to derivative notation. Steps 1, 3, and 5 are mechanical; however, steps 2 and 4 require the exercise of judgment as the final result depends on these decisions. Because only the temperature derivative of the entropy is a measurable quantity, it is usually desirable in step 2 to eliminate a Jacobian containing the entropy, either c or n. The choice of Jacobian denominator in step 4 is guided by the Jacobian numerators and by the desired result.

5-9 THE GENERALITY OF THE THERMODYNAMIC METHOD

In the preceding section we saw how relationships among thermodynamic properties are derived for systems of constant composition capable of performing only PV work. Although such systems constitute the major engineering application of thermodynamics, the first and second laws apply to all types of systems, and the thermodynamic method is applicable to any system capable of existing in equilibrium states.[15] While systems will differ in the variables used to describe them, temperature is a variable common to all systems, and internal energy and entropy are properties common to all systems.

In the thermodynamic sense, systems are distinguished by the manner in which work is exchanged with the surroundings. Starting with the definition of work in terms of a force producing a displacement, we write

$$dW = \mathscr{F}\,dr \qquad (5\text{-}68)$$

or

$$W = \int \mathscr{F}\,dr \qquad (5\text{-}69)$$

where $\mathscr{F}$ is an intensive variable which we have previously identified as a generalized force and r is an extensive variable we have designated as a generalized displacement. With this

---

[15] The concept of a property has no thermodynamic meaning unless the system is in an equilibrium state.

generalized statement of work, the derivation of the fundamental equation (see Sec. 5-2) can be repeated to obtain a more general statement:[16]

$$dU = T\,dS + \mathscr{F}\,dr \tag{5-70}$$

In Jacobian notation the Maxwell relation is

$$J(T, S) = -J(\mathscr{F}, r) = J(\mathscr{F}, -r) \tag{5-71}$$

Comparison of this with the original Maxwell Jacobian [Eq. (5-67)] reveals the direct correspondence between P and $\mathscr{F}$ and V and $-r$. The method of Jacobians can therefore be used to establish the interrelationships among the thermodynamic variables for other types of systems, if in Table 5-1, P is replaced by $\mathscr{F}$ and V is replaced by $-r$.[17] Because they are defined in terms of PV, the enthalpy and Gibbs free energy are of little value in the general treatment, and those columns and rows in Table 5-1 can be ignored. However, analogs of these functions could be developed by replacing PV with $\mathscr{F}r$ in their defining equations.

We now consider several specific systems and identify the generalized forces and displacements. For a fluid we have already seen that

$$\mathscr{F}\,dr = -P\,dV \tag{5-72}$$

For the elastic stretching or compression of a solid

$$\mathscr{F}\,dr = \mathbf{f}\,dl \tag{5-73}$$

where $\mathbf{f}$ is force and l is length. For a change in surface area,

$$\mathscr{F}\,dr = \sigma\,d\mathbf{A} \tag{5-74}$$

where σ is the surface tension and $\mathbf{A}$ is the surface area. For the magnetization of a paramagnetic solid

$$\mathscr{F}\,dr = \mathbf{H}\,dI \tag{5-75}$$

where $\mathbf{H}$ is the magnetic field strength and I is the intensity of magnetization.

---

[16] This formulation is for a system that can perform only one kind of work. Otherwise, an $\mathscr{F}dr$ term is required for each kind of work.

[17] Because $J(\mathscr{F}, -r) = J(-\mathscr{F}, r)$ an alternative procedure would be to replace P by $-\mathscr{F}$ and V by r in Table 5-1.

EXAMPLE 5-6

Determine the internal energy change associated with the isothermal extension of the surface of a liquid film. Also derive an expression relating change of temperature with change of surface area for an adiabatic process.

Solution 5-6

We desire to express $(\partial U/\partial \mathbf{A})_T$ in terms of measurable quantities. Using the Jacobian method, we write

$$\left(\frac{\partial U}{\partial \mathbf{A}}\right)_T = \frac{J(U, T)}{J(\mathbf{A}, T)} = -\frac{J(U, T)}{J(-\mathbf{A}, T)}$$

In Table 5-1 we replace V by $-\mathbf{A}$ and P by σ and obtain

$$\left(\frac{\partial U}{\partial \mathbf{A}}\right)_T = -\left(\frac{-Tb - \sigma a}{a}\right) = \sigma + T\frac{J(\sigma, -\mathbf{A})}{J(-\mathbf{A}, T)} \qquad (5\text{-}76)$$

$$= \sigma - T\left(\frac{\partial \sigma}{\partial T}\right)_{\mathbf{A}}$$

As the surface tension is known to be independent of area, the partial derivative may be replaced with $d\sigma/dT$ and the desired internal energy change can be readily calculated from surface tension measured at several temperatures. Langmuir[18] did this and found that all aliphatic hydrocarbons have essentially the same value of $(\partial U/\partial \mathbf{A})_T$ and that it is also the same as that of aliphatic alcohols. The molecular interpretation of this is that the alcohols bury their OH group in the liquid and expose their hydrocarbon part to the vapor phase. Because the surface tension of hydrocarbons is much less than that of water, aliphatic alcohols are good detergents.

For the adiabatic change of surface area we desire $(\partial T/\partial \mathbf{A})_S$ and write

$$\left(\frac{\partial T}{\partial \mathbf{A}}\right)_S = \frac{J(T, S)}{J(\mathbf{A}, S)} = \frac{J(\mathbf{A}, \sigma)}{J(\mathbf{A}, S)}$$

where Eq. (5–71) has been applied.

---

[18] For his work on surfaces, Langmuir received the Nobel prize. This work is summarized in I. Langmuir, *Phenomena, Atoms and Molecules*, Philosophical Library, New York, 1950.

Dividing by $J(\mathbf{A}, T)$, we obtain

$$\left(\frac{\partial T}{\partial \mathbf{A}}\right)_S = \frac{J(\mathbf{A}, \sigma)/J(\mathbf{A}, T)}{J(\mathbf{A}, S)/J(\mathbf{A}, T)} = \frac{(\partial \sigma/\partial T)_\mathbf{A}}{(\partial S/\partial T)_\mathbf{A}}$$

The denominator can be expressed in terms of heat capacity—the heat capacity at constant surface area $C_\mathbf{A}$:

$$\left(\frac{\partial S}{\partial T}\right)_\mathbf{A} = \frac{C_\mathbf{A}}{T}$$

And we obtain

$$\left(\frac{\partial T}{\partial \mathbf{A}}\right)_S = \frac{T}{C_\mathbf{A}}\frac{d\sigma}{dT}$$

Now heat capacities are generally positive and surface tension is known to decrease with increasing temperature, $d\sigma/dT < 0$; therefore we can say that

$$\left(\frac{\partial T}{\partial \mathbf{A}}\right)_S < 0$$

For example, a coalescence of drops results in a decrease in surface area and hence an increase in temperature.

EXAMPLE 5-7

For a paramagnetic solid, determine (a) the isothermal heat effect accompanying an increase in intensity of magnetization and (b) the temperature change accompanying an adiabatic magnetization.

Solution 5-7

(a) We require $(\partial S/\partial I)_T$ from which we may obtain $(\partial Q_{rev}/\partial I)_T$ by employing the defining equation $dS = Q_{rev}/T$.

$$\left(\frac{\partial S}{\partial I}\right)_T = \frac{J(S, T)}{J(I, T)}$$

For this system the Maxwell Jacobian is

$$J(T,S) = -J(\mathbf{H},I)$$

and is used to obtain

$$\left(\frac{\partial S}{\partial I}\right)_T = \frac{J(\mathbf{H},I)}{J(I,T)}$$

Dividing by $J(\mathbf{H},\,T)$ gives

$$\left(\frac{\partial S}{\partial I}\right)_T = \frac{J(\mathbf{H},I)/J(\mathbf{H},T)}{J(I,T)/J(\mathbf{H},T)} = \frac{(\partial I/\partial T)_\mathbf{H}}{(\partial I/\partial \mathbf{H})_T} \tag{5-78}$$

Thus

$$\left(\frac{\partial Q_{\text{rev}}}{\partial I}\right)_T = \frac{T(\partial I/\partial T)_\mathbf{H}}{(\partial I/\partial \mathbf{H})_T} \tag{5-79}$$

From the study of this type of system it is found that $(\partial I/\partial T)_\mathbf{H}$ is negative and $(\partial I/\partial \mathbf{H})_T$ is positive, a result that seems intuitively obvious. We may therefore state that

$$\left(\frac{\partial Q_{\text{rev}}}{\partial I}\right)_T < 0$$

which requires the removal of heat for an isothermal increase in magnetization.

(b) For the adiabatic process we desire to relate $(\partial T/\partial I)_S$ to measurable quantities and write

$$\left(\frac{\partial T}{\partial I}\right)_S = \frac{J(T,S)}{J(I,S)} = \frac{J(I,\mathbf{H})}{J(I,S)}$$

Dividing by $J(I,\,T)$ gives

$$\left(\frac{\partial T}{\partial I}\right)_S = \frac{J(I,\mathbf{H})/J(I,T)}{J(I,S)/J(I,T)} = \frac{(\partial \mathbf{H}/\partial T)_I}{(\partial S/\partial T)_I} = \frac{T}{C_I}\left(\frac{\partial \mathbf{H}}{\partial T}\right)_I \tag{5-80}$$

where C_I is the heat capacity at constant magnetization intensity. Both C_I and $(\partial \mathbf{H} / \partial T)_I$ are positive quantities,[19] and

$$\left(\frac{\partial T}{\partial I}\right)_S > 0$$

Therefore, an adiabatic decrease in magnetization intensity results in a decrease in temperature. Extremely low temperatures approaching absolute zero are attained by employing this process.

The preceding examples illustrate that the applicability of thermodynamics is broad and that its methods are general, yet this aspect of the subject often seems unappreciated. One wonders why we can appreciate the generality of a theory but fail to perceive this quality in thermodynamics. Probably the answer is that a theory is usually based on a model that provides physical insight and allows us to understand the phenomenon and make a priori predictions of behavior. Also, the ramifications of a theory are usually specific and concrete. On the other hand, thermodynamics is not a theory but is better described as a method. It provides a framework for viewing and correlating the behavior of a system based on experimentally gained knowledge of the system. While its methods are general, the question of whether or how they can be applied to any given type of system cannot be determined a priori but depends on our experimental knowledge of the system. From the examples just studied it is observed that, as opposed to those of a theory, the ramifications of the thermodynamic method are many and diverse.

While the methods of thermodynamics are general, we do not know whether or how they can be applied to a new type of system until we have verified the application. It is probably this a posteriori nature that tends to obscure the generality of the thermodynamic method and also gives the appearance of a certain amount of circularity in its application. In each successful application of thermodynamics we must possess sufficient knowledge of the system to know the type of work exchanged (and how to express it) and the number of variables required to specify an equilibrium state. Suppose, for example, that the change in surface area of a liquid were accompanied by a change in volume of which we were unaware. The total work in an infinitesimal change would be $\sigma\, d\mathbf{A} - P\, dV$, but we have considered it to be only $\sigma\, d\mathbf{A}$. Likewise, the system would require three variables, say T, $\mathbf{A}$, and V, to specify an equilibrium state as opposed to our selection of only two. The application of the thermodynamic method to this improperly described system would produce erroneous results which would be contrary to our experience. We would then be alerted to our incomplete description and could study the system more thoroughly to discern the omitted factors.

---

[19] An increase in temperature results in increased kinetic or vibrational energy, which must be overcome by an increased magnetic field strength to preserve a constant magnetization intensity. Hence, we expect $(\partial \mathbf{H}/\partial T)_I > 0$.

There is some degree of circularity associated with every theory or model in the sense that to be accepted they must yield results that conform to experimental observation. In this way, the application of the thermodynamic method is no different. However, because an understanding of the behavior of the system, at least in an empirical way, is essential to the application of the thermodynamic method, and experimental information is needed for its implementation, a greater degree of circularity is suggested.

We have just said that it is necessary to know a great deal about the behavior of a system before the methods of thermodynamics can be applied. If this is so, then it is natural to ask the following: What profit is derived from the application of thermodynamics? The answer to this is that we are able to extend our experimentally gained knowledge of a system, or, correspondingly, information gained through the thermodynamic method reduces the experimental effort required to describe a system. The thermodynamic method also provides a means of testing the consistency of experimental data.

PROBLEMS

5-1. It is desired to calculate and tabulate the thermodynamic properties (namely h and s) of isotootenol using the pure saturated liquid at 77°F where $h = 0$ and $s = 0$. Show in detail how you would calculate h and s of the saturated vapor at the temperature 200°F. Write the necessary equations and be specific about the data needed to make the calculations.

5-2. Show how h and s could be obtained as functions of T and v. Develop the necessary equations, specify the type of experimental data required, and show how the data would be utilized in making the calculations.

5-3. The equation which describes the path of a reversible, adiabatic process involving an ideal gas is

$$\frac{T_2}{T_1} = \left(\frac{P_2}{P_1}\right)^{(\gamma-1/\gamma)}$$

A student argues that this equation is not restricted to reversible processes because it is simply an equation relating state properties. Do you agree? If not, give a reasoned argument as to why is the student incorrect.

5-4. In Sec. 5-5 it was shown that h and s could be calculated for a gas as functions of T and P from its heat capacity and PVT behavior. The Joule-Thomson coefficient, $(\partial T/\partial P)_h$ is easily determined in a throttling experiment, and we wish to know if the calculation of h and s could be made from the PVT behavior and a knowledge of the Joule-Thomson coefficient as a function of T and P. Is this possible? Explain.

5-5. A gas obeys the following equation of state:

$$P(v - B) = RT + \frac{aP^2}{T}$$

where B and a are constants.

    (a) Calculate the entropy change involved when the gas changes from state 1 ($P = 4$ atm, $T = 300$ K) to state 2 ($P = 12$ atm, $T = 400$ K). The mean c_p at atmospheric pressure is 8 cal/g mol · K. The values of a and B are as follows: $a = 1.0$ liter · K/atm · g mol; $B = 0.080$ liter/g mol.

    (b) Estimate the mean c_p at 12 atm.

5-6. Determine whether the gas described in Prob. 5-5 behaves realistically with respect to the pressure dependency of its enthalpy. Is the behavior of this gas qualitatively consistent with Fig. 5-3? Explain.

5-7. Ott, Goates, and Hall [*J. Chem. Educ.*, **48**, 515 (1971)] present the following equation of state obtained by fitting the compressibility factor in the range $0 < P_r \leq 0.6$, $1.0 \leq T_r \leq 2.0$:

$$Pv = RT\left[1 + \frac{PT_c}{17P_cT}\left(1 - \frac{15T_c^2}{2T^2}\right)\right]$$

At the conditions $T = 450$ K, $P = 20$ atm, determine the entropy of a gas obeying this equation of state if s is taken to be zero for the gas at $T = 300$ K and $P = 1$ atm. At 1 atm and over the temperature range 300–450 K, the mean constant-pressure heat capacity is 7 cal/g mol · K. For this gas $T_c = 300$ K and $P_c = 40$ atm.

5-8. A gas obeys the following equation of state:

$$P(v - b) = RT$$

where the constant b was found to be 0.2 m³/kmol. Over the temperature range 300–400 K the heat capacity may be taken to be constant at 40 kJ/kmol · K. Calculate h and s for the following change of state:

    State 1: $P = 10$ bar $T = 300$ K

    State 2: $P = 20$ bar $T = 400$ K

Also for this gas show that

    (a) c_p is independent of pressure.

    (b) the temperature increases in an adiabatic throttling process.

5-9. Comment on the thermodynamic consistency of the following data:

P\T	700	800
500		
v	1.305	1.442
h	1356	1411
s	1.611	1.656

P\T	700	800
600		
v	1.074	1.192
h	1350	1407
s	1.587	1.634

Units are T, °F; P, psia; v, ft³/lb; h, Btu/lb; and s, Btu/lb °R.

5-10. A manuscript is submitted for publication to a journal. The author claims to have determined the equation of state of a pure solid to be

$$v = V_0 - AP + BT$$

and the internal energy to be

$$u = CT - BPT$$

where A, B, C, and V_0 are constants. Check these equations for thermodynamic consistency. Would you recommend publication?

5-11. You and your office mate are working late on an assignment that is due the next day. In the course of your calculations you require thermodynamic data for isotootenol, but to your dismay you find that crucial entries in the property table for this substance have been obliterated due to someone's carelessness. No other data source is available; however, your office mate suggests that you may be able to estimate the missing data from the extant data. Can you obtain a reasonable value for the enthalpy of isotootenol at 120°C and 1.80 MPa from the data shown below? If so, make the calculation.

	P = 1.60 MPa	P = 1.80 MPa
$t = 120°C$	$v = 0.014608$	$v = 0.012697$
	$h = 257.035$	_____
	$s = 0.8059$	$s = 0.7944$
$t = 130°C$	$v = 0.015195$	$v = 0.013244$
		$h = 263.094$
	$s = 0.8253$	$s = 0.8141$

Units are: v, m³/kg; h, kJ/kg; s, kJ/kg · K.

5-12. Using volume-temperature data from the steam tables, determine the enthalpy change (Δh) of superheated steam between $P_1 = 7$ MPa and $P_2 = 8$ MPa at a temperature of 370°C. Compare with tabulated enthalpy values.

5-13. Show that all lines of constant pressure on a Mollier (hs) chart must have the same slope at the same temperature. Further, show that lines of constant pressure must be concave upward.

5-14. A mass of saturated liquid water at a pressure of 1 bar fills a container. The saturation temperature is 99.63°C. Heat is added to the water until its temperature reaches 120°C. If the volume of the container does not change, what is the final pressure?

5-15. For a gas obeying the van der Waals equation of state show that

(a) c_V depends only on T.
(b)   isochores (lines of constant volume) are linear.

5-16. (a)   Isochores of pure liquids are often represented by

$$P = bT - a$$

where a and b are functions only of v. Show that this relationship leads to the conclusion that

$$\left(\frac{\partial c_V}{\partial v} \right)_T = 0$$

(b)   Show that at the point of maximum density for water (4°C)

$$c_P = c_V$$

5-17. Use the following information about a liquid to estimate its isothermal compressibility, β, where

$$\beta = -\frac{1}{V}\left(\frac{\partial V}{\partial P}\right)_T$$

At 325 K the density is 799.4 kg/m³ and at 375 K it is 753.0 kg/m³.

When the liquid was heated from 345 K to 355 K in a container of constant volume, the pressure increase was 148.8 atm.

5-18. It is well-known that reduced isometrics for most gases are linear. Use this information to show that the entropy change of an isothermal process between the two reduced volumes V_{r_1} and V_{r_2} is always the same regardless of the temperature.

5-19. $(\partial V / \partial T)_P$ is positive for real gases. Show that the temperature always decreases for a reversible adiabatic expansion and that this temperature decrease is greater than that resulting from a corresponding decrease of pressure by a Joule-Thomson expansion (constant enthalpy).

5-20. The density of liquid water passes through a maximum at 4°C.

   (a)   Show that in the region $0 < t < 4°C$, liquid water is cooled by an adiabatic compression and not heated as other liquids and gases.

   (b)   Can the temperature 4°C be reached via an adiabatic process?

5-21. It is claimed that the heat capacity, c_p, can be determined from measurements leading to the evaluation of $(\partial T / \partial P)_s$ and $(\partial V / \partial T)_P$. Test this claim by deriving a relationship between these quantities.

5-22. It is known that the thermal coefficient of expansion, α, approaches zero as the absolute temperature approaches zero. Show that this is a confirmation of the third law of thermodynamics.

$$\alpha = \frac{1}{v}\left(\frac{\partial v}{\partial T}\right)_P$$

5-23. What is the maximum work (in Btu/lb) obtainable from the adiabatic expansion of ethylene (C_2H_4) gas through a turbine from 1000 psia and 340°F to a pressure of 100 psia?

   (a)   Assume ideal-gas behavior.

   (b)   Use the generalized enthalpy and entropy departures.

5-24. A gas is compressed from 10 atm and 300 K to 50 atm and 400 K at the rate of 1 kmol/s. Cooling water flows through the compressor at the rate of 110 lb/s entering at 80°F and leaving at 100°F. Calculate the work supplied to the compressor per kmol of gas compressed. Do not assume ideal-gas behavior. The gas has the following properties: M.W. = 28; c_p = 30 kJ/kmol · K; T_c = 300 K; P_c = 50 atm; ω = 0.10.

5-25. For the compressor in Prob. 5-24, calculate the minimum work required to compress the gas to the same final state when the cooling water again enters at 80°F and leaves at 100°F. Do not assume ideal-gas behavior.

5-26. A gas is compressed adiabatically from 10 atm and 300 K to 50 atm and 450 K at the rate of 1 kmol/s. Calculate the work required to compress a kmol of the gas. Do not assume ideal-gas behavior. Some properties of the gas are: c_p = 40 kJ/kmol · K; M.W. = 28; T_c = 300 K; P_c = 50 atm; ω = 0.10.

5-27. Calculate the minimum work to adiabatically compress the gas of Prob. 5-26 from the same initial state to the same final pressure. Do not assume ideal-gas behavior.

5-28. We wish to cool a gas in a steady-state throttling process. High-pressure gas at 360 K and 30 atm will pass through a well-insulated throttle valve to a pressure of 1 atm. Estimate the temperature of the low-pressure gas.

The gas has the following properties: c_p constant and equal to 7 cal/gmol · K; T_c = 300 K; P_c = 30 atm; M.W. = 32; ω = 0.108.

5-29. A gas (T_c = 250 K, P_c = 50 bar, M.W. = 36, ω = 0.11) is to be compressed adiabatically from 50 bar and 300 K to a pressure of 100 bar. Assume reversible operation and calculate the work required per kmol of gas. The low-pressure heat capacity of the gas is

$$c'_P = 30 + 0.02\,T$$

Where c'_P is in kJ/kmol · K and T is in K.

5-30. Nitrogen gas at a high pressure flows through a pipe where its temperature is measured to be 200 K. A small stream of the gas is bled out of the pipe through an insulated throttle valve and into the atmosphere. The temperature of the gas leaving the throttle valve is 180 K. Estimate the pressure of the nitrogen gas in the pipe. The heat capacity, c_p, can be assumed constant at 7 Btu/lb mol °R.

5-31. A gas (T_c = 300 K, P_c = 35 atm, M.W. = 30, ω = 0.18) is stored in a bomb at 360 K and 70 atm. The bomb is connected to a large tank that is initially evacuated. A valve is opened and the pressures in the bomb and tank equalize at 0.37 atm. Estimate the final temperature assuming that the process is adiabatic, the final temperature is uniform throughout, and c_p is constant and equal to 7 cal/gmol · K.

5-32. A Joule-Thomson expansion is irreversible and adiabatic and thus occurs at constant enthalpy. The behavior of a gas undergoing such an expansion is characterized by the Joule-Thomson coefficient $(\bar{\partial} T / \bar{\partial} P)_H$. We wish to generalize the behavior of gases with respect to a Joule-Thomson (J-T) expansion and wonder if Fig. 5-3 could be used for guidance. Can this figure ($\Delta h^* / T_c$ vs. T_r and P_r) be used to estimate the J-T coefficient? Show how this could be done and what additional data are needed. Over the reduced temperature and pressure range of the figure, is the J-T coefficient ever negative? What happens to it at the critical point?

5-33. For a gas in which heat capacity is known, show how Fig. 5-4 (Δs^* vs. T_r and P_r) might be used to construct a line of constant entropy on pressure-temperature coordinates. Specifically, for a gas with $T_c = 300$ K, $P_c = 50$ bar, and

$$c_P' = 30.0 + 0.02T$$

(where c_P' is in kJ/kmol $\cdot$ K and T is in K), graph a constant-entropy line passing through $T = 300$ K, $P = 5$ bar.

5-34. Ethylene gas at 1000 psia and 340°F is throttled adiabatically to 100 psia. Estimate the final temperature using Fig. 5-3. The heat capacity may be assumed constant at 13.5 Btu/lb mol °F.

5-35.

(a) Show how the corresponding states correlation for $\Delta h^* / T_c$ and/or Δs^* could be used to determine c_P for a gas at a high pressure when c_P' is known.

(b) For a gas with $T_c = 320$ K, $P_c = 40$ atm, estimate the difference between c_P' and c_P at 60 atm and 400 K.

(c) Based on the corresponding states correlations, are there any generalizations that can be made about the sign of $c_P - c_P'$?

5-36. Figure 15-4 shows a constant-enthalpy line, an isenthalp, for methane. Point A represents a pressure of 10,300 psia and a temperature of 480°R. Using Fig. 5-3 and $c_P = 8.0$ Btu/lb mol °R,

(a) Calculate the pressure on this isenthalp when the temperature is 450°R.

(b) Check the position of the maximum against that predicted by Fig. 5-3.

Note that the Joule-Thomson inversion curve shown on the figure is merely the locus of isenthalp maxima.

5-37. Show that the residual volume, $RT / P - v$, does not approach zero as the pressure goes to zero.

5-38. The velocity of sound, **C**, is a thermodynamic property defined as

$$\mathbf{C}^2 = \left(\frac{\partial P}{\partial \rho} \right)_S$$

where ρ is density.

(a)   show that

$$\mathbf{C}^2 = \left(\frac{\partial P}{\partial \rho}\right)_S = \frac{C_P}{C_V}\left(\frac{\partial P}{\partial \rho}\right)_T$$

(b)   derive an expression for the velocity of sound in terms of measurable properties with volume and temperature as independent variables

5-39.   For certain types of nozzle calculations it is necessary to know the rate at which the enthalpy of the fluid changes with density. Since nozzle flow is usually approximated as isentropic, the derivative in question is $(\partial h / \partial \rho)_S$

(a)   Express this derivative in terms of only P, v, T, c_P, and c_V and their derivatives.
(b)   Evaluate for an ideal gas.

5-40.   A fluid has an isothermal compressibility

$$\beta = K\left[1 + b(T - T_0)\right]$$

and a thermal coefficient of expansion

$$\alpha = A(1 - aP)$$

where

$$K = 2.52(10^{-5})\,\text{atm}^{-1}$$

$$b = 2(10^{-3})\text{K}^{-1}$$

$$T_0 = 273\ \text{K}$$

$$a = 1.2(10^{-4})\,\text{atm}^{-1}$$

$$A = 4.2(10^{-4})\,\text{K}^{-1}$$

(a)   Test the consistency of these data.
(b)   Determine the equation of state for the fluid.

5-41.   The equation of state of a rubber band is

$$\mathbf{f} = aT\left[\frac{l}{l_0} - \left(\frac{l_0}{l}\right)^2\right]$$

where $\mathbf{f}$ = tension, l = length, $a = 1.3(10^3)$ dyne K^{-1}, and $l_0 = 1$ m.

(a)   Show that the internal energy is a function only of temperature.

(b)   The band is stretched reversibly and isothermally from a length of 1 m to a length of 2 m at 300 K. Find the heat and work effects.

(c)   If the band is stretched adiabatically and reversibly between 1 and 2 m with an initial temperature of 300 K, what would be the final temperature? The heat capacity at constant length is $C_l = 1.2$ J K^{-1}.

(d)   Consider the operation of a Carnot cycle using this material as a working substance, and sketch the cycle on the proper coordinates, indicating the nature of each step.

(e)   Show that the efficiency of such a Carnot cycle is $(T_2 - T_1)/T_2$.

5-42.   A constrained metal bar is heated under conditions of constant length. Derive an expression for the force necessary to constrain the bar as a result of a moderate increase in temperature. Express this force in terms of measurable quantities such as the linear coefficient of expansion, α, and Young's modulus, Y. Where

$$\alpha = \frac{1}{l}\left(\frac{\partial l}{\partial T}\right)_{\mathbf{f}}$$

$$Y = \frac{l}{A}\left(\frac{\partial \mathbf{f}}{\partial l}\right)_{T}$$

l = length, $\mathbf{f}$ = force, A = cross sectional area

5-43.   The tensile force, X, of a rubber band constrained to constant length has the following dependency on temperature

$$X = \alpha T$$

where T is absolute temperature and α is a constant depending only on length. Show that

(a)   at constant temperature the entropy decreases with increasing length.

(b)   the temperature rises on adiabatic stretching.

(c)   heat is rejected during isothermal stretching.

5-44.   (a)   Assume that the surface tension of a liquid depends only on temperature. Then a Clapeyron-like equation should be valid. Show that

$$\frac{d\sigma}{dT} = -\frac{\Delta S}{\Delta \mathbf{A}} = -\frac{Q}{T}$$

where Q is the heat absorbed when the surface area $\mathbf{A}$ is increased by a unit amount isothermally.

(b)   The surface tension of water is given as follows:

t	0	10	20	30	40	50	60	70	80	100
σ	75.6	74.22	72.75	71.18	69.56	67.91	66.18	64.4	62.6	58.9

t in °C; σ in dynes/cm

Does Q change appreciably with T? If not, plot the data so that a straight line consistent with the integrated equation would be obtained.

5-45. For a paramagnetic solid undergoing magnetization or demagnetization, Curie's law is valid at moderate temperatures but fails at low temperatures. Show that a substance obeying Curie's law cannot also obey the third law of thermodynamics. Curie's law is

$$I = c\frac{\mathbf{H}}{T}$$

where $\mathbf{H}$ is the magnetic field strength, I is the intensity of magnetization, T is the absolute temperature, and c is a constant which is always positive.

5-46. It has been stated that a paramagnetic solid obeying Curie's law (see Prob. 5-45) has an internal energy that depends only on temperature. Is this statement correct? Explain.

5-47. Use the program PREOS.EXE to construct a line of constant entropy for methane gas on pressure-temperature coordinates. The line should pass through the point $T = 500$ K, $P = 100$ bar.

5-48. Rework Prob. 5-23 using the program PREOS.EXE.

5-49. Rework Prob. 5-34 using the program PREOS.EXE.

5-50. Rework Prob. 5-36(a) using the program PREOS.EXE.

5-51. An evacuated, insulated cylinder is connected to a source of chlorine gas at 350 K and 30 atm. A valve is opened and chlorine flows into the cylinder until its pressure reaches 20 atm. How much chlorine has flowed into the cylinder?

5-52. Methane is stored in a bomb at 330 K and 90 bar. The bomb is connected to an evacuated tank. A valve is opened and the bomb and tank presssures equalize at 0.37 bar. Estimate the final temperature of the methane assuming that the process is adiabatic and the temperature is uniform throughout.

CHAPTER 6

Heat Effects

The calculation of heat effects for various types of processes is a task that regularly confronts the practicing engineer. While this is an important application of thermodynamics for which one naturally expects equations and techniques which have direct utility, an understanding of the subject is best developed from the perspective of the experimentalist who acquires data and presents them in a usable form. These data, which will often be used routinely, will take on physical meaning only if we understand how they came into existence. In this chapter our goal is to see how the methods of thermodynamics can be employed for the most effective use of experimentally derived information.

The chapter begins with the concept of a computational path which allows us to calculate property changes for a complex process by identifying and considering separately the various phenomena which combine to produce the overall process. These phenomena are then discussed and their thermodynamic treatment developed. Finally, several examples are presented to illustrate the delineation of the computational path and the execution of the attendant calculations.

6-1 THE COMPUTATIONAL PATH

The calculation of heat effects is based on application of the first law and therefore requires values of ΔH and ΔU for the process under study. Because these are state properties, their changes corresponding to any particular change of state can be calculated by any convenient computational path rather than the actual path taken by the system. For example, if SO_3 reacts with an excess of water, sulfuric acid, H_2SO_4, will form but will be dissolved in the excess water. Thus, there are heat effects due to the chemical reaction

$$SO_3 + H_2O = H_2SO_4$$

and the formation of a solution of sulfuric acid in water and changes in temperature. While all the processes giving rise to heat effects actually occur simultaneously, it is convenient to calculate ΔH or ΔU by considering a series of simple steps. For calculational purposes a reasonable sequence of elementary steps would be the following: (1) React SO_3 with the stoichiometric quantity of water, (2) dissolve the acid in the remaining water, and (3) heat the solution to its final temperature.

All the separate steps in a computational path require some type of experimentally derived information for the determination of property changes, and each is chosen to take advantage of the available data. Usually, the steps are chosen so that only one type of process occurs per step. Any overall process we may wish to consider can always be decomposed into a series of simple steps corresponding to either change in temperature, pressure, or phase, isothermal mixing or unmixing, and isothermal chemical reaction. Accordingly, the results of experimental measurement are always presented as one of the preceding simple steps even though the experiment might not correspond to a single simple step. In many instances this requires the delineation of a computational path for the observed state change for which the heat effect is measured. Known data are then used to evaluate property changes for all but the desired simple step. The property change for this step is then obtained from the measured heat effect and the property changes for the other steps in the computational path. A little reflection should convince you that this is the only logical approach for depositing and withdrawing information needed for the calculation of the limitless variety of changes which could be encountered. Without the thermodynamic approach the literature would contain a chaotic collection of measured heat effects which would be useful only in the unlikely event that the considered change in state exactly matched that of a reported value.

6-2 HEAT EFFECTS DUE TO CHANGE OF TEMPERATURE

From Chap. 2 we have seen that through thermodynamics the internal energy and enthalpy are defined and their changes can be determined from experimental measurements. Here we are interested in how experimentally derived information is deposited in the literature and how in turn it may be used to calculate enthalpy or internal energy changes. In this section we will consider only pure substances and will treat mixtures in a later section.

Information relating change of enthalpy with temperature at constant pressure can be acquired calorimetrically or calculated from spectroscopic data through the methods of quantum and statistical mechanics. The latter type of data is calculated for the ideal-gas state and can usually be applied to processes involving gases near atmospheric pressure. In fact, the accuracy of the spectroscopically determined data is considered equal to, or perhaps better than, calorimetrically determined data. The various forms in which one finds these data in the literature and methods of using them will now be described.

Enthalpy Tables. With the enthalpy set equal to zero in a datum state, values may be tabulated at various conditions. A tabulation of enthalpies of combustion gases in the ideal-gas state at various temperatures may be found in Table 6-1. In the thermochemical data sources listed in Table 12-1 we may find $H° - H_0°$ or $(H° - H_0°)/T$ tabulated against temperature. These are spectroscopically determined and are the enthalpy differences in the ideal-gas state between absolute zero and the tabulated temperature.

Heat Capacity. From Chap. 2 we have seen that for a change in temperature at constant pressure the enthalpy change is found from the heat capacity:

$$\Delta h = \int_{T_1}^{T_2} c_P \, dT$$

Thus, a knowledge of heat capacity as a function of temperature is equivalent to a table of enthalpies. Most heat capacity data are available at atmospheric pressure and are often fitted to polynomial equations in temperature. A tabulation of the parameters in these equations for a variety of compounds may be found in the Appendix. In addition to the various handbooks, heat capacity data can be found in the references for thermochemical data listed in Table 12-1. When experimentally determined heat capacities are unavailable, methods of estimating this property may be employed.[1]

---

[1] See R. C. Reid, J. M. Prausnitz, and B. E. Poling, *The Properites of Liquids and Gases*, 4th ed., McGraw-Hill, New York, 1978.

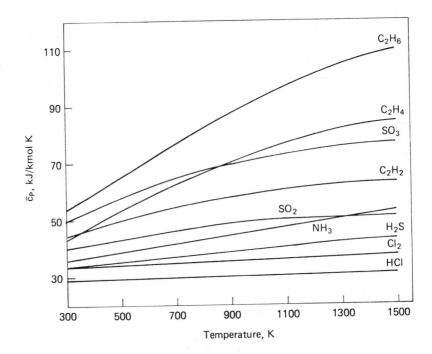

Figure 6-1   Mean heat capacities of selected gases. Reference temperature 298 K. [Based mainly on data from D. D. Wagman, ed., Selected Values of Chemical and Thermodynamic Properties, *Natl. Bur. Stand. Circ., 500* (1952).]

Mean Heat Capacity.   Often changes in enthalpy are more conveniently calculated from the mean heat capacity $\bar{c}_P$, defined as

$$\bar{c}_P = \frac{\Delta h}{\Delta T} = \frac{\displaystyle\int_{T_0}^{T} c_P\, dT}{T - T_0} \tag{6-1}$$

From Eq. (6-1) $\bar{c}_P$ is seen to be the mean heat capacity between T_0 and T, where T_0 is a reference temperature. A convenient reference temperature is 298 K. Mean heat capacities referenced to 298 K for several common gases are plotted in Fig. 6-1.

TABLE 6-1
ENTHALPIES OF COMBUSTION GASES IN THE IDEAL-GAS STATE

T(K)	N_2	O_2	H_2	CO	CO_2	H_2O	CH_4
			Enthalpy (kJ/kmol)				
273	0	0	0	0	0	0	0
298	728.2	732.4	718.1	728.6	911.5	838.3	859.2
400	3,696	3,753	3,656	3,700	4,905	4,285	4,721
500	6,646	6,813	6,591	6,654	9,207	7,755	9,061
600	9,630	9,973	9,521	9,667	13,810	11,330	14,850
700	12,660	13,230	12,460	12,750	18,660	15,020	17,410
800	15,760	16,570	15,420	15,900	23,720	18,830	25,540
900	18,970	19,980	18,390	19,130	28,940	22,770	32,070
1000	22,180	23,440	21,390	22,420	34,320	26,830	39,050
1100	25,480	26,950	24,430	25,770	39,810	31,020	46,410
1200	28,830	30,500	27,520	29,160	45,420	35,320	54,150
1300	32,220	34,090	30,630	32,600	51,100	39,730	62,190
1400	35,650	37,700	33,800	36,080	56,870	44,250	70,480
1500	39,150	41,350	37,000	39,590	62,690	48,860	79,010
1750	47,950	50,570	45,290	48,470	77,460	60,770	101,280
2000	56,920	59,930	53,690	57,500	92,490	73,150	124,500
2250	66,000	69,470	63,360	66,580	107,760	85,880	148,440
2500	75,080	79,140	71,230	75,790	123,310	98,890	172,880
2750	84,290	88,930	80,310	85,040	138,730	112,120	197,780
3000	93,530	98,850	89,480	94,290	154,380	125,550	222,980
3500	112,160	119,060	108,060	113,000	185,940	152,840	274,160
4000	130,910	139,670	127,560	131,830	217,830	180,460	326,010

Source: references 2 and 8 in Table 12-1.

To use mean heat capacities for calculating enthalpy changes between two temperatures T_1 and T_2, where neither is the reference temperature, we devise a two-step computational path—a change from T_1 to T_0 followed by a change from T_0 to T_2. This results in

$$\Delta h = \int_{T_1}^{T_0} c_P \, dT + \int_{T_0}^{T_2} c_P \, dT$$

$$= (T_0 - T_1)\bar{c}_P\big|_{T_1} + (T_2 - T_0)\bar{c}_P\big|_{T_2} \tag{6-2}$$

where $\overline{c}_P\big|_{T_1}$ and $\overline{c}_P\big|_{T_2}$ are mean heat capacities evaluated at T_1 and T_2.

6-3 HEAT EFFECTS DUE TO CHANGE OF PRESSURE

For the most part we have neglected changes of enthalpy with pressure for liquids and solids and saw in Chap. 5 that this approximation is justified. An ideal gas shows no enthalpy change with pressure, and in Chap. 5 we have seen how this enthalpy change may be calculated or estimated for a real gas. In comparison with the other steps in a computational path, that due to a change in pressure can usually be neglected in calculating process heat effects. For precise thermochemical calculations it, of course, should be included.

6-4 HEAT EFFECTS DUE TO CHANGE OF PHASE

A change of phase occurs at constant temperature and pressure. The enthalpy change accompanying a phase change, often referred to as a latent heat, may be determined calorimetrically or from vapor pressure data as shown in Chap. 8. Latent heats are, in general, temperature dependent. This dependence can be determined by the computational path shown in Fig. 6-2. The latent heat for the phase change at T_2, shown by the heavy arrow, may be expressed in terms of the latent heat at T_1 and the heat capacities of the phases

$$\left(h^{\beta} - h^{\alpha}\right)_{T_2} = \int_{T_2}^{T_1} c_{P\alpha}\, dT + \left(h^{\beta} - h^{\alpha}\right)_{T_1} + \int_{T_1}^{T_2} c_{P\beta}\, dT \tag{6-3}$$

For a differential change this becomes

$$\frac{d\left(h^{\beta} - h^{\alpha}\right)}{dT} = c_{P\beta} - c_{P\alpha} \tag{6-4}$$

These expressions ignore the usually negligible effect of pressure change which accompanies the temperature change.

Tabulations of latent heats are found in the same sources cited for heat capacity data. For calculation of process heat effects, the heat of vaporization is the most important latent heat effect, and therefore much effort has been devoted to establishing estimation methods for this property.[2]

---

[2] See R. C. Reid, *ibid.*

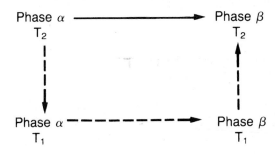

Figure 6-2 Computational path for determining temperature dependency of latent heat.

6-5 MIXING HEAT EFFECTS

From our experience in the chemistry laboratory we know that a heat effect often accompanies a mixing process. The best known example of this would be the mixing of sulfuric acid and water where the effect is easily perceived. If the mixing occurs batchwise at constant pressure or in a steady-state flow process, the heat effect is the enthalpy change. Therefore, experimental measurement of the heat effect results in the determination of the change in a state property. The experimental measurement is usually made in a calorimeter, although it is possible to calculate the heat of mixing from other types of measurements.[3] As in the determination of many other thermodynamic properties, the concept behind the experiment is simple, but its execution requires sophisticated apparatus and painstaking technique if data of high quality are to be obtained.[4] Most measurements are made batchwise in a constant-pressure adiabatic calorimeter. A simplified representation of the experiment is shown in Fig. 6-3. On adiabatic mixing of the components originally at the temperature T_1, the mixture achieves a temperature T_a. If the process is endothermic, $T_a < T_1$ and the quantity of electrical energy required to bring the mixture to T_1 is measured. Application of the first law results in

$$\Delta H = W_e$$

If the mixing process is exothermic, $T_a > T_1$. The mixture is allowed to cool to T_1, and the amount of electrical energy required to return the system to T_a is measured. From the first law we obtain

---

[3] See Sec. 11-2.

[4] For a thorough, yet lucid, description of the experimental approach to calorimetry, see M. L. McGlashan, *Chemical Thermodynamics*, Academic Press, New York, 1979, Chap. 4.

$$\Delta H = -W_e$$

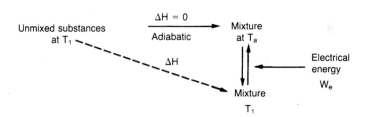

Figure 6-3   Simplified representation of adiabatic mixing calorimetry.

Experimentally determined heat effects are reduced to intensive property changes and can be presented in either of two ways: as heats of mixing Δh based on a mol of solution or as heats of solution (often called "integral heats of solution"), Δh_s based on a mol of solute. When the components are liquids and solutions of all proportions are possible, the heat of mixing Δh is the preferred method of presentation. Figure 6-4 shows the heat of mixing for the ethanol-water system plotted vs. mol fraction of ethanol at several temperatures. The process is seen to be exothermic at low temperatures and low ethanol concentrations, changing to endothermic at higher temperatures and ethanol concentrations. When one component is a gas or solid, it is not possible to form solutions of all proportions, and it is convenient to represent data as heats of solution Δh_s. Figure 6-5 shows the heats of solution for several substances in water plotted against the number of mols of water per mol of solute.

The heat of mixing can be expressed as the difference between the solution enthalpy h and the pure component enthalpies h_1 and h_2:

$$\Delta h = h - x_1 h_1 - x_2 h_2 \tag{6-5}$$

Dividing by x_1 gives

$$\Delta h_s = \frac{\Delta h}{x_1} = \frac{h}{x_1} - h_1 - n h_2 \tag{6-6}$$

where the heat of solution Δh_s is expressed in terms of the enthalpy of a quantity of solution containing a mol of solute, component 1, and $n = x_2 / x_1$ is the mols of solvent per mol of solute. The value of Δh or Δh_s can be known through experimentation, while absolute values of enthalpies can never be known. Whether experimental information is presented as heats of mixing or heats of solution, the enthalpy change always represents the isothermal

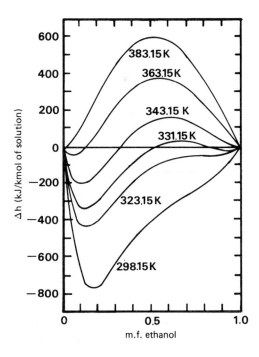

Figure 6-4 Heat of mixing ethanol and water. [Based on the data of J. A. Larkin, *J. Chem. Thermodyn.*, *7*, 137 (1975); J. A. Boyne and A. G. Williamson, *J. Chem. Eng. Data*, *12*, 318 (1967); and R. F. Lama and B. C. Lu, *J. Chem. Eng. Data*, *10*, 216 (1965).]

formation of a solution at a pressure of 1 atm from pure substances at 1 atm. Because these standard states[5] have been specified, the heat effect depends only on temperature.

A comparison of Figs. 6-4 and 6-5 shows that the heat effect of dissolving strong electrolytes, especially acids and bases, in water is much larger than for dissolving a nonelectrolyte, ethanol. This is generally true. It is found that the heat of mixing nonelectrolyte liquids seldom exceeds 1000 kJ/kmol, and except for where specific interactions such as hydrogen bonding exist, the process is endothermic. For many industrial process calculations a heat effect of this magnitude can usually be ignored. Also, there is no heat effect on mixing ideal gases,[6] and hence real-gas mixtures formed at low to moderate pressures should exhibit a negligible heat of mixing.

EXAMPLE 6-1

Calculate the temperature attained by a solution resulting from the adiabatic mixing of 0.80 mol of water with 0.20 mol of ethanol. Both pure liquids are originally at 298 K.

---

[5] The subject of standard states will be more fully developed in Secs. 6-7 and 12-3. Strictly speaking, the standard state for a gas is the ideal-gas state at 1 atm; however, for most applied thermochemical calculations involving the enthalpy this refinement is unnecessary.

[6] See Sec. 11-3.

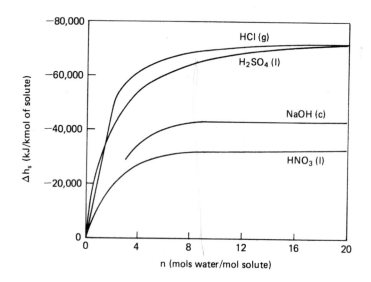

Figure 6-5
Heat of
Solution of
selected
electrolytes at
298 K. [Based
on data from
D. D. Wag-
man, ed., Se-
lected Values
of Chemical
and Thermo-
dynamic Prop-
erties, *Natl.
Bur. Stand.
Circ., 500*
(1952).]

Solution 6-1

The final temperature is obtained through the use of the two-step computational path shown in Fig. 6-6. However, to execute the calculations,
the heat capacity of the 20 mol % solution is required. This property
may be obtained from the computational path shown in Fig. 6-7, which
requires the heat of mixing at two temperatures and enthalpy or heat capacity data for pure water and ethanol. From Fig. 6-4 the heat of mixing
to obtain a 20 mol % ethanol solution is -758 J/mol at 298.15 K
and -415 J/mol at 323.15 K. The enthalpy change for pure liquid water ΔH_1 can be obtained from the steam tables in the Appendix:

$$\Delta H_1 = 0.8(18.02)(104.8 - 209.3) = -1506 \text{ J}$$

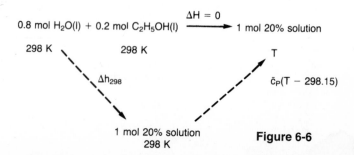

Figure 6-6

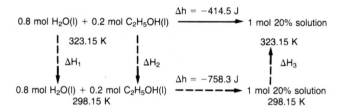

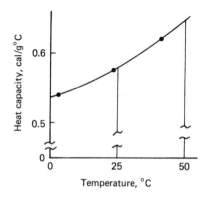

Figure 6-7

Figure 6-8 Heat capacity of $C_2H_5OH(l)$. [Data from R. H. Perry and C. H. Chilton. *Chemical Engineer's Handbook*, 5th ed., McGraw-Hill, New York, 1973.]

The heat capacity of pure liquid ethanol is plotted vs. temperature in Fig. 6-8. The enthalpy change ΔH_2 for ethanol is

$$\Delta H_2 = -0.2(46.07) \int_{25°C}^{50°C} c_P \, dt$$

where the integral is simply the area under the curve in Fig. 6-8 between 25 and 50°C. We find $\Delta H_2 = -594\,\text{J}$. From the computational path of Fig. 6-7 we have

$$\Delta h_{323.15\,K} = \Delta H_1 + \Delta H_2 + \Delta h_{298.15\,K} + \Delta H_3$$

$$-415 = -1506 - 594 - 758 + \Delta H_3$$

$$\Delta H_3 = \bar{c}_P(323.15 - 298.15) = 2444\,\text{J}$$

$$\bar{c}_P = 97.8 \ \text{J / mol·K}$$

This is the mean heat capacity of a 20% solution between 25 and 50°C. This value will be used to evaluate the enthalpy change for the temperature change from 25 to t °C shown in Fig. 6-6. For the computational path of Fig. 6-6 we write

$$0 = \Delta h_{298.15} + \bar{c}_P(T - 298.15)$$

$$0 = -758 + 97.8(T - 298.15)$$

$$T = 305.91 \text{ K}$$

A temperature rise of 7.76 K is expected for this adiabatic mixing process.

As shown in Ex. 6-1, the heat capacity of a solution can be obtained from heats of mixing (or heats of solution) at two temperatures. It is also possible to determine the heat capacity of a solution in the same way as for a pure substance. Heat capacities of solutions are found throughout the literature. Usually data are available for solutions exhibiting strong heat effects. When experimental data are unavailable, the heat capacity of nonelectrolyte solutions can be approximated by setting $\Delta h = 0$ in Eq. (6-5) and differentiating with respect to temperature at constant pressure to obtain

$$c_P = x_1 c_{P1} + x_2 c_{P2} \tag{6-7}$$

This approximation should be satisfactory for many types of process heat effect calculations.

EXAMPLE 6-2

Calculate the heat effect when 1 kmol of water is added to a solution containing 1 kmol of sulfuric acid and 3 kmol of water. The process is isothermal at 25°C.

Solution 6-2

Using the heat of solution data of Fig. 6-5 and the two-step computational path shown in Fig. 6-9, we write

$$\Delta H = \Delta H_1 + \Delta H_2$$

ΔH_1 is the negative of the heat of solution for 1 mol of acid in 3 mols of water. Four mols of water and 1 mol of acid are subsequently mixed in step 2 to form the desired final solution. Therefore, ΔH_2 is the heat of solution for 1 mol of acid in 4 mols of water. Both ΔH_1 and ΔH_2, as well as ΔH, are based on a kmol of acid:

$$\Delta H = -(-49{,}000) + (-54{,}100) = -5100 \text{ kJ}$$

To maintain a constant temperature, it is necessary to remove 5100 kJ during the mixing process.

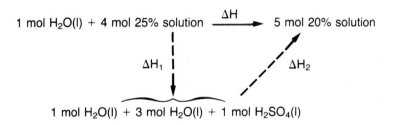

Figure 6-9

6-6 ENTHALPY-CONCENTRATION DIAGRAMS

Calculations involving the mixing of binary solutions can be carried out quite conveniently if an enthalpy-concentration diagram has been constructed for the system of interest. These diagrams can be found throughout the literature for commercially important systems with large mixing heat effects.[7] An enthalpy-concentration diagram for sulfuric acid and water is shown in Fig. 6-10. The utility of these diagrams derives from the fact that an adiabatic mixing of two solutions can be represented on them by a straight line connecting the points corresponding to the mixture and the original solutions. The lever-arm principle also applies.

This graphical representation of the mixing process may be verified by writing material balances and applying the first law to an adiabatic mixing process as illustrated in Fig. 6-11. The overall material balance is

$$A + B = C \tag{6-8}$$

where A, B, and C are either quantities of solution or flow rates expressed in either mass or mol units. A material balance over one of the components is

$$Ax_A + Bx_B = Cx_c \tag{6-9}$$

---

[7] Several diagrams are available in R. H. Perry and C. H. Chilton, *Chemical Engineers' Handbook*, 5th ed., McGraw-Hill, New York, 1973, Chap. 3.

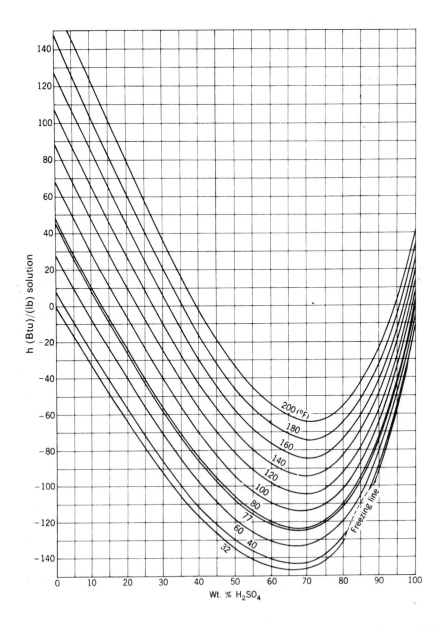

Figure 6-10 Enthalpy-concentration diagram for sulfuric acid and water. [With permission from W. D. Ross, *Chem. Eng. Prog., 48*, 314 (1952), as adapted in J. M. Smith and H. C. Van Ness, *Introduction to Chemical Engineering Thermodynamics*, 3rd. ed., McGraw-Hill, New York, 1975.]

where the x's are either mass or mol fractions. For adiabatic mixing $\Delta H = 0$ and

$$Ah_A + Bh_B = Ch_C \qquad (6\text{-}10)$$

where, consistent with the material balances, either molar or specific enthalpies may be used. These three equations can be rearranged to

$$-\frac{A}{B} = \frac{x_C - x_B}{x_C - x_A} \qquad (6\text{-}11)$$

$$= \frac{h_C - h_B}{h_C - h_A} \qquad (6\text{-}12)$$

and combined to yield

$$\frac{h_B - h_C}{x_B - x_C} = \frac{h_C - h_A}{x_C - x_A} \qquad (6\text{-}13)$$

A graphical representation of the mixing process is shown in Fig. 6-12. The state of each solution is specified by its composition and enthalpy[8] with the points A, B, and C representing the initial solutions and the final solution, respectively. From Fig. 6-12 it is seen that $(h_B - h_C)/(x_B - x_C)$ is the slope of line segment $\overline{BC}$ and $(h_C - h_A)/(x_C - x_A)$ is the slope of

Figure 16-11 Adiabatic mixing process.

line segment $\overline{AC}$. These segments have equal slopes and the common point C, and hence ABC is a straight line. Thus, the mixing of solutions A and B results in the solution C which lies on a straight line connecting A and B. Equations (6-11) and (6-12) show the applicability of the lever-arm principle.

As illustrated in Fig. 6-13, a nonadiabatic mixing process (enclosed by the dashed envelope) may be considered as an adiabatic mixing followed by heat addition or removal at constant composition. On an enthalpy-concentration diagram A, B, and C' are collinear, and C is located from C' by application of the first law to the second step:

$$Q = \Delta H = C(h_C - h_{C'}) \qquad (6\text{-}14)$$

$$h_C = \frac{Q}{C} + h_{C'}$$

---

[8] While the phase rule requires specification of three intensive variables for a two-component, single-phase system, the pressure, tacitly assumed atmospheric, has a negligible effect on liquid properties.

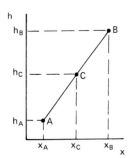

Figure 6-12   Graphical representation of an adiabatic mixing process.

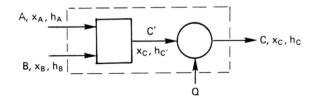

Figure 6-13
Nonadiabatic mixing
process.

EXAMPLE 6-3

Use the enthalpy-concentration diagram of Fig. 6-10 to work Ex. 6-2.

Solution 6-3

For this nonadiabatic process we first consider the adiabatic mixing to produce the desired composition but at a temperature different from the desired value. Essential portions of Fig. 6-10 are sketched in Fig. 6-14 to illustrate this graphical solution technique.

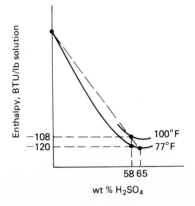

Figure 6-14

Pure water and the 25 mol % acid solution (65 wt %) at 25°C (77°F) are located and connected by a straight line. The resulting 20 mol % solution (58 wt %) could be located by application of the lever-arm principle or by material balance calculation as in Ex. 6-2. This adiabatic mixing is seen to produce a solution at 100°F with an enthalpy of −108 Btu/lb. The enthalpy of this solution at 77°F is −120 Btu/lb as determined from Fig. 6-10, and the enthalpy change on cooling to 77°F is therefore −12 Btu/lb of solution. A pound of 58 wt % solution contains

$$\frac{0.58(0.454)}{98.08} = 2.68(10^{-3}) \text{ kmol of } H_2SO_4$$

and per kmol of H_2SO_4 the heat effect is[9]

$$\frac{-12(1.055)}{2.68(10^{-3})} = -4723 \text{ kJ / kmol}$$

Considering the uncertainty of reading enthalpies from Fig. 6-10, the agreement with the value of −5100 kJ/kmol obtained from Ex. 6-2 is as good as can be expected.

In contemplating the use and construction of an enthalpy-concentration diagram, the question of assigning a value to the enthalpy of a solution arises. The enthalpy of the solution must be referenced to some datum condition for the pure components, but what datum states are permissible? To answer this, we begin by writing the enthalpy of the solution h in terms of pure component enthalpies h_1 and h_2, the heat of mixing Δh, and the mol fractions of components 1 and 2 in the solution:

$$h = \Delta h + x_1 h_1 + x_2 h_2 \tag{6-5}$$

For an adiabatic process where n' mol of solution of enthalpy h' are mixed with n'' mol of solution of enthalpy h'' to produce $n' + n''$ mol of solution of enthalpy h^F, we write

$$(n' + n'')h^F - n'h' - n''h'' = 0 \tag{6-15}$$

Substituting Eq. (6-5) for the solution enthalpies, and rearranging one obtains

$$0 = (n' + n'')\Delta h^F - n'\Delta h' - n''\Delta h'' + x_1 n'(h_1^F - h_1')$$

$$+ x_1 n''(h_1^F - h_1'') + x_2 n'(h_2^F - h_2') + x_2 n''(h_2^F - h_2'') \tag{6-16}$$

---

[9] 1 Btu = 1.055 kJ

The terms Δh^F, $\Delta h'$, and $\Delta h''$ are heat effects when pure components at 1 atm are mixed isothermally to form a solution at 1 atm. These are quantities which could be determined from experimental measurement and are obviously unaffected by the choice of datum levels for pure components. The remaining terms are seen to be enthalpy differences for the pure components and are likewise unaffected by choice of datum. Similarly, for a change in temperature at constant composition the solution enthalpy change is expressed in terms of heats of mixing and pure-component enthalpy differences and is also independent of choice of datum. Thus, any datum is permissible for pure-component enthalpies.

The construction of an enthalpy-concentration diagram requires heat capacities of the pure components and either heats of solution at several temperatures or heats of solution at a single temperature along with heat capacities of several solutions. Equation (6-5) or (6-6) is used for determining isotherms at temperatures where heat of solution data are available. Isotherms at other temperatures may be determined from these isotherms and heat capacities at several solution concentrations. Details of construction may be found elsewhere.[10]

6-7 CHEMICAL HEAT EFFECTS

Continuing with the approach of separating the various types of heat effects, we now consider chemical heat effects. To isolate these effects, it is necessary to define the standard heat of reaction $\Delta H°$. This is the enthalpy change when reactants in their standard states react isothermally and completely to form products in their standard states. We have already seen that for purposes of calculating enthalpy changes the standard states are pure substances at a pressure of 1 atm. Thus, a standard heat of reaction depends only on temperature and may be determined at any temperature, although a great deal of thermochemical data are tabulated at 298 K.

Very few reactions can actually be carried out under conditions which correspond to those for the standard heat of reaction. However, through an appropriate computational path, $\Delta H°$ may be calculated from a measured heat effect. We must recognize that while $\Delta H°$ is associated with an artificial process, it represents a heat effect due solely to chemical reaction. It is a thermodynamic property change corresponding to a well-defined change of state and thus can be employed in any consistent computational path.

Chemical heat effects can be measured either under conditions of constant volume or constant pressure. In the former case a mixture is made to react adiabatically to produce another mixture at a different temperature, and a computational path similar to that shown in Fig. 6-3 is employed to determine the chemical heat effect. Flow calorimeters, operating at constant pressure with a steady flow of reactants, can achieve a steady-state condition with reaction heat being removed by a flow of cooling water. These calorimeters are used mainly for combustion reactions, and the heat of combustion is determined from the measured flow rates of reactants and cooling water, the cooling water temperature rise, the mass of collected

---

[10] See W. L. McCabe, *Trans. AIChE, 31, 129* (1935).

water resulting from combustion, and the temperature and analysis of the exiting combustion gas. Measurements accurate to within a few percent can be obtained with off-the-shelf calorimeters such as shown in Fig. 6-15. More accurate determinations require considerably more complex equipment, elaborate technique, and detailed calculations.[11]

Compiled experimentally determined information about chemical heat effects is most effectively utilized by considering the reaction giving rise to $\Delta H°$ to proceed as follows:

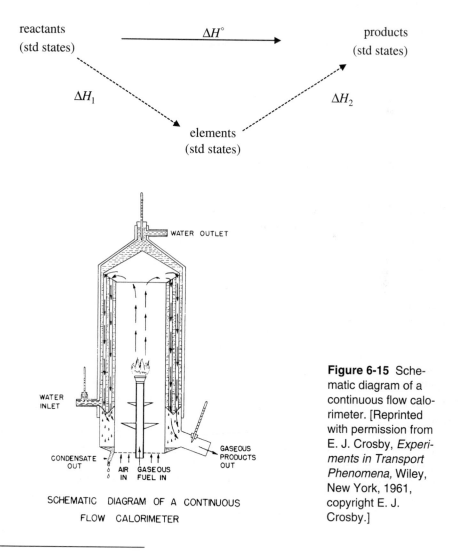

SCHEMATIC DIAGRAM OF A CONTINUOUS
FLOW CALORIMETER

Figure 6-15 Schematic diagram of a continuous flow calorimeter. [Reprinted with permission from E. J. Crosby, *Experiments in Transport Phenomena,* Wiley, New York, 1961, copyright E. J. Crosby.]

---

[11] See F. D. Rossini, ed., *Experimental Thermochemistry*, Wiley-Interscience, New York, 1956, or M. L. McGlashan, *ibid.*

Reactants in their standard states are dissociated isothermally into elements in their standard states followed by an isothermal recombination of elements to form products in their standard states. To exploit this computational path, it is necessary to define the heat of formation ΔH^f of a chemical species. This is simply the standard heat of reaction for the formation of 1 mol of the species from its elements in their naturally occurring states.[12]

$$\text{elements} \quad \xrightarrow{\Delta H^\circ = \Delta H_i^f} \quad \text{species } i$$
$$\text{(std states)} \qquad\qquad\qquad \text{(std state)}$$

In the preceding computational path it is now possible to express ΔH_1 and ΔH_2 in terms of heats of formation:

$$\Delta H_1 = - \sum_{\substack{\text{all} \\ \text{reactants}}} n_j \, \Delta H_j^f$$

$$\Delta H_2 = \sum_{\substack{\text{all} \\ \text{products}}} n_i \, \Delta H_i^f$$

For the reaction we obtain

$$\Delta H^\circ = \sum_{\substack{\text{all} \\ \text{products}}} n_i \, \Delta H_i^f - \sum_{\substack{\text{all} \\ \text{reactants}}} n_j \, \Delta H_j^f \qquad\qquad (6\text{-}17)$$

Consistent with the chosen computational path, the preceding summations are made over all products and all reactants which are not elements in their naturally occurring states. The omission of these elements from the summations is equivalent to setting heats of formation of elements in their naturally occurring states equal to zero. It should be noted that Eq. (6-17) can be applied to a single chemical reaction or where many reactions are occurring. The former case is routine and requires only that the n's be replaced by stoichiometric coefficients. In the latter case it is not necessary to delineate the reaction mechanism by identifying individual reactions and the extent to which each progresses. Because we are calculating the change in a thermodynamic property, we need only know the initial and final states. The application of Eq. (6-17) to a complex reacting mixture is illustrated in Ex. 6-11.

Thus, it is seen that the concept of a standard heat of reaction, although somewhat artificial, allows chemical heat effects to be isolated. Because it represents only chemical heat effects, the standard heat of reaction is expressible from heats of formation of the reaction participants. This is a tremendous accomplishment! Instead of needing to determine the heat effect for every possible reaction, the task has been considerably reduced to that of determining the heat of formation of compounds of interest. Thermodynamics still requires ex-

---

[12] While species such as H(g), O(g) and Cl(g) exist under certain conditions, the naturally occurring states of these elements are $H_2(g)$, $O_2(g)$, and $Cl_2(g)$. The monatomic species therefore are assigned heats of formation.

experimentally derived information for its application, but here is an outstanding example of how its methods have allowed us to make very effective use of our experimental efforts.

In writing a chemical reaction to which Eq. (6-17) will be applied or in delineating a computational path, the phase in which the substance exists should be noted by (g), (l), or (c), which represent gas, liquid, or crystalline states. Consistency between these states and those of the heats of formation should be observed. For example, we must distinguish between the heat of formation of liquid water and gaseous water:

$$H_2(g) + \frac{1}{2}O_2(g) = H_2O(l) \qquad \Delta H^f = -286.0 \text{ MJ / kmol}$$

$$H_2(g) + \frac{1}{2}O_2(g) = H_2O(g) \qquad \Delta H^f = -242.0 \text{ MJ / kmol}$$

It will be noted that the state to which the heat of formation of $H_2O(g)$ refers is physically impossible at 298 K because the standard state pressure of 1 atm exceeds water's vapor pressure. However, through an appropriate computational path it is possible to calculate this enthalpy change from experimental data. This is little different from calculating ΔH° for an artificial process, which was proved a useful tactic. The heat of formation of $H_2O(g)$ represents an enthalpy change for a well-defined, albeit hypothetical, change in state, and thus no problem arises as long as it is used in a consistent computational path.

A table of heats of formation can be constructed by determining ΔH° for simple formation reactions which may be readily carried out, such as

$$H_2(g) \;+\; \frac{1}{2}O_2(g) \;=\; H_2O(l) \qquad \Delta H^\circ = \Delta H^f_{H_2O(l)} \tag{A}$$

$$C(c) \;+\; O_2(g) \;=\; CO_2(g) \qquad \Delta H^\circ = \Delta H^f_{CO_2(g)} \tag{B}$$

Through the use of Eq. (6-17) heats of formation of other species may be determined from measured values of ΔH° for reactions in which all but one heat of formation is known. For example, ΔH° for the reaction

$$CH_4(g) + 2O_2(g) = CO_2(g) + 2H_2O(l) \tag{C}$$

can be experimentally determined, and application of Eq. (6-17) yields

$$\Delta H^\circ = \Delta H^f_{CO_2(g)} \;+\; 2\Delta H^f_{H_2O(l)} \;-\; \Delta H^f_{CH_4(g)}$$

From previously determined values of the heats of formation of $H_2O(l)$ and $CO_2(g)$, the heat of formation of $CH_4(g)$ can be determined. If we desired to experimentally determine ΔH^f for a newly synthesized compound, we would need to find a reaction involving that compound which goes essentially to completion and involves other species for which heats of formation are known. We would also need to know that there were no competing reactions which might occur. One type of reaction which satisfies these criteria is a combustion reaction, and consequently these reactions have been thoroughly studied. In fact, $\Delta H°$ for a combustion reaction is called the heat of combustion and is often a tabulated thermochemical quantity. For example, $\Delta H°$ for reaction (C) is the heat of combustion of methane. While heats of combustion are obviously convenient for calculating combustion heat effects, it would be redundant to tabulate both the heat of formation of a substance and its heat of combustion. For fuels of undetermined composition (e.g., coal) it is only possible to determine the heat of combustion on a mass basis, and hence it is customary to report data in this manner.

Heats of formation at 298 K for several substances are given in Tables D-1 and D-2 of the Appendix. Thermochemical data for other substances can be found in the references listed in Table 12-1.

Although most handbooks tabulate heats of formation only at 298 K, in the more extensive tabulations[13] this property is available at other temperatures and can be used with Eq. (6-17) to determine $\Delta H°$ at other temperatures. The heat of formation of an element is taken to be zero at any temperature and the heat of formation of a species at the temperature T may be calculated by a three-step computational path: (1) take elements from their standard states at T to their standard states at 298 K, (2) react elements at 298 K to form one mol of the species in its standard state (ΔH_{298}^f), (3) take the species from 298 K to its standard state at T.

For the calculation of chemical heat effects at temperatures removed from 298 K, another property, the *total enthalpy*, proves useful. This quantity is defined as $(H_T° - H_{298}° + \Delta H_{298}^f)$ and is seen to be the enthalpy of a species at the temperature T relative to its elements at the reference temperature 298 K. The usefulness of the total enthalpy will be demonstrated by consideration of the general gas-phase reaction

$$aA(g) + bB(g) = cC(g) + dD(g)$$

where for the sake of generality the reactants and products will be at different temperatures. The computational path needed to calculate ΔH for this reaction is shown in Fig. 6-16. The value of ΔH for the reaction as shown is obtained by summing the enthalpy changes for each step and on rearrangement becomes[14]

$$\Delta H = c\left(H_{T_P}° - H_{T_O}° + \Delta H_{T_O}^f\right)_C + d\left(H_{T_P}° - H_{T_O}° + \Delta H_{T_O}^f\right)_D$$

$$- a\left(H_{T_A}° - H_{T_O}° + \Delta H_{T_O}^f\right)_A - b\left(H_{T_B}° - H_{T_O}° + \Delta H_{T_O}^f\right)_B$$

Further generalization casts this equation into the form of Eq. (6-17)

---

[13] For example, reference 1-D of Table 12-1.

[14] While all reaction participants are in their standard states, here the superscript ° is not used on the enthalpy change because it refers to a reaction in which the temperature remains constant.

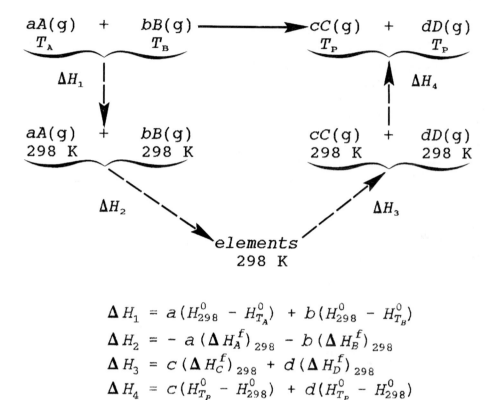

$$\Delta H_1 = a(H^0_{298} - H^0_{T_A}) + b(H^0_{298} - H^0_{T_B})$$

$$\Delta H_2 = -a(\Delta H^f_A)_{298} - b(\Delta H^f_B)_{298}$$

$$\Delta H_3 = c(\Delta H^f_C)_{298} + d(\Delta H^f_D)_{298}$$

$$\Delta H_4 = c(H^0_{T_P} - H^0_{298}) + d(H^0_{T_P} - H^0_{298})$$

$$\Delta H = \Delta H_1 + \Delta H_2 + \Delta H_3 + \Delta H_4$$

Figure 16-16

$$\Delta H = \sum_{products} n_i \left(H^\circ_{T_i} - H^\circ_{T_0} + \Delta H^f_{T_0}\right)_i - \sum_{reactants} n_j \left(H^\circ_{Tj} - H^\circ_{T_0} + \Delta H^f_{T_0}\right)_j \quad (6\text{-}18)$$

The total enthalpy will be useful for reactions at temperatures exceeding 298 K and can be calculated at any temperature for many of the species in Tables D-1 and D-2 from the spreadsheet LOGKF(T).WQ1(or.XLS). This spreadsheet also calculates ΔH^f at any temperature.[15]

---

[15] For details of this calculation see Ex. 12-1.

EXAMPLE 6-4

Use data from the spreadsheet LOGKF(T).WQ1(or .XLS) to determine the heat to be removed from the oxidation of ammonia with the stoichiometric amount of oxygen as per the reaction:
$NH_3(g) + 2O_2(g) = NO(g) + 1.5H_2O(g) + 0.75O_2(g)$
        Gaseous ammonia enters the reactor at 400 K, oxygen enters at 320 K, and products leave at 500 K.

Solution 6-4

The solution will be obtained in two ways: (a) use of the total enthalpy $(H_T^\circ - H_{298}^\circ + \Delta H_{298}^f)$ and (b) use of the heat of formation. The following data was obtained from the spreadsheet

Species	ΔH_{500}^f (cal / g mol)	$H_T^\circ - H_{298}^\circ + \Delta H_{298}^f$ (cal / g mol)		
		$T = 320$ K	$T = 400$ K	$T = 500$ K
NH_3	−11,962.3	——	−10,128.7	−9161.3
O_2	0.0	154.1	——	1468.9
NO	21,617.8	——	——	23,060.9
H_2O	−58,268.8	——	——	−56,139.4

(a) Using Eq. (6-18) with the data in the table we have

$$Q = \Delta H = 23,060.9 + 1.5(-56,139.4) + 0.75(1468.9) - (-10,128.7) - 2(154.1)$$

$$Q = -50,226 \text{ cal / g mol } NH_3$$

(b) A two-step path will be used: (1) heat reactants to 500 K, (2) carry out reaction at 500 K.

    It is easily seen that the difference of any two values of the total enthalpy, $\Delta H_T^\circ - H_{298}^\circ + \Delta H_{298}^f$, is simply the enthalpy change between the two temperatures. The enthalpy change for heating reactants to 500 K is therefore

$$\Delta H_1 = -9161.3 - (-10,128.7) + 2(1468.9 - 154.1)$$

$$\Delta H_1 = 3597$$

From Eq. (6-17) and the data of the table, $\Delta H°$ at 500 K is

$$\Delta H_2 = \Delta H° = 21,617.8 + 1.5(-58,268.8) - (-11,962.3)$$

$$\Delta H_2 = -53,823.1$$

The total enthalpy change is

$$\Delta H = \Delta H_1 + \Delta H_2 = 3597 + (-53,823) = -50,226 = Q$$

As expected, results obtained from the two paths are identical.

6-8 HEATS OF FORMATION IN SOLUTION

Often heat of solution data are reported as heats of formation of the solute in solutions of specified strengths. For example, the heat of formation of Na_2CO_3 in an aqueous solution containing 15 mols of water per mol of Na_2CO_3 refers to the following isothermal reaction:

$$2Na + C + \frac{3}{2}O_2 + 15H_2O(l) = Na_2CO_3(15H_2O)$$

where everything on the left-hand side is in the standard state of pure substance at 1 atm of pressure. The notation on the right-hand side signifies 1 mol of Na_2CO_3 in solution with 15 mol of water which is also at a pressure of 1 atm. This reaction can be considered to occur in two steps, as shown on the computational path of Fig. 6-17. This leads to

$$\Delta H^f_{Na_2CO_3(15H_2O)} = \Delta H^f_{Na_2CO_3(c)} + \Delta h_s$$

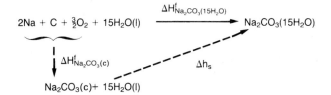

$$\Delta H^f_{Na_2CO_3(15H_2O)}$$

$2Na + C + \frac{3}{2}O_2 + 15H_2O(l) \xrightarrow{\hspace{3cm}} Na_2CO_3(15H_2O)$

$\downarrow \Delta H^f_{Na_2CO_3(c)}$

Δh_s

$Na_2CO_3(c) + 15H_2O(l)$

Figure 6-17 computational path for the heat of formation of $Na_2CO_3(15H_2O)$.

By rearrangement the heat of solution Δh_s is seen to be

$$\Delta h_s = \Delta H^f_{Na_2CO_3(15H_2O)} - \Delta H^f_{Na_2CO_3(c)}$$

Thus, tabulation of the heat of formation of the solute and the heats of formation of the solute in solutions of specified strengths is equivalent to a tabulation of heats of solution. This type of data for several compounds and solutions is presented in Table 6-2. In using this type of data, it should be noted that the heat of formation of a component in a solution does not include the heat of formation of the water in the solution.

TABLE 6-2
HEATS OF FORMATION AT 298 K[a]

Compound	ΔH^f (kcal / mol)	Compound	ΔH^f (kcal / mol)
$Na_2CO_3(c)$	−270.3	$NaOH(5H_2O)$	−111.015
$Na_2CO_3 \cdot H_2O(c)$	−341.8	$NaOH(7H_2O)$	−111.836
$Na_2CO_3(15H_2O)$	−278.13	$NaOH(10H_2O)$	−112.148
$Na_2CO_3(20H_2O)$	−277.91	$NaOH(15H_2O)$	−112.228
$Na_2CO_3(25H_2O)$	−277.72	$NaOH(20H_2O)$	−112.235
$Na_2CO_3(50H_2O)$	−277.09	$NaOH(25H_2O)$	−112.221
$Na_2CO_3(100H_2O)$	−276.57	$NaOH(50H_2O)$	−112.154
$NaOH(c)$	−101.99	$H_2O(l)$	−68.317
$NaOH(3H_2O)$	−108.894	$CO_2(g)$	−94.051

[a] Source: reference 1A in Table 12-1.

EXAMPLE 6-5

From the data of Table 6-2, determine the heat of solution of 1 mol of Na_2CO_3 in 15 mol of water.

Solution 6-5

As indicated by the computational path shown in Fig. 6-17, the heat of solution Δh_s is

$$\Delta h_s = -278.13 - (-270.3) = -7.83 \text{ kcal}$$

EXAMPLE 6-6

From the data of Table 6-2, determine the heat of solution of 1 mol of sodium carbonate monohydrate, $Na_2CO_3 \cdot H_2O(c)$, in 14 mol of water to form a solution with 15 mol of water per mol of sodium carbonate.

Solution 6-6

For the solution of this problem the computational path shown in Fig. 6-18 is used, where it is seen that

$$\Delta h_s = -\Delta H^f_{Na_2CO_3 \cdot H_2O(c)} + \Delta H^f_{H_2O(l)} + \Delta H^f_{Na_2CO_3(15H_2O)}$$

$$= -(-341.8) + (-68.317) + (-278.13)$$

$$= -4.647 \, kcal$$

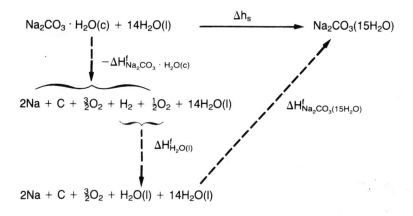

Figure 6-18

EXAMPLE 6-7

Calculate the heat effect when CO_2 is reacted with aqueous NaOH to produce a solution of 1 mol of Na_2CO_3 in 15 mol of water.

Solution 6-7

The computational path for this example is shown in Fig. 6-19, from which we write

$$\Delta H = -2\Delta H^f_{NaOH(7H_2O)} - \Delta H^f_{CO_2} + \Delta H^f_{H_2O(l)} + \Delta H^f_{Na_2CO_3(15H_2O)}$$

$$= -2(-111.836) - (-94.051) + (-68.317) + (-278.13)$$

$$= -28.724 \, kcal$$

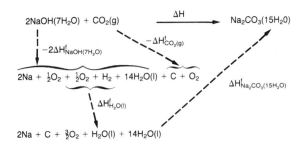

Figure 6-19

These examples show that the use of heats of formation of substances in solution facilitates somewhat the calculation of heat effects in aqueous solution. However, care must be taken in determining when the heat of formation of water must be included. In Ex. 6-6 the heat of formation of a hydrate is required. This is a definite chemical compound, and thus its heat of formation is defined in the usual manner with all elements in their naturally occurring states including the hydrogen and oxygen comprising the water of hydration. In Ex. 6-7 it was necessary to include the mol of water formed from the reaction of CO_2 with NaOH. With these points in mind no trouble should be encountered as long as the computational path is consistent.

6-9 APPLIED THERMOCHEMISTRY

We have seen how enthalpy changes can be determined for each type of possible state change which would correspond to a separate step in a computational path. Here a few examples will be worked to illustrate how computational paths are delineated and calculations executed for problems dealing with process heat effects. While it is possible to write and tempting to use *working equations* which may be applied to certain common classes of problems, we will forego this temporary convenience in favor of the more basic and unfettered approach . Equipped with an understanding of the basic phenomena and an acquired facility in delineating computational paths, you should be able to tackle any type of heat-effect problem with confidence.

EXAMPLE 6-8

Sulfuric acid is produced in a packed column fed with SO_3 and water and provided with cooling coils for heat removal. The operation is carried out at atmospheric pressure and is sketched in Fig. 6-20. A 60 wt % product is produced from 134.4 m^3(STP)/s of a gas containing 16.7%

EXAMPLE 6-8 CON'T

SO_3, 16.7% oxygen, and 66.6% nitrogen that enters at 298 K and 1 atm. Sufficient water to react with the SO_3 and produce the 60 wt % acid enters the top of the column at 290 K and flows counter to the gas. The 60 wt % acid leaves at 355.2 K (180°F). Essentially all the SO_3 is removed, and the gas exits at 320 K. Determine the rate of heat removal in the column.

Solution 6-8

The stoichiometric calculations based on an interval of 1 s and assuming ideal-gas behavior are

$$\frac{134.4}{22.4} = 6 \text{ kmol}$$

$$6 \times 0.167 = 1 \text{ kmol of } SO_3$$

$$6 \times 0.167 = 1 \text{ kmol of } O_2$$

$$6 \times 0.666 = 4 \text{ kmol of } N_2$$

For the chemical reaction

$$SO_3(g) + H_2O(l) = H_2SO_4(l)$$

1 kmol of H_2O is required. The 60 wt % solution will contain 3.63 mol of H_2O per mol of H_2SO_4, and hence the total inflow of water should be 4.63 kmol.

Application of the first law to this steady-state flow system results in

$$\Delta H = Q$$

The task of calculating ΔH is accomplished through the computational path shown in Fig. 6-21. In this figure the process represented by the heavy arrow corresponds to the change of state occurring within the column for which we require ΔH, and the computational path is represented by dashed arrows. Considering each separate step in the computational path, we have the following.

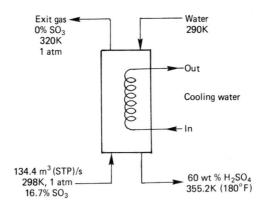

Figure 6-20

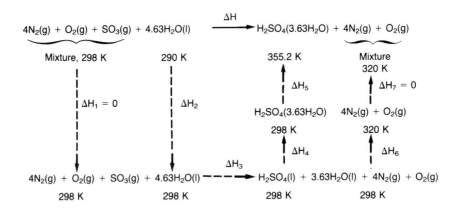

Figure 6-21

ΔH_1. Because the chemical reaction will occur between liquid water and pure gaseous SO_3 both at 1 atm, the entering gas mixture is separated into its pure components isothermally and isobarically. The gas mixture will be assumed ideal, and hence there is no enthalpy change for this step. $\Delta H_1 = 0$.

ΔH_2. The reaction and solution steps will occur at 298 K, and we must therefore heat the incoming water from 290 to 298 K. By taking enthalpies from the steam table in the Appendix, this is

$$\Delta H_2 = 4.63(18.02)(104.8 - 74.1) = 2561 \text{ kJ}$$

ΔH_3. Water and SO_3, now in their standard states, are reacted to form 1 kmol of pure $H_2SO_4(l)$:

$$\Delta H_3 = \Delta H°$$

$$\Delta H° = \Delta H^f_{H_2SO_4(l)} - \Delta H^f_{H_2O(l)} - \Delta H^f_{SO_3(g)}$$

Using heats of formation from Table D-2 results in

$$\Delta H° = -814.12 - (-286.0) - (-395.4)$$

$$= -132.72 \text{ MJ} = -132,720 \text{ kJ}$$

ΔH_4. In this step, 1 kmol of $H_2SO_4(l)$ is mixed with 3.63 kmol of $H_2O(l)$ to form the 60 wt % solution. The process is isothermal at 298 K. The heat of solution Δh_s, as read from Fig. 6-5, is $-52,000$ kJ/kmol of H_2SO_4 and is equal to ΔH_4.

ΔH_5. The 60 wt % acid is heated from 298 to 355.2 K (77 to 180°F). Rather than locate heat capacity data for aqueous sulfuric acid solutions, we will use Fig. 6-10 and obtain the following enthalpies:

Enthalpy of 60 wt % acid at 180°F: -65 Btu/lb of solution
Enthalpy of 60 wt % acid at 77°F: -121 Btu/lb of solution

A pound of solution contains

$$\frac{1(0.60)(0.454)}{98} = 2.78(10^{-3}) \text{ kmol of } H_2SO_4$$

and we find that

$$\Delta H_5 = \frac{[-65 - (-121)]1.055}{2.78(10^{-3})} = 21,252 \text{ kJ}$$

It should be noted that the enthalpy change corresponding to $\Delta H_4 + \Delta H_5$ could have been obtained from Fig. 6-10. By connecting pure water and acid, each at 77°F, with a straight line, the enthalpy of the adiabatically formed 60 wt % solution is found to be 20 Btu/lb of solution. We have already found the enthalpy of the solution at 180°F to be -65 Btu/lb of solution and write

$$\Delta H_4 + \Delta H_5 = \frac{[-65 - (20)]1.055}{2.78(10^{-3})} = -32,260 \text{ kJ}$$

When compared to the sum of the previously obtained ΔH_4 and ΔH_5, $-30,948$ kJ, the agreement is seen to be reasonable.

ΔH_6.   The 1 kmol of O_2 and 4 kmol of N_2 which have remained at 298 K until now are heated to 320 K. Using the enthalpies of Table 6-1, we obtain

$$\Delta H_6 = (1384 - 732) + 4(1368 - 728) = 3213 \text{ kJ}$$

ΔH_7.   The isothermal and isobaric mixing of oxygen and nitrogen results in zero enthalpy change.

The desired enthalpy change for the total process may now be found:

$$\Delta H = \Delta H_1 + \Delta H_2 + \Delta H_3 + \Delta H_4 + \Delta H_5 + \Delta H_6 + \Delta H_7$$

$$= 0 + 2561 - 132,720 - 52,200 + 21,252 + 3212 + 0 = -157,895 \text{ kJ}$$

$$Q = -158,000 \text{ kJ / s}$$

The calculations have been carried out using tabulated data without regard to the number of justifiable significant figures. The enthalpy change for step 5 was obtained by reading the enthalpy-concentration diagram where each enthalpy can be determined to ± 1 Btu/lb of solution, which is equivalent to an uncertainty of ± 760 kJ in ΔH_5. The most uncertainty is associated with step 5, and it therefore seems reasonable to round our answer to the nearest 1000 kJ. The resulting three significant figures is certainly of adequate precision for process design purposes.

EXAMPLE 6-9

Propane is used for drying stored grain. Propane gas at room temperature is burned with enough air so that combustion is complete and gases leave the burner at 1400 K. The combustion gas is then mixed with sufficient air so that the resulting gas mixture for drying is at 400 K. How many mols of gas are available for drying per mol of propane burned?

Solution 6-9

A sketch of the system is shown in Fig. 6-22. For this steady-state flow process, application of the first law around the dashed envelope, which includes burner and mixer, results in

$$\Delta H = 0$$

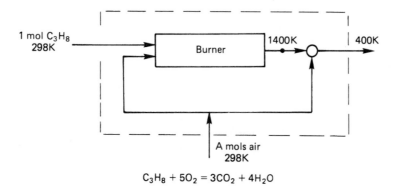

1 mol C3H8
298K
Burner
1400K
400K
A mols air
298K
C3H8 + 5O2 = 3CO2 + 4H2O

Figure 6-22

We let A represent the total kmol of air required per kmol of propane and note that complete combustion of propane requires 5 kmol of O_2:

$$C_3H_8(g) + 5O_2(g) = 3CO_2(g) + 4H_2O(g)$$

The computational path expressed in terms of the unknown A consists of two steps and is shown in Fig. 6-23.

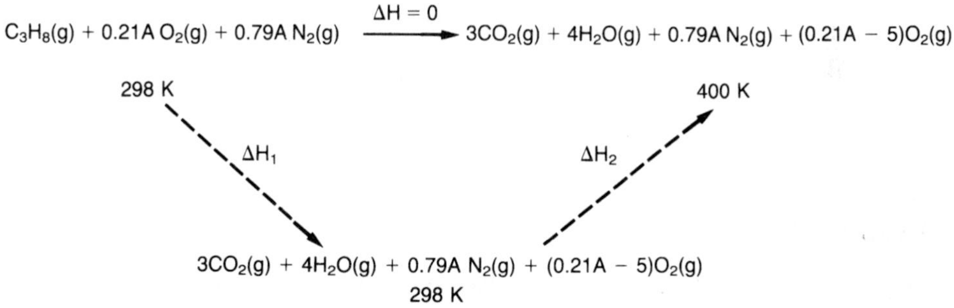

ΔH = 0
C3H8(g) + 0.21A O2(g) + 0.79A N2(g)
3CO2(g) + 4H2O(g) + 0.79A N2(g) + (0.21A − 5)O2(g)
298 K
400 K
ΔH1
ΔH2
3CO2(g) + 4H2O(g) + 0.79A N2(g) + (0.21A − 5)O2(g)
298 K

Figure 6-23

ΔH_1**.** This enthalpy change is ΔH° for the preceding reaction:

$$\Delta H^\circ = 3\Delta H^f_{CO_2(g)} + 4\Delta H^f_{H_2O(g)} - \Delta H^f_{C_3H_8(g)}$$

From Tables D-1 and D-2 of the Appendix we have

$$\Delta H^\circ = 3(-393.8) + 4(-242.0) - (-103.9)$$

$$\Delta H_1 = \Delta H^\circ = -2045.5 \text{ MJ} = -2{,}045{,}500 \text{ kJ}$$

ΔH_2.   This enthalpy change corresponds to heating the gas mixture from 298 K to 400 K. Enthalpies obtained from Table 6-1 are included in the following table:

Compound	Mol	h_{400}	h_{298}	Δh	$n\Delta h$
CO_2	3	4905	912	3993	11,979
H_2O	4	4285	838	3447	13,788
O_2	$0.21A - 5$	3754	732	3022	$635A - 15{,}110$
N_2	$0.79A$	3696	728	2968	$2345A$
	$A + 2$				$2980A + 10{,}657$

$$\Delta H_2 = \sum n\Delta h = 2980A + 10{,}657$$

For the total process we write

$$\Delta H = 0 = \Delta H_1 + \Delta H_2$$

$$0 = -2{,}045{,}500 + 2980A + 10{,}657$$

$$A = 682 \text{ mol of air}$$

The 682 mol of air supply 684 mol of gas for drying.

EXAMPLE 6-10

Use the software POLYMATH on the CD-ROM accompanying this text to generalize the combustion problem of Ex. 6-9.

Solution 6-10

Because T is specified, the solution of this problem in Ex. 6-9 is easily obtained from a single linear algebraic equation; however, if we were given A and required T for adiabatic operation we would need to solve a nonlinear equation. The solution would be messy or require a trial-and-error approach. Such a problem could be solved using the nonlinear algebraic equation feature of POLYMATH. This problem has been set up in POLYMATH; it is stored in the library as EX6-10. A facsimile of the POLYMATH worksheet follows.

POLYMATH PROBLEM EX6-10

Equations:

$f(T) = delH1 + delH2$

$delH1 = -2045.5*238.85$

$delH2 = 3*delhCO2 + 4*delhH2O + (0.21*A - 5)*delhO2 + 0.79*A*delhN2$

$delhCO2 = 5.316*(T - 298) + 1.4285*(T^2 - 298^2)/200 -0.8362*(T^3-298^3)/(3*10^5)+1.784*(T^4- 298^4)/(4*10^9)$

$delhH2O = 7.7*(T - 298) + 0.04594*(T^2 - 298^2)/200 + 0.2521*(T^3 - 298^3)/(3*10^5)- 0.8587*(T^4 - 298^4)/(4*10^9)$

$delhO2 = 6.085*(T - 298) + 0.3631*(T^2 - 298^2)/200 - 0.1709*(T^3 - 298^3)/(3*10^5) + 0.3133*(T^4 - 298^4)/(4*10^9)$

$delhN2 = 6.903*(T - 298) - 0.03753*(T^2 - 298^2)/200 + 0.193*(T^3 - 298^3)/(3*10^5)- 0.6961*(T^4 - 298^4)/(4*10^9)$

$A = 682$

$T_{min} = 350; T_{max} = 500$

For this problem there is only one nonlinear equation, the first law statement that $\Delta H = 0$ which is linear in A but nonlinear in T. In POLYMATH all nonlinear equations are written so that the expression on the right-hand side will be zero for a valid solution and the left-hand side is expressed as a function of the variable. The first listed equation is of this type and the remaining statements are auxiliary equations. The molar enthalpy changes, the 4th through the 7th equations, are simply integrations of the heat capacity equations listed in Table B-1 for the substances CO_2, H_2O, O_2, and N_2 respectively. Specification of T_{min} and T_{max} defines the search range for the numerical solution. The solution for $A = 682$ is $T = 399.6$ K.

EXAMPLE 6-11

The determination of equilibrium compositions for the partial oxidation of a sulfur-containing gas is the subject of Ex. 13-12. For a specified air to feed ratio, a pressure of 1 atm, and an exit gas temperature of 1478 K (2200°F) combustion gas compositions were calculated. These compositions and the feed gas compositions are tabulated for an air to feed ratio of 3.77:

Feed		Combustion gas	
Component	Mol %	Component	Mol %
H_2S	43.2	N_2	61.2
H_2O	17.7	H_2	2.5
NH_3	11.9	S_2	3.1
CO_2	11.4	CO_2	6.6
CH_3SH	7.1	SO_2	2.2
C_2H_5SH	0.9	CO	2.1
CH_4	0.7	H_2O	20.3
C_2H_6	0.5	H_2S	1.9
C_3H_8	0.5		
C_3H_6	4.9		
C_4H_{10}	0.2		
C_5H_{12}	1.0		

We desire to find the heat liberated in this oxidation when the feed gas enters at 333.3 K (140° F) and air enters at 298 K.

Solution 6-11

A nitrogen balance based on 1 kmol of feed gas is used to find the kmol of combustion gas G:

$$\frac{0.119}{2} + 0.79(3.77) = 0.612G$$

$$G = 4.96$$

For a steady-state flow process, application of the first law results in

$$\Delta H = Q$$

A three-step computational path shown in Fig. 6-24 will be used to calculate ΔH based on a kmol of feed gas.

ΔH_1. The feed gas is cooled from 333.3 to 298 K. For this change,

$$\Delta H_1 = (298 - 333.3)\sum n_j \bar{c}_{Pj}$$

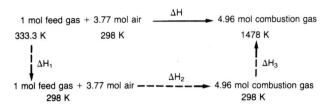

Figure 6-24

The mean heat capacities, the mol numbers, and their products are tabulated:

REACTANTS TABLE

Component	n_j	$\overline{c}_{Pj}$ [a]	$n_j \overline{c}_{Pj}$	ΔH_j^f [b]	$n_j \Delta H_j^f$
H_2S	0.432	34.1	14.73	−20.2	−8.583
H_2O	0.177	33.6	5.95	−242.0	−42.804
NH_3	0.119	36.0	4.28	−46.2	−5.497
CO_2	0.114	37.7	4.30	−393.8	−44.859
CH_3SH	0.071	51.8	3.68	−87.38	−6.204
C_2H_5SH	0.009	80.7	0.73	−110.5	−0.995
CH_4	0.007	36.3	0.25	−74.9	−0.524
C_2H_6	0.005	54.3	0.27	−84.72	−0.423
C_3H_8	0.005	76.2	0.38	−103.9	−0.519
C_3H_6	0.049	65.9	3.23	+20.4	+1.000
C_4H_{10}	0.002	102.2	0.20	−124.2	−0.249
C_5H_{12}	0.010	124.3	1.24	−146.5	−1.464
	1.000		39.3		−111.121 MJ

[a] Calculated from $H° − H_0° + \Delta H_0^f$ data of reference 7 in Table 12-1.

[b] From Tables D-1 and D-2 of the Appendix. ΔH^f for CH_3SH and C_2H_5SH from reference 2 in Table 12-1.

With $\Sigma n_j \overline{c}_{Pj} = 39.3$ we obtain

$$\Delta H_1 = 1(298 - 333.3)(39.3) = -1387 \text{ kJ}$$

ΔH_2. The reactions which actually occur need not be identified, be-cause the derivation of Eq. (6-17) is based on the artificial, but thermo-dynamically legitimate, mechanism of dissociating reactants to elements with subsequent recombination of elements to form products. Values ΔH^f at 298 K and the product $n_j \Delta H_j^f$ are tabulated in the reactants ta-ble, and $\Sigma n_j \Delta H_j^f$ is found to be $-111{,}121$ kJ. These quantities for prod-ucts are tabulated in the accompanying products table, where it is seen that $\Sigma n_i \Delta H_i^f = -398{,}053$ kJ. For the reaction step we have

$$\Delta H_2 = \Delta H^\circ = \sum n_i \, \Delta H_i^f - \sum n_j \, \Delta H_j^f$$

$$\Delta H_2 = -398{,}053 - (-111{,}121) = -286{,}932 \text{ kJ}$$

PRODUCTS TABLE

Components	n_i	$\bar{c}_{Pi}$ [a]	$n_i \bar{c}_{Pi}$	ΔH_i^f [b]	$n_i \Delta H_i^f$
N_2	3.036	31.9	96.85	0	0
H_2	0.124	30.1	3.73	0	0
S_2	0.154	36.2	5.57	+129.07	19.877
CO_2	0.327	51.3	16.78	−393.8	−128.675
SO_2	0.109	51.4	5.60	−297.1	−32.362
CO	0.104	32.3	3.36	−110.6	−11.494
H_2O	1.007	40.0	40.28	−242.0	−243.523
H_2S	0.094	43.1	4.05	−20.2	−1.876
	4.955		176.2		-398.053 MJ

[a] Calculated from $H^\circ - H_0^\circ + \Delta H_0^f$ data of reference 7 in Table 12-1.

[b] From Table D-2 of the Appendix. ΔH^f for S_2 from reference 1-D in Table 12-1.

ΔH_3. For heating the combustion gas from 298 to 1478 K,

$$\Delta H_3 = (1478 - 298) \sum n_i \, \bar{c}_{Pi}$$

From the products table, $\Sigma n_i \bar{c}_{Pi} = 176.2$, and hence

$$\Delta H_3 = (1478 - 298)176.2 = 207{,}916 \text{ kJ}$$

For the total process,

$$\Delta H = \Delta H_1 + \Delta H_2 + \Delta H_3$$

$$= -1387 - 286{,}932 + 207{,}916 = -80{,}403 \text{ kJ}$$

$$Q = -80{,}403 \text{ kJ}$$

This result suggests that this partial oxidation might be carried out profitably in a waste heat boiler. Note that no unmixing or mixing steps were included in the computational path. This is because we have tacitly assumed ideal-gas behavior and are therefore justified in omitting these steps. Also note that the heat of formation for $S_2(g)$ is not zero. At 298 K the naturally occurring form of this element is crystalline (rhombic), and heats of formation of sulfur-containing compounds at 298 K are based on this elemental state.

Example 6-12

The dehydrogenation of butene to butadiene is an endothermic reaction:

$$C_4H_8(g) = C_4H_6(g) + H_2(g).$$

This reaction is to be carried out adiabatically and at atmospheric pressure, and in order to minimize the temperature drop, the reactor feed will consist of 10 mol of steam per mol of butene. The steam is nonreactive. The feed mixture enters the reactor at 900 K and at atmospheric pressure. What will be the reactor effluent temperature when 20% of the butene has been converted?

Solution 6-12

For this steady-state adiabatic flow process the first law reduces to $\Delta H = 0$. As shown on the computational path in Fig. 6-25, there are three enthalpy changes which add to zero. ΔH_3 is a function of final temperature, which can be found by solving

$$\Delta H = 0 = \Delta H_1 + \Delta H_2 + \Delta H_3(T)$$

The following data are available:[16]

---

[16] $\bar{c}_P$ is calculated from $H° - H_0°$ + ΔH_0^f data of reference 7 in Table 12-1. ΔH^f is from Table D-1 of the Appendix.

	$\bar{c}_P$, 298 – 900 K	ΔH_{298}^f
	(kJ/kmol · K)	(kJ/kmol)
Butene	148.7	−130
Butadiene	131.2	110,200
Steam	36.6	————
Hydrogen	29.4	0

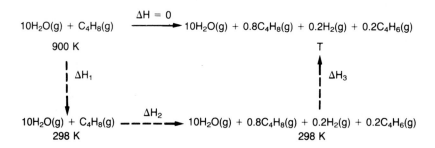

Figure 6-25

ΔH_1. For a cooling from 900 to 298 K,

$$\Delta H_1 = [10(36.6) + 148.7](298 - 900)$$

$$= -309{,}849 \text{ kJ}$$

ΔH_2. The standard heat of reaction at 298 K is

$$\Delta H^\circ = \Delta H^f_{C_4H_6(g)} - \Delta H^f_{C_4H_8(g)}$$

$$= 110{,}200 - (-130) = 110{,}330 \text{ kJ}$$

For the reaction of 0.2 kmol of C_4H_8,

$$\Delta H_2 = 0.2(110{,}330) = 22{,}066 \text{ kJ}$$

ΔH_3. For the heating from 298 K to T of 10 kmol of $H_2O(g)$, 0.8 kmol of $C_4H_8(g)$, 0.2 kmol of $H_2(g)$, and 0.2 kmol of $C_4H_6(g)$,

$$\Delta H_3 = (T - 298)[10(36.6) + 0.8(148.7) + 0.2(29.4) + 0.2(131.2)]$$

$$= 517.1(T - 298)$$

For $\Delta H = 0$, we have

$$-309,849 + 22,066 + 517.1(T - 298) = 0$$

$$T = 854.5 \text{ K}$$

The temperature drops 45.5 K through the reactor.

PROBLEMS

6-1. Hydrogen gas is fed to a burner, where it is combusted with 500% excess air. If the hydrogen and air are at room temperature and atmospheric pressure, and the burner operates adiabatically, estimate the flame temperature.

6-2. One lb-mol/h of carbon monoxide and 1 lb-mol/h of oxygen at 25°C are fed continuously into a reactor where carbon monoxide is oxidized to carbon dioxide. When heat is removed from the reactor at 70,000 Btu/h, the reaction products leave the reactor at 500°C. What is the percentage conversion of carbon monoxide?

6-3. Methane is burned with 100% excess air with liquid water injected into the combustion space so that the combustion gases leave the furnace at 1000°F. The methane, air, and liquid water all enter the furnace at 77°F, and the combustion is assumed to be complete. Calculate the mols of water injected per mol of methane burned.

6-4. Methane reacts with steam to produce hydrogen via the following reaction

$$CH_4(g) + H_2O(g) = 3H_2(g) + CO(g)$$

Methane and steam in the ratio 3 mol steam per mol methane are fed to a reactor operating at 1200 K where the methane is completely reacted. With reactants entering at 1200 K and products leaving at 1200 K, find the amount of heat that must be added or removed in the reactor per kmol of methane reacted.

6-5. Methane and air, both at room temperature and pressure, are fed to a burner which operates adiabatically and combustion is complete. If a combustion gas temperature of 1200 K is desired, what ratio of air to methane must be used?

6-6. Hydrogen is produced by reforming methane via the following endothermic reaction

$$CH_4(g) + H_2O(g) = 3H_2(g) + CO(g)$$

Methane and low-pressure steam, each at 1000 K, are fed in a 1:1 ratio to a reactor operating adiabatically. If the product stream is to leave the reactor at 800 K, what percentage of the methane should react?

6-7. The production of hydrogen from methane proceeds at high temperature by the following reaction

$$CH_4(g) + H_2O(g) = 3H_2(g) + CO(g)$$

This reaction is endothermic, but instead of supplying heat to the reactor, it is possible to add oxygen to the reactor so that the following exothermic reaction occurs

$$CH_4(g) + 2O_2(g) = CO_2(g) + 2H_2O(g)$$

If the reactor is operated at 1000 K, what ratio of oxygen to methane should be used if the net heat of reaction is to be zero?

6-8. The experimental value for the heat evolved when liquid benzene is burned to carbon dioxide and liquid water in a constant-volume bomb calorimeter at 25°C is 780,090 cal/g mol of benzene. Calculate the standard enthalpy of formation of liquid benzene at 25°C. Assume all gases behave ideally.

6-9. Calculate the heat effect for a process in which 1 lb-mol of liquid SO_3 at 77°F is reacted with water at 77°F to give a 65 wt % H_2SO_4 solution at 180°F.

6-10. The dissolution of NH_4NO_3 in water produces a cooling effect which we believe might be utilized to cool beer. If we desire to cool a 12-oz can of beer initially at 70°F to a delightfully cool temperature of 40°F, how much NH_4NO_3 and water should we use? Heat of formation at 298 K is as follows:

$NH_4NO_3(c)$:	−87.27 kcal/g mol
$NH_4NO_3(3H_2O)$:	−83.30 kcal/g mol
$(4H_2O)$:	−83.20 kcal/g mol
$(5H_2O)$:	−83.03 kcal/g mol
$(10H_2O)$:	−82.46 kcal/g mol
$(20H_2O)$:	−81.95 kcal/g mol
$(50H_2O)$:	−81.49 kcal/g mol
∞:	−81.11 kcal/g mol

The heat capacity of $NH_4NO_3(c)$ is 43.5 cal/g mol °C. The heat capacity of aqueous solutions of NH_4NO_3 is as follows:

wt.% NH_4NO_3	4.25	8.16	18.18	35.71	59.70
c_p (cal/ g °C)	0.962	0.931	0.860	0.750	0.652

The solubility of NH_4NO_3 is as follows:

t (°C)	0	25	35
wt % NH_4NO_3	54.94	68.19	72.21

6-11. Methane and air at room temperature and pressure enter a burner which operates adiabatically. If a combustion gas temperature of 2200°F is desired, what ratio of air to methane must be used?

6-12. It is proposed to manufacture hydrochloric acid (HCl) from chlorine via the following reaction:

$$Cl_2 + H_2O = 2HCl + \frac{1}{2} O_2$$

The reaction is to be carried out at 400°F with saturated steam at 400°F and chlorine gas at 400°F entering the reactor with a ratio of 3 mol of steam per mol of chlorine gas.

(a) If the reaction is 100% complete and products leave the reactor at 400°F, how much heat must be added or removed for every lb-mol of HCl obtained?

(b) The gas stream leaving the reactor is to be cooled and the steam and HCl condensed so that leaving the cooler will be an aqueous solution of HCl and a pure oxygen stream each at 77°F. How much heat must be removed in the cooler per lb-mol of HCl produced in the reactor?

6-13. $LiCl \cdot H_2O(c)$ is dissolved isothermally in enough water to form a solution containing 5 mol of water per mol of LiCl. What is the heat effect? The following enthalpies of formation are given:

$$LiCl(c): \quad -97{,}700 \text{ cal}$$

$$LiCl \cdot H_2O(c): \quad -170{,}310 \text{ cal}$$

$$LiCl(5H_2O): \quad -104{,}431 \text{ cal}$$

$$H_2O(l): \quad -68{,}317 \text{ cal}$$

6-14. One kmol of ammonia gas, $NH_3(g)$, reacts adiabatically with one kmol of nitric acid in 20 kmols of water, $HNO_3(20 \ H_2O)$, each at 298 K, to form a solution of ammonium nitrate. What is the final temperature of the solution? See problem 6-10 for ammonium nitrate data.

6-15. One kmol of sodium carbonate monohydrate, $Na_2CO_3 \cdot H_2O(c)$, reacts with aqueous HCl at 298 K. The reaction goes to completion and all the CO_2 formed in the reaction escapes the liquid phase so that in the final state there is a solution of NaCl in 10 mols of water. Calculate $\Delta H°$ for the reaction. See problem 6-17 for NaCl data.

6-16. Ammonia and air are fed to a reactor where the ammonia is oxidized. If the temperature of reactants entering the reactor is 77°F, how much heat has to be added or removed per lb-mol of NH_3 fed if the product gas has the following composition and leaves at 500°F?

H_2O:	17.40 mol %
NO:	2.90 mol %
NO_2:	5.80 mol %
N_2O:	1.45 mol %
NH_3:	2.90 mol %
O_2:	1.45 mol %
N_2:	68.20 mol %

Assume that the heat capacity of the product gas is the same as that of N_2.

6-17. What is the heat effect in calories when 6 g mol of $NaOH(5H_2O)$ solution is reacted with 5 g mol of $HCl(4H_2O)$ solution, each at 25°C, to form 11 g mol of $NaCl(10H_2O)$ solution at 25°C? The heat of formation of $NaCl(10H_2O)$ is $-97,768$ cal/g mol.

6-18. Hydrogen necessary for ammonia production is usually produced from natural gas via steam reforming:

$$CH_4 + H_2O = CO + 3H_2$$

If a reactor operating adiabatically is fed with an equimolar mixture of methane and steam and 10% conversion is obtained, calculate the temperature of the product stream leaving the reactor. Methane enters at 1 atm and 700°F, and steam enters at 100 psia and 800°F. The reactor is operated at atmospheric pressure.

6-19. Hydrogen gas at 500°F and chlorine gas at 77°F are fed to a reactor in a 1:1 mol ratio to produce HCl. The reaction is assumed to be complete. If the HCl leaving the reactor is at a temperature of 1000°F, how much heat must be added or removed from the reactor per lb-mol of HCl produced?

6-20. One lb of ice at 32°F is mixed adiabatically with 1 lb of pure $H_2SO_4(l)$ at 77°F. What is the final temperature of the solution? The heat of fusion of ice is 143 Btu/lb, and the heat capacity of ice is 0.44 Btu/lb °F.

6-21. Concentrated sulfuric acid solutions are sometimes used for drying gases. Nitrogen gas containing 2% water vapor is dried by passing it upwards through a packed column where it contacts a downward-flowing stream of concentrated sulfuric acid solution. The column is operated adiabatically. The gas enters the column at 298 K and 1 atm at a rate of 50 kmol/min. The gas leaves the column at 330 K, and the acid enters at 298 K. The acid entering the column is 90 wt %, and the effluent acid is 80 wt %. Essentially, the effluent nitrogen stream is completely dry. Find the temperature of the effluent acid stream.

6-22. Nitric acid is produced by passing an NO_2–air mixture into the bottom of a packed column where it flows upward and contacts a downward-flowing stream of water. The following reactions occur:

$$3NO_2 + H_2O = 2HNO_3 + NO$$

$$NO + \frac{1}{2}O_2 = NO_2$$

which can be combined to yield

$$2NO_2 + \frac{1}{2}O_2 + H_2O = 2HNO_3$$

A column is fed with 1 kmol/s of a gas containing 10% NO_2 and 90% air at 298 K and 1 atm. Sufficient water at 298 K is supplied to the top of the column to produce a 60 wt % HNO_3 solution. Essentially all the NO_2 is recovered as nitric acid, and leaving streams are at 298 K. Calculate the rate of heat removal expressed in kJ/s.

6-23. Hydrogen is produced by reforming methane via the following reaction:

$$CH_4 + H_2O = 3H_2 + CO$$

This reaction is endothermic and is carried out at 1200 K. Methane and steam, each at 800 K and 1 atm, are fed to the reactor in a 1:1 ratio. The reaction essentially goes to completion, and the hydrogen and carbon monoxide leave at 1200 K. How much heat must be supplied per mol of methane reacted?

6-24. From data found only in this book, calculate a value for the heat of formation of

    (a) $H_2SO_4(5H_2O)$
    (b) $HCl(9H_2O)$
    (c) $HNO_3(20H_2O)$

6-25. A process for the manufacture of ethylene oxide by the air oxidation of ethylene is based on passing the reactants over a silver catalyst at temperatures from 200 to 260°C. Suppose that a 5 vol % ethylene–95% air mixture enters the reactor at 200°C and that 50% of the ethylene is converted to ethylene oxide and 40% completely burned to carbon dioxide. How much heat must be removed from the reactor per mol of ethylene if the exit temperature of gases is not to exceed 260°C?

6-26. Determine the error which would result from using Eq. (6-7) to estimate the heat capacity of an equimolar solution of ethanol and water.

6-27. Recently considerable interest has arisen over the use of farm-produced ethanol as a fuel. Although pure anhydrous ethanol is difficult to produce, aqueous ethanol of between 160 and 190 proof can be obtained via distillation. Several farmers claim to have operated their cars and pickups with this crude fuel but report no gas mileage data. You are to estimate the gas mileage of 160-proof alcohol (80 vol %) relative to gasoline based on the following premise: The gas mileage is proportional to the lower heat of combustion (water in the gaseous state) of a gallon of fuel. Assume for purposes of calculation that gasoline is n-octane (n-C_8H_{18}) and has a density of 0.72 g/ml. Proof is two times the volume percent alcohol. The density of ethanol is 0.789 g/ml. The heat of formation of n-octane is

$$C_8H_{18}(l): \quad \Delta H^f = -59,740 \text{ cal / g mol}$$

$$C_8H_{18}(g): \quad \Delta H^f = -49,820 \text{ cal / g mol}$$

6-28. Ammonium nitrate solution is formed by contacting a gas containing NH_3 with a nitric acid solution in a packed column equipped with cooling coils. The entering nitric acid solution is at 298 K and contains 3 mols of water per mol HNO_3. The entering gas is at 298 K and contains 25% NH_3 and 75% N_2. Essentially all the NH_3 is removed and the N_2 leaves at 298 K and essentially all the HNO_3 is reacted. The resulting ammonium nitrate (NH_4NO_3) solution leaves at 298 K also. Calculate the heat which must be added or removed per mol of entering ammonia. See Prob. 6-10 for ammonium nitrate data.

6-29. One hundred (100) lb/min of 20 wt % sulfuric acid solution at 100°F are continuously mixed with 100 lb/min of an 80 wt % sulfuric acid solution at 120°F. At what rate must heat be removed from the mixer if the exit solution is to be at 80°F?

6-30. A mixture containing 50% n-butane and 50% i-butane is to be dehydrogenated to butenes in an adiabatic reactor

$$C_4H_{10} = C_4H_8 + H_2$$

Because the reaction is endothermic and we do not want the temperature to drop excessively, a diluent gas, steam, will be added to the feed. The butane—steam mixture is preheated to 800°F before entering the reactor, and we desire 10% conversion of butane with no more than 100°F temperature drop in the reactor. It is expected that the three butene isomers (1-butene, cis-2-butene, $trans$-2-butene) will be present in approximately equal amounts. Calculate the molar ratio of steam to butane for the desired reactor performance.

Use the following c_p data

butanes: 35.0 Btu/lb mol °F $H_2O(g)$: 8.44 Btu/lb mol °F

butenes: 30.0 Btu/lb mol °F   hydrogen: 7.00 Btu/lb mol °F

Assume equal extents of reaction for the butane isomers.

6-31. It is well-known that sodium metal reacts violently with water to produce NaOH and hydrogen. In an insulated container 1 g mol of sodium is added to 1 liter of water, each initially at 25°C, in such a way that the generated hydrogen escapes the container. Estimate the temperature reached by the resulting NaOH solution. List your assumptions and comment on their possible consequences.

6-32. Estimate the maximum temperature to which a tank of pure water can be heated at atmospheric pressure by the submerged combustion of butane with the stoichiometric amount of dry air. Combustion occurs below the surface of the water and combustion gases leave saturated with water vapor. How many pounds of water can be evaporated per std. cubic foot of butane burned?

6-33. For the combustion problem of Exs. 6-9 and 6-10 make the necessary calculations and construct a curve relating exit gas temperature, T, to the mols of air per mol of fuel, A. Use methane as the fuel with methane and air entering at 333 K. Is there any limit beyond which your results may not be reliable?

6-34. Rework the combustion problem of Exs. 6-9 and 6-10 for a system that loses 1000 kcal per kmol of propane. At an air rate of 682 mol/mol fuel, what will be the exit gas temperature?

6-35. Develop a POLYMATH program to calculate the adiabatic reaction temperature when a fuel of composition $C_aH_bO_c$ is combusted with A mols of air per mol of fuel. Calculate the minimum air requirement and base your calculations on using excess air so that combustion products are CO_2 and H_2O. Fuel and air will enter at 298 K. Use your program to calculate T for the combustion of one mol of liquid methanol with 100 mols of air.

Equilibrium and Stability

Our definition of the condition of equilibrium and of an equilibrium state was made within an experimental context and contained as essential elements the concepts of reproducibility and time invariance. Thermodynamic properties are defined for equilibrium states and therefore have meaning only for such states. In Chap. 5, methods of evaluating thermodynamic property changes were delineated, and many of the important relationships among these properties were developed for closed systems of unchanging composition.

In this chapter the thermodynamic property relationships will be extended to open systems or closed systems undergoing composition changes. Also, mathematical statements characterizing the equilibrium state and referred to as conditions of equilibrium will be derived. These developments require the introduction of a new thermodynamic property, an intensive property called the chemical potential. Finally, conditions will be derived which ensure that a system is in a stable equilibrium state.

7-1 CRITERIA OF EQUILIBRIUM

The Clausius inequality serves as the starting point for development of thermodynamic criteria of equilibrium. As we have seen, this inequality is an embodiment of the first and second laws. It is valid for any process occurring within a closed system for which any heat is exchanged with the surroundings at a constant temperature. It can be written for either a finite or infinitesimal change:

$$\Delta U - T\Delta S - W \leq 0 \tag{5-5}$$

$$dU - T\,dS - dW \leq 0 \tag{5-5a}$$

The delineation of the several equilibrium criteria proceeds by applying this inequality under various constraints. Each set of constraints considered corresponds to a physically realistic or commonly encountered situation.

The Isolated System. By definition, an isolated system does not exchange mass, heat, or work with its surroundings. We expect that enclosure within a well-insulated vessel of constant volume would closely approximate these conditions and for any process occurring under such constraints. We may state that

$$Q = 0$$

$$W = 0$$

$$\Delta V = 0$$

$$\Delta U = 0$$

The statement that $\Delta U = 0$ results from application of the first law with the conditions $Q = 0$ and $W = 0$. The preceding conditions applied to the Clausius inequality results in

$$\Delta S_{U,V} \geq 0 \tag{7-1}$$

Here the constraint of isolation is expressed in terms of the thermodynamic variables V and U. The interpretation of this result is that the entropy can never decrease for any process occurring in a system of constant U and V. The inequality sign refers to a spontaneous, or irreversible, process. Thus, spontaneous processes, which always move the system toward an equilibrium state, increase the entropy with the result that the entropy has its maximum value at equilibrium when no further spontaneous processes are possible.

Constant T and V. For a process occurring in a closed system capable of only PV work and constrained under conditions of constant temperature and volume, the Clausius inequality becomes

$$\Delta U - T\Delta S \leq 0$$

From the defining equation for the Helmholtz free energy [Eq. (5-1)] it is seen that the left-hand side of the preceding inequality is the isothermal Helmholtz free energy change, and therefore we may write

$$\Delta A_{T, V} \leq 0 \tag{7-2}$$

Thus, at constant temperature and volume, spontaneous processes are accompanied by a decrease in the Helmholtz free energy which reaches its minimum value at equilibrium.

Constant T and P.   In dealing with systems constrained to constant temperature and pressure, it is convenient to use the Clausius inequality expressed in terms of net work W':

$$\Delta U - T\Delta S + P\Delta V - W' \leq 0 \tag{5-6}$$

If the system exchanges no work other than that due to changes in the system volume against the surroundings at the constant pressure P, then $W' = 0$, and

$$\Delta U - T\Delta S + P\Delta V \leq 0$$

The left-hand side is the Gibbs free energy change under conditions of constant temperature and pressure as can be seen from Eq. (5-2). Thus, we may write

$$\Delta G_{T, P} \leq 0 \tag{7-3}$$

Thus, in a closed system constrained to constant temperature and pressure, spontaneous processes result in negative Gibbs free energy changes, and at equilibrium the Gibbs free energy has its minimum value.

Application of the Clausius inequality with other sets of constraints gives rise to statements similar to Eqs. (7-1)–(7-3). These are seldom used and are stated simply for the sake of completeness:

$$\Delta U_{S, V} \leq 0 \tag{7-4}$$

$$\Delta S_{H, P} \geq 0 \tag{7-5}$$

$$\Delta H_{S, P} \leq 0 \tag{7-6}$$

The statements derived so far [Eqs. (7-1)–(7-6)], which we may call criteria of spontaneity, have resulted from our focus on the process which brings the system to a state of equilibrium. In most applications we are rarely interested in verifying whether the system is in a state of equilibrium because this is generally known or assumed. Of more interest will be the thermodynamic statements which characterize the equilibrium state. From this perspective let us consider the system to be in an equilibrium state and consider infinitesimal variations away from this state. From our previous statements regarding extremal conditions, for these variations we may write

$$dS_{U,V} = 0 \tag{7-7}$$

$$dA_{T,V} = 0 \tag{7-8}$$

$$dG_{T,P} = 0 \tag{7-9}$$

$$dU_{S,V} = 0 \tag{7-10}$$

$$dS_{H,P} = 0 \tag{7-11}$$

$$dH_{S,P} = 0 \tag{7-12}$$

Equations (7-7)–(7-12) are simply differential equations stating that the change in a state property is zero for an infinitesimal change along a specified path. These statements may be called criteria of equilibrium. The coefficients in these equations will be evaluated at the initial equilibrium state, and each of these equations will be applied to the same initial state. Thus, the information extracted from any one of these equations concerns the nature of the equilibrium state regardless of the path the system is considered to take in moving from that state. This point will be illustrated in Sec. 7-3 for the case of equilibrium between two phases of a single component.

7-2 THE CHEMICAL POTENTIAL

The equilibrium criteria we have obtained so far are quite general; however, if we wish to use them to obtain information concerning equilibrium in systems capable of undergoing composition changes, it will first be necessary to relate the thermodynamic properties to composition. Let us first consider the internal energy, beginning with the fundamental equation

$$dU = T\, dS - P\, dV \tag{5-8}$$

If we are dealing with a single-phase system which experiences no composition change, its state can be specified by two variables. Therefore, Eq. (5-8) can be regarded as an equation of state relating changes in U to changes in the independent variables S and V. We may write this in functional notation as

$$U = U(S,V)$$

When a composition change is possible, we must now consider U to depend also on the number of mols of each of the C components, n_i and make the following functional state-ment:[1]

$$U = U(S, V, n_1, n_2, \ldots, n_i, \ldots, n_C)$$

From the calculus it follows that

$$dU = \left(\frac{\partial U}{\partial S}\right)_{V, n_i} dS + \left(\frac{\partial U}{\partial V}\right)_{S, n_i} dV + \sum \left(\frac{\partial U}{\partial n_i}\right)_{S, V, n_j} dn_i \qquad (7\text{-}13)$$

The first two coefficients[2] refer to conditions of constant composition—exactly the same conditions which apply to Eq. (5-8). From inspection of this equation we see that

$$\left(\frac{\partial U}{\partial S}\right)_{V, n_i} = T \qquad (7\text{-}14)$$

$$\left(\frac{\partial U}{\partial V}\right)_{S, n_i} = -P \qquad (7\text{-}15)$$

Additionally, we will simplify Eq. (7-13) by defining the chemical potential μ_i:

$$\mu_i = \left(\frac{\partial U}{\partial n_i}\right)_{S, V, n_j} \qquad (7\text{-}16)$$

Equation (7-13) now becomes

$$dU = T\, dS - P\, dV + \sum \mu_i\, dn_i \qquad (7\text{-}17)$$

If the quantity $d\,(PV)$ is added to each side of this equation, we obtain

$$dH = V\, dP + T\, dS + \sum \mu_i\, dn_i \qquad (7\text{-}18)$$

Subtracting $d\,(TS)$ from each side of Eq. (7-17) yields

$$dA = -S\, dT - P\, dV + \sum \mu_i\, dn_i \qquad (7\text{-}19)$$

Similarly, subtracting $d\,(TS)$ from each side of Eq. (7-18) results in

---

[1] Note that this statement is consistent with the phase rule which specifies that $C + 1$ variables are needed to deter-mine the intensive properties of a single-phase, C-component system. An additional variable, the quantity of the phase, would be needed to determine the extensive properties.

[2] On a partial derivative, the subscript n_i means that all n's are held constant, while the subscript n_j reads all n's except n_i are held constant.

$$dG = V \, dP - S \, dT + \sum \mu_i \, dn_i \tag{7-20}$$

Equations (7-17)–(7-20) are the various statements of the fundamental equation applicable to systems in which the number of mols of one or more components changes during a process. They can be applied to closed systems where the change is caused by a chemical reaction or open systems (e.g., a single phase) where a transfer of matter occurs. These equations are the most general forms of the fundamental equation for it is readily seen that they reduce to Eqs. (5-8)–(5-11) when no change in mol numbers occurs (dn_i's = 0).

The chemical potential, which was originally defined in terms of a partial derivative of U, can also be expressed in terms of H, A, and G. From comparison of Eqs. (7-16)–(7-20) we may write

$$\mu_i = \left(\frac{\partial U}{\partial n_i} \right)_{S, V, n_j} = \left(\frac{\partial H}{\partial n_i} \right)_{P, S, n_j}$$

$$= \left(\frac{\partial A}{\partial n_i} \right)_{T, V, n_j} = \left(\frac{\partial G}{\partial n_i} \right)_{T, P, n_j} \tag{7-21}$$

These apparently disparate statements are not conflicting definitions but can be regarded as alternative ways of determining a thermodynamic property. They all allow the determination of the chemical potential in one and the same state. With the system in some initial equilibrium state each relation specifies the path of a displacement from the equilibrium state, but each relation yields the value of the chemical potential in the initial state. Obviously, a path of constant temperature and pressure is experimentally the most convenient, and therefore the last term in Eq. (7-21) will be more meaningful physically. From its definition the chemical potential is seen to be an intensive property.

A physical interpretation of the chemical potential may be obtained when Eq. (7-17) is applied to a process. Since the equation applies only to changes between equilibrium states, such a process would be reversible, and hence $T \, dS$ is the heat effect. Thus, in comparison with the first law ($dU = dQ + dW$) we see that

$$-dW = P \, dV - \sum \mu_i \, dn_i$$

The chemical potential therefore provides a measure of the work a system is capable of when a change in mol numbers occurs (e.g., a chemical reaction or a transfer of mass).

7-3 APPLICATION OF THE EQUILIBRIUM CRITERIA

Consider two phases of a pure substance in equilibrium at a specified condition. We wish to characterize this equilibrium by the application of one or more of Eqs. (7-7)–(7-12), recognizing that the selection of a particular equation specifies only the path taken in moving away from the equilibrium state. Let us first employ Eq. (7-7):

$$dS_{U,V} = 0$$

This statement applies to the total system comprised of both phases. This will be a closed system, although each phase will be open, and we can write

$$\left(dS^\alpha + dS^\beta\right)_{U,V} = 0 \tag{7-22}$$

where the superscripts identify the phases. Equation (7-17) can be applied to each phase:

$$dU^\alpha = T^\alpha\, dS^\alpha - P^\alpha\, dV^\alpha + \mu^\alpha\, dn^\alpha \tag{7-23}$$

$$dU^\beta = T^\beta\, dS^\beta - P^\beta\, dV^\beta + \mu^\beta\, dn^\beta \tag{7-24}$$

In employing Eq. (7-7), we must visualize the system as moving from the initial equilibrium state along a path prescribed by constant U and V. Thus,

$$dU = dU^\alpha + dU^\beta = 0 \tag{7-25}$$

$$dV = dV^\alpha + dV^\beta = 0 \tag{7-26}$$

and because the system is closed,

$$dn = dn^\alpha + dn^\beta = 0 \tag{7-27}$$

If Eqs. (7-23) and (7-24) are solved for dS^α and dS^β and substituted into Eq. (7-22), on applying the conditions of Eqs. (7-25)–(7-27) we obtain

$$dS = 0 = \left(\frac{1}{T^\alpha} - \frac{1}{T^\beta}\right)dU^\alpha + \left(\frac{P^\alpha}{T^\alpha} - \frac{P^\beta}{T^\beta}\right)dV^\alpha - \left(\frac{\mu^\alpha}{T^\alpha} - \frac{\mu^\beta}{T^\beta}\right)dn^\alpha \tag{7-28}$$

Because dS must equal zero for any changes of the independent variables U^α, V^α, and n^α, the coefficients of each of these differentials must be zero. Thus, we may state that

$$T^\alpha = T^\beta \tag{7-29}$$

$$P^\alpha = P^\beta \tag{7-30}$$

$$\mu^\alpha = \mu^\beta \tag{7-31}$$

These three conditions characterize the initial equilibrium state and apply to the equilibrium state regardless of the criterion employed with its prescribed path away from the equilibrium state. This will be shown by using Eq. (7-12):

$$dH_{S,P} = 0$$

As before,

$$\left(dH^\alpha + dH^\beta\right)_{S,P} = 0 \tag{7-32}$$

and

$$dn^\alpha + dn^\beta = 0 \tag{7-33}$$

We now apply Eq. (7-18) to each phase with the intensive property P remaining constant $dP = 0$:

$$dH^\alpha = T^\alpha dS^\alpha + \mu^\alpha dn^\alpha \tag{7-34}$$

$$dH^\beta = T^\alpha dS^\beta + \mu^\beta dn^\beta \tag{7-35}$$

The variation away from the initial equilibrium state is constrained to constant entropy, and

$$dS = 0 = dS^\alpha + dS^\beta \tag{7-36}$$

Combination of Eqs. (7-12) and (7-32)–(7-36) results in

$$dH = 0 = \left(T^\alpha - T^\beta\right)dS^\alpha + \left(\mu^\alpha - \mu^\beta\right)dn^\alpha \tag{7-37}$$

Because dS^α and dn^α are independent variations, Eq. (7-37) is satisfied only when Eqs. (7-29) and (7-31) hold. Since the process is constrained to constant pressure, the condition of Eq. (7-30) is, of course, implied.

Because of the convenience of experimentally imposing the constraints of constant temperature and pressure, the Gibbs free energy criterion [Eq. (7-9)] is usually thought to be most valuable. However, because it is not required that the variation from the equilibrium state actually be carried out, this is not an impelling reason to select a criterion. In fact, in a few simple steps paralleling those preceding, it can be easily shown that only Eq. (7-31) results when the Gibbs free energy criterion is used. Thus, in terms of supplying explicit information characterizing the equilibrium state, for this system the criterion $dS_{U,V} = 0$ is to be preferred.

The result just obtained for equilibrium between two phases of a single component can be generalized to include any number of components and phases. Proceeding along the lines of the previous derivation, we write

$$\left(\sum_j dS^j \right)_{U,V} = 0 \tag{7-38}$$

where the superscript j indexes the phases and the summation is over all phases. For each phase,

$$dU^j = T^j \, dS^j - P^j \, dV^j + \sum_i \mu_i^j \, dn_i^j \tag{7-39}$$

or

$$dS^j = \frac{dU^j}{T^j} + \frac{P^j}{T^j} dV^j - \sum_i \frac{\mu_i^j}{T^j} dn_i^j \tag{7-39a}$$

where μ_i^j is the chemical potential and dn_i^j is the change in mol number of the ith component in the jth phase. The constraints of isolation are

$$dU = \sum_j dU^j = 0 \tag{7-40}$$

$$dV = \sum_j dV^j = 0 \tag{7-41}$$

$$dn_i = \sum_j dn_i^j = 0 \qquad (i = 1 \; to \; C) \tag{7-42}$$

Substitution of Eq. (7-39a) into Eq. (7-38) produces

$$0 = \sum_j \frac{dU^j}{T^j} + \sum_j \frac{P^j}{T^j} dV^j - \sum_j \sum_i \frac{\mu_i^j}{T^j} dn_i^j \tag{7-43}$$

If all the variables, U's, V's, and n's, were independent, the equation would be satisfied when each differential coefficient was zero. However, because of the constraints stated in Eqs. (7-40)–(7-42), all the variations are not independent, and this is an example of a constrained extremum which can be treated by the use of Lagrangian multipliers. Equation (7-40) is multiplied by λ, Eq. (7-41) is multiplied by π, and Eqs. (7-42) are each multiplied by a ρ_i. The resulting equations are added to Eq. (7-43) to give

$$0 = \sum_j \left(\frac{1}{T^j} + \lambda \right) dU^j + \sum_j \left(\frac{P^j}{T^j} + \pi \right) dV^j - \sum_i \sum_j \left(\frac{\mu_i^j}{T^j} + \rho_i \right) dn_i^j \qquad (7\text{-}44)$$

All variations may now be considered independent, and Eq. (7-44) is satisfied when the differential coefficients are zero. This leads to

$$T^j = -\frac{1}{\lambda} \qquad (7\text{-}45)$$

$$P^j = -\pi \, T^j = \frac{\pi}{\lambda} \qquad (7\text{-}46)$$

$$\mu_i^j = -\rho_i \, T^j = \frac{\rho_i}{\lambda} \qquad (7\text{-}47)$$

Although undetermined, λ and π are constant, and Eqs. (7-45) and (7-46) state that at equilibrium the temperature and pressure each have the same value in each phase. Because ρ_i depends only on i, Eq. (7-47) states that at equilibrium the chemical potential of each component is equal in each phase.

7-4 THE ESSENCE OF THERMODYNAMICS

All the basic thermodynamic tools have now been developed. Before going on to the major chemical engineering applications, it is well to step back and put into perspective the various laws, functions, and relationships we have thus far encountered. Figure 7-1 may be helpful in doing this; it illustrates both the conceptual development of the thermodynamic tools and the application of these tools for the solution of practical problems. The sequence of rectangles connected by heavy arrows traces the development of the laws, functions, and relationships that constitute the tools of thermodynamics. The sequence of circles connected by light arrows shows the steps in applying thermodynamics to practical problems, and the dashed arrows indicate where the various thermodynamic tools are employed in the problem-solving process.

Conceptual Development. Each law was fashioned in mathematical language from statements of experience and served to define a function: U and S. These functions, which were previously unknown, were shown to be state properties. Combination of the first and second law statements resulted in the Clausius inequality and subsequently the fundamental equation. Application of the methods of calculus to this equation yielded the network of relationships among the state properties. Thus, the laws which define U and S lead eventually to the prescriptions for their evaluation from experimental data.

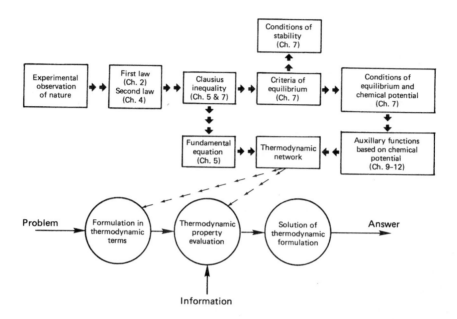

Figure 7-1   Road map for the conceptual development and application of thermodynamics.

Application of the Clausius inequality produced criteria of equilibrium for systems under various constraints. These criteria subsequently led to the conditions of equilibrium requiring uniformity of temperature, pressure, and chemical potential throughout the system. The latter quantity was shown to be a well-defined thermodynamic property essential to the description of systems that change size or composition. It, and other functions derived from it, forms the basis of the thermodynamic treatment of phase and chemical equilibrium. Through the application of the methods of calculus to these functions (as will be shown in later chapters), the number of relationships forming the thermodynamic network is increased. This expanded network, which also includes the first and second laws, contains all the tools for problem solving and in Fig. 7-1 is shown to be the link between the conceptual development of thermodynamics and its applications.

Application of Thermodynamic Tools.   We have already seen how problems are formulated in terms of the first and second laws and have observed that the solution usually depends on knowing the values of property changes. This procedure is formalized as a three-step process in Fig. 7-1. For most of the problems we have worked, the thermodynamic property evaluation consisted of reading from tables or graphs or performing relatively simple calculations involving ideal gases of constant heat capacity. While convenient, this rather routine step tends to obscure the fact that this information was obtained by processing experimental data.

The thoughtful reader will observe that the problem-solving process such as shown in Fig. 7-1 is not unique to thermodynamics but could also apply to other subjects such as fluid mechanics. While this is true, there are two features which distinguish thermodynamic problems: the nature of the variables and the manner in which they are evaluated. While the problem is formulated in terms of thermodynamic variables, little physical significance can be associated with these variables, and we are seldom interested in knowing their values per se. Moreover, their evaluation is often complicated and requires multistep paths (which may include hypothetical steps) and several types of data. As we saw in Chap. 6 especially, this sometimes complicated property-evaluation process is designed to obtain the maximum use of experimental information. Therefore, the applications of thermodynamics may be characterized by an emphasis on information processing.

7-5 STABILITY

The results just obtained regarding the conditions of equilibrium were obtained by applying Eq. (7-7). While this equation was written for a variation from the equilibrium state where the entropy was shown to be a maximum, it should be recognized that it could also apply if instead the entropy were a minimum. To ensure that we have a maximum, and hence a stable equilibrium state, it is necessary that the inequality of Eq. (7-1) be satisfied. The same will be true of the other criteria of equilibrium [Eqs. (7-8)–(7-12)] and their corresponding inequalities [Eqs. (7-2)–(7-6)]. In testing for conformance to these inequalities, we will discover additional information about the system which ensures stable equilibrium. These statements are referred to as stability conditions. We will first employ Eq. (7-4) to obtain the conditions of mechanical and thermal stability and later use Eq. (7-3) to investigate stability with regard to phase equilibrium.

Because we are considering systems in an initial equilibrium state, the inequality signs of Eqs. (7-1)–(7-6) should be reversed when considering changes away from the initial state to another equilibrium state. This type of variation is termed a *virtual variation* and must be conceptualized in terms of constraints imposed on the system. In Sec. 7-6 it will be shown that a system will move spontaneously from a more constrained to a less constrained equilibrium state. These changes are visualized as occurring when devices such as partitions or adiabatic walls dividing the original system into separate parts are removed. A virtual variation is therefore a change from a less constrained equilibrium state to a more constrained equilibrium state and may be visualized as being accomplished through the agency of devices such as partitions and adiabatic walls.

As our system, let us consider a pure fluid originally at equilibrium and constrained to constant volume and entropy.[3] We now divide the system into two parts, designated A and B,

---

[3] The condition of constant entropy may be maintained by the addition or removal of heat.

by means of a suitable partition, and we alter the volume and entropy of the separate parts. These variations are small and are denoted δV^A, δV^B, δS^A, and δS^B; however, they will be subject to constraints of constant total volume and entropy so that

$$\delta V^A + \delta V^B = 0 \tag{7-48}$$

$$\delta S^A + \delta S^B = 0 \tag{7-49}$$

Because variations are small, a Taylor series expansion will be used to obtain the corresponding change in internal energy. For part A we write

$$\delta U^A = \left(\frac{\partial U}{\partial V}\right)_S^A \delta V^A + \left(\frac{\partial U}{\partial S}\right)_V^A \delta S^A + \frac{1}{2}\left(\frac{\partial^2 U}{\partial V^2}\right)_S^A \left(\delta V^A\right)^2$$

$$+ \frac{1}{2}\left(\frac{\partial^2 U}{\partial S^2}\right)_V^A \left(\delta S^A\right)^2 + \left(\frac{\partial^2 U}{\partial V \partial S}\right)^A \left(\delta V^A\right)\left(\delta S^A\right) + \cdots \tag{7-50}$$

Because we are considering small variations from the original equilibrium state, the series expansion is truncated after second-order terms. From an examination of the fundamental equation [Eq. (5-8)], the following substitutions can be made in Eq. (7-50):

$$\left(\frac{\partial U}{\partial V}\right)_S^A = -P^A \tag{7-51}$$

$$\left(\frac{\partial U}{\partial S}\right)_V^A = T^A \tag{7-52}$$

$$\left(\frac{\partial^2 U}{\partial V^2}\right)_S^A = \frac{\partial}{\partial V}\left(\frac{\partial U}{\partial V}\right)_S^A = -\left(\frac{\partial P}{\partial V}\right)_S^A \tag{7-53}$$

$$\left(\frac{\partial^2 U}{\partial S^2}\right)_V^A = \frac{\partial}{\partial S}\left(\frac{\partial U}{\partial S}\right)_V^A = \left(\frac{\partial T}{\partial S}\right)_V^A \tag{7-54}$$

$$\left(\frac{\partial^2 U}{\partial V \partial S}\right)^A = \frac{\partial}{\partial S}\left(\frac{\partial U}{\partial V}\right)_S^A = -\left(\frac{\partial P}{\partial S}\right)_V^A \tag{7-55}$$

For what follows it will be convenient to express the last three of these relations in terms of intensive variables. Because $V = nv$ and $S = ns$, we may write

$$\left(\frac{\partial^2 U}{\partial V^2}\right)_S^A = -\frac{1}{n^A}\left(\frac{\partial P}{\partial v}\right)_S^A \tag{7-53a}$$

$$\left(\frac{\partial^2 U}{\partial S^2}\right)_V^A = \frac{1}{n^A}\left(\frac{\partial T}{\partial s}\right)_V^A \tag{7-54a}$$

and

$$\left(\frac{\partial^2 U}{\partial V \partial S}\right)^A = -\frac{1}{n^A}\left(\frac{\partial P}{\partial s}\right)_V^A \tag{7-55a}$$

From Eq. (5-28) it is seen that Eq. (7-54a) can be further simplified to

$$\left(\frac{\partial^2 U}{\partial S^2}\right)_V^A = \frac{T^A}{n^A c_V^A} \tag{7-54b}$$

Using these simplifications and writing an identical expansion for δU^B, we obtain

$$\delta U_{V,S} = \delta U^A + \delta U^B = T\left(\delta S^A + \delta S^B\right) - P\left(\delta V^A + \delta V^B\right)$$

$$-\frac{1}{2}\left(\frac{1}{n^A}+\frac{1}{n^B}\right)\left(\frac{\partial P}{\partial v}\right)_S \left(\delta V^A\right)^2 + \frac{1}{2}\left(\frac{1}{n^A}+\frac{1}{n^B}\right)\frac{T}{c_V}\left(\delta S^A\right)^2 \tag{7-56}$$

$$-\left(\frac{1}{n^A}+\frac{1}{n^B}\right)\left(\frac{\partial P}{\partial s}\right)_V \delta V^A \, \delta S^A$$

The coefficients in the expansions are evaluated for the initial state where all intensive properties of parts A and B are identical. Additional simplification of the first two second-order terms was obtained from the relations $(\delta V^B)^2 = (\delta V^A)^2$ and $(\delta S^B)^2 = (\delta S^A)^2$ which result from Eqs. (7-48) and (7-49). Also, from Eqs. (7-48) and (7-49) it is seen that the first-order terms are zero. For this small, but finite, variation from the original equilibrium state it is required that $\delta U_{V,S} > 0$. To ensure this, it is required that the remaining terms in the expansion be greater than zero:

$$-\frac{1}{2}\left(\frac{1}{n^A}+\frac{1}{n^B}\right)\left[\left(\frac{\partial P}{\partial v}\right)_S (\delta V^A)^2 - \frac{T}{c_V}(\delta S^A)^2 + 2\left(\frac{\partial P}{\partial s}\right)_V \delta V^A \, \delta S^A\right] > 0 \qquad (5\text{-}57)$$

or

$$\left(\frac{\partial P}{\partial v}\right)_S (\delta V^A)^2 - \frac{T}{c_V}(\delta S^A)^2 + 2\left(\frac{\partial P}{\partial s}\right)_V \delta V^A \, \delta S^A < 0 \qquad (5\text{-}58)$$

Now V^A and S^A are independent variables, and the preceding inequality must hold for all possible values of V^A and S^A. A variation in V^A when S^A remains constant requires that

$$\left(\frac{\partial P}{\partial v}\right)_S < 0 \qquad (7\text{-}59)$$

Similarly, a variation in S^A at a constant V^A requires that

$$\frac{T}{c_V} > 0$$

or

$$c_V > 0 \qquad (7\text{-}60)$$

Other related conditions of stability result from the application of the other equilibrium criteria. Using the enthalpy criterion [Eqs. (7-6) and (7-12)] and the same procedure as before, one obtains the additional condition

$$c_P > 0 \qquad (7\text{-}61)$$

while the Helmholtz free energy criterion [Eqs. (7-2) and (7-8)] yields

$$\left(\frac{\partial v}{\partial P}\right)_T < 0 \qquad (7\text{-}62)$$

The stability conditions, Eqs. (7-59)–(7-62), are requirements placed on the type of systems we have studied (those capable only of PV work) which ensure that stable equilibrium states exist. There is certainly nothing startling about these apparently obvious results. In fact, they might be deduced intuitively since they simply state that an external forcing produces a response in the system which opposes that forcing. Thus, the heat which flows into a system by virtue of an imposed higher temperature causes a rise in the system temperature (for $c_P > 0$), thereby lessening the potential for further heat flow and permitting the system to move toward a condition of thermal equilibrium.

In considering systems which undergo a change in mol numbers, it is more convenient to work with the Gibbs free energy criterion. The system is a single-phase mixture of components 1 and 2 constrained to constant temperature and pressure. Initial equilibrium is presumed, and a variation is devised whereby the system is divided into two parts, again designated A and B, and a transfer of mass is made to occur between the separate parts.[4] The total system is closed, and therefore

$$dn_1^A + dn_1^B = 0 \qquad (7\text{-}63)$$

$$dn_2^A + dn_2^B = 0 \qquad (7\text{-}64)$$

As before, a Taylor series expansion will be used to represent the property change for each phase. They will be combined to express the total Gibbs free energy change, which must then satisfy

$$\delta G_{T,P} > 0 \qquad (7\text{-}65)$$

For $\delta G_{T,P}$, on omitting terms higher than second order, we have

$$\delta G_{T,P} = \mu_1\left(dn_1^A + dn_1^B\right) + \mu_2\left(dn_2^A + dn_2^B\right)$$

$$+\frac{1}{2}\left[\left(\frac{\partial^2 G}{\partial n_1^2}\right)^A_{T,P,n_2} + \left(\frac{\partial^2 G}{\partial n_1^2}\right)^B_{T,P,n_2}\right]\left(\delta n_1^A\right)^2$$

$$+\frac{1}{2}\left[\left(\frac{\partial^2 G}{\partial n_2^2}\right)^A_{T,P,n_1} + \left(\frac{\partial^2 G}{\partial n_2^2}\right)^B_{T,P,n_1}\right]\left(\delta n_2^A\right)^2 \qquad (7\text{-}66)$$

$$+\left[\left(\frac{\partial^2 G}{\partial n_1\,\partial n_2}\right)^A_{T,P} + \left(\frac{\partial^2 G}{\partial n_1\,\partial n_2}\right)^B_{T,P}\right]\left(\delta n_1^A\right)\left(\delta n_2^A\right)$$

All intensive properties of parts A and B are equal, and the two first-order terms are zero by virtue of Eqs. (7-63) and (7-64). To satisfy Eq. (7-65), the sum of second-order terms in Eq. (7-66) must be positive; however, by considering the special case of a variation in n_1^A with n_2^A held constant, it will be necessary to include only the first second-order term to obtain the desired stability condition. Thus we must have

$$\left(\frac{\partial^2 G}{\partial n_1^2}\right)^A_{T,P,n_2} + \left(\frac{\partial^2 G}{\partial n_1^2}\right)^B_{T,P,n_2} > 0 \qquad (7\text{-}67)$$

---

[4] For example, in an electrolyte solution, a concentration cell can be established by imposing a voltage.

or

$$\left(\frac{\partial \mu_1}{\partial n_1}\right)_{T,P,n_2}^{A} + \left(\frac{\partial \mu_1}{\partial n_1}\right)_{T,P,n_2}^{B} > 0 \qquad (7\text{-}67a)$$

These derivatives may be expressed in terms of intensive variables by utilizing

$$x_1 = \frac{n_1}{n_1 + n_2} \qquad (7\text{-}68)$$

and for constant n_2 writing

$$dn_1 = \frac{n_1 + n_2}{x_2} dx_1 \qquad (7\text{-}69)$$

Equation (7-67a) now becomes

$$\frac{x_2^A}{n_1^A + n_2^A}\left(\frac{\partial \mu_1}{\partial x_1}\right)_{T,P}^{A} + \frac{x_2^B}{n_1^B + n_2^B}\left(\frac{\partial \mu_1}{\partial x_1}\right)_{T,P}^{B} > 0 \qquad (7\text{-}70)$$

and because the intensive properties of parts A and B are identical, we write

$$\left(\frac{\partial \mu_1}{\partial x_1}\right)_{T,P}\left(\frac{x_2^A}{n_1^A + n_2^A} + \frac{x_2^B}{n_1^B + n_2^B}\right) > 0 \qquad (7\text{-}71)$$

Because the second term in parentheses will always be positive, the stability condition is

$$\left(\frac{\partial \mu_1}{\partial x_1}\right)_{T,P} > 0 \qquad (7\text{-}72)$$

To reduce the mathematical complexity, we have considered only a binary system; however, it should be obvious that the derivation could have been broadened to include any number of components. The stability condition stated in Eq. (7-72) therefore would apply to each and every component in the mixture. Because diffusion occurs from a higher to a lower chemical potential, the mass flux entering a system by virtue of exposure to a medium of higher chemical potential will increase the mol fraction of a component and hence its chemical potential. Thus, compositional differences, which lead to chemical potential differences, give rise to diffusion, which levels the differences and leads to equilibrium.

7-6 Constraints, Equilibrium, and Virtual Variations

In applying the Clausius inequality, we found that under various sets of constraints the sign of a thermodynamic property change accompanying a spontaneous process was specified. Also, in determining the conditions of equilibrium and stability, property changes for processes originating from an equilibrium state were specified. Because a thermodynamic property has meaning only for an equilibrium state, these specified property changes must refer to a process which connects two equilibrium states. This leads us to inquire as to how processes between two equilibrium states are to be carried out. For example, if our system is a pure, nonreacting gas under conditions of isolation where the volume and internal energy can be used to specify an equilibrium state, what type of changes do we visualize occurring at constant volume and internal energy?

The answer is that such processes must result in a change in the number of constraints on the system. As an example, consider a gas confined in a cylinder fitted with a free-floating piston which divides the gas into two parts. The piston is initially locked into position, and the pressure of the gas on either side may not be the same. The volume of the cylinder remains constant during any process, and the cylinder is well insulated—conditions which define an isolated system. We now consider the spontaneous process which follows the unlocking of the piston. The pressures will equalize, and so will the temperatures if heat may be transferred across the dividing piston. The system has moved from a more constrained equilibrium state to a less constrained equilibrium state. Because the process occurs between two equilibrium states, an entropy change has meaning, and the Clausius inequality requires it to be positive.

As another example, consider an ionic reaction occurring in an aqueous solution contained in an electrochemical cell. The imposition of a definite voltage across the cell will result in a definite equilibrium state for the system. When the applied voltage is removed, the system will spontaneously move to a new equilibrium state characterized by different compositions. If the temperature and pressure were maintained constant, the Clausius inequality would require $\Delta G < 0$ for this process.

Yet another example would be the introduction of a spark or catalyst to a mixture of hydrogen gas, oxygen gas, and water vapor in a well-insulated container of constant volume. Initially an equilibrium state exists because of the large activation energy which effectively prevents the reaction and acts as a constraint on the system. The spark or catalyst removes the constraint by lowering the activation energy. This process is known to be spontaneous and results in a positive entropy change.

In application of the Clausius inequality to obtain Eqs. (7-1)–(7-6), we consider spontaneous processes which move the system to a less constrained state. As can be seen from the examples, these are changes which can be easily visualized and could be closely approximated in a well-designed experiment.

Processes which move a system to a more constrained equilibrium state are called virtual processes or virtual variations. Such variations from the least constrained equilibrium state[5] are represented by the various forms of the fundamental equation. In Secs. 7-3 and 7-5 virtual variations were employed to determine the conditions which characterize the least constrained equilibrium state.

As the name implies, virtual processes may not correspond to a readily executable experiment but may rather be conceived as occurring in a "thought experiment." Nevertheless, it is only necessary to know that such processes could be made to occur, and we need not furnish details of their execution. A virtual variation is simply regarded as the reverse of a process that brought the system to the least constrained equilibrium state. If needed, any thermodynamic property changes could be determined as the negative of those representing the original spontaneous process.

Because the coefficients in the differential or difference equations that represent virtual variations [e.g., Eqs. (7-28), (7-37), (7-44), and (7-56) and all forms of the fundamental equation employed in Chap. 5] are evaluated at the least constrained equilibrium state, specification of the exact nature of the final state is not required. It is only required that the virtual variation be made to occur under the appropriate constraints. For example, we may consider the restoration of the gas separated by the free-floating piston to occur by somehow returning the piston to its original position. This, of course, would require work, but if heat is removed, it should be possible, at least in principle, to maintain constant internal energy throughout the return path. Such a virtual variation could therefore be visualized as occurring at constant U and V, and we would therefore expect the entropy to decrease.[6]

PROBLEMS

7-1. Show that

$$C_P \geq C_V$$

and

$$\beta_T \geq \beta_S$$

where

---

[5] The least constrained equilibrium state should not be difficult to recognize as it will correspond to the least number of specified variables. The phase rule will usually provide the necessary guidance for this determination.

[6] Note that an isolated system is characterized by constant U and V but that conditions of constant U and V are not restricted to isolated systems.

$$\beta_T = -\frac{1}{V}\left(\frac{\partial V}{\partial P}\right)_T \text{ and } \beta_S = -\frac{1}{V}\left(\frac{\partial V}{\partial P}\right)_S$$

7-2. An isotherm for the van der Waals equation in the two-phase region is shown in Fig. 3-4. Show that roots lying on segment AB are unstable.

7-3. For the van der Waals isotherm shown in Fig. 3-4, show that the saturation pressure can be determined by locating the horizontal, two-phase segment of the isotherm so that two equal areas are enclosed between it and the van der Waals curve.

7-4. For an ideal gas, the internal energy depends only on temperature. For any system described by $\mathscr{F}$, r, and T, what information results from the statement $U = U(T)$?

7-5. Using the condition $U = U(T)$ and the stability conditions, can an equation of state for a gas be obtained? Is it unique?

7-6. Determine the conditions for thermal and mechanical stability in terms of the generalized fundamental equation [Eq. (5-60)].

7-7. Starting with the Clausius inequality show that for a spontaneous process

    (a) $\Delta U_{S,V} \le 0$

    (b) $\Delta S_{H,P} \ge 0$

    (c) $\Delta H_{S,P} \le 0$

Write the corresponding equilibrium criteria and apply them to a single-component, two-phase system in equilibrium to obtain conditions of equilibrium.

Thermodynamics of Pure Substances

8-1 THE PHASE DIAGRAM

Phase equilibrium information for a pure substance is best displayed on pressure-temperature coordinates and is referred to as a phase diagram. Such a diagram for sulfur is shown in Fig. 8-1.

The areas labeled S_R, S_M, S_l and S_g are regions where a single phase exists. They are, respectively, the two crystalline forms, rhombic and monoclinic, and liquid and vapor. Both temperature and pressure must be specified to determine the state of pure sulfur in one of these regions as is consistent with the phase rule.

Separating the single-phase areas are lines which represent the state of two coexisting phases of pure sulfur. The solid lines represent equilibrium states, and the dashed lines represent metastable states which are reproducible but not necessarily time invariant. Line AO represents equilibrium between rhombic sulfur and its vapor and is thus the vapor pressure curve for rhombic sulfur. Because the transition from the rhombic to the monoclinic crystalline form occurs slowly, it is possible to obtain the vapor pressure of rhombic sulfur at temperatures higher than O. The dashed line Ob represents this metastable state. Line OB represents equilibrium between the monoclinic crystalline form and vapor, and line BE represents equilibrium between liquid and vapor with point E as the critical temperature above which no liquid can exist. Similarly, line OC represents equilibrium between the rhombic and monoclinic crystalline forms, and line BC represents equilibrium between crystalline monoclinic sulfur and liquid. Subcooling of the liquid is possible and is represented by the dashed line Bb, while Oa represents metastable states of subcooled monoclinic crystals. Again, consistent with the dictates of the phase rule, we see that when two phases of a pure substance are in equilibrium the state of the system is specified by a single intensive variable. For example, a system comprised of liquid and vapor is determined when the specified temperature locates the equilibrium point on the line BE.

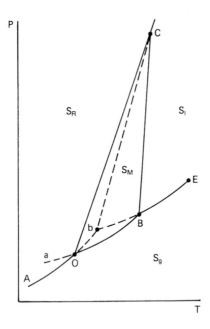

Figure 8-1 Phase diagram for sulfur.

On this particular phase diagram there are three places where two-phase coexistence lines intersect: O, B, and C. They represent equilibrium between three coexisting phases and are called triple points. At O rhombic crystals, monoclinic crystals, and vapor coexist; at B monoclinic crystals, liquid, and vapor coexist; and at C rhombic crystals, monoclinic crystals, and liquid coexist. While the phase rule demands that triple points are invariant states, it places no limit on the number of such states. It merely states that the experimenter can specify no intensive variables in preparing a three-phase system in equilibrium.

This phase diagram, while sketched simply here for instructional purposes, must be determined from experimental measurements. Vapor pressure measurements can be used to construct the liquid-vapor and solid-vapor lines, while observation of the melting temperature and the crystal-crystal transition temperature at various pressures is required to construct the solid-liquid and solid-solid lines.

A greater physical appreciation for the vapor pressure-temperature relationship may be obtained by a brief review of the experimental methods of obtaining such data. Two general methods exist: the static method and the dynamic method. In the static method a pressure is measured at a known temperature, while in the dynamic method a temperature is measured at a known pressure.

An isoteniscope, pictured in Fig. 8-2a, is the device used to determine vapor pressure by the static method. The bulb A is filled about two-thirds and the U-tube B to about one-

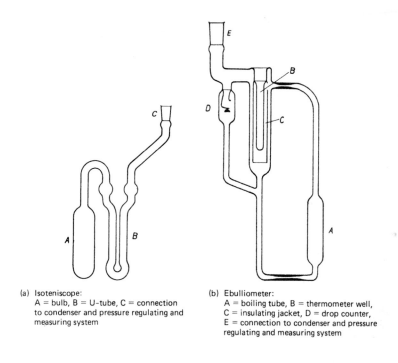

(a) Isoteniscope:
A = bulb, B = U-tube, C = connection
to condenser and pressure regulating and
measuring system

(b) Ebulliometer:
A = boiling tube, B = thermometer well,
C = insulating jacket, D = drop counter,
E = connection to condenser and pressure
regulating and measuring system

Figure 8-2 Devices for determination of vapor pressure of a liquid. [Reprinted from Hála, J. Pick, V. Fried, and O. Vilím, *Vapor-Liquid Equilibrium*, 2nd ed., 1967, with permission of Pergamon Press Ltd., Elmsford, NY.]

third of its height with the liquid to be tested. The isoteniscope is immersed in a thermostated bath of the desired temperature, and at joint C is connected through a condenser to a pressure regulating and measuring system. The pressure is then lowered until the liquid boils and all air is expelled from the system. The pressure is then raised until the liquid levels in the U-tube equalize. The measured pressure is the vapor pressure at the bath temperature.

An ebulliometer, such as pictured in Fig. 8-2b, is used to determine the boiling temperature of a pure liquid at a fixed pressure. Heat is added to the boiling tube A, and a mixture of liquid and vapor is pumped onto the thermometer well B. Equilibrium between liquid and vapor is assumed to be attained as the two-phase mixture flows from A to B. Errors in temperature measurement due to heat loss are minimized by the inclusion of an insulating jacket. In the space between this jacket and the outer tube vapors will condense because of heat loss, but the temperature of the outside of the insulating jacket should be close to the condensing (boiling) temperature. The ebulliometer is connected to a pressure regulating and measuring system through a condenser connected at joint E.

8-2 THE CLAPEYRON EQUATION

We will now be concerned with applying the thermodynamic tools previously developed in order to determine the relationship between temperature and pressure when two phases of a pure substance are in equilibrium. For such a system we have seen that the conditions of equilibrium are

$$T^\alpha = T^\beta \tag{7-29}$$

$$P^\alpha = P^\beta \tag{7-30}$$

$$\mu^\alpha = \mu^\beta \tag{7-31}$$

The first two statements are the rather obvious conditions for thermal and mechanical equilibrium within the system. Because the chemical potential of a phase is an intensive property that can be expressed as a function of two intensive properties, say temperature and pressure, it can be represented as a surface on a three-dimensional plot. Such surfaces are sketched in Fig. 8-3 for the α and β phases.[1] The intersection of these surfaces results in a line (shown dashed) where the three conditions of equilibrium are met. This line is the relationship between T and P that represents all the possible equilibrium states in this particular two-phase system; it corresponds to a line on the phase diagram of Fig. 8-1. Thus, when the chemical potentials in Eq. (7-31) are expressed as functions of T and P, this equation can be stated functionally as

$$f(T, P) = 0 \tag{8-1}$$

Our object now is to obtain an explicit statement of this relationship. Let us now consider the infinitesimal displacement of the system from its original equilibrium state described by Eqs. (7-29)–(7-31) to another equilibrium state. This change will move the system along one of the coexistence lines of Fig. 8-1 to a new T and P where

---

[1] See Prob. 8-15.

$$T \rightarrow T + dT$$

$$P \rightarrow P + dP$$

Because equilibrium again prevails, we state that

$$\mu^\alpha + d\mu^\alpha = \mu^\beta + d\mu^\beta$$

or simply that

$$d\mu^\alpha = d\mu^\beta$$

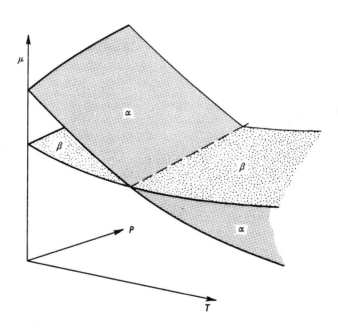

Figure 8-3 Surfaces of chemical potential for α and β phases. [Reprinted from K. Denbigh, *Principles of Chemical Equilibrium*, 3rd ed., 1971, with permission of Cambridge University Press, New York.]

The differentials of μ for each phase can be expressed in terms of T and P according to

$$\left(\frac{\partial \mu^\alpha}{\partial P} \right)_T dP + \left(\frac{\partial \mu^\alpha}{\partial T} \right)_P dT = \left(\frac{\partial \mu^\beta}{\partial P} \right)_T dP + \left(\frac{\partial \mu^\beta}{\partial T} \right)_P dT \qquad (8-2)$$

We must now determine the coefficients in this equation and begin by recalling the fundamental equation

$$dG = V\,dP - S\,dT + \sum \mu_i\,dn_i \tag{7-20}$$

where, as we have previously seen,

$$\mu_i = \left(\frac{\partial G}{\partial n_i}\right)_{T,P,n_j} \tag{7-21}$$

Because Eq. (7-20) is exact, the reciprocal relationship may be used to generate a new set of Maxwell relations:

$$\left(\frac{\partial \mu_i}{\partial P}\right)_{T,n_i} = \left(\frac{\partial V}{\partial n_i}\right)_{T,P,n_j} \tag{8-3}$$

$$\left(\frac{\partial \mu_i}{\partial T}\right)_{P,n_i} = -\left(\frac{\partial S}{\partial n_i}\right)_{T,P,n_j} \tag{8-4}$$

These relations are quite general and apply to both mixtures and pure substances. For a pure substance the extensive and intensive properties are related by

$$V = nv$$

$$S = ns$$

$$G = ng$$

and Eqs. (7-21), (8-3), and (8-4) simplify to

$$\left(\frac{\partial \mu}{\partial P}\right)_T = v \tag{8-5}$$

$$\left(\frac{\partial \mu}{\partial T}\right)_P = -s \tag{8-6}$$

$$\mu = g \tag{8-7}$$

As they are no longer required, the subscript i on μ and the subscripts n_i on the partial derivatives have been dropped. For a pure substance we see that the chemical potential is merely the molar Gibbs free energy.

We now return to Eq. (8-2) and utilize Eqs. (8-5) and (8-6) to obtain

$$v^\alpha \, dP - s^\alpha \, dT = v^\beta \, dP - s^\beta \, dT$$

Rearrangement leads to

$$\frac{dP}{dT} = \frac{s^\alpha - s^\beta}{v^\alpha - v^\beta} \tag{8-8}$$

It is desirable to eliminate the entropy from this equation. This can be done by restating the equilibrium condition in terms of the molar Gibbs free energy:

$$g^\alpha = g^\beta$$

$$h^\alpha - Ts^\alpha = h^\beta - Ts^\beta$$

or

$$s^\alpha - s^\beta = \frac{h^\alpha - h^\beta}{T} \tag{8-9}$$

Substitution of Eq. (8-9) into Eq. (8-8) yields

$$\frac{dP}{dT} = \frac{h^\alpha - h^\beta}{(v^\alpha - v^\beta)T} \tag{8-10}$$

Equation (8-10) is referred to as the Clapeyron equation and is a perfectly general relationship among pressure, temperature, volume change, and enthalpy change for a single-component, two-phase system at equilibrium. All these properties or property changes are experimentally measurable, and the validity of the equation has been tested and is well established. A test such as this increases our confidence in the laws and methods of thermodynamics. The actual utility of Eq. (8-10), or its subsequent simplifications, is that it may be employed with easily available information about the system to calculate information that is unknown or which may be difficult to obtain.

8-3 SOLID-LIQUID EQUILIBRIUM

When considering solid-liquid equilibrium, we are usually interested in the change of freezing point with pressure dT/dP, and therefore we invert Eq. (8-10):

$$\frac{dT}{dP} = \frac{T(v^L - v^S)}{h^L - h^S} \tag{8-10a}$$

For most substances dT/dP is small and positive; however, for the ice-water transition at 0°C $dT/dP = -0.007$ K/atm as calculated from data available in the steam tables.[2] This unusual behavior for water occurs because the solid phase is less dense than the liquid phase, thus making $v^L - v^S < 0$.

EXAMPLE 8-1

The temperature at which pure ice is in equilibrium with air-saturated water at 1 atm is 0.0100°C lower than the triple point where ice, liquid, and vapor are all in equilibrium at a pressure of 4.6 mmHg. Can this temperature difference be explained by the effect of pressure on the melting temperature? Data:
  Density of ice: 0.917 g/cm³
  Density of liquid water: 1.000 g/cm³
  Latent heat of fusion: 79.6 cal/g

Solution 8-1

The effect of pressure on the freezing point of pure water can be determined from Eq. (8.10a):

$$\frac{dT}{dP} = \frac{273(1 - 1/0.917)}{79.6} = -0.311 \text{ K} \cdot \text{cm}^3 / \text{cal}$$

$$\frac{dT}{dP} \doteq \frac{\Delta T}{\Delta P} = -0.00752 \text{ K} / \text{atm}$$

For

$$\Delta P = \frac{(760 - 4.6)}{760} = 0.994 \text{ atm}$$

we obtain

$$\Delta T = -(0.00752)(0.994) = -0.00746 \text{ K}$$

The pressure effect is seen to account for a large part of the observed difference; however, considering the quality of the data used for the calculation, we cannot attribute the lack of agreement to data of insufficient accuracy. Instead we must consider other factors, the largest of which is the lowering of the freezing point due to air solubility in liquid water. This will be dealt with in Ex. 11-2.

---

[2] J. H. Keenan and F. G. Keyes, *Thermodynamic Properties of Steam*, Wiley, NY, 1936.

EXAMPLE 8-2

The transition temperature of rhombic to monoclinic sulfur varies with pressure, as shown in the following table:

P(atm)	1	100	360	610	850
t(°C)	95.5	100	110	120	130

At 1 atm the volume change of this transition was determined to be 13.8 cm$^3 \cdot$ kg^{-1}. Calculate the heat of transition from rhombic to monoclinic sulfur at 1 atm. Compare this with the value of 401.7 ±2 J/g atom determined calorimetrically.[3]

Solution 8-2

A plot of the PT data is essentially linear. A least-squares fit of these data yields a slope, or dP/dT, equal to 24.8 atm K^{-1} with a correlation coefficient of 0.99986. From Eq. (8.10) we have

$$\Delta h = \left(\frac{dP}{dT} \right) \Delta v T$$

and

$$\Delta h = 24.8(13.8)(95.5 + 273.15) = 126{,}170 \text{ cm}^3 \cdot \text{atm} / \text{kg}$$

$$= 126{,}170 \text{ cm}^3 \cdot \text{atm} / \text{kg} \times 0.03206 \text{ kg} / \text{g atom} = 4045 \text{ cm}^3 \cdot \text{atm} / \text{g atom}$$

$$= 4045 \text{ cm}^3 \cdot \text{atm} / \text{g atom} \times 0.1013 \text{ J} / \text{cm}^3 \cdot \text{atm} = 410 \text{ J} / \text{g atom}$$

The agreement with the calorimetrically determined value is seen to be quite good.

8-4 SOLID-VAPOR AND LIQUID-VAPOR EQUILIBRIUM

When applied to equilibrium involving a vapor phase, the pressure is referred to as the vapor pressure and will be denoted by $P°$. For this type of phase equilibrium the major application of Eq. (8-10) is in the calculation of heats of vaporization for vapor-liquid equilibrium and

---

[3] E . D. West, *J. Am. Chem. Soc.*, *81*, 29 (1959).

heats of sublimation for solid-liquid equilibrium from the slope of the vapor pressure curve $dP°/dT$ and the densities of the phases. Here it is a matter of utilizing data which can be determined fairly easily to calculate data which would be more difficult to obtain.

An equation relating T and $P°$, which is thermodynamically less rigorous than Eq. (8-10) but which is more useful, can be derived if the vapor volume is much larger than the liquid or solid volume. The volume difference in Eq. (8-10) is then nearly equal to the vapor volume, and the further assumption of ideal-gas behavior for the vapor results in replacing Δv with $RT/P°$:

$$\frac{dP°}{dT} = \frac{P°\Delta h}{RT^2}$$ (8-11)

Rearrangement yields

$$\frac{dP°/P°}{dT/T^2} = -\frac{d\ln P°}{d(1/T)} = \frac{\Delta h}{R}$$ (8-12)

If Eq. (8-12) is integrated under the assumption of constant Δh, the result is

$$\ln P° = c - \frac{\Delta h}{RT}$$ (8-13)

where c is a constant. Equations (8-12) and (8-13) are often referred to as the Clausius-Clapeyron equation and find use in correlation, interpolation, and extrapolation of vapor pressure data and also in estimating latent heats from vapor pressure data. Equation (8-13) suggests that a plot of logarithm of $P°$ vs. the reciprocal of the absolute temperature should be a straight line over the region where Δh is constant. Actually, it has been found that such plots are linear over a surprisingly wide range of temperature; undoubtedly this can be attributed to a fortunate cancellation of errors introduced by the various assumptions.

The Antoine equation, used for the precise analytical representation of vapor pressure data,[4] can be regarded as an empiricized version of the Clausius-Clapeyron equation:

$$\log P° = A - \frac{B}{C+t}$$ (8-14)

Except for the use of 10 base logarithms, the only difference between the Antoine equation [Eq. (8-14)] and the Clausius-Clapeyron equation [Eq. (8-13)] lies in allowing the parameter C to assume values other than 273.15.

If the latent heat of vaporization is assumed to depend linearly on temperature,[5] integration of Eq. (8-11) leads to an equation of the following form:

---

[4] The spreadsheet ANTOINE.WQ1 (.XLS) in the COMPUTE directory of the accompanying CD-ROM uses the Antoine equation to calculate vapor pressures of several common substances.
[5] See Sec. 6-4.

$$\log P° = A - \frac{B}{T} + C \log T \tag{8-15}$$

With a sound thermodynamic basis and the flexibility afforded by three adjustable parameters this equation should be capable of fitting vapor pressure data quite well; however, the Antoine equation is more widely accepted.

EXAMPLE 8-3

From the following vapor pressure data for carbon tetrachloride, evaluate the mean heat of vaporization in this range:

t (°C)	25	35	45	55
P° (mmHg)	113.8	174.4	258.9	373.6

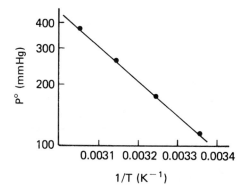

Figure 8-4 Clausius-Clapeyron plot for carbon tetrachloride.

Solution 8-3

The data, plotted as $P°$ vs. $1/T$ on semilog paper, are shown in Fig. 8-4, where it is seen that a good linear relationship results. From a least-squares fit, the slope $d \ln P°/d(1/T)$ is found to be -3923 K, and from Eq. (8-12) the heat of vaporization averaged over this temperature range is

$$\Delta h = -R \frac{d \ln P°}{d(1/T)} = -1.987(-3923) = 7795 \text{ cal / gmol}$$

$$= \frac{7795}{154} = 50.6 \text{ cal / g}$$

This value compares well with the reported values[6] of 50.78 cal/g at 21.11°C and 48.83 cal/g at 48.89°C.

EXAMPLE 8-4

The vapor pressures of solid and liquid hydrogen cyanide expressed in mmHg are given by

Solid (from 243 to 258 K): $\log P° = 9.33902 - (1864.8 / T)$

Liquid (from 265 to 300 K): $\log P° = 7.74460 - (1453.06 / T)$

Calculate (1) the heat of sublimation, (2) the heat of vaporization, (3) the heat of fusion, (4) the triple point, and (5) the normal boiling point.

Solution 8-4

The equations are in the form of Eq. (8-13), which allows the heat of sublimation $(h^G - h^S)$ to be determined from the solid vapor-pressure equation and the heat of vaporization $(h^G - h^L)$ to be determined from the liquid vapor pressure equation.

1. $(h^G - h^S) = 2.303(1.987)(1864.8) = 8640 \text{ cal / g mol}$

2. $(h^G - h^L) = 2.303(1.987)(1453.06) = 6650 \text{ cal / g mol}$

3. The heat of fusion can be obtained from the heats of sublimation and vaporization as follows:

$$(h^L - h^S) = (h^G - h^S) - (h^G - h^L)$$

$$h^L - h^S = 8640 - 6650 = 1990 \text{ cal / g mol}$$

4. The triple point is located by the intersection of the two vapor pressure curves. Simultaneous solution of the solid and liquid vapor pressure equations yields

$$T = 258 \text{ K}$$

$$P° = 135 \text{ mmHg}$$

5. The normal boiling point is obtained by simply substituting $P° = 760$ mmHg into the liquid vapor pressure equation. This yields

$$T = 299 \text{ K}$$

---

[6] *Handbook of Chemistry and Physics*, 49th ed., The Chemical Rubber Co., Cleveland, OH, 1969, p. E-26.

8-5 PRESENTATION OF THERMODYNAMIC PROPERTY DATA

Chapter 5 dealt with the calculation of thermodynamic properties for a single phase of a substance, and the previous parts of this chapter have dealt with the calculation of property changes accompanying phase changes. Therefore, if the necessary experimental data are available, thermodynamic properties of a pure substance may be evaluated over a wide range of temperature and pressure, including two-phase regions. In the past, whenever sufficient data to enable thermodynamic property calculations accumulated in the literature, invariably someone would perform the useful service of executing the calculations and publishing the information in the form of tables or property diagrams. This type of information is available in the literature for many substances. Today, with the development of serviceable generalized equations of state and the widespread availability of computing facilities, the data base is becoming the preferred method of storing and retrieving thermodynamic information.

Property Diagrams. The properties of major interest are temperature, pressure, volume, enthalpy, and entropy. For a pure substance or a mixture of unchanging composition (e.g., air) a maximum of two of these properties need to be specified in order to fix the others. When presenting data in tabular form, T and P are the most convenient independent variables and are therefore almost always used. However, diagrams involving several different combinations of independent variables are commonly used: hT, Ph, Ts, and hs diagrams on which the remaining properties are plotted as lines of constant property values—isotherms, isobars (constant pressure), isochores (constant volume), isenthalps (constant enthalpy), and isentropes (constant entropy). Certain diagrams are convenient for specific types of calculations. For example, the Carnot cycle with its two isothermal and two isentropic steps plots as a rectangle on Ts coordinates.

Programs using interactive computer graphics to construct three-dimensional views of PVT surfaces, including paths of constant P, V, T, s, u, or h, are available.[7] EQUATIONS OF STATE, a version of this software dedicated to running tutorials, is included on the accompanying CD-ROM.

Data Bases. Most data bases utilize generalized cubic equations of state and heat capacity equations of the type found in Table B-1 to calculate thermodynamic properties. The use of the Peng-Robinson equation of state to calculate fluid density and the residual properties Δh^* and Δs^* for chlorine gas was illustrated in Examples 3-2 and 5-2. Values of h and s can be obtained from Δh^* and Δs^*, together with the ideal-gas h' and s' values calculated from a heat capacity equation. With h, s, and v known at any specified temperature and pressure, one can determine the value of any other thermodynamic property for a pure fluid. These calculations can be executed with the program PREOS.EXE on the accompanying CD-ROM. See App. E.

---

[7] K. R. Jolls, *Chem. Eng. Prog.*, *85*, 64(1989).

The Peng-Robinson equation of state can be used for either the liquid or gas phase and also for the determination of vapor pressure (see Sec. 9-12). Thus, the calculation of all the thermodynamic properties requires only T_c, P_c, ω, and the constants in a heat capacity equation.

Data bases are available as commercial software;[8] they also play a major role in process simulation programs. Algorithms and sample computer code are available to those who would construct their own data base.[9]

EXAMPLE 8-5

Illustrate the use of the Peng-Robinson equation and a heat capacity equation for the calculation of h and s for a pure substance.

Solution 8-5

Values of h and s are calculated from Eqs. (5-37) and (5-38):

$$\Delta h = h - h_0 = h = \Delta h_0^* + \int_{T_0}^{T} c_P \, dT - \Delta h^* \qquad (5\text{-}37)$$

$$\Delta s = s - s_0 = s = \Delta s_0^* + \int_{T_0}^{T} \frac{c_P}{T} \, dT - R \ln \frac{P}{P_0} - \Delta s^* \qquad (5\text{-}38)$$

In these equations Δh_0^* and Δs_0^* are evaluated at T_0 and P_0, where h and s have been set equal to zero and Δh^* and Δs^* are evaluated at the T and P where the property values are desired. With the c_p equations of Table B-1, the integrals in Eqs. (5-37) and (5-38) are

$$\int_{T_0}^{T} c_P \, dT = a(T - T_0) + \frac{b}{2}(T^2 - T_0^2) + \frac{c}{3}(T^3 - T_0^3) + \frac{d}{4}(T^4 - T_0^4)$$

and

$$\int_{T_0}^{T} \frac{c_P}{T} \, dT = a \ln \frac{T}{T_0} + b(T - T_0) + \frac{c}{2}(T^2 - T_0^2) + \frac{d}{3}(T^3 - T_0^3)$$

---

[8] The annual Software Directory accompanying the January issue of *Chemical Engineering Progress* contains brief descriptions of commercially available data bases and data estimation programs. The directory is scheduled for inclusion on *AIChE*'s Web site (www.aiche.org).
[9] P. Benedek and F. Olti, *Computer Aided Chemical Thermodynamics of Gases and Liquids*, Wiley, New York 1985.

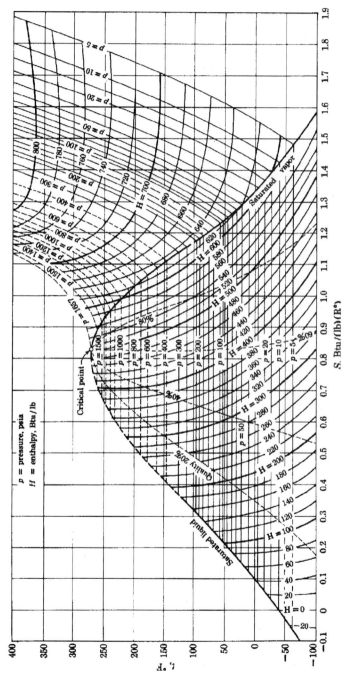

Figure 8-5 Temperature-entropy diagram for ammonia. [From O. A. Hougen, K. M. Watson, and R. A. Ragatz, *Chemical Process Principles*, Part II, Wiley, 1959, with permission of John Wiley & Sons, Inc.]

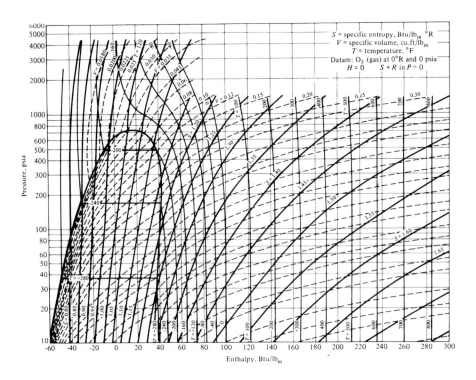

Figure 8-6 Oxygen pressure-enthalpy diagram. (From L. N. Canjar and F. S. Manning, *Thermodynamic Properties and Reduced Correlations for Gases*, copyright Gulf Publishing Co., Houston, TX, 1967. Reproduced by permission.)

The needed values of Δh^*, Δh_0^*, Δs^*, and Δs_0^* are calculated from Eqs. (5-58) and (5-59):

$$\frac{\Delta h^*}{RT_c} = 2.078(1+\kappa)[(1+\kappa(1-T_r^{1/2})]\ln\left[\frac{Z+\left(1+\sqrt{2}\right)B}{Z+\left(1-\sqrt{2}\right)B}\right] - T_r(Z-1) \qquad (5\text{-}58)$$

$$\frac{\Delta s^*}{R} = 2.078\kappa\left[\frac{1+\kappa}{T_r^{1/2}} - \kappa\right]\ln\left(\frac{Z+\left(1+\sqrt{2}\right)B}{Z+\left(1-\sqrt{2}\right)B}\right) - \ln(Z-B) \qquad (5\text{-}59)$$

These equations require values of $T_r, \kappa, B,$ and Z. In accordance with Eq. (3-5h), κ depends only on ω and is therefore a constant characteristic of the substance.

$$\kappa = 0.37464 + 1.5422\omega - 0.26992\omega^2 \qquad (3\text{-}5h)$$

The parameters A and B depend on T_r and P_r according to

$$A = 0.45724\frac{P_r}{T_r^2}[1 + \kappa(1 - T_r^{1/2})]^2 \qquad (3\text{-}5l)$$

$$B = 0.07780\frac{P_r}{T_r} \qquad (3\text{-}5m)$$

and Z depends on A and B as follows:

$$Z^3 + (B-1)Z^2 + (A - 3B^2 - 2B)Z + (B^3 + B^2 - AB) = 0 \qquad (3\text{-}5i)$$

The task of solving these equations is complicated but straightforward and could be accomplished through the following algorithm.

Sample algorithm
1. Evaluate A and B at T_0 and P_0 using Eqs. (3-5h), (3-5l), and (3-5m).
2. Calculate Z from Eq. (3-5i). Take the smallest root for the liquid phase and the largest root for the vapor phase. Newton's method can be used to find the roots.
3. Using these values of $B, Z, T_r,$ and κ, calculate Δh_0^* and Δs_0^* via Eqs. (5-58) and (5-59).
4. At the specified T and P, evaluate A and B as in step 1.
5. Calculate Z as in step 2.
6. Calculate Δh^* and Δs^* as in step 3.
7. Evaluate the heat capacity integrals at T.
8. Evaluate $\ln P/P_0$ at P.
9. Calculate h and s via Eqs. (5-37) and (5-38).
10. For a different T and P, return to step 4.

PROBLEMS

8-1. Use data from the steam tables (Tables C-1) in the Appendix to demonstrate validity of the

(a) Clapeyron equation.
(b) Clausius-Clapeyron equation.

8-2.

(a) Does the fact that ice floats on water guarantee that we can ice skate (apart from the obvious fact that there would be no surface to skate on if ice did not float)? Use thermodynamic reasoning in developing your argument.

(b) Do you expect that there would be a temperature below which ice skating would not be possible? Explain.

8-3. The following vapor pressure data is available for ice and liquid water

	$P°$ (psia)	
t (°F)	Ice	Water
30	0.0808	———
32	0.08854	0.08854
34	———	0.09603

Estimate the

(a) heat of vaporization of water.
(b) heat of sublimation of ice.
(c) heat of melting.

8-4. The heat of fusion of ice is 80 cal/g at 0°C and 1 atm, and the ratio of the specific volume of water to that of ice is 1.000:1.091. The saturated vapor pressure and the heat of vaporization of water at 0°C are 4.58 mmHg and 600 cal/g, respectively. Estimate the triple point using these data.

8-5. Below the triple point (−56.2°C) the vapor pressure of solid carbon dioxide may be expressed by the relation

$$\log P° = \frac{-1353}{T} + 9.832$$

where $P°$ is in mmHg and T in K. The latent heat of fusion is 1990 cal/g mol. Make an estimate of the vapor pressure of liquid carbon dioxide at 0°C.

8-6. A liquid (mercury) normally boils at 357°C, and its heat of vaporization is 68 cal/g. It is required to distill the liquid at 100°C; estimate the approximate pressure that would be used.

8-7. The melting point of benzene is found to increase from 5.50 to 5.78°C when the external pressure is increased by 100 atm. The heat of fusion of benzene is 30.48 cal/g. What is the change of volume per gram accompanying the fusion of benzene?

8-8. At 125°C there is a transition between the two forms of selenium, the vitreous form and the gray form. The entropy of the former is 7.40 cal/g mol · K, while that of the latter is 10.04 cal/g mol · K. Assuming that the entropy change is independent of temperature, calculate the free energy change the conversion of 1 mol of vitreous selenium to gray selenium at 25°C. Which is the stable form at 25°C?

8-9. What percentage of the volume of a tube must be occupied by liquid water at 70°F (the remainder being water vapor) so that when the tube is sealed and heated the contents will pass through the critical state? How much heat must be added to the contents of the tube between 70°F and the critical state if the enclosed volume is 1 ft^3?

8-10. The free energy change for the following change in state is +685 cal/g atom at 25°C:

$$\text{carbon (graphite, 1 atm)} \rightarrow \text{carbon (diamond, 1 atm)}$$

Estimate the pressure at which these two forms of carbon are in equilibrium at 25°C. The densities of graphite and diamond are 2.26 and 3.51 g/cc, respectively, at 25°C.

8-11.

    (a) Show that for a single phase

$$\left(\frac{\partial P}{\partial T}\right)_S = \frac{c_P}{Tv\alpha}$$

where

$$\alpha = \frac{1}{v}\left(\frac{\partial v}{\partial T}\right)_P$$

    (b)   Ice originally at –3°C and 1 atm is compressed adiabatically until the melting point is reached. What is the T and P of this melting point? The following data are given for water in this range of conditions: The heat of fusion is 79.6 cal/g.

Ice	Liquid water
$c_p = 0.48$ cal/g °C	$c_p = 1.0$ cal/g °C
$v = 1.09$ cc/g	$v = 1.00$ cc/g
$\alpha = 158(10^{-6})$ K^{-1}	$\alpha = -67(10^{-6})$ K^{-1}

8-12. Show that the constant-volume heat capacity for a system consisting of two phases in equilibrium can be expressed as

$$C_V = -T\left(\frac{\partial V}{\partial P}\right)_S\left(\frac{dP}{dT}\right)^2$$

8-13. A saturated vapor is compressed adiabatically and reversibly. In terms of readily available properties, determine the condition for which condensation will occur. Assume the vapor to behave as an ideal gas.

8-14. The transition point for rhombic to monoclinic sulfur is 95.5°C, and the melting point of monoclinic sulfur is 119.3°C at atmospheric pressure where the latent heat for the transition rhombic → monoclinic is 2.78 cal/g and the latent heat for the transition monoclinic → liquid is 13.2 cal/g. The densities of rhombic, monoclinic, and liquid sulfur are 2.07, 1.96, and 1.90 g/cc, respectively. At 120°C the latent heat of vaporization of the liquid is 84.8 cal/g, and the vapor pressure of liquid sulfur is $4.0(10^5)$ atm. Estimate the coordinates (P, T) of all three triple points in this system: rhombic-monoclinic-vapor, rhombic-monoclinic-liquid, and monoclinic-liquid-vapor. What assumptions are made, and which ones are apt to cause the largest error?

8-15. There is a slight error in the representation of the $\mu - T - P$ surfaces of Figure 8-3. Close scrutiny reveals that the curvature is such that

$$\left(\frac{\partial^2 \mu}{\partial T^2} \right)_P < 0$$

Show that this curvature is incorrect.

8-16. On Figure 8-1 at point O the two-phase curves of rhombic and monoclinic sulfur intersect. Note that the unstable portions of these curves (the dashed curves) lie above the curves for the stable forms. Is there a thermodynamic explanation for this behavior?

8-17. Trouton's rule states that at the normal boiling point, nbp, the entropy of vaporization is 21 cal/g mol · K. This rule works for very few substances, but the following improved version by Kistyakowsky applies to a wide variety of nonpolar substances.

$$\frac{\Delta h}{T_{nbp}} = 8.75 + 4.571 \log T_{nbp} \qquad \text{where } \Delta h / T_{nbp} \text{ has the units cal/g mol · K.}$$

Using the Kistyakowsky relation, derive an equation relating vapor pressure to temperature that contains only one parameter, T_{nbp}.

8-18. The acentric factor, ω, is defined in terms of the reduced vapor pressure, $P°/P_c$, at a reduced temperature of 0.7:

$$\omega = -1.000 - \log \left(\frac{P°}{P_c} \right)_{T_r = 0.7}$$

Use the acentric factor definition and the fact that the vapor pressure curve terminates at the critical point to determine a

  (a) generalized reduced vapor pressure equation.
  (b) generalized heat of vaporization.

8-19. Express the acentric factor ω in terms of T_c, P_c, and the constants in the Antoine equation.

8-20. Titanium can exist in three different crystalline phases, designated β, ε, and ω. The following Gibbs free energy changes have been determined[10] for the three binary transitions.

$$\varepsilon \to \beta: \quad \Delta G = 1050 - 0.91T - 1.43P$$

$$\omega \to \varepsilon: \quad \Delta G = -360 - 0.08T + 4.54P$$

$$\omega \to \beta: \quad \Delta G = 690 - 0.99T + 3.11P$$

where ΔG is in cal/g atom, T is in K, and P is in kbars. Plot the phase diagram in the solid phase region and show the triple point. Can these equations be checked for thermodynamic consistency?

8-21. Tabulations of v, h, and s in the superheated vapor region are available for several substances.[11] Choose a substance from the following list and make a comparison of these properties with those calculated from the Peng-Robinson equation of state and the heat capacity equation of App. B via the program PREOS.EXE. Set the datum level for h and s such that your calculated values coincide with the tabulated data at the highest temperature and lowest pressure reported—where the gas is most likely to behave ideally. Present your results in such a way that trends may be easily discerned.

  Substances n-butane, carbon dioxide, ethane, hydrogen, methane, and methyl chloride

8-22. A well-insulated tank contains 10 lb-mols of liquid ammonia initially at 80°F. Ammonia is removed continuously from the tank and throttled to a pressure of 15 psia.

  (a) Estimate the temperature in the tank after one lb-mol of liquid ammonia has been removed.
  (b) Estimate the temperature in the tank after one lb-mol of ammonia vapor has been removed.

Assume no heat change.

8-23. A tank with heating coils contains 10 lb-mols of liquid ammonia initially at 80°F. Ammonia is removed continuously from the tank and throttled to a pressure of 15 psia. Heat is added to the tank so that the temperature remains at 80°F.

  (a) Estimate the heat added to the tank after one lb-mol of liquid ammonia has been removed.
  (b) Estimate the heat added to the tank after one lb-mol of ammonia vapor has been removed.

---

[10] L. Kaufman and H. Bernstein, *Computer Calculation of Phase Diagrams*, Academic Press, New York, 1970, Chap. 2.

[11] *Perry's Chemical Engineers' Handbook,* 6th ed., J. H. Perry, D. W. Green, and J. O. Maloney, eds., McGraw-Hill, New York, 1984

8-24.   Oxygen gas at 100°F and 100 psia is throttled adiabatically to a pressure of 15 psia. What is its final temperature? What can we conclude from this result?

8-25.   Oxygen gas at 14.7 psia and 100°F is to be compressed adiabatically to 100 psia. What is the reversible work of compression and the final temperature? If the actual work is 50% greater than the reversible work what will be the final temperature for an adiabatic compression to 100 psia?

Principles of Phase Equilibrium

*A*s we now wish to apply the principles of thermodynamics to phase equilibrium, let us consider the meaning of an equilibrium state as applied to a multicomponent, multiphase system comprised of nonreacting substances. As an example, suppose we introduce liquid water and liquid ethanol into a vessel equipped with means of measuring pressure and temperature and of withdrawing samples of liquid and vapor for analyses. We now place the vessel in a bath maintained at a specified temperature. Within the vessel water and ethanol will begin to distribute themselves between the liquid and vapor phases. This distribution process involves vaporization and diffusion; it is usually slow but can be accelerated by mechanical agitation. When diffusion has ceased, the system will be in an equilibrium state which can be verified when successive readings of temperature, pressure, liquid mol fractions, and vapor mol fractions are identical. This equilibrium state is reproducible, and according to the phase rule, this is possible for our two-component, two-phase system when any two of its intensive properties are duplicated. Here we are saying that all the intensive properties of the system are duplicated, although the quantities of the phases, and thus the extensive properties, may differ. In applying thermodynamics to phase equilibrium, we are mainly interested in intensive properties and the relationships between them; therefore, the phase rule specifies the conditions necessary to determine the equilibrium state adequately for our purposes.

As we have already seen, the successful application of the methods of thermodynamics requires an empirical understanding of the system under study. To assist in cultivating this understanding, we first consider the various ways in which vapor-liquid equilibrium data may be presented and thereby seek to develop a physical appreciation of the variables of phase equilibrium and how they are interrelated. Often, however, the concept of phase equilibrium and the variables expressing it do not acquire physical meaning until one considers the experimental determination. The chapter therefore begins with a descriptive account of the presentation and experimental determination of phase equilibrium data. Tools for the thermodynamic treatment of phase equilibrium are then developed. Finally, the thermodynamic approach to phase equilibrium is delineated.

9-1 PRESENTATION OF VAPOR-LIQUID EQUILIBRIUM DATA

In studying vapor-liquid equilibrium in a binary system, there are four intensive variables which will concern us: temperature, pressure, a single liquid mol fraction,[1] and a single vapor mol fraction. Application of the phase rule to a binary system shows that the maximum number of intensive variables which can be specified is three—this occurs when only a single phase is present. Therefore, to graphically represent the phase behavior of a binary system, a three-dimensional plot, such as shown in Fig. 9-1, is necessary. This figure is worthy of considerable study.[2] It shows that each phase is represented by a surface on $PTxy$ coordinates. The upper surface represents the liquid phase, and the lower surface represents the vapor phase. When there is equilibrium between the two phases, we know that the temperatures and pressures must be equal. Therefore, the compositions of the equilibrium phases are determined by the intersection of a constant-temperature plane with a constant-pressure plane. In Fig. 9-1 a constant-temperature plane is shown by the lens-shaped envelope AB which lies in the vertical plane DAV_aL_aBE. This envelope is shown to intersect a constant-pressure envelope KV_aL_aH which lies in the horizontal plane. The intersection of the envelopes occurs in two places: on the liquid surface at the point L_a and on the vapor surface at the point V_a. These points represent values of x_1 and y_1 that are in equilibrium at the temperature and pressure we have selected. Equilibrium liquid and vapor compositions are also shown as L_b and V_b where the isothermal plane $C^{IV}V_bL_bNP$ corresponding to T_b intersects the isobaric plane $MLV_bL_bC^I$ corresponding to P_b.

---

[1] Only a single mol fraction is required to describe the composition of a phase in a binary system because of the condition that mol fraction in a phase must sum to unity.

[2] Visualization and understanding of these diagrams can be greatly enhanced through the use of interactive computer graphics. See G. N. Charos, P. Clancy, and K. E. Gubbins, *Chem. Eng. Educ., 20,* 80 (1986) and K. R. Jolls, *Chem. Eng. Educ., 32,* 113 (1998).

Because three variables are required to describe the state of a single phase, the representation of phase behavior on a two-dimensional plot can be accomplished by holding one variable constant; three such plots are possible. These three types of plots corresponding to the three-dimensional diagram of Fig. 9-1 are shown on Figs. 9-2, 9-3, and 9-4. Unless we are concerned with equilibrium near the critical region, we will have little use for the PT plot and we will find that the envelopes on the Pxy and Txy diagrams will span the entire range of mol fractions. Usually we are interested in systems at low to moderate pressure where the constant-temperature envelope will resemble T_a in Fig. 9-2 and the constant-pressure envelope will resemble P_a in Fig. 9-3.

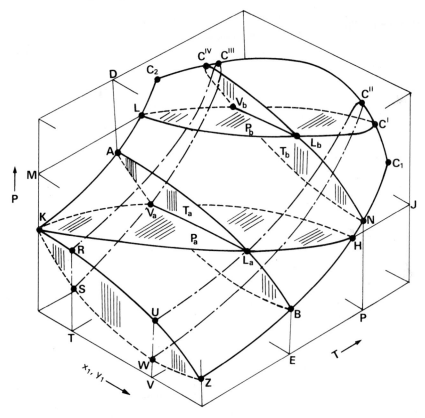

Figure 9-1 *PTxy* diagram.

Consider equilibrium at a constant temperature of T_a as represented on Fig. 9-2. At equilibrium the pressure must be equal for each phase and a horizontal line ties together the saturated liquid and vapor curves. Such lines determine the equilibrium liquid and vapor composition at a given pressure and are called *tie lines*. The phase rule requires specification of two variables to determine a two-component, two-phase system. As we have already specified T_a, only one other variable may be chosen. If we choose P, then x_1 and y_1 are found at the ends of the tie line. If we choose either x_1 or y_1, we have fixed a point on either saturated curve and a tie line determines P and the mol fraction of the other phase. In Fig. 9-3, a constant-pressure diagram, tie lines are characterized by temperature. Figure 9-4 shows the curves formed by the intersection of the two constant-composition planes, $TSRV_aC^{III}$ and $WVUL_aC^{II}$, with the liquid and vapor surfaces of Fig. 9-1. Note that it is possible for mixtures to have critical temperatures, C^{II} and C^{III}, above those of either pure component. The intersection of the projections of the vapor portion of one curve with the liquid portion of another locates an equilibrium point, L_a, V_a.

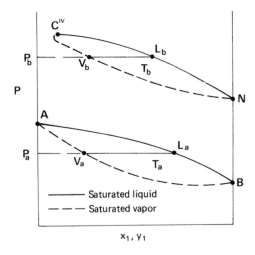

x_1, y_1 **Figure 9-2** *Pxy* diagram.

As the temperature increases there is a narrowing of the range of composition over which a liquid phase can exist. For example, from Fig. 9-1 it is seen that for the isothermal plane $C^{IV}V_bL_bNP$ corresponding to T_b component 2 is above its critical temperature and therefore cannot exist as a pure liquid—in fact, there is a maximum mol fraction of component 2 above which no liquid phase can form at this temperature. The composition range also narrows as the pressure is increased. Fig. 9-1 shows that the lens-shaped envelope formed by the intersection of the constant-pressure plane, $MLV_bL_bC^I$ at P_b with the liquid and vapor surfaces does not extend to $x_1 = y_1 = 1$. This limited composition range can also be discerned from Fig. 9-4 where it is observed that the pressure P_b lies above the vapor pressure curve for component 1. As the temperature or pressure continues to increase, the composition range steadily diminishes. Often, a point is reached where neither component can exist as pure conjugate liquid and vapor phases.

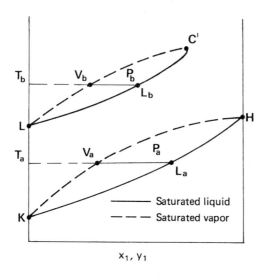

Figure 9-3 *Txy* diagram.

The preponderance of experimentally determined vapor-liquid equilibrium data is reported at either constant temperature or pressure as either of these constraints can be easily imposed on an experimental determination. Unless we are dealing with systems following ideal solution behavior where the phase behavior may be calculated from pure component properties, experimental data are needed to construct the phase diagrams.

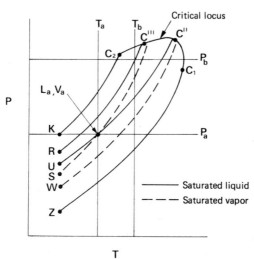

Figure 9-4 *PT* diagram.

9-2 DETERMINATION OF VAPOR-LIQUID EQUILIBRIUM DATA

The complexities of apparatus and refinements of technique required for obtaining high-quality phase equilibrium data have been thoroughly discussed elsewhere.[3] Here the objective is merely to outline the general approach and to identify some of the salient factors involved in the experimental determinations.

Static Apparatus.   A schematic representation of an apparatus for the static determination of vapor-liquid equilibrium data is shown in Fig. 9-5. The principle is simply to allow vapor and liquid to come to equilibrium at a fixed temperature and then measure the pressure and sample the phases. The samples are then analyzed to determine compositions. The approach to equilibrium is usually slow but can be hastened by some type of mechanical agitation. The condition of equilibrium may be determined by a constancy of the pressure reading. This type of apparatus is well suited to obtaining isothermal data. A major difficulty lies in degassing the system so that reliable pressure readings can be obtained.

Recirculating Stills.   A schematic representation of a recirculating still is shown in Fig. 9-6. Vapor is generated in the reboiler by the addition of heat, it flows to a condenser, and totally condensed vapor returns by gravity flow to the reboiler. Thus, there is a continuous circulation of vapor. The pressure is regulated, and constancy of temperature indicates attainment of steady-state conditions. Samples of liquid and condensed vapor are then withdrawn for analysis.

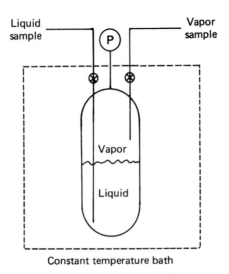

Figure 9-5 Static vapor-liquid equilibrium apparatus.

[3] E. Hála, J. Pick, V. Fried, and O. Vilím, *Vapor-Liquid Equilibrium,* 2nd ed., Pergamon, Elmsford, NY, 1967.

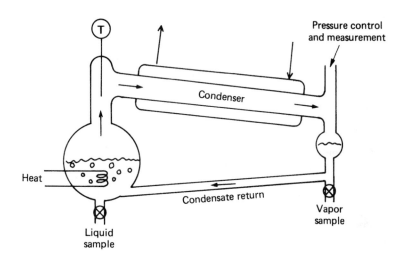

Figure 9-6 Recirculating vapor-liquid equilibrium still.

To ensure equilibrium, the generated vapor is usually forced to travel through a tube where it entrains slugs of liquid. After traveling the length of the tube, the liquid-vapor mixture, assumed at equilibrium, then impinges on a thermometer or thermocouple which determines the equilibrium temperature. All entrained liquid must be removed from the vapor before it is condensed.

Data Acquisition. With either type of apparatus it is necessary to be able to analyze the collected liquid and vapor samples. Whether this is done by measurement of a physical property such as refractive index, chromatography, or chemical analysis, considerable effort is usually required to establish and calibrate the method.

The achievement of equilibrium in the static apparatus or steady state in the recirculating still and the subsequent measurements produce a single *PTxy* data point. The apparatus must then be recharged so that data points at other compositions may be obtained. To construct a *Pxy* diagram similar to Fig. 9-2 or a *Txy* diagram similar to Fig. 9-3 would require acquisition of several data points. The construction of diagrams of more complexity, such as shown in Figs. 10-1 and 10-2, requires more data points.

9-3 THE THERMODYNAMIC BASIS FOR THE PHASE RULE

A correct mathematical description of any system requires that the number of applicable equations be equal to the number of unknown variables. The phase rule is merely an accounting device which allows us to quickly make the determination of mathematical correctness for systems we desire to treat by the methods of thermodynamics.

Consider a system consisting of C nonreacting components distributed among π phases. To describe each phase, we need $C + 1$ intensive variables, $C - 1$ independent composition variables, and temperature and pressure. For the system of π phases there will be a total of $\pi (C + 1)$ intensive variables.

The conditions of equilibrium [Eqs. (7-45)–(7-47)] provide the describing equations:

Number of
independent equations

$$T^\alpha = T^\beta = \cdots = T^\pi \qquad\qquad \pi - 1$$

$$P^\alpha = P^\beta = \cdots = P^\pi \qquad\qquad \pi - 1$$

$$\mu_1^\alpha = \mu_1^\beta = \cdots = \mu_1^\pi \qquad\qquad \pi - 1$$

$$\mu_2^\alpha = \mu_2^\beta = \cdots = \mu_2^\pi \qquad\qquad \pi - 1$$

$$\vdots \qquad \vdots \qquad \vdots$$

$$\mu_C^\alpha = \mu_C^\beta = \cdots = \mu_C^\pi \qquad\qquad \pi - 1$$

From the definition [Eq. (7-21)] the chemical potential is seen to depend on temperature, pressure, and composition, and therefore this set of $(C + 2)(\pi - 1)$ independent equations relates the $\pi(C + 1)$ intensive variables. The difference between the number of variables and the number of independent equations is called the degrees of freedom and is designated by $\mathscr{F}$:

$$\mathscr{F} = \pi(C + 1) - (C + 2)(\pi - 1)$$

$$= C + 2 - \pi \qquad\qquad\qquad (1\text{-}8)$$

The degrees of freedom $\mathscr{F}$ are the number of variables which must be specified in order that the system be defined in a correct mathematical sense. When this condition is assured, we may then proceed to determine the values of the unknown variables using the principles and tools to be developed in this chapter.

9-4 THE FUGACITY

It is again convenient to define another thermodynamic property: the fugacity f. Although we have seen that the chemical potential provides a concise condition of equilibrium, the fugacity will also be shown to be useful for that purpose and in addition will facilitate certain calculations involved with phase and chemical equilibrium. The fugacity is defined in terms of the chemical potential, and for a pure substance the relationship is

$$\mu = \mu° + RT \ln f \qquad (9\text{-}1)$$

where $\mu°$ is a function only of temperature and is seen to be the chemical potential of the substance when its fugacity is unity.[4] An alternative definition in terms of changes occurring at constant temperature is

$$[d\mu = RT d \ln f]_T \qquad (9\text{-}2)$$

Neither equation provides a complete definition, and a boundary condition is needed. The equations are adequate for calculating changes in fugacity or chemical potential and apply to pure gases, liquids, or solids; however, the following boundary condition applies only to gases:

$$\lim_{P \to 0} \frac{f}{P} = 1 \qquad (9\text{-}3)$$

The fugacity is completely defined with Eq. (9-3) and either Eq. (9-1) or (9-2).

Some insight into the reason the fugacity is so defined can be obtained by considering the chemical potential of an ideal gas. If Eq. (8-5) is applied to an ideal gas under conditions of constant temperature, we have

$$d\mu = v \, dP = \frac{RT}{P} \, dP = RT d \ln P \qquad (9\text{-}4)$$

Integration gives

$$\mu = \mu° + RT \ln P \qquad (9\text{-}5)$$

where $\mu°$, the constant of integration, can depend only on temperature. The similarity between Eqs. (9-1) and (9-2) and Eqs. (9-4) and (9-5) is obvious, and one could therefore consider the fugacity as an effective pressure. Equation (9-3) is the statement that the fugacity of an ideal gas is equal to its pressure and that all gases approach ideal-gas behavior as the pressure is reduced. The fugacity has the units of pressure.

[4] One atm.

The fugacity of a substance in a solution (either gaseous, liquid, or solid) is defined by analogy with Eqs. (9-1) and (9-2):

$$\hat{\mu}_i = \mu_i^\circ + RT \ln \hat{f}_i \qquad (9\text{-}6)$$

or

$$[d\hat{\mu}_i = RT\, d \ln \hat{f}_i]_T \qquad (9\text{-}7)$$

The circumflex is used to distinguish the chemical potential and fugacity of a component in solution from those of the pure substance. In dealing with solutions, the indexing subscript i is also necessary. Again, μ_i° is a function only of temperature and is the chemical potential of pure component i at a fugacity of unity, because Eq. (9-6) must apply to all compositions and hence applies as the mol fraction of the substance approaches unity. The boundary condition needed to complete the definition of fugacity again applies at low pressure. We assert that for a *perfect gas mixture* the fugacity of a component is equal to its partial pressure and that all gaseous mixtures become perfect on approaching zero pressure:[5]

$$\lim_{P \to 0} \frac{\hat{f}_i}{p_i} = 1 \qquad (9\text{-}8)$$

The partial pressure is defined as the product of total pressure P and mol fraction:

$$p_i = P y_i \qquad (9\text{-}9)$$

The fugacity's role as a condition of equilibrium can be seen by considering two multicomponent phases in equilibrium at a specified temperature and pressure. We have already seen that the chemical potentials must be equal, and from an integration of the identity [Eq. (9-7)] we can write

$$\hat{\mu}_i^\alpha - \hat{\mu}_i^\beta = 0 = RT \ln \frac{\hat{f}_i^\alpha}{\hat{f}_i^\beta}$$

which requires equality of fugacity of each component in each phase:

---

[5] Other characteristics of a perfect gas mixture are no volume or enthalpy change on forming the mixture, and the mixture equation of state is $PV = \Sigma n_i RT$.

$$\hat{f}_i^{\alpha} = \hat{f}_i^{\beta} \tag{9-10}$$

Either the chemical potential or the fugacity may be used as a condition of equilibrium; however, for vapor-liquid equilibrium the fugacity is more convenient.

9-5 DETERMINATION OF FUGACITIES OF PURE SUBSTANCES

Pure Gases. To determine how fugacities are related to measurable properties, we begin with our definition,

$$[d\mu = RT\, d\ln f]_T \tag{9-2}$$

and recall also that

$$[d\mu = v\, dP]_T \tag{8-5}$$

Eliminating $d\mu$ produces

$$RT\, d\ln f = v\, dP \tag{9-11}$$

We now add $-RT\, d\ln P$ to each side of Eq. (9-11) and rearrange to obtain

$$RT\, d\ln\frac{f}{P} = \left(v - \frac{RT}{P}\right) dP \tag{9-12}$$

This equation is then integrated with the boundary condition $P = 0; f\,/\,P = 1$ as prescribed by Eq. (9-3):

$$RT \ln\frac{f}{P} = \int_0^P \left(v - \frac{RT}{P}\right) dP \tag{9-13}$$

Equation (9-13) provides the means of calculating values of f corresponding to a particular temperature and pressure. This can be done if the molar volume v of the gas is known as a function of pressure P at the specified temperature. This knowledge could be in the form of an equation of state with parameters determined from experimental measurements or simply as a set of experimentally determined values of P and v.

As you have probably noted, the ratio f/P appears regularly. For convenience this ratio is called the fugacity coefficient and is denoted by ϕ:

$$\phi = \frac{f}{P} \tag{9-14}$$

We will now show that the fugacity coefficient is a reduced property which can be correlated by means of the corresponding states principle. We begin with Eq. (9-13) and replace v with ZRT/P:

$$\ln \phi = \int_0^P (Z-1) \frac{dP}{P} \qquad (9\text{-}15)$$

Equation (9-15) provides a means of estimating fugacities where insufficient experimental data exist for application of Eq. (9-13). Values of the compressibility factor Z estimated from a corresponding states plot can be used via graphical integration of Eq. 9-15 to obtain ϕ. Values of ϕ so determined can then be correlated in terms of reduced variables, because ϕ is also a reduced property, as can be seen from rearrangement of Eq. (9-15):

$$\ln \phi = \int_0^{P_r} (Z-1) \frac{dP_r}{P_r} \qquad (9\text{-}15a)$$

Because Z is a reduced property, the right-hand side of Eq. (9-15a) depends only on reduced variables, and hence ϕ should also depend only on these reduced variables. Figure 9-7 is a corresponding states correlation of ϕ.

EXAMPLE 9-1

Calculate the fugacity of carbon dioxide at 100°F and 200 psia from volumetric data.

Solution 9-1

The following Pv data are available[6] at 100°F:

P (psia)	1	10	20	40	80
v (ft^3/lb)	136.6	13.61	6.778	3.363	1.657

P (psia)	120	160	200	240
v (ft^3/lb)	1.088	0.8033	0.6376	0.5237

---

[6] J. H. Perry, *Chemical Engineer's Handbook*, 3rd ed., McGraw-Hill, New York, 1950, p. 255.

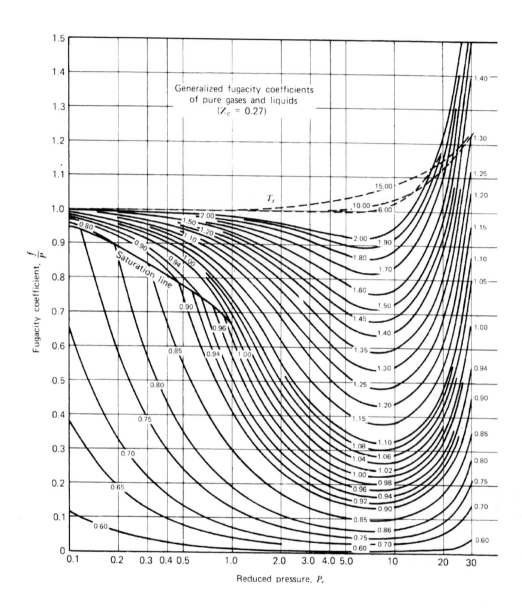

Figure 9-7 [From O. A. Hougen, K. M. Watson, and R. A. Ragatz, *Chemical Process Principles Charts,* 2nd ed., Wiley, 1960 as adapted in *Chemical and Engineering Thermodynamics* by S. I. Sandler, Wiley, 1977, with permission of John Wiley & Sons, Inc.]

After converting from specific volume to molar volume, the quantity $v - (RT / P)$ is evaluated for each pressure and is plotted vs. pressure in Fig. 9-8. The integral

$$\int_0^P \left(v - \frac{RT}{P} \right) dP$$

will be evaluated graphically from this figure after the question of how to draw the $v - (RT / P)$ vs. P curve in the low-pressure region has been answered. There is considerable uncertainty associated with $v - (RT / P)$ at low pressures because it involves the difference between two large, almost-equal numbers. At low pressures the virial equation truncated after the second coefficient should adequately represent the volumetric behavior:

$$Z = 1 + \frac{B}{v}$$

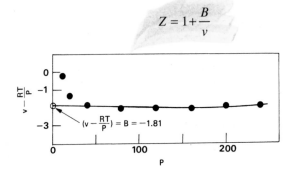

Figure 9-8

Now

$$v = \frac{ZRT}{P}$$

and we write

$$v - \frac{RT}{P} = \frac{RT}{P}\left(1 + \frac{B}{v}\right) - \frac{RT}{P}$$

$$v - \frac{RT}{P} = \frac{BRT}{Pv} = \frac{B}{Z}$$

In the limit as $P \to 0$, $Z \to 1.0$, and we write

$$\lim_{P \to 0} \left(v - \frac{RT}{P} \right) = B$$

and observe that the second virial coefficient B is the limiting value of the quantity $v - (RT / P)$.

The second virial coefficient is now needed and will be obtained by fitting experimental Pv data at 100°F to the following linearized version of the virial equation:

$$(Z - 1)v = B + \frac{C}{v} \tag{3-3a}$$

For this parameter estimation the following additional[7] data will be used:

P (psia)	300	360	440	520	600	800
v (ft^3/lb)	0.4100	0.3341	0.2652	0.2174	0.1823	0.1196

Converting to molar volume and computing $(Z - 1)v$ allows plotting of

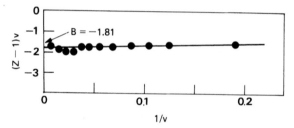

Figure 9-9

this quantity vs. $1/v$, as shown in Fig. 9-9. With the exception of a few points at low pressure, the linear relationship suggested by Eq. (3-3a) is observed. The value of B determined from the intercept is -1.81 ft^3/lb mol. This is the value of $v - (RT / P)$ at $P = 0$ and is shown in Fig. 9-8. A smooth curve can be drawn through this point and all but the two low-pressure points. Graphical integration based on this curve yields

---

[7] Also from J. H. Perry, *ibid.*

$$\int_0^{200} \left(v - \frac{RT}{P} \right) dP = -389.5 \ \text{ft}^3 \cdot \text{psi} / \text{lb mol}$$

and from Eq. 9-13 we have

$$\ln \frac{f}{P} = \ln \phi = \frac{-389.5}{10.72(560)} = -0.0649$$

$$\phi = \frac{f}{P} = 0.937$$

$$f = 0.937(200) = 187.4 \ \text{psia}$$

EXAMPLE 9-2

The source of Pv data for carbon dioxide also contains values of h and s. Use them to calculate the fugacity of carbon dioxide at 100°F and 200 psia.

Solution 9-2

We integrate Eq. (9-2) between states I and II to obtain

$$\mu_{\text{II}} - \mu_{\text{I}} = RT \ln \frac{f_{\text{II}}}{f_{\text{I}}} \qquad (9\text{-}16)$$

Recalling that $\mu = g$ [Eq. (8-7)] and $g = h - Ts$ [Eq. (5-2)], we may write

$$h_{\text{II}} - h_{\text{I}} - T(s_{\text{II}} - s_{\text{I}}) = RT \ln \frac{f_{\text{II}}}{f_{\text{I}}} \qquad (9\text{-}17)$$

We find h and s tabulated at 100°F for pressures of 200 and 1 psia:

$$P_{\text{I}} = 1 \ \text{psia} \qquad h_{\text{I}} = 318.0 \ \text{Btu} / \text{lb} \qquad s_{\text{I}} = 1.5506 \ \text{Btu} / \text{lb} \ °\text{R}$$
$$P_{\text{II}} = 200 \ \text{psia} \quad h_{\text{II}} = 310.6 \ \text{Btu} / \text{lb} \qquad s_{\text{II}} = 1.3038 \ \text{Btu} / \text{lb} \ °\text{R}$$

A pressure of 1 psia is low enough for us to safely assume ideal-gas behavior where $P_1 = f_{\text{I}} = 1 \ \text{psia}$. Substitution in the preceding equation yields

$$44.1[(310.6-318.0)-560(1.3038-1.5506)] = 1.987(560) \ln \frac{f}{1}$$

$$f = 178.4 \text{ psia}$$

Much closer agreement with the fugacity calculated in Ex. 9-1 was expected; therefore an error analysis is warranted. Use of Eq. (9-13) requires only v as a function of P. However, h and s needed in Eq. (9-17) require the more uncertain quantity $(\partial v / \partial T)_P$ for their calculation.[8] Further, the use of Eq. (9-17) involves subtraction, which can also increase the uncertainty. Equation (9-13) involves the most direct use of the experimental Pv data and therefore is expected to yield the more reliable value.

EXAMPLE 9-3

Estimate the fugacity of carbon dioxide at 100°F and 200 psia using
(a) Figure 9-7.
(b) the Peng-Robinson equation of state.

Solution 9-3

From Table A-1 we find

$$T_c = 304.2 \text{ K} \qquad P_c = 72.9 \text{ atm} \qquad \omega = 0.239$$

The reduced temperature and pressure are

$$T_r = \frac{560}{1.8(304.2)} = 1.023$$

$$P_r = \frac{200}{14.7(72.9)} = 0.1866$$

(a) From Fig. 9-7 we find $\phi = 0.955$ and

$$f = 0.955(200) = 191 \text{ psia}$$

(b) We may interpret ϕ as the ratio of the fugacity of a gas in the real state divided by its fugacity in the ideal-gas state. Thus, utilizing Eqs. (9-16) and (8-7) we can write

---

[8] See Sec. 5-5.

$$RT \ln \phi = RT \ln \frac{f}{P} = -\Delta g^*$$

where Δg^* is the residual Gibbs free energy, expressible as

$$\Delta g^* = \Delta h^* - T \Delta s^*$$

We now write

$$-RT \ln \phi = \Delta g^* = \Delta h^* - T \Delta s^*$$

or

$$-T_r \ln \phi = \frac{\Delta h^*}{RT_c} - T_r \frac{\Delta s^*}{R}$$

By the method illustrated in Ex. 5-2, Eqs. (5-58) and (5-59) yield

$$\frac{\Delta h^*}{RT_c} = 0.2208 \quad \text{and} \quad \frac{\Delta s^*}{R} = 0.1488$$

which lead to

$$-1.023 \ln \phi = 0.2208 - 1.023(0.1488)$$

$$\phi = 0.935$$

$$f = 0.935(200) = 187.0 \text{ psia}$$

Here we see that Fig. 9-7 provides a good estimate and the Peng-Robinson equation, an excellent estimate to the actual value of 187.4 psia determined in Ex. 9-1. A more direct derivation of ϕ from the Peng-Robinson equation is given in Sec. 9-12.

Pure Liquids.   To calculate the fugacity of a pure liquid, use is made of the condition

$$f^L = f^G \tag{9-18}$$

which applies when liquid and vapor are in equilibrium. At a specified temperature when liquid and vapor are in equilibrium, the pressure is referred to as the vapor pressure or saturation pressure. If we can calculate the fugacity of the vapor at saturation, we then know via Eq. (9-18) the fugacity of the liquid at the desired temperature and the saturation pressure.

Equation (9-11) is readily integrated for a liquid assumed incompressible, and the use of the preceding boundary condition results in

$$RT \ln \frac{f}{f^{\text{sat}}} = v(P - P^{\text{sat}})$$

(9-19)

where f is the liquid fugacity at the pressure P, f^{sat} is the liquid (or vapor) fugacity at saturation where the pressure is P^{sat}, and v is the liquid molar volume. As with many other properties of liquids, the fugacity is not strongly dependent on pressure. What has been presented here for the calculation of liquid fugacity applies equally well to the calculation of the fugacity of a solid with a measurable saturation pressure.

EXAMPLE 9-4

Calculate the fugacity of liquid water at 30°C (303.15 K) and at the following pressures: the saturation pressure, 10 bar, and 100 bar.

Solution 9-4

From the steam tables we find $P^{\text{sat}} = 0.0424$ bar at 30°C. This is a pressure low enough that we may safely state that

$$f^{\text{sat}} = P^{\text{sat}} = 0.0424 \text{ bar}$$

To calculate the fugacity at other pressures via Eq. (9-19), we require the liquid molar volume. The liquid specific volume at saturation is $0.001004 \text{ m}^3/\text{kg}$, and the corresponding molar volume is

$$0.001004(18.02) = 0.01809 \text{ m}^3 / \text{kmol}$$

Equation (9-19) becomes

$$8314.3(303.15) \ln \frac{f}{0.0424} = 0.01809(P - 0.0424)(10^5)$$

and we obtain

$$f = 0.0427 \text{ bar for } P = 10 \text{ bar}$$
$$f = 0.0455 \text{ bar for } P = 100 \text{ bar}$$

It is seen that over a moderate pressure range there is little variation of the fugacity of a pure liquid.

9-6 DETERMINATION OF FUGACITIES IN MIXTURES

Gaseous Solutions.   For the ith component of a gaseous solution we defined the fugacity through Eq. (9-7):

$$d\hat{\mu}_i = RT d \ln \hat{f}_i \tag{9-7}$$

The preceding definition applies to isothermal changes. We have previously determined the dependence of the chemical potential of a component in a mixture on pressure under conditions of constant temperature and composition [Eq. (8-3)]. Rearranging Eq. (8-3) gives

$$d\hat{\mu}_i = \left(\frac{\partial V}{\partial n_i}\right)_{T,P,n_j} dP$$

Eliminating $d\hat{\mu}_i$ between these two equations leaves

$$RT d \ln \hat{f}_i = \left(\frac{\partial V}{\partial n_i}\right)_{T,P,n_j} dP \tag{9-20}$$

which applies to a change in pressure occurring at constant temperature and composition. From each side of Eq. (9-20) we will subtract $RT d \ln p_i$ and obtain

$$RT d \ln \frac{\hat{f}_i}{p_i} = \left(\frac{\partial V}{\partial n_i}\right)_{T,P,n_j} dP - RT d \ln p_i \tag{9-21}$$

From the definition of p_i we may write the last term as

$$RT d \ln p_i = RT d \ln P + RT d \ln y_i$$

Because we are considering a change at constant composition, $d \ln y_i = 0$, and Eq. (9-21) becomes

$$RT d \ln \frac{\hat{f}_i}{p_i} = \left(\left(\frac{\partial V}{\partial n_i}\right)_{T,P,n_j} - \frac{RT}{P}\right) dP \tag{9-22}$$

Equation (9-8) provides the boundary condition for the integration of Eq. (9-22), which results in

$$RT \ln \frac{\hat{f}_i}{p_i} = \int_0^P \left(\left(\frac{\partial V}{\partial n_i}\right)_{T,P,n_j} - \frac{RT}{P}\right) dP \tag{9-23}$$

The ratio of fugacity to pressure is again referred to as the fugacity coefficient and is denoted by $\hat{\phi}_i$:

$$\hat{\phi}_i = \frac{\hat{f}_i}{p_i} \tag{9-24}$$

Equation (9-23) provides the means of calculating the fugacity of a component in a gaseous solution. This calculation requires the evaluation of $(\partial V/\partial n_i)_{T,P,n_j}$ as a function of pressure, which in turn requires at each pressure the knowledge of how the solution volume depends on composition. This type of data must be obtained by experimentation and is rarely available, thus prohibiting direct and rigorous calculation of fugacity via Eq. (9-23).

Because the calculation of fugacities via Eq. (9-23) is rarely possible, it will be necessary to adopt a different strategy. We will use a model which allows us to calculate a value for the fugacity, and we must determine the limits of applicability of the model and devise means of correcting the model when it is applied outside these limits. The model is referred to as the *ideal solution model*; it specifies that the fugacity of component i in an ideal gaseous solution is[9]

$$\hat{f}_i^G = f_i^G y_i \tag{9-25}$$

where f_i^G is the vapor fugacity of pure component i at the temperature and pressure of the solution, and y_i is its mol fraction.

Previously, reference was made to a *perfect gas mixture* in which component fugacities were equal to partial pressures

$$\hat{f}_i = p_i = Py_i$$

The similarity to an *ideal gaseous solution* is apparent on comparing the preceding with Eq. (9-25). The *perfect gas mixture* model is used at low pressure, while the *ideal gaseous solution* model can be used at increasingly higher pressures. Both are models and should be used only under conditions where they have been shown to represent actual mixture behavior. The following guidelines are offered in determining the useful range of Eq. (9-25):[10]

1. It is always a good approximation at low pressure.
2. It is always a good approximation at any pressure whenever i is present in large excess and becomes exact in the limit as y_i approaches unity.
3. It is often a fair approximation over a wide range of composition and pressure whenever the physical properties of all the components are nearly the same.

---

[9] Equation (9-25) is often referred to as the Lewis and Randall rule.
[10] J. M. Prausnitz, R. N. Lichtenthaler, and E. Gomes de Azevedo, *Molecular Thermodynamics of Fluid-Phase Equilibria*, 2nd ed., Prentice-Hall, Englewood Cliffs, NJ, 1986, Chap. 5.

4. It is almost always a poor approximation at moderate and high pressures whenever the molecular properties of the other components are significantly different from those of i and when i is not present in excess.

Another approach to calculating fugacities in vapor mixtures is to use an equation of state. The virial equation is widely used for this purpose because the composition dependence of the parameters is prescribed by theory. Based on the virial equation, the fugacity coefficient of component i in an m-component gas mixture is[11]

$$\ln \hat{\phi}_i = \frac{2}{v}\sum_{j=1}^{m} y_j B_{ij} + \frac{3}{2v^2}\sum_{j=1}^{m}\sum_{k=1}^{m} y_i y_k C_{ijk} - \ln Z \qquad (9\text{-}26)$$

where Z is the compressibility factor of the mixture. At low to moderate pressures this expression can be simplified by setting the C's equal to zero. In this pressure range a more convenient expression resulting from the use of Eq. (3-3a) in Eq. (9-23) is

$$RT \ln \hat{\phi}_i = \left[2\sum_{j=1}^{m} y_j B_{ij} - B_m \right] P \qquad (9\text{-}27)$$

where the mixture coefficient is

$$B_m = \sum_{i=1}^{m}\sum_{j=1}^{m} y_i y_j B_{ij}$$

For binary mixtures Eq. (9-27) reduces to

$$RT \ln \hat{\phi}_1 = [2(y_1 B_{11} + y_2 B_{12}) - B_m]P \qquad (9\text{-}28a)$$

$$RT \ln \hat{\phi}_2 = [2(y_2 B_{22} + y_1 B_{12}) - B_m]P \qquad (9\text{-}28b)$$

where

$$B_m = y_1^2 B_{11} + 2y_1 y_2 B_{12} + y_2^2 B_{22}$$

B_{11} and B_{22} are the second virial coefficients for pure 1 and 2 and can be measured experimentally or estimated from corresponding states correlations. B_{12} is called the cross coefficient and is estimated from corresponding states correlations based on critical properties determined from empirical mixing rules. A thorough treatment of fugacities from the virial

---

[11] *Ibid.*
[12] *Ibid.*

equation is given by Prausnitz et al.[13] In Sec. 9-12 the Peng-Robinson equation is used to calculate fugacity coefficients in vapor and liquid mixtures and the program PRVLE.EXE computes them.

Liquid Solutions. For liquid solutions a modeling approach is again used. The concept of an ideal liquid solution analogous to an ideal gaseous solution is used, where

$$\hat{f}_i^L = f_i^L x_i \tag{9-29}$$

$\hat{f}_i^L$ the fugacity of component i in the ideal liquid solution, f_i^L is the fugacity of pure liquid i at the temperature and pressure of the mixture, and x_i is the liquid mol fraction of i. While we have seen that for a reasonable range of systems and pressure the ideal gaseous solution model is adequate, the same cannot be said for the ideal liquid solution model. Very few solutions follow Eq. (9-29) over the entire composition range—only those in which components are isomers or members of the same chemical family. Additionally, the magnitude of deviations of real solutions from this model can often be enormous. We will deal first with that limited, but important, class of systems which do conform to the model, and then we will see how the inadequacy of the model can be surmounted.

9-7 IDEAL SYSTEMS

In this section we will treat systems which conform to the ideal solution model in both the gas and liquid phases. As mentioned before, most systems conform reasonably well to the model in the gas phase at low to moderate pressures, but few systems conform in the liquid phase. Only systems comprised of very closely related compounds such as isomers or members of a homologous series (e.g., paraffin hydrocarbons) have been found to closely follow the ideal liquid solution model. In spite of this limited applicability, this type of system is encountered often in the petroleum industry, but, more importantly, in dealing first with these systems which are easy to treat you should develop a physical appreciation for the concepts of phase equilibrium without becoming overwhelmed with calculation details.

If vapor and liquid phases are in equilibrium, we may apply the fugacity condition

$$\hat{f}_i^G = \hat{f}_i^L$$

When the ideal solution model is applied to each phase, we obtain

$$y_i f_i^G = x_i f_i^L \tag{9-30}$$

For any component the fugacities f_i^G and f_i^L depend only on temperature and pressure, as does their ratio

---

[13] *Ibid.*

$$K_i = \frac{f_i^L}{f_i^G}$$

where

$$y_i = K_i x_i \qquad\qquad (9\text{-}31)$$

The quantity K_i is referred to as the *equilibrium ratio* for a component and can be easily calculated and correlated as a function of temperature and pressure. Such a correlation for several hydrocarbons appears in Fig. 9-10. While it is possible to evaluate the pure component fugacities directly and use Eq. (9-30), when K's are available, the use of Eq. (9-31) is more expeditious. At low pressures the fugacities in Eq. (9-30) can be replaced by the appropriate pressure term. The fugacity f_i^G of the pure gas simply becomes the system pressure P. When both P and P^{sat} are low, Eq. (9-19) shows that $f^L = f^{sat}$, and f^{sat} becomes the vapor pressure P°. Therefore, the low-pressure statement of Eq. (9-30) is

$$Py_i = P_i^\circ x_i \qquad\qquad (9\text{-}32)$$

This is often referred to as Raoult's law. This law can be written in the form of Eq. (9-31) with

$$K_i = \frac{P_i^\circ}{P} \qquad\qquad (9\text{-}33)$$

In using equilibrium ratios, K_i's, a minor computational problem arises because K depends on temperature and pressure, and usually only one is known from the problem statement. This is a restriction imposed by the phase rule. In a two-phase, C-component system there will be C variables to be specified. Usually, the complete composition of a phase will be specified requiring $C-1$ mol fractions and leaving only one other variable which can be fixed—either temperature or pressure. Therefore, computations for ideal systems usually involve trial and error. For a system where the temperature and the liquid mol fractions are known, the following computation scheme may be followed:

Step 1. Assume P.

Step 2. Evaluate K's.

Step 3. Calculate y's from $y_i = K_i x_i$.

Step 4. Test: Does $\Sigma y_i = 1$?

A similar scheme may be followed where either y_i's and T, x_i's and P, or y_i's and P are known.

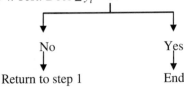

No   Yes

Return to step 1   End

Ideal solutions only?

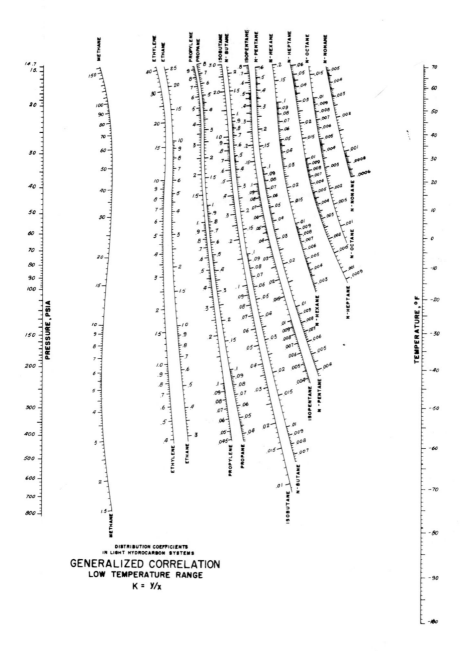

Figure 9-10 Equilibrium ratios for light hydrocarbons. [Reproduced by permission from C. L. DePriester, *Chem. Eng. Prog. Symp. Ser., 7,* 49 (1953).]

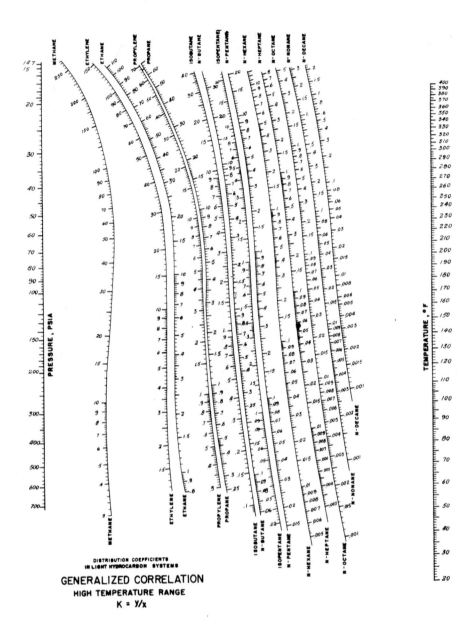

DISTRIBUTION COEFFICIENTS
IN LIGHT HYDROCARBON SYSTEMS

GENERALIZED CORRELATION
HIGH TEMPERATURE RANGE
K = y/x

Figure 9-10 (Con't)

EXAMPLE 9-5

Because they are members of the same chemical family, benzene and toluene are expected to form ideal liquid solutions. Calculate the data needed to construct a Txy diagram for this system at a pressure of 1 bar (750.1 mmHg). Constants in the Antoine equation are given. For the equation:

$$\log P^{\circ} = A - \frac{B}{t + C}$$

with P° in mmHg and t in $^{\circ}$C, the parameters are

	A	B	C
Benzene	6.90565	1211.033	220.790
Toluene	6.95334	1343.943	219.377

Solution 9-5

A trial and error calculation can be avoided if we choose a temperature and solve for x_B and y_B. At a temperature of 90°C we have

$$P_B^{\circ} = 1021 \text{ mm} \qquad P_T^{\circ} = 406.7 \text{ mm}$$

and

$$K_B = \frac{1021}{750.1} = 1.361$$

$$K_T = \frac{406.7}{750.1} = 0.5422$$

We now write

$$y_B = 1.361 x_B$$

$$1 - y_B = 0.5422(1 - x_B)$$

and solve to find

$$x_B = 0.559 \qquad y_B = 0.761$$

The following table is obtained by repeating the calculation for other temperatures:

$t(°C)$	x_B	y_B
85	0.755	0.887
90	0.559	0.761
95	0.390	0.612
100	0.244	0.439
105	0.116	0.239

Construction of the Txy diagram is aided by a knowledge of the saturation temperatures of pure benzene and toluene at 1 bar. These are 79.7 and 110.7°C, respectively, as found from the Antoine equation. The Txy diagram for this system is shown in Fig. 9-11.

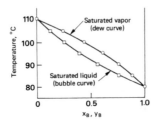

Figure 9-11 Calculated Txy diagram or benzene and toluene.

EXAMPLE 9-6

A vapor mixture of the composition

C_2H_4: 20%
C_2H_6: 20%
C_3H_8: 40%
C_4H_{10}: 20%

is to be condensed at 80°F. Find the dew point pressure and the pressure required to totally condense the vapor.

Solution 9-6

At the dew point the vapor composition is unchanged, and by writing Eq. (9-31) for each component, we have

$$0.2 = K_1 x_1$$

$$0.2 = K_2 x_2$$

$$0.4 = K_3 x_3$$

$$0.2 = K_4 x_4$$

The procedure is to assume a pressure, evaluate the K's from Fig. 9-10, calculate x's, and check for $\sum x_i = 1$. A summary of the calculations is shown in the following table:

	Trial		
	1	2	3
Pressure (psia)	160	120	130
K_1	4.55	5.9	5.5
K_2	2.98	3.75	3.6
K_3	0.97	1.25	1.18
K_4	0.29	0.358	0.340
$x_1 = 0.2/K_1$	0.044	0.034	0.036
$x_2 = 0.2/K_2$	0.067	0.053	0.056
$x_3 = 0.4/K_3$	0.412	0.320	0.339
$x_4 = 0.2/K_4$	0.690	0.559	0.588
$\sum x_i$	1.213	0.966	1.019

The first assumed pressure of 160 psia yielded $\sum x_i > 1$, indicating that K's should be larger. This corresponds to lower pressure; therefore, 120 psia was the second guess. A $\sum x_i < 1$ indicated that the pressure had been bracketed, and linear interpolation indicated that 130 psia should be reasonable. Calculations performed at this pressure yielded $\sum x_i = 1.019$, which can be considered reasonable in light of the uncertainties of reading K's from Fig. 9-10. The dew point pressure is approximately 130 psia. Liquid compositions can be normalized by dividing by 1.019.

When the vapor is totally condensed, the liquid has the same composition as the original vapor, and we write

$$y_1 = 0.2K_1 \qquad\qquad y_3 = 0.4K_3$$
$$y_2 = 0.2K_2 \qquad\qquad y_4 = 0.2K_4$$

The results of the trial and error procedure are summarized as follows:

	Trial		
	1	2	3
Pressure (psia)	300	400	365
K_1	2.65	2.08	2.25
K_2	1.80	1.41	1.51
K_3	0.58	0.47	0.51
K_4	0.187	0.154	0.165
$y_1 = 0.2K_1$	0.530	0.416	0.450
$y_2 = 0.2K_2$	0.360	0.282	0.302
$y_3 = 0.4K_3$	0.230	0.188	0.204
$y_4 = 0.2K_4$	0.037	0.031	0.033
Σy_i	1.159	0.917	0.989

The pressure needed for total condensation is seen to be approximately 365 psia.

For the operations of flash vaporization or partial condensation the system description requires material balance equations as well as equilibrium relationships. Consider the case of partial condensation as illustrated in Fig. 9-12.

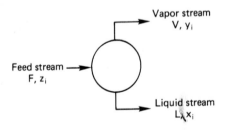

Figure 9-12 Flash vaporization or partial condensation

At the pressure P there are F mol of vapor entering with component mol fractions denoted by z_i. Leaving will be L mol of liquid with mol fractions x_i and V mol of vapor with mol fractions y_i. Equilibrium between the exit streams is assumed, and

$$y_i = K_i x_i$$

Material balances are

$$F = L + V$$

$$Fz_i = Lx_i + Vy_i$$

Solving these equations for x_i gives

$$x_i = \frac{z_i}{(L/F) + [1 - (L/F)]K_i}$$

(9-34)

This same equation applies to the partial vaporization of a liquid feed of flow rate F and mol fractions z_i's. Computations for this type of problem very often also involve a trial and error procedure.

EXAMPLE 9-7

The overhead vapor from a fractionating column has the following analysis:
  Ethane: 15%
  Propane: 20%
  Isobutane: 60%
  n-Butane: 5%
It is desired to condense 75% of this vapor with the condenser temperature at 80°F. What pressure is required?

Solution 9-7

For this problem Eq. (9-34) becomes

$$x_i = \frac{z_i}{0.75 + 0.25\, K_i}$$

(9-34a)

The solution will be accomplished by trial and error using the following calculation scheme:

Step 1. Assume P.
Step 2. Evaluate K_i's at 80°F and P.
Step 3. Calculate x_i's from Eq. (9-34a).

Step 4. Test: Does $\Sigma x_i = 1$?

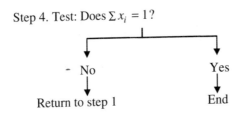

- No Yes

Return to step 1 End

The results are tabulated for two assumed pressures:

Component	z_i	$P = 100$ psia		$P = 110$ psia	
		K_i	x_i	K_i	x_i
Ethane	0.15	4.53	0.080	4.15	0.084
Propane	0.20	1.46	0.180	1.35	0.184
i-Butane	0.60	0.59	0.669	0.54	0.677
n-Butane	0.05	0.42	0.058	0.38	0.059
			0.987		1.004

Because we have bracketed $\Sigma x_i = 1$, the pressure lies between 100 and 110 psia and closer to 110 psia. Actually, either of these pressures provides reasonable results in view of the uncertainties in reading K's from Fig. 9-10. The x_i's are seen to be essentially the same for the two pressures and if normalized to $\Sigma x_i = 1$ would be in better agreement. Values of y_i's can, of course, be obtained from these x_i's and Eq. (9-31).

9-8 THE ACTIVITY COEFFICIENT

We have seen that the ideal solution model represents gas phase behavior reasonably well up to moderate pressure but that the model is generally inadequate in regard to the liquid phase. To overcome this deficiency, we will introduce a correction to the ideal liquid solution model in the following way:

$$\hat{f}_i^L = f_i^L x_i \gamma_i \qquad (9\text{-}35)$$

where γ_i is the correction by which the ideal solution fugacity $f_i^L x_i$ is multiplied so that it equals the actual fugacity. In a way this is similar to using the ideal gas model and correcting it with the compressibility factor Z, except that while Z was found to depend only on T and P, it can be expected that γ_i will depend on liquid phase composition as well as T and P. Because γ_i is defined in terms of fugacities, it is a well-defined thermodynamic property and is given the name of activity coefficient. We now wish to use the available thermodynamic tools to determine how the activity coefficient depends on the system variables.

We first rearrange Eq. (9-35) and take the logarithm of both sides:

$$RT \ln \frac{\hat{f}_i^L}{f_i^L} = RT \ln x_i + RT \ln \gamma_i \qquad (9\text{-}36)$$

Integration of Eq. (9-7) yields

$$\hat{\mu}_i^L - \mu_i^L = RT \ln \frac{\hat{f}_i^L}{f_i^L} \qquad (9\text{-}37)$$

Combining Eqs. (9-36) and (9-37) gives

$$\hat{\mu}_i^L - \mu_i^L = RT \ln x_i + RT \ln \gamma_i \qquad (9\text{-}38)$$

We now differentiate Eq. (9-38) with respect to pressure, holding temperature and liquid-phase composition constant:

$$\left(\frac{\partial \hat{\mu}_i^L}{\partial P} \right)_{T,x} - \left(\frac{\partial \mu_i^L}{\partial P} \right)_T = RT \left(\frac{\partial \ln \gamma_i}{\partial P} \right)_{T,x} \qquad (9\text{-}39)$$

Utilizing Eqs. (8-3) and (8-5) yields

$$RT \left(\frac{\partial \ln \gamma_i}{\partial P} \right)_{T,x} = \left(\frac{\partial V^L}{\partial n_i} \right)_{T,P,n_j} - v_i^L \qquad (9\text{-}40)$$

As the data needed to evaluate the first term on the right-hand side are seldom available, Eq. (9-40) is rarely used. However, Eq. (9-40) suggests and the weight of experimental evidence shows that, like many other liquid phase properties, the activity coefficient is not strongly dependent on pressure. It is customary to ignore pressure effects up to moderate pressures.

A similar approach can be used to determine a rigorous relation for the dependence of activity coefficient on temperature.[14] As with Eq. (9-40) the needed data are usually lacking; however, this effect cannot be neglected. The usual procedure is to determine the temperature dependence empirically in a manner we shall discuss later.

To determine the composition dependence of γ_i, it will first be necessary to derive a basic thermodynamic relation: the Gibbs-Duhem equation. We begin with the fundamental equation applied to a single phase,

---

[14] See Sec. 11-2.

$$dU = T \, dS - P \, dV + \sum \hat{\mu}_i \, dn_i \tag{7-17}$$

and consider its integration for the special case where temperature, pressure, and mol fractions are held constant while the quantity of the phase is increased. This means that the various dn_i's are adjusted so that the relative proportions of the n_i's remain constant. Under these conditions all intensive variables, such as μ_i's, which depend on intensive properties will remain constant. This integration yields

$$\Delta U = T \Delta S - P \Delta V + \sum \hat{\mu}_i \, \Delta n_i \tag{9-41}$$

Let the original values of the extensive variables be U, S, V, and n_i. If the system is increased to k times its original quantity, the final values will be kU, kS, kV, and kn_i so that the changes will be $(k-1)U$, $(k-1)S$, $(k-1)V$, and $(k-1)n_i$, and Eq. (9-41) becomes

$$(k-1)U = T(k-1)S - P(k-1)V + \sum \hat{\mu}_i (k-1)n_i$$

and thus

$$U = TS - PV + \sum \hat{\mu}_i \, n_i \tag{9-42}$$

which can be written as

$$G = \sum \hat{\mu}_i n_i \tag{9-43}$$

Although it was derived under what might appear to be somewhat restricted conditions, Eq. (9-43) is a very useful relationship. It relates at any given temperature and pressure the Gibbs free energy of the solution to the chemical potentials of each component. This equation is now differentiated:

$$dG = \sum \hat{\mu}_i \, dn_i + \sum n_i \, d\hat{\mu}_i \tag{9-44}$$

At constant T and P the Gibbs fundamental equation [Eq.(7-20)] becomes

$$\left[dG = \sum \hat{\mu}_i \, dn_i \right]_{T, P}$$

Comparison of this equation with Eq. (9-44) reveals that

$$\left[\sum n_i \, d\hat{\mu}_i = 0 \right]_{T, P} \tag{9-45}$$

Equation (9-45), known as the Gibbs-Duhem equation, places a restriction on changes in chemical potential in a solution at constant temperature and pressure. Alternatively, this equation can be expressed in terms of fugacity by invoking Eq. (9-7):

$$\left[\sum n_i \, d \ln \hat{f}_i = 0\right]_{T,P} \tag{9-46}$$

Mol numbers n_i's can be changed to mol fractions x_i's by dividing by $\sum n_i$:

$$\left[\sum x_i \, d \ln \hat{f}_i = 0\right]_{T,P} \tag{9-47}$$

Expression in terms of activity coefficients is possible if we write the defining equation [Eq. (9-35)] in logarithmic form and differentiate under conditions of constant T and P:

$$d \ln \hat{f}_i^L = d \ln f_i^L + d \ln x_i + d \ln \gamma_i$$

The first right-hand term is zero because f_i^L depends only on T and P. We now substitute the preceding into Eq. (9-47) to obtain

$$\sum x_i \, d \ln x_i + \sum x_i \, d \ln \gamma_i = 0$$

Now

$$\sum x_i \, d \ln x_i = \sum x_i \frac{dx_i}{x_i} = \sum dx_i = 0$$

and hence

$$\left[\sum x_i \, d \ln \gamma_i = 0\right]_{T,P} \tag{9-48}$$

Equations (9-45)–(9-48) are each different forms of the Gibbs-Duhem equation.

Let us now consider the application of Eq. (9-48) to a liquid phase consisting of two components:

$$x_1 \, d \ln \gamma_1 + x_2 \, d \ln \gamma_2 = 0$$

Although the notation has been dropped, it should be remembered that this equation applies at conditions of constant T and P. If T and P are fixed, we recognize that the changes in γ_1 and γ_2 must result from changes in composition of the phase, and if we choose x_1 as our composition variable, we may write

$$x_1 \frac{d \ln \gamma_1}{dx_1} + x_2 \frac{d \ln \gamma_2}{dx_1} = 0 \tag{9-49}$$

From Eq. (9-49) we see that γ_1 and γ_2 are not independent of each other and that at constant T and P changes in γ_1 are related to changes in γ_2.

We have seen that the activity coefficient is a thermodynamic property and that its dependence on pressure and temperature is prescribed precisely by thermodynamic relationships. For example, changes in the activity coefficient due to changes in pressure could be determined via Eq. (9-40) if the necessary experimental data were available. This is not the case for changes in γ with respect to composition because while Eq. (9-49) is a necessary condition it is not sufficient to specify a unique relationship between γ and x. While the thermodynamic treatment of phase equilibrium would be much simpler and more direct if such a unique relationship existed, as we shall see later, this does not present an insurmountable problem.

9-9 EXPERIMENTAL DETERMINATION OF ACTIVITY COEFFICIENTS

After defining a thermodynamic property, it has been our custom, whenever possible, to show how the property may be determined from experimental measurements. First, this reminds us of the empirical basis for thermodynamics, which is extremely important to keep in mind when studying phase equilibrium. Second, we are often able to attach physical significance to the property when it is viewed in this context.

It should be borne in mind that the activity coefficient, while essential to the calculation of equilibrium between phases, has been defined as a property of the liquid phase. We have viewed it as a correction to the oversimplified ideal liquid solution model which in corrected form is

$$\hat{f}_i^L = f_i^L x_i \gamma_i \tag{9-35}$$

Also, we have seen that the fugacity can be regarded as an *equivalent pressure*, although this is more obvious when considering gas-phase fugacities. For the gas phase we have written

$$\hat{f}_i^G = f_i^G y_i \tag{9-25}$$

which at low pressures reduces to

$$\hat{f}_i^G = p_i = P y_i \tag{9-26}$$

These equations allow us to develop a sense of physical significance for $\hat{f}_i^G$ because it is approximated by partial pressure which in turn is determined by P and y_i, directly measurable quantities.

Thus, the fugacity of a component in a gas phase can be determined by measurements performed on, or knowledge of, the gas phase alone. The same cannot be said for the liquid phase.

The fugacity of a component in the liquid phase must be determined in a situation where the liquid phase is in equilibrium with a vapor phase. Thus, for each component we can write

$$\hat{f}_i^G = \hat{f}_i^L$$

Substitution of Eqs. (9-25) and (9-35) yields

$$f_i^G y_i = f_i^L x_i \gamma_i \tag{9-50}$$

which at low pressure becomes

$$p_i = P y_i = P_i^\circ x_i \gamma_i \tag{9-51}$$

Thus, the conception of the fugacity of a component in a liquid phase may be formed only in terms of the partial pressure of that component in a vapor phase which is in equilibrium with the liquid. At low pressure Eq. (9-51) allows values of γ_i to be calculated from P, y_i, and x_i resulting from phase equilibrium measurements and P_i°, which is a property of each substance and depends only on temperature. At higher pressures Eq. (9-50), which requires the determination of pure liquid and vapor fugacities, would be used. This calculation would be more tedious, but we have already seen how these fugacities may be evaluated. Let us again examine the behavior of binary vapor-liquid systems at equilibrium.

Experimental data for the system carbon disulfide–acetone determined at a constant temperature of 35.17°C are shown in Fig. 9-13 as a plot of total pressure P vs. liquid mol fraction of carbon disulfide. This curve is plotted from direct measurement of P and x, while the partial pressure curves are calculated from the measured values of P, y, and x. Dashed lines representing partial pressures and total pressure calculated from the ideal solution model are also shown. While it is obvious that the ideal solution model does a poor job of modeling the behavior of this system, it should be noted that actual partial pressure curves become asymptotic to the ideal partial pressure lines as the mol fraction of that component approaches unity. A plot of activity coefficients vs. liquid mol fraction for this system is shown in Fig. 9-14. These we have regarded as correction factors for the ideal solution model, and Fig. 9-14 shows that the correction can be quite large and is greatest when the mol fraction of a component approaches zero. We also see that activity coefficients approach unity as the mol fraction approaches unity. This system is said to exhibit positive deviations from ideal behavior because actual pressures exceed those predicted from the ideal solution model.

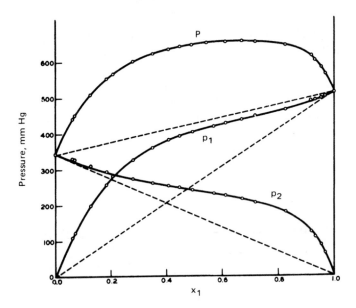

Figure 9-13 Partial and total pressures of carbon disulfide–acetone solutions at 35.17°C. [From J. Hildebrand and R. Scott, *The Solubility of Nonelectrolytes*, copyright © 1950 by Reinhold Publishing. Reprinted by permission of Dover Publications Inc, New York.]

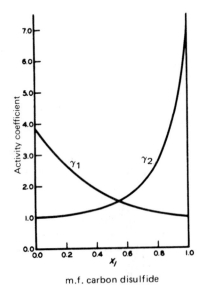

m.f. carbon disulfide

Figure 9-14 Activity coefficients for carbon disulfide–acetone solutions at 35.17°C. [From J. Hildebrand and R. Scott, *The Solubility of Nonelectrolytes*, copyright © 1950 by Reinhold Publishing. Reprinted by permission of Dover Publications Inc, New York.]

Shown in Fig. 9-15 is a plot of equilibrium data for the chloroform–acetone system at a temperature of 35.17°C. Here it is seen that actual pressures are always less than those predicted by the ideal solution model, and systems of this type are said to exhibit negative deviations. Note again that as the mol fraction of a component approaches unity the partial pressure curve asymptotically approaches the ideal partial pressure line. This is also manifested by an asymptotic approach of activity coefficients to unity in Fig. 9-16. Again, as shown in Fig. 9-16, deviations can be quite large and are largest as the mol fraction of the component approaches zero.

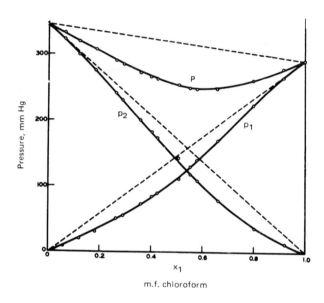

Figure 9-15 Partial and total pressures of chloroform-acetone solutions at 35.17°C. [From J. Hildebrand and R. Scott, *The Solubility of Nonelectrolytes*, copyright © 1950 by Reinhold Publishing. Reprinted by permission of Dover Publications Inc., New York.]

The observation that the ideal solution model becomes valid for a component at mol fractions approaching unity can be made for the two systems just examined. This observation has been made for many systems and is accepted as an empirical fact. Thus, we may say that the ideal liquid solution model is valid for only a few systems over the entire composition range but is valid for all systems in this limit. Also, this means that the activity coefficient will always approach unity as the mol fraction of the component approaches unity.

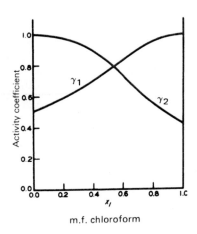

Figure 9-16 Activity coefficients for chloro-
form–acetone solutions at 37.17°C. [From J.
Hildebrand and R. Scott, *The Solubility of
Nonelectrolytes*, copyright © 1950 by Rein-
hold Publishing. Reprinted by permission of
Dover Publications Inc., New York.]

9-10 HENRY'S LAW

We have just seen that for all systems

$$\lim_{x_i \to 1.0} \hat{f}_i^L = f_i^L x_i \tag{9-52}$$

At this limit and at constant T and P we use the ideal solution model to write

$$d \ln \hat{f}_i^L = d \ln f_i^L + d \ln x_i = 0 + d \ln x_i \tag{9-53}$$

We recall now the Gibbs-Duhem equation written for a binary system [Eq.(9-47)]:

$$x_1 d \ln \hat{f}_1^L + x_2 d \ln \hat{f}_2^L = 0 \tag{9-54}$$

The substitution of Eq. (9-53) into Eq. (9-54) with the subscript 1 referring to the component in excess yields

$$x_1 d \ln x_1 = -x_2 d \ln \hat{f}_2^L$$

Recalling that $dx_1 = -dx_2$ and rearranging, we obtain

$$d \ln \hat{f}_2^L = d \ln x_2$$

which integrates to

$$\hat{f}_2^L = k_2 x_2 \qquad (9\text{-}55)$$

where k_2 is a characteristic constant for component 2 and depends only on temperature and pressure. Equation (9-55) is called Henry's law, and we see that over whatever range component 1 follows the ideal liquid solution model component 2 must follow Henry's law. Here we have taken an established experimental observation concerning one component and used a thermodynamic tool to prescribe the behavior of the other component. *This is typical of the type of victories we win through thermodynamics.*

9-11 ACTIVITY COEFFICIENT EQUATIONS

We have already remarked that thermodynamics provides insufficient guidance to allow the formulation of a unique relationship between activity coefficient and liquid composition. Instead there are any number of algebraic functions relating γ and x which will satisfy the Gibbs-Duhem equation and are therefore thermodynamically acceptable. However, over many years of the study of phase equilibrium several equations have been identified as having rather broad applicability and consequently have enjoyed wide usage. Following are three of these equations:

The Margules equation:

$$\ln \gamma_1 = x_2^2 \left[A_{12} + 2(A_{21} - A_{12})x_1 \right] \qquad (9\text{-}56\text{a})$$

$$\ln \gamma_2 = x_1^2 \left[A_{21} + 2(A_{12} - A_{21})x_2 \right] \qquad (9\text{-}56\text{b})$$

The van Laar equation:

$$\ln \gamma_1 = \frac{B_{12}}{\left[1 + (B_{12}x_1/B_{21}x_2) \right]^2} \qquad (9\text{-}57\text{a})$$

$$\ln \gamma_2 = \frac{B_{21}}{\left[1 + (B_{21}x_2/B_{12}x_1) \right]^2} \qquad (9\text{-}57\text{b})$$

The Wilson equation:

$$\ln \gamma_1 = -\ln (x_1 + x_2 G_{12}) + x_2 \left(\frac{G_{12}}{x_1 + x_2 G_{12}} - \frac{G_{21}}{x_2 + x_1 G_{21}} \right) \qquad (9\text{-}58\text{a})$$

$$\ln \gamma_2 = -\ln(x_2 + x_1 G_{21}) - x_1 \left(\frac{G_{12}}{x_1 + x_2 G_{12}} - \frac{G_{21}}{x_2 + x_1 G_{21}} \right) \qquad (9\text{-}58b)$$

In these equations the A's, B's, and G's are adjustable parameters which must be determined from an experimentally derived set of γx data. Some type of curve-fitting procedure or parameter-estimation routine can be used for this purpose. Because each equation contains only two parameters, a minimum of one Vapor-Liquid Equilibrium (VLE) data point is sufficient to determine them. This one data point provides T, P, x_1, and y_1 and allows the calculation of γ_1 and γ_2, which along with the values of x_1 and x_2 may be substituted into the equations to provide a set of two equations and two unknown parameters.

For the purpose of parameter estimation, the Margules and van Laar equations can be rearranged into linear form. The Margules equation becomes

$$\frac{\ln \gamma_1}{x_2^2} = A_{12} + 2(A_{21} - A_{12})x_1 \qquad (5\text{-}56c)$$

$$\frac{\ln \gamma_2}{x_1^2} = A_{21} + 2(A_{12} - A_{21})x_2 \qquad (5\text{-}56d)$$

and the van Laar equation becomes

$$\frac{1}{\sqrt{\ln \gamma_1}} = \frac{1}{\sqrt{B_{12}}} + \frac{\sqrt{B_{12}}}{B_{21}} \frac{x_1}{x_2} \qquad (5\text{-}57c)$$

$$\frac{1}{\sqrt{\ln \gamma_2}} = \frac{1}{\sqrt{B_{21}}} + \frac{\sqrt{B_{21}}}{B_{12}} \frac{x_2}{x_1} \qquad (5\text{-}57d)$$

Margules parameters can be obtained from an experimentally determined isothermal set of γx data by fitting straight lines to plots of $\ln \gamma_1 / x_2^2$ vs. x_1 and $\ln \gamma_2 / x_1^2$ vs. x_2. Similarly, plots of $1/\sqrt{\ln \gamma_1}$ vs. x_1/x_2 and $1/\sqrt{\ln \gamma_2}$ vs. x_2/x_1 yield the van Laar parameters. This procedure is useful when data for only one component are available,[15] but its disadvantage is that it yields a set of parameters from the plot for each component. As the two sets of parameters will rarely be in agreement, some type of averaging procedure must be used to obtain a single set. This problem can be avoided if the data for both components are combined through the function $\mathbf{Q}$:[16]

$$\mathbf{Q} = x_1 \ln \gamma_1 + x_2 \ln \gamma_2$$

---

[15] For example, when one component is essentially nonvolatile.
[16] $\mathbf{Q}$ is related to the excess Gibbs free energy of mixing. See Eq. (10-14).

Now, instead of two sets of γx data, there is a single set of Qx data. With the Margules and van Laar equations Q can be expressed in a linear form. For the Margules equation:

$$\frac{Q}{x_1 x_2} = A_{12} + \left(A_{21} - A_{12}\right)x_1 \tag{9-56e}$$

For the van Laar equation:

$$\frac{x_1 x_2}{Q} = \frac{1}{B_{12}} + \frac{B_{12} - B_{21}}{B_{12}B_{21}}x_1 \tag{9-57e}$$

EXAMPLE 9-8

Use VLE data for the ethanol–benzene system at 40°C to determine Margules and van Laar parameters.

Solution 9-8

The experimental data are shown in the first three left-hand columns of Table 9-1. With the vapor pressures

$$P_1^\circ = 134.02\,\text{mmHg}; \quad P_2^\circ = 182.78\,\text{mmHg}$$

and the experimental xyP data, activity coefficients can be calculated from Eq. (9-51). These are tabulated in Table 9-1 along with corresponding values of Q.

TABLE 9-1
COMPARISON OF CALCULATED AND EXPERIMENTAL VLE DATA (EX. 9-8)[a, b]

x_1	y_1	P	γ_1	γ_2	Q	Calculated y_1 Margules	Calculated y_1 van Laar
0.020	0.145	208.4	11.27	0.995	0.0432	0.105	0.105
0.095	0.280	239.8	5.274	1.044	0.1968	0.287	0.283
0.204	0.332	249.1	3.025	1.144	0.3327	0.362	0.354

TABLE 9-1 CON'T

x_1	y_1	P	γ_1	γ_2	Q	Calculated y_1	
						Margules	van Laar
0.378	0.362	252.3	1.803	1.416	0.4391	0.381	0.377
0.490	0.384	248.8	1.455	1.644	0.4373	0.384	0.385
0.592	0.405	245.7	1.254	1.960	0.4086	0.396	0.401
0.702	0.440	237.3	1.110	2.440	0.3390	0.429	0.437
0.802	0.507	219.4	1.035	2.989	0.2444	0.496	0.503
0.880	0.605	196.3	1.007	3.535	0.1577	0.596	0.600
0.943	0.747	169.5	1.002	4.116	0.0835	0.744	0.745
0.987	0.912	145.6	1.004	5.392	0.0258	0.924	0.924

[a] System: ethanol (1)–benzene (2) at 40°C, pressure in mmHg.
[b] Source: reference B8 in Table 10-8.

Margules equation. According to Eq. (9-56e), a linear relationship between Q/x_1x_2 and x_1 results if the system can be represented by the Margules equation. The xyP data in Table 9-1 were used with the spreadsheet MARGULES.WQ1(or .XLS) on the accompanying CD-ROM to determine the least-squares fit of Q/x_1x_2 vs. x_1 and the goodness of fit $\Delta\bar{y}$ defined as

$$\Delta\bar{y} = \frac{\sum_{i=1}^{n}\left|y_1^{exp} - y_1^{cal}\right|}{n}$$

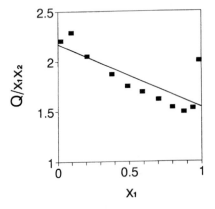

Figure 9-17 Determination of Margules parameters via Eq. (9-56e). System: ethanol (1)–benzene (2) at 40°C.

The spreadsheet displays a plot of experimental values of Q/x_1x_2 vs. x_1 showing the least-squares line and also a plot of experimental xy data compared to calculated values using the parameters determined by the least squares fit of Q/x_1x_2 vs. x_1. There is also the option of dropping any number of experimental data points from the least-squares fit. These two plots are shown as Figs. 9-17 and 9-18. The best value of $\Delta\bar{y}$ was obtained by using all of the data even though the fit as shown in Fig. 9-17 seemed better when data points at the composition extremes were dropped. Because our objective in fitting the data will usually be calculation or correlation of VLE, it seems fitting to use the criterion of a minimum $\Delta\bar{y}$ in selecting the

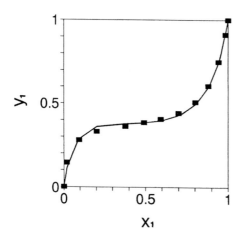

Figure 9-18
Goodness of fit of Margules equation to xy data. Squares are experimental data; line is calculated from Margules equation with fitted parameters. System: ethanol (1)– benzene (2) at 40°C.

parameters. Accordingly, a value of $\Delta\bar{y} = 0.014$ was otained for

$$\text{intercept} = A_{12} = 2.173$$

$$\text{slope} = A_{21} - A_{12} = -0.634$$

$$A_{12} = 2.173; \ A_{21} = 1.539$$

The calculation of the y's needed to evaluate $\Delta\bar{y}$ is illustrated for the case of $x_1 = 0.490$ where the Margules activity coefficients are

$$\ln \gamma_1 = (1 - 0.490)^2 [2.173 + 2(1.539 - 2.173)(0.490)]$$

$$\ln \gamma_1 = 1.50$$

$$\ln \gamma_2 = (0.490)^2[1539 + 2(2.173 - 1539)(1 - 0.490)]$$

$$\ln \gamma_2 = 1.69$$

The partial pressures are

$$p_1 = 0.490(134.02)(1.50) = 98.3$$

$$p_2 = (1 - 0.490)(182.78)(1.69) = 157.5$$

and the calculated vapor mol fraction is

$$y_1^{cal} = \frac{p_1}{p_1 + p_2} = 0.384$$

Values of y calculated in the manner are tabulated in Table 9-1 and were used to calculate $\Delta\bar{y}$.

   van Laar equation. The spreadsheet VANLAAR.WQ1(or .XLS) on the accompanying CD-ROM was used to determine the best fit of the van Laar equation to the xyP data by applying least squares to the data plotted as x_1x_2/Q vs. x_1. This arrangement, according to Eq. (9-57e), should produce a straight line when the van Laar equation fits the data. The spreadsheet also determines the value of $\Delta\bar{y}$ and displays a plot of x_1x_2/Q vs. x_1 and a graphical comparison of experimental and calculated values of y_1. These plots are shown in Figs. 9-19 and 9-20.

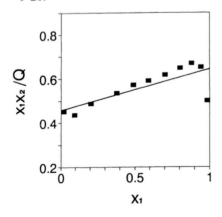

Figure 9-19 Determination of the van Laar parameters via Eq. (9-57e). System: ethanol (1)–benzene (2) at 40°C.

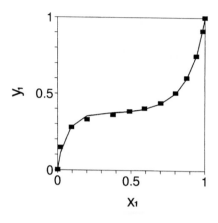

Figure 9-20 Goodness of fit of van Laar equation to *xy* data. Squares are experimental data; line is calculated from van Laar equation with fitted parameters. System: ethanol (1)–benzene (2) at 40°C.

As with the Margules equation, the lowest value of $\Delta\bar{y}$ was obtained when all of the data points were used in the least-squares calculation. This value, $\Delta\bar{y} = 0.0100$, corresponds to

$$\text{intercept} = 1/B_{12} = 0.4575$$

$$\text{slope} = \left(B_{12} - B_{21}\right)/B_{12}B_{21} = 0.1857$$

$$B_{12} = 2.186;\ B_{21} = 1.555$$

Again, the calculation of y_1 corresponding to $x = 0.490$ begins with the determination of activity coefficients.

$$\ln \gamma_1 = \frac{2.186}{\left[1 + \dfrac{2.186(0.490)}{1.555(0.510)}\right]^2}$$

$$\ln \gamma_1 = 1.49$$

$$\ln \gamma_2 = \frac{1.555}{\left[1 + \dfrac{1.555(0.510)}{2.186(0.490)}\right]^2}$$

$$\gamma_2 = 1.67$$

$$p_1 = 0.490(134.02)(1.49) = 97.5$$

$$p_2 = 0.510(182.78)(1.67) = 155.7$$

$$y_1^{cal} = 0.385$$

Calculated values of y_1 are listed in Table 9-1.

Both the Margules and van Laar equations seem to represent activity coefficients in this system reasonably well, with the van Laar equation being somewhat better.

Unfortunately, the Wilson equation cannot be recast into linear form, and therefore relatively simple graphical techniques, as illustrated above, are inapplicable and parameter estimation methods are required. These methods are based on the selection of parameters in the activity coefficient equation which minimize the deviations between measured and calculated properties. Several objective functions can be used to measure the extent of deviations. Hirata et al.[17] have evaluated several objective functions (OF's) which had been previously suggested in the literature and recommend the following.

$$OF_Q = \sum_{i=1}^{n} (Q^{exp} - Q^{cal})_i^2$$

Q^{exp} is evaluated from an experimental data point; Q^{cal} is calculated from the Wilson equation corresponding to x_1 and x_2 of the data point and its value is, of course, dependent on the Wilson parameters. The summation is over all data points. There are several computational techniques for determining the parameters which minimize the objective function. Hirata et al. tested the techniques

Nonlinear least squares
Gradient search
Pattern search
Complex search

and found the nonlinear least-squares technique to be most suitable for using the preceding OF_Q to obtain parameters in the Wilson equation. They have determined Wilson equation parameters in this manner for 800 binary systems and have reported them along with a listing of the experimental data and a measure of the goodness of fit of the Wilson equation.

---

[17] M. Hirata, S. Ohe, and K. Nagahama, *Computer-Aided Book of Vapor-Liquid Equilibria*, Kodansha/Elsevier, Tokyo/New York, 1975.

EXAMPLE 9-9

Use VLE data for the methanol–carbon tetrachloride system at 20°C to determine parameters in the Wilson equation using the following techniques:
(a) a G_{12}–G_{21} contour plot,
(b) nonlinear regression (POLYMATH4), and
(c) simplex pattern-search technique using OFQ (WEQ-ISOT.EXE).

Solution 9-9

The experimental data are shown in the first three left-hand columns of Table 9-2. With these data and the vapor pressures

$$P_1^\circ = 96.87 \text{ mmHg} \qquad P_2^\circ = 92.08 \text{ mmHg}$$

activity coefficients are calculated from Eq. (9-51). These are tabulated in Table 9-2 along with corresponding values of Q.

(a) To illustrate the nature of the parameter estimation process, a direct, brute-force procedure, which is easily executed with a spreadsheet and a contour plotting routine, will be used. Values of OFQ are calculated for predetermined sets of G_{12} and G_{21}. These parameter sets are points of intersection on an 11×11 grid established with G_{12} running from 0.0680 to 0.0720 in increments of 0.0004 and G_{21} running from 0.3000 to 0.3140 in increments of 0.0014. Figure 9-21 displays the results of this calculation as OFQ contours projected onto the G_{12}–G_{21} plane. All points (parameter sets) on a contour will represent the data equally well and, practically speaking, we should not expect to find a unique parameter set. Figure 9-21 indicates that in a three-dimensional projection the surface for this system in the region of the minimum in OFQ is very flat. For this reason we should not be too concerned with finding the exact minimum.

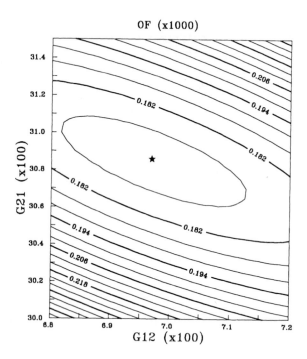

OF (x1000)

Figure 9-21 Contour plot for Wilson equation. Contour parameters are $OF_Q \times 10^3$. System: methanol (1)–carbon tetrachloride (2) at 20°C.

(b) For the Wilson equation the relationship between Q and x is

$$Q = -x_1 \ln (x_1 + x_2 G_{12}) - x_2 \ln (x_2 + x_1 G_{21}) \tag{9-58c}$$

The nonlinear regression feature of POLYMATH (on the accompanying CD-ROM) was used with the Qx data of Table 9-2 to obtain the following parameters in Eq. (9-58c).

$$G_{12} = 0.06971; \quad G_{21} = 0.3085$$

These parameters minimized the sum of squares (OF_Q) at a value of $0.1762(10^{-3})$. This problem is stored in the POLYMATH library as EX9-9.

TABLE 9-2

EXPERIMENTAL DATA FOR DETERMINATION

OF WILSON PARAMETERS (EX.9-9)[a, b]

						Calculated data	
x_1	y_1	P	γ_1	γ_2	Q	y_1	P
0.020	0.344	136.6	24.3	0.993	0.0569	0.279	125.6
0.040	0.391	146.4	14.8	1.01	0.117	0.358	139.7
0.100	0.430	154.8	6.87	1.06	0.245	0.424	153.1
0.200	0.449	158.0	3.66	1.18	0.392	0.449	157.7
0.300	0.462	159.5	2.54	1.33	0.479	0.459	159.1
0.600	0.494	159.9	1.36	2.20	0.500	0.492	159.2
0.700	0.512	157.5	1.19	2.78	0.429	0.515	157.0
0.800	0.546	152.0	1.07	3.75	0.318	0.558	151.3
0.900	0.662	135.5	1.03	4.97	0.187	0.653	136.7
0.960	0.792	117.2	0.998	6.62	0.0737	0.791	118.0

[a] System: methanol (1)–carbon tetrachloride (2) at 20°C, pressure in mmHg.
[b] Source: reference B8 in Table 10-8.

TABLE 9-3

RESULTS OF WILSON PARAMETER SEARCH (EX.9-9)[a, b, c]

Run	First Guess			End Point		
No.	G_{12}	G_{21}	Precision	G_{12}	G_{21}	OFQ $\times (10^3)$
1	2	2	10^{-8}	0.06972	0.3085	0.17620
2	2	2	10^{-5}	0.06972	0.3085	0.17620
3	1	1	10^{-4}	0.06972	0.3085	0.17620
4	1	0.05	10^{-5}	0.06969	0.3085	0.17620
5	0.05	1	10^{-5}	0.06967	0.3085	0.17620
6	0.05	0.05	10^{-5}	0.06970	0.3085	0.17620

[a] System: methanol (1)–carbon tetrachloride (2) at 20°C.
[b] Source: reference B8 in Table 10-8.
[c] WEQ-ISOT.EXE operating parameters
  Scaling factor for step-size reduction: 0.25
  Scaling factor for step-size increment: 2
  Initial step size : 2

(c) The program WEQ-ISOT.EXE on the accompanying CD-ROM uses the simplex pattern-search technique to determine the G_{12}–G_{21} set which minimizes the objective function OFQ. The program was executed for five different starting points and three levels of precision to yield the results in Table 9-3. For all six runs the end points are seen to be essentially identical and identical with the values obtained by nonlinear regression. This parameter set, plotted on Fig. 9-21 as a star, was used to calculate values of P^{cal} and y_1^{cal} in the manner of Ex. 9-8. These values are tabulated in Table 9-2; $\Delta\bar{y}$ was found to be 0.014.

In determining the parameters (A's, B's, and G's) in Eqs. (9-56)–(9-58), it has been tacitly assumed that the γx data used are isothermal or, at least, cover only a narrow temperature range. The parameters are expected to depend on temperature as we have earlier noted the dependence of the activity coefficient on temperature. In the case of the Margules and van Laar equations, the temperature dependence has to be determined empirically from γx data sets at different temperatures. However, the Wilson equation, developed through the consideration of molecular behavior, prescribes the dependence of the parameters on temperature as

$$G_{12} = \frac{v_2}{v_1}\exp\left(-\frac{a_{12}}{RT}\right) \tag{9-58d}$$

$$G_{21} = \frac{v_1}{v_2}\exp\left(-\frac{a_{21}}{RT}\right) \tag{9-58e}$$

where v_1 and v_2 are the molar volumes of the pure liquids at the absolute temperature T and a_{12} and a_{21} are parameters which characterize molecular interactions but are regarded simply as empirical constants.

If the parameters G_{12} and G_{21} are known at only one temperature, the preceding equations along with known values of v_1 and v_2 can be used to evaluate a_{12} and a_{21}, which in turn can be used to estimate G_{12} and G_{21} at other temperatures. The form of Eqs. (9-58d) and (9-58e) suggests that a plot of the logarithm of G vs. $1/T$ will be linear. When the parameters G_{12} and G_{21} are known at two or more temperatures, this linearity can be utilized for obtaining temperature dependence. It should be remembered that while Eqs. (9-58d) and (9-58e) have a basis in molecular theory, they have little thermodynamic basis and therefore cannot be expected to be applicable in all cases.

9-12 PHASE EQUILIBRIUM VIA AN EQUATION OF STATE

Another approach for using experimentally gained information for phase equilibrium calculations is based on the use of an equation of state. For this, an equation capable of accurately representing the *PVT* behavior of both the liquid and vapor phases is needed. The early work and the development of this approach were based on the Benedict-Webb-Rubin (BWR) equation or one of its modifications.[18] However, more recently, generalized cubic equations of state have been employed with increasing frequency. While it is obvious that the fairly simple Peng-Robinson (PR) equation does not possess the accuracy of the BWR equation, it nevertheless is suitable in certain situations and has the advantage of requiring less experimental information and computational effort. The equation-of-state approach to phase equilibrium will be illustrated with this simpler equation.

Single-Component Equilibrium. For single-component equilibrium this approach is founded on Eq. (9-10) and Eq. (9-11) or their equivalents:

$$f^L = f^G \tag{9-10}$$

$$RT d \ln f = v d P \tag{9-11}$$

If an equation of state capable of accurately representing v as a function of T and P were available, Eq. (9-11) could be integrated to allow f^L and f^G to be obtained as functions of these variables. Application of Eq. (9-10) would then permit the desired relationship to be established between T and P—the vapor pressure curve.

The PR equation, like most effective equations of state, is explicit in P, and use of Eq. (9-11) is inconvenient; however, the following equivalent equation[19] is useful:

$$\ln \phi = Z - 1 - \ln Z + \frac{1}{RT} \int_v^\infty \left[P - \frac{RT}{v} \right] dv \tag{9-59}$$

Using the PR equation, Eq. (3-5), to express the dependence of pressure on volume, Eq. (9-59) integrates to

$$\ln \phi = Z - 1 - \ln (Z - B) - \frac{A}{2\sqrt{2}B} \ln \left[\frac{Z + \left(1 + \sqrt{2}\right)B}{Z + \left(1 - \sqrt{2}\right)B} \right] \tag{9-60}$$

---

[18] For a detailed treatment of this approach using the BWR equation, see R. V. Orye, *Ind. Eng. Chem. Process Des. Dev.*, **8**, 579 (1969) or K. E. Starling, *Fluid Thermodynamic Properties for Light Petroleum Systems*, Gulf Pub. Co., Houston, 1973, Chap. 18.

[19] J. A. Beattie, *Thermodynamics and Physics of Matter* (F. D. Rossini, ed.), Princeton University Press, Princeton, NJ, 1955, Chap. 3.

At a specified temperature a trial and error procedure is used to find the pressure at which the liquid and vapor phases are in equilibrium—the vapor pressure. A pressure is assumed and Eq. (3-5i) is solved for Z:

$$Z^3 + (B-1)Z^2 + (A - 3B^2 - 2B)Z + (B^3 + B^2 - AB) = 0 \qquad (3\text{-}5i)$$

The smallest root, Z^L, is identified with the liquid phase and the largest, Z^G, with the vapor phase. Using these in Eq. (9-60) we can calculate ϕ^L and ϕ^G and test for the equilibrium condition $\phi^L = \phi^G$. When the test is met, we have the correct pressure, the vapor pressure P° at the selected temperature. A study has shown the PR equation is capable of representing vapor pressures of paraffin hydrocarbons, N_2, CO_2, and H_2S, with an average deviation of 0.60%.[20]

Multicomponent Equilibrium. To apply the equation-of-state approach to multicomponent phase equilibrium calculations, it is necessary to incorporate composition dependence into the equation of state. This is done by specifying mixing rules for the parameters. The following mixing rules are used for the PR equation:

$$b = \sum_{i=1}^{C} \chi_i b_i$$

$$a = \sum_{i=1}^{C} \sum_{j=1}^{C} \chi_i \chi_j a_{ij}$$

with

$$a_{ij} = a_{ji} = (1 - k_{ij})\sqrt{a_{ii} a_{jj}}$$

which on applying Eqs. (3-5j) and (3-5k) leads to

$$B = \sum_{i=1}^{C} \chi_i B_i$$

$$A = \sum_{i=1}^{C} \sum_{j=1}^{C} \chi_i \chi_j A_{ij}$$

$$A_{ij} = A_{ji} = (1 - k_{ij})\sqrt{A_{ii} A_{jj}}$$

In these equations, χ is replaced by x for the liquid phase and by y for the vapor phase. The binary interaction parameter k_{ij} is obtained by fitting the equation of state to experimental phase equilibrium data.

---

[20] D. Y. Peng and D. B. Robinson, *Ind. Eng. Chem. Fundam.*, *15*, 59 (1976).

The counterpart of Eq. (9-23) for use with pressure-explicit equations of state is[21]

$$\ln \hat{\phi}_i = \frac{1}{RT} \int_v^\infty \left[\left(\frac{\partial P}{\partial n_i} \right)_{T,V,n_j} - \frac{RT}{v} \right] dv - \ln Z \qquad (9\text{-}61)$$

When evaluated with the PR equation of state, this equation yields

$$\ln \hat{\phi}_i = \frac{A}{2\sqrt{2}B} \left[\frac{B_i}{B} - \frac{2\sum\limits_{j=1}^{C} \chi_j A_{ij}}{A} \right] \ln \left[\frac{Z + \left(1 + \sqrt{2}\right)B}{Z + \left(1 - \sqrt{2}\right)B} \right] \qquad (9\text{-}62)$$

$$+ \frac{B_i}{B}(Z - 1) - \ln\,(Z - B)$$

For vapor-liquid equilibrium, the compositions of the phases will generally be different, giving rise to a set of A and B for each phase. With the liquid-phase parameters Eq. (3-5i) is solved for the smallest root, Z^L, while with vapor-phase parameters the largest root, Z^G, is found. Thus, Eq. (3-5i) is written for each phase and Eq. (9-62) is written for each component in each phase. The computational difficulty encountered in Sec. 9-7 with the use of K's also applies here and complicates the calculational procedure. The procedure is illustrated in Example 9-10.

EXAMPLE 9-10

Use the PR equation to calculate the temperature and vapor composition in equilibrium with a liquid containing 50 mol % C_3H_6 and 50 mol % $i\text{-}C_4H_{10}$ at a pressure of 20 atm.

Solution 9-10

From Table A-1 we have the following:

Component	Component number	T_c (K)	P_c (atm)	ω
C_3H_6	1	365	45.6	0.148
$i\text{-}C_4H_{10}$	2	408	36.0	0.176

---

[21]J. A. Beattie, *op. cit.*

From Eq. (3-5h) values of κ are computed: $\kappa_1 = 0.597; \kappa_2 = 0.638$
From Eqs. (3-5l) and (3-5m) we write, for the pure-component parameters,

$$A_{11} = \frac{0.45724(20/45.6)}{(T/365)^2}\left[1+0.597\left(1-\sqrt{\frac{T}{365}}\right)\right]^2$$

$$A_{22} = \frac{0.45724(20/36.0)}{(T/408)^2}\left[1+0.638\left(1-\sqrt{\frac{T}{408}}\right)\right]^2$$

$$B_1 = \frac{0.0778(20/46.5)}{T/365}$$

$$B_2 = \frac{0.0778(20/36.0)}{(T/408)}$$

For this system a binary interaction parameter of -0.014 has been reported,[22] thus allowing A_{12} to be determined from

$$A_{12} = 1.014\sqrt{A_{11}A_{22}}$$

These five parameters are seen to depend only on temperature. Composition dependence is introduced when the parameters for each phase are written. For the liquid phase we have

$$B^L = 0.50B_1 + 0.50B_2$$

$$A^L = (0.50)^2 A_{11} + 2(0.50)(0.50)A_{12} + (0.50)^2 A_{22}$$

and for the vapor phase,

$$B^G = y_1B_1 + (1-y_1)B_2$$

$$A^G = y_1^2 A_{11} + 2y_1(1-y_1)A_{12} + (1-y_1)^2 A_{22}$$

Note that A^L and B^L depend only on temperature, while A^G and B^G depend on temperature and vapor composition. Thus, in calculating fugacity coefficients from Eqs. (9-62) and (3-5i) we will find that $\hat{\phi}_1^L$ and $\hat{\phi}_2^L$ depend on T, while $\hat{\phi}_1^G$ and $\hat{\phi}_2^G$ depend on y_1 and T. The correct values

[22] S. I. Sandler, *Chemical and Engineering Thermodynamics*, 2nd ed., Wiley, New York, 1989, p. 319.

of y_1 and T are found when the condition of equal fugacities is satisfied:

$$20(0.5)\hat{\phi}_1^L(T) = 20y_1\hat{\phi}_1^G(T,\ y_1)$$

$$20(0.5)\hat{\phi}_2^L(T) = 20(1-y_1)\hat{\phi}_2^G(T,\ y_1)$$

Normally, a computer program using an iterative procedure is employed, but to illustrate the functional relationships the calculated fugacities are plotted vs. temperature in Fig. 9-22. Single curves

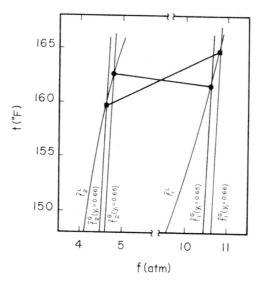

Figure 9-22 Graphical representation of fugacities in binary VLE.

represent $\hat{f}_1^L$ and $\hat{f}_2^L$, while curves for $\hat{f}_1^G$ and $\hat{f}_2^G$ are shown for two values of y_1. For the correct T and y_1 the intersection of the $\hat{f}^L$ and $\hat{f}^G$ curves for each component should occur at the same temperature. These intersections are shown in Fig. 9-22 as circles for y_1 = 0.65 and as squares for y_1 = 0.66. A line connecting these intersections should be horizontal, but we see that it has a negative slope for y_1 = 0.65 and a positive slope for y_1 = 0.66. We therefore expect that y_1 lies between these values and that t lies close to the average value of 162°F.

While the algorithm employed in Ex. 9-10 may be instructive, it would not be computation-ally effective. Other, more efficient, algorithms are used in computer codes. For systems containing up to 10 components, the program PRVLE.EXE on the accompanying CD-ROM executes calculations for the following sets of fixed variables: T and x's; P and x's; T and y's; and P and y's. The algorithm for the calculation where T and the x's are specified is shown in Fig. 9-23 where superscripts in parentheses index the iterations and the equations labeling the flows show the values of P and y_i passed to and from the calculation block.[23] The first estimate is based on an ideal-liquid solution and a perfect gas mixture with the vapor pressure estimated from Eq. (8-12) integrated for constant Δh between $T_r = 0.7$ where ω is defined and the critical point which is the last point on the vapor pressure curve.[24] Values of y_i are normalized before being passed to the calculation block. After the first iteration the value of P to be used in the second iteration is obtained by a special adjustment;[25] thereafter linear interpolation is used. With $\varepsilon = 10^{-4}$ six or less iterations are usually required.

VLE Data Reduction.   The determination of parameters in an activity coefficient equation provides a means of reducing experimentally gained VLE information. An alterna-tive reduction procedure is the determination of the binary interaction parameter k_{12} from a set of experimental VLE data. The program PRKFIT.EXE on the CD-ROM accompanying this text is capable of finding the value of k_{12} that fits the Peng-Robinson equation to a set of up to fifteen binary, isothermal xyP data points. This is done by minimizing the following objective function

$$OF = \sum \left[\alpha - \frac{K_1}{K_2} \right]^2$$

where the sum is taken over all data points, α is the relative volatility defined as

$$\alpha = \frac{y_1/x_1}{y_2/x_2}$$

and

$$\frac{K_1}{K_2} = \frac{\hat{\phi}_1^L \hat{\phi}_2^G}{\hat{\phi}_1^G \hat{\phi}_2^L}$$

---

[23] These equations should be read: The left-hand side takes the value of the right-hand side.
[24] See Prob. 8-18.
[25] See Prob. 9-26.

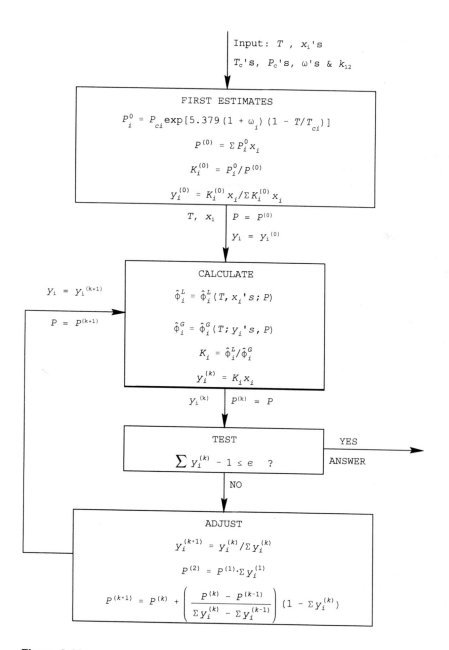

Figure 9-23

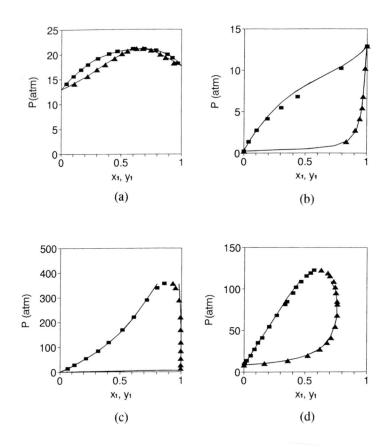

Figure 9-24 *Pxy* diagrams at constant temperature: (a) carbon dioxide (1)–ethane (2) at 250 K, $k_{12} = 0.1355$; (b) propane (1)–benzene (2) at 37.8°C, $k_{12} = 0.0536$; (c) methane (1)–decane (2) at 37.8°C, $k_{12} = 0.0490$; (d) methane (1)–*n*-butane (2) at 71.1°C, $k_{12} = 0.0648$. Solid lines are fitted curves; experimental x's are squares and y's are triangles. Source: reference B2 in Table 10-8.

with ϕ's determined from Eq. (9-62). Each VLE data point yields a value of α and at the specified T, P, x_1, and y_1 the ratio K_1/K_2 is a function only of k_{12}, hence OF depends only on k_{12}. Figure 9-24 shows four rather disparate systems in which the program PRKFIT.EXE was used to obtain the optimum value of k_{12} from experimental isothermal VLE data and the program PRVLE.EXE was used to calculate the *xyP* data from this value of k_{12}. The figure compares the experimental data points to curves constructed from calculated values[26] and

---

[26] The data needed for plotting these systems can be found on the spreadsheets FG9-24A.WQ1 through FG9-24D.WQ1 in the BACKUP directory of the CD-ROM.

shows that reasonable agreement was obtained. The goodness of fit shown in Fig. 9-24 is comparable to that found by Ohe in a study 700 systems.[27]

It is customary to take the binary interaction parameter k_{12} to be independent of composition and temperature as we have done in Ex. 9-10. The ability to reasonably fit isothermal xyP data as shown by Ohe and in Fig. 9-24 seems sufficient justification for the non-dependence of k_{12} on composition, however, Ohe fitted different isotherms of the same system and usually found a temperature dependence for k_{12} that was often systematic.

The equation-of-state approach can be used to calculate VLE data over a wide range of temperatures and pressures and in addition can be used to calculate mixture properties (e.g., v, h, s) and property changes on mixing (e.g., Δv, Δh, Δs). It has been customary to use the equation-of-state approach for systems comprised of simple or relatively nonpolar molecules and to use the activity coefficient approach for systems containing larger and more complex molecules. While this is still the usual practice, recent work has focused on combining the two approaches.[28]

9-13 THE THERMODYNAMIC APPROACH TO PHASE EQUILIBRIUM

The thermodynamic approach to phase equilibrium begins by stating the condition of equilibrium

$$\hat{f}_i^{\,G} = \hat{f}_i^{\,L}$$

From here we have developed two different approaches which we will call the activity-coefficient approach and the equation-of-state approach. Broadly speaking, the equation-of-state approach is used for systems comprised of relatively simple, low-molecular-weight species and the activity-coefficient approach is used for systems containing larger, more complex species. The activity-coefficient approach is the traditional approach while equation-of-state approach has only recently achieved wide usage. This is mainly due to the development of the improved van der Waals equations of state and the availability of the personal computer (PC). These equations of state are not overly complex mathematical expressions, they do a good job of representing PVT behavior, and their parameters are easily evaluated from readily available data. With its readily accessible, enormous computing power the PC has changed a daunting, computationally intensive approach into routine calculations.

The Activity-Coefficient Approach. Using the relations previously developed for liquid and vapor fugacities we write

---

[27] S. Ohe, *Vapor-Liquid Equilibrium Data at High Pressure*, Elsevier, Amsterdam, 1990.
[28] See Sec. 17-1.

$$Py_i \hat{\phi}_i = f_i^L x_i \gamma_i \qquad (9\text{-}63)$$

This equation embodies all the concepts we have developed; it is thermodynamically rigorous and has been called the basic equation of vapor-liquid equilibrium. It provides a systematic means of viewing and analyzing phase equilibrium by separating effects attributable to each phase: $\hat{\phi}_i$ is a gas-phase property, γ_i is a liquid-phase property, and f_i^L is a pure-component property.

The properties—fugacity, fugacity coefficient, and activity coefficient—are abstractions which may acquire meaning only after considerable study. While they have been presented here in a manner intended to provide them physical significance, it must be remembered that they were developed as tools to be used in achieving the ultimate result of finding the relationship among the measurable properties T, P, x, and y. This we show by considering their functionality. From our previous developments we may state that

$$\hat{\phi}_i = \phi(T, P, y_i's)$$

$$f_i^L = f(T, P)$$

$$\gamma_i = \gamma(T, P, x_i's)$$

and therefore the functionality of Eq. (9-63) may be stated as

$$F(T, P, x_i's, y_i's) = 0$$

and the abstractions, $\hat{\phi}_i$, f_i^L, and γ_i may be seen as a means of achieving the desired goal of relating the physical variables.

It is important to remember that implied in these functionality statements is the need for experimentally determined property data and that thermodynamics should be regarded as a tool which shows us how to effectively use our available knowledge. This available knowledge may be in the form of physical data for a specific system or a generalization based on considerable experimental evidence. In the latter category are observations which concern the applicability of the ideal solution models and the use of the principle of corresponding states to estimate fugacities. The former category would include equilibrium data for the specific system needed to evaluate parameters in an activity coefficient equation.

At this point the thoughtful reader will probably recognize that some circularity appears to exist in this approach and might ask the following: *Of what value is an approach which requires equilibrium data for a system in order to evaluate parameters in an activity coefficient equation which is then to be used in calculating equilibrium data for the system? Why bother with this approach?* Only a brief response to these questions will be offered now, but we will return to them after the detailed treatment of vapor-liquid and liquid-liquid equilibria presented in the following chapter.

The answer is as follows: In dealing with phase equilibrium, the activity coefficient approach offers the following advantages:

1. It provides a framework for viewing and analyzing phase equilibrium. This is especially useful in (a) extrapolating and interpolating data and establishing correlations and (b) evaluating and applying the results of theories based on molecular considerations.

2. It facilitates experimentation by (a) reducing the amount of data required and (b) allowing simpler measurements to be substituted for more difficult measurements.

3. It provides a means of testing the consistency of experimental data.

The Equation-of-State Approach.   Using Eq. (9-24) to express the fugacities, we may write for the vapor phase

$$\hat{f}_i^G = \hat{\phi}_i p_i = \hat{\phi}_i y_i P$$

and similarly for the liquid phase

$$\hat{f}_i^L = \hat{\phi}_i^L x_i P$$

Equating fugacities and substituting these expressions results in

$$y_i \hat{\phi}_i^G = x_i \hat{\phi}_i^L \tag{9-64}$$

where

$$\hat{\phi}_i^G = \phi(T, P, y_i's)$$

$$\hat{\phi}_i^L = \phi(T, P, x_i's)$$

and therefore Eq. (9-64) is functionally is equivalent to

$$f(T, P, x_i's, y_i's) = 0$$

and again the $\hat{\phi}$'s are seen to be tools that allow us to relate the desired intensive variables of phase equilibrium.

Here there is only one set of parameters that applies to both phases and because the implementation of Eq. (9-64) is usually by means of a computer program, we seldom note the values of the ϕ_i's. Therefore, we do not attempt to identify separately the effects attributable to each phase as expressed by the ϕ_i's. For this reason the equation-of-state approach seems less circuitous than the activity coefficient approach. The approaches are similar, however, because the implementation of each requires experimental equilibrium data for the system under study.

PROBLEMS

9-1. A gas-phase activity coefficient is sometimes used when dealing with gas mixtures. This coefficient $\hat{\gamma}$ is defined by

$$\hat{f}_i = \hat{\gamma}_i y_i f_i$$

where f_i is the pure component fugacity at the mixture conditions.

(a)   Show that

$$RT \ln \hat{\gamma}_i = \int_0^P (\overline{V}_i - v_i) dP$$

where

$$\overline{V}_i = \left(\frac{\partial V}{\partial n_i} \right)_{T,P,n_j}$$

(b)   What restriction does the Gibbs-Duhem equation place on $\hat{\gamma}$'s?

9-2. The following relationship between fugacity and pressure has been proposed:

$$f = P + \alpha P^2$$

where α is a function only of temperature. Find the equation of state of a gas conforming to this equation. Is the equation of state realistic? Explain.

9-3. J. B. Ott, J. R. Goates, and H. T. Hall [*J. Chem. Educ., 48*, 515 (1971)] determined the following equation of state by fitting the compressibility factor in the range $0 < P_r \leq 0.6$ and $1.0 \leq T_r \leq 2.0$:

$$Pv = RT \left[1 + \frac{PT_c}{17 P_c T} \left(1 - \frac{15 T_c^2}{2 T^2} \right) \right]$$

Based on this equation, derive an expression for the fugacity of a gas in this range of T_r and P_r.

9-4. Show that an alternative statement of the ideal-gaseous solution model is

$$\hat{\phi}_i = \phi_i$$

What restrictions does the Gibbs-Duhem equation place on the $\hat{\phi}_i$'s?

9-5. Determine under what circumstances the following equation is a reasonable approximation:

$$f = \frac{P^2 v}{RT}$$

9-6. Two compounds A and B are known to form ideal-liquid solutions. A vapor mixture containing 50 mol % of A and 50 mol % of B is initially at 100°F and 1 atm. This mixture is compressed isothermally until condensation occurs. At what pressure does condensation occur, and what is the composition of the liquid that forms? The vapor pressures of A and B at this temperature are 1.20 and 1.40 atm respectively.

9-7. Two homologs, a and b, form an ideal solution and have the following vapor pressures at 330 K

$$P_a^\circ = 900 \text{ mmHg}, \quad P_b^\circ = 800 \text{ mmHg}$$

For a equimolar vapor mixture,

(a) what is the dew point pressure?
(b) what pressure is necessary for complete condensation?

9-8. Compounds A and B are members of the same chemical family and have the following vapor pressures.

$t(^\circ F)$	P_A° (mmHg)	P_B° (mmHg)
150	600	500
200	1000	950

Assume that these compounds form ideal solutions and calculate the vapor mol fraction of A and the total pressure for VLE when the liquid mol fraction of A is 0.500 and the temperature is 175°F.

9-9. One mol of n-butane and 1 mol of n-pentane are charged into a container. The container is heated to 180°F where the pressure reads 100 psia. Determine the quantities and compositions of the phases in the container.

9-10. A vessel initially containing propane at 30°C is connected to a nitrogen cylinder, and the pressure is increased to 300 psia. Assuming that the nitrogen is insoluble in liquid propane, what is the mol fraction of propane in the vapor phase? List all assumptions. The vapor pressure of propane at 30°C is 10.5 atm.

9-11. What is the dew point temperature of a vapor mixture of 50 mol % of n-pentane (C_5H_{12}) and 50 mol % of n-butane (C_4H_{10}) at 200 psia? At what temperature is the vapor mixture completely condensed if the pressure is maintained constant at 200 psia?

9-12. Liquefied petroleum gas (LPG) used for heating, cooking, etc., in rural homes is stored outside in above-ground tanks. This fuel is a mixture of propane and butane the composition of which depends on weather conditions. In cold weather more propane is needed to produce a reasonable pressure in the tank, and in hot weather the amount of propane is reduced to prevent excessive pressures within the tank. If the maximum and minimum pressures are 100 and 30 psia, the maximum summertime temperature is 120°F, and the minimum wintertime temperature is 0°F, specify the maximum amount of propane allowable in the summer and the minimum allowable in winter. There will be a vapor phase in the tank, but on a mass basis very little of the tank contents are vapor.

9-13. A vapor containing 50 mol % propane and 50 mol % n-butane is held at 100°F and the pressure is increased until the dew point is reached. What is the dew point pressure?

9-14. Find the pressure and the vapor and liquid compositions when an equimolar mixture of propane and n-butane is 50% condensed at 100°F.

9-15. A liquid containing 50 mol % benzene and 50 mol % toluene is subjected to a flash vaporization process at a pressure of 1 bar where the temperature inside the flash drum is 95°C. Use any information in Ex. 9-5 that you wish and answer the following.

    (a)   What is the exit vapor composition?
    (b)   What is the exit liquid composition?
    (c)   What mol % of the liquid feed is flashed?

9-16. A liquid stream containing ethane and propane enters a flash chamber operated at 80°F and 300 psia. Leaving the chamber, the vapor stream flow rate is twice that of the liquid on a mol basis. Calculate the feed composition.

9-17. For the design of an absorption refrigeration system using propane as the working fluid, we require equilibrium data for mixtures of propane and a nonvolatile hydrocarbon solvent. Ideal-liquid solutions are expected and we need propane partial pressure (total pressure, P) vs. mol fraction of propane in the liquid solution, x_p. Prepare a plot of P vs. x_p for a temperature of 200°F. Cover the range $0 < x_p \leq 0.5$.

9-18. A liquid stream of 4.9 mol % propane and 95.1% butane at 100 psia and 152°F is throttled to a pressure of 50 psia. Calculate, for the resulting stream,

    (a)   temperature.
    (b)   liquid and vapor compositions.
    (c)   vapor-to-liquid ratio.

Enthalpy data is available in J. H. Perry, D. W. Green, and J. O. Maloney, eds., *Perry's Chemical Engineers' Handbook*, 6th ed., McGraw-Hill, New York, 1984, Chap. 3.

9-19. Vapor-liquid equilibrium for the system A-B at a constant temperature of 80°C shows that component B obeys Henry's law in the range $0 < x_B \leq 0.020$. The partial pressure of B in this range is given by

$$P_B = 500x_B$$

The vapor pressures at 80°C are $P_A^\circ = 1000$ mmHg and $P_B^\circ = 250$ mmHg. For $x_B = 0.01$ calculate the equilibrium pressure and vapor composition.

9-20. For a binary system at constant temperature, show that where Henry's law is followed the total pressure is linear in mol fraction. Use the following data to determine the Henry's law constant for component B.

VAPOR-LIQUID EQUILIBRIA FOR *A–B* SYSTEM AT 75°C

m. f. of *B* in liquid	Total Pressure (mmHg)
0.005	1005
0.010	1010
0.015	1015
0.020	1020

9-21. Vapor-liquid equilibrium data for the *A–B* system at a constant temperature of 350 K shows that component *B* follows Henry's law in the range $0 < x_B \leq 0.05$. At this temperature the following data point has been reported

$$x_A = 0.975, \quad y_A = 0.942, \quad P = 1035 \text{ mmHg}$$

and the vapor pressures are

$$P_A^\circ = 1000 \text{ mmHg}, \quad P_B^\circ = 800 \text{ mmHg}$$

Calculate y_B and P when $x_B = 0.040$

9-22. Aqueous emulsions of perfluorochemicals are being considered as "artificial bloods" because of their high oxygen solubility. At 25°C and an oxygen pressure of 1 atm, 384 ml of oxygen gas (measured at 25°C and 1 atm) dissolve in 1 liter of perfluorotributylamine, $(C_4F_9)_3N$, which has a liquid density of 1.883 g/ml. Determine the Henry's law constant, in units of atmospheres, for oxygen dissolved in perfluorotributylamine. The corresponding value for oxygen dissolved in water is 4.38 (10^4) atm.[29]

---

[29]Suggested by Prof. E. A. O'Rear III, University of Oklahoma.

The blood substitute *Oxypherol* is an emulsion of 20% perfluorotributylamine and 80% water by volume. Estimate the volume of oxygen gas (measured at 25°C and 1 atm) dissolved in a liter of liquid when *Oxypherol* is equilibrated with air at 25°C.

9-23. In a binary system the relative volatility α can be represented as a power series in $x_1 - x_2$:

$$\log \alpha = a + b(x_1 - x_2) + c(x_1 - x_2)^2 + \ldots$$

This series can be used to generate an expression for activity coefficients as a function of composition. Truncate the series after the second term, and determine the activity coefficient equations. The relative volatility is defined as

$$\alpha = \frac{y_1/x_1}{y_2/x_2}$$

where y is vapor-phase mol fraction and x is liquid-phase mol fraction.

9-24. Table 9-2 contains no entries for the liquid mol fraction of component 1 in the region 0.3 to 0.6. Use any information found in Example 9-9 to calculate P and y_1 for $x_1 = 0.5$.

9-25. Use the vapor-liquid equilibrium data listed in Ex. 9-9 for the methanol–carbon tetrachloride system at 20°C to determine the

(a) parameters in the van Laar equation.
(b) parameters in the Margules equation.

9-26. In the program PRVLE.EXE first estimates are needed for the iterative calculation algorithm.[30] These are based on the assumption of ideal liquid solutions and perfect gas mixtures where $K_i = P_i^\circ / P$. The first iteration yields new values which must be adjusted and used in the second iteration and we seek thermodynamic guidance as to the nature of this adjustment. Two cases will be considered:

(a) T and the x_i's are fixed and values of P and the y_i's are desired
(b) P and the x_i's are fixed and values of T and y_i's are desired.

For case *a* justify the following new estimate for P, $P^{(2)}$, based on values obtained from the first iteration [superscript(1)]

$$P^{(2)} = P^{(1)} \cdot \sum y_i^{(1)}$$

For case *b* justify the following new estimate for T

---

[30]See Fig. 9-23.

$$T^{(2)} = T^{(1)} - \frac{R(T^{(1)})^2}{\Delta h^V} \ln \sum y_i^{(1)}$$

where Δh^V is the average heat of vaporization of the mixture.

9-27. Use fugacity coefficients calculated from the program PREOS.EXE to determine the vapor pressures of one of the following substances over a range of temperature. Compare with values found in *Perry's Handbook*.

   Substances *n*-butane, carbon dioxide, ethane, ethylene, methyl chloride, and propane

9-28. Use the program PRVLE.EXE to work the following problems and compare the results with those obtained with K values.

   (a) 9-9
   (b) 9-11
   (c) 9-12
   (d) 9-13
   (e) 9-14

9-29. To show the effect of the value of k_{12} on calculated VLE data, calculate and plot a *Pxy* diagram for the ethane–propane system at 270 K using k_{12} values of 0, 0.1, & –0.1.

9-30. For a pure substance obeying the van der Waals equation of state at a temperature below T_c and at a pressure greater than the saturation pressure—the liquid state—show that the liquid fugacity calculated from Eq. (9-59) is equal to the value calculated from the integration of Eq. (9-11) from P^{sat} to P where ϕ at P^{sat} is evaluated by Eq. (9-59).

9-31. Use the condition of equal areas demonstrated in Prob. 7-3 and the van der Waals equation of state to determine the saturation pressure for CO_2 at 260 K. An analytical solution is not required.

9-32. We wish to use the Peng-Robinson equation of state along with a heat capacity equation from App. B-1 to construct a data base for several pure substances. We desire values of v, h, and s for the saturated liquid and vapor phases and for the superheated vapor region and will set h and s to zero for saturated liquid at the normal boiling point. Write the necessary equations and outline a method of computation.

Applied Phase Equilibrium

There are many operations performed in the chemical, petroleum, and related industries which involve the contact between phases. The analysis or design of these operations requires an understanding of the principles of phase equilibrium and an availability of equilibrium data. Such equilibrium data are seldom available at the exact conditions prevailing in the process and must usually be obtained by interpolating or extrapolating data, often fragmentary, which are available at other conditions. For this task the engineer again needs an understanding of the principles of phase equilibrium. This chapter is intended to develop that understanding and is focused on the approach to and the rationale behind the application of thermodynamics to practical problems involving vapor-liquid and liquid-liquid equilibrium; no attempt is made to be encyclopedic.

We have previously developed Eq. (9-63) which we have called the basic equation of vapor-liquid equilibrium and have seen that it provides a systematic means of treating this subject. In this chapter we will consider the application of this equation and other thermodynamic tools to specific topics in vapor-liquid and liquid-liquid equilibrium, and we should always keep in mind the question raised previously in regard to the practical utility of the thermodynamic approach. We can best understand and appreciate the thermodynamic approach if we try to visualize how a situation might be handled without the aid of thermodynamics. However, to facilitate the development of this understanding and appreciation, we must remember that when applied to phase equilibrium thermodynamics usually requires some information which is specific to the system under study and which must be determined experimentally. For this reason a familiarity with the experimental determination of equilibrium data is helpful, and you should always try to visualize how such required data might be obtained.

In this chapter our major concern will be with systems at low to moderate pressures where ideal gaseous solutions are expected, subject to the exceptions previously mentioned. Where ideal gaseous solutions are found, $\hat{\phi}_i$ in Eq. (9-63) may be replaced with ϕ_i, the pure-component fugacity coefficient, and γ_i is the only remaining composition-dependent quantity in the equation. Therefore, when we know the γ_i's as a function of temperature and liquid composition, we may calculate the relationship between the variables $PTxy$ for the desired conditions. The major thrust of this chapter will be directed toward obtaining this information about the γ_i's and utilizing it under various conditions for the solution of practical problems.

The low-pressure form of Eq. (9-63) is

$$Py_i = p_i = P_i^\circ x_i \gamma_i \qquad (9\text{-}63a)$$

and it is seen that the behavior of vapor-liquid systems can be visualized and analyzed in terms of the pure-component vapor pressures and the activity coefficients. Consider, for example, the four vastly different systems displayed at constant temperature in Fig. 10-1 and at constant pressure in Fig. 10-2. These systems share little in common in regard to their phase equilibrium behavior as manifested on these plots. This is typical of the diversity of phase equilibria which can be expected, and yet all these systems are amenable to the thermodynamic approach. Because it is mainly the behavior of the activity coefficient which gives rise to such diverse behavior, the study of this property is the key to understanding phase equilibrium.

Because it is applicable to a wide variety of systems, the activity-coefficient approach is awarded major coverage here, however, current research in phase equilibrium is focusing on ways of extending the applicability of the equation-of-state approach to higher-molecular-weight systems.[1] As shown in Fig. 9-24, the Peng-Robinson equation with a single parameter, k_{12}, is capable of representing a wide variety of behaviors for systems consisting of relatively simple molecules.

---

[1] See Section 17-1.

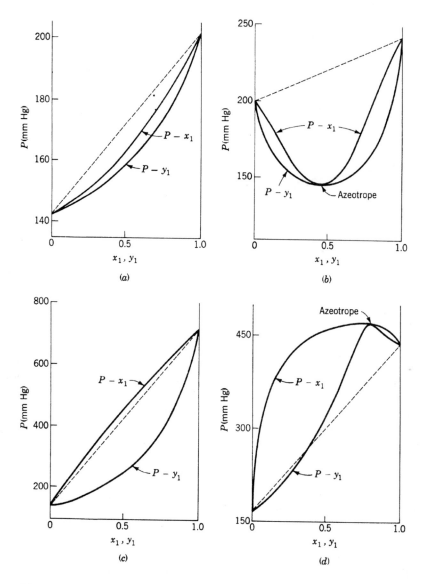

Figure 10-1 *Pxy* diagrams at constant temperature: (a) tetrahydrofuran (1)–carbon tetrachloride (2) at 30°C, (b) chloroform (1)–tetrahydrofuran (2) at 30°C, (c)
furan (1)–carbon tetrachloride (2) at 30°C, (d) ethanol (1)–toluene (2) at
65°C. Dashed line: *Px* relation for Raoult's law. [From J. M. Smith and H.
C. Van Ness, *Introduction to Chemical Engineering Thermodynamics*, 3rd
ed., copyright © 1975 by McGraw-Hill. Used with permission of McGraw-
Hill Book Company, New York.]

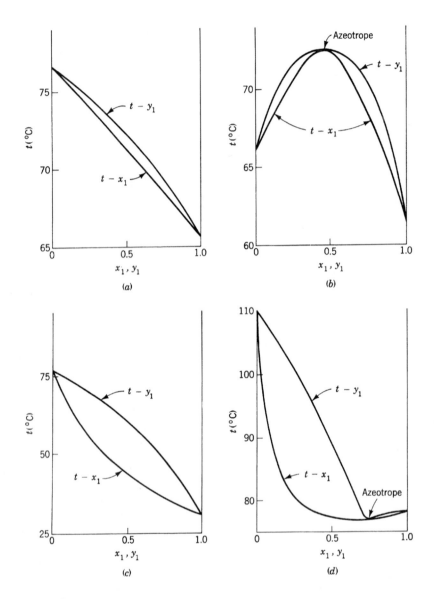

Figure 10-2 *txy* diagrams at a constant pressure of 1 (atm): (a) tetrahydrofuran (1)–carbon tetrachloride (2), (b) chloroform (1)–tetrahydrofuran (2), (c) furan (1)–carbon tetrachloride (2), (d) ethanol (1)–toluene (2). [From J. M. Smith and H. C. Van Ness, *Introduction to Chemical Engineering Thermodynamics*, 3rd ed., copyright © 1975 by McGraw-Hill. Used with permission of McGraw-Hill Book Company.]

10-1 THE CONSUMMATE THERMODYNAMIC CORRELATION OF VAPOR-LIQUID EQUILIBRIUM

Let us begin by considering the consummate application of thermodynamics to the correlation of vapor-liquid equilibrium for a binary system. We ask the following question: What is the minimum amount of experimental data we need in order to be confident that we can represent the behavior of this system over the desired ranges of temperature, pressure, and composition? The answer, of course, will depend on the desired accuracy, but let us require reasonably accurate representation typical of most engineering design needs.

The Activity-Coefficient Approach. In the most rigorous sense the consummate thermodynamic approach would employ Eq. (9-63) to calculate the desired information. This would require f_i^L as a function of temperature and pressure as well as a knowledge of the dependence of $\hat{\phi}_i$ and γ_i on temperature, pressure, and composition.

The calculation of f_i^L from pure-component properties as outlined in Sec. 9-5 is straightforward and should present no problems. Here our attention will be focused on systems near atmospheric pressure where it will be reasonable to replace f_i^L with the vapor pressure P_i°.

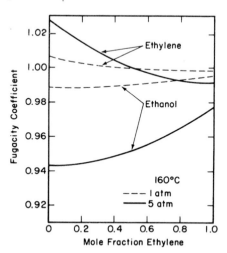

Figure 10-3 Fugacity coefficients for the system ethylene–ethanol. [Prausnitz, Anderson, Grens, Eckert, Hsieh, and O'Connell, *Computer Calculations for Multicomponent Vapor-Liquid and Liquid-Liquid Equilibria*, 1980, p. 32. Reprinted by permission of Prentice-Hall, Inc., Englewood Cliffs, NJ.]

As can be seen in Fig. 10-3, the assumption $\hat{\phi}_i = 1$ or $\hat{\phi}_i = \phi_i$ may be subject to slight error even at atmospheric pressure. While it would be desirable to include $\hat{\phi}_i$ in the calculations through the use of Eqs. (9-26) or (9-62), this would require knowing the cross coefficient in the virial equation or k_{12} in the Peng-Robinson equation. Except for systems in which all components are nonpolar and correlations of virial coefficients are available,[2] this will require some experimental *PVTy* data for the specific gaseous system. The lack of this information severely limits the use of Eqs. (9-26) or (9-62), and by necessity one is usually forced to assume $\hat{\phi}_i = 1$ at low pressure and $\hat{\phi}_i = \phi_i$ at moderate pressures. This will be the procedure followed here.

Near atmospheric pressure Eq. (9-63a) and the representation of activity coefficients as functions of temperature and liquid composition will allow the correlation of experimental vapor-liquid equilibrium data and subsequently the calculation of equilibria at other conditions. The correlation is based on fitting activity coefficients determined from experimental isothermal vapor-liquid equilibrium data to a selected activity coefficient equation. The parameters so obtained apply at that temperature, and the procedure must be repeated at other temperatures if the temperature dependence of the parameters is to be discerned. This fitting procedure could be a simple graphical technique such as plotting linearized versions of the Margules or van Laar equations (e.g., MARGULES.WQ1 or VANLAAR.WQ1) or a computer-based parameter estimation method (e.g., WEQ-ISOT.EXE).

As mentioned previously, there are many possible activity coefficient equations which satisfy the thermodynamic requirement imposed by the Gibbs-Duhem equation. Therefore, if the equation we have chosen does not fit the experimental data well enough, we should try another equation. Unqualified generalizations are not possible, but evaluations based on the comparison of many systems have shown[3] the Wilson equation to be superior to the Margules or van Laar equation for a vast majority of the systems studied. While the Wilson equation is more difficult to manipulate, its proven efficacy explains its wide acceptance and frequent current usage.

After we have found an activity coefficient equation which has been shown to fit our experimental data and have determined its parameters at enough temperatures to allow establishment of their temperature dependence,[4] we possess sufficient information to completely describe the system under any conditions. For example, we might wish to calculate *Pxy* data at a given temperature or *Txy* data at a specified pressure.

EXAMPLE 10-1

For the system chloroform–*n*-hexane, the Wilson parameters have been evaluated[5] at temperatures of 35, 45, and 55°C and are tabulated here. Vapor pressure data are available as constants in the Antoine equation,

$$\log P^\circ = A - \frac{B}{C + t}$$

---

[2] See Sec. 9-6.

[3] For example, R. V. Orye and J. M. Prausnitz, *Ind. Eng. Chem.*, *57*, 19 (1965).

[4] Just as the temperature dependence of vapor pressure allowed us to determine heats of vaporization, the temperature dependence of the activity coefficients will allow us to determine the heat of mixing. We will deal with this subject in Chap. 11.

[5] M. Hirata, S. Ohe, and K. Nagahama, *Computer-Aided Book of Vapor-Liquid Equilibria*, Kodansha/Elsevier, Tokyo/New York, 1975.

EXAMPLE 10-1 CON'T

Calculate the equilibrium temperature and vapor mol fraction of chloroform corresponding to a chloroform liquid mol fraction of 0.500 when the total pressure is 760 mmHg.

	Antoine Constants		
	A	B	C
chloroform	6.90328	1163.030	227.400
n-hexane	6.87760	1171.530	224.366

Wilson parameters: 1 = chloroform; 2 = n-hexane.

$t(°C)$	G_{12}	G_{21}
35	1.020	0.398
45	1.233	0.378
55	1.300	0.378

The temperature dependence of the Wilson parameters is shown in Fig. 10-4, where it is seen that a reasonably good straight line results when $\ln G$ is plotted vs. $1/T$.

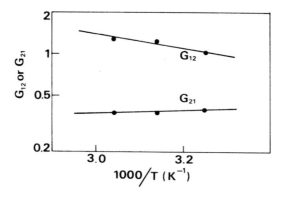

Figure 10-4 Wilson parameters for the chloroform–n-hexane system.

Solution 10-1

Because the equilibrium temperature is unknown, the calculation of vapor-liquid equilibria at a specified pressure involves a trial and error procedure. For the first trial we will pick a temperature of 61°C. At this temperature we find from the Antoine equation that the vapor pressures are

$$P_1^\circ = 742.3 \text{ mmHg}$$

$$P_2^\circ = 591.9 \text{ mmHg}$$

From Fig. 10-4 at 61°C we find that

$$G_{12} = 1.40$$

$$G_{21} = 0.371$$

Using these parameters in Eq. (9-58a) and (9-58b) at $x_1 = 0.5$, $x_2 = 0.5$ gives

$$\gamma_1 = 1.139 \qquad \gamma_2 = 1.067$$

Partial pressures are then calculated:

$$p_1 = x_1 P_1^\circ \gamma_1 = 0.5(742.3)(1.139) = 422.7$$

$$p_2 = x_2 P_2^\circ \gamma_2 = 0.5(591.9)(1.067) = \dfrac{315.8}{738.5}$$

The calculated total pressure of 738.5 is too low; therefore for the next trial we will use a temperature of 62.5°C. The vapor pressures are more strongly dependent on temperature than the Wilson parameters; therefore, only the vapor pressures will be recalculated. At 62.5°C we obtain

$$P_1^\circ = 778.8 \text{ mmHg}$$

$$P_2^\circ = 621.9 \text{ mmHg}$$

The partial pressures at 62.5°C are then

$$p_1 = 0.5(778.8)(1.139) = 443.5$$

$$p_2 = 0.5(621.9)(1.067) = \frac{331.8}{775.3}$$

We have now bracketed the desired pressure of 760 mmHg and can determine the equilibrium temperature by interpolation. The vapor mol fraction can be found from the calculated partial pressures:

$$y_1 = \frac{p_1}{p_1 + p_2} \qquad P_1 + P_2 = 760$$

Both sets of calculated partial pressures yield the same value of y_1. The calculated t and y_1 are

$$t = 61.2°C$$

$$y_1 = 0.572$$

TABLE 10-1
COMPARISON OF CALCULATED AND EXPERIMENTAL
VAPOR-LIQUID EQUILIBRIUM DATA (Ex. 10-1)[a, b]

	y_1		$t(°C)$	
x_1	Exptl.	Calc.	Exptl.	Calc.
0.106	0.159	0.168	66.5	66.4
0.206	0.292	0.297	64.8	64.6
0.295	0.383	0.394	63.5	63.3
0.397	0.484	0.489	62.3	62.1
0.500	0.566	0.572	61.4	61.2
0.594	0.641	0.642	60.9	60.7
0.700	0.720	0.717	60.5	60.3
0.800	0.796	0.791	60.4	60.3
0.900	0.884	0.878	60.7	60.6

[a] System: chloroform (1) – hexane (2) at 760 mmHg.
[b] Source: reference B8 in Table 10-8.

We have just calculated a single vapor-liquid equilibrium data point. This procedure has been repeated at other selected liquid compositions, and the calculated results are compared with experimental data in Table 10-1 where it is observed that the agreement is quite good. Perhaps the most uncertainty in the calculation arises from the determination of the temperature dependence of the Wilson parameters. Figure 10-4 shows that while the data can be fit by a straight line, the fit is not perfect as all points do not fall exactly on this line. Another cause of uncertainty might have arisen from the inability of the Wilson equation to exactly fit the activity coefficients for this system; however, Hirata et al. have evaluated this from the data at 35, 45, and 55°C and have found an excellent fit.

In Ex. 10-1 we have used one type of information about the system (isothermal *Pxy* data) to calculate another type of information (isobaric *Txy* data). Visualized in terms of the thermodynamic problem-solving process of Fig. 7-1, the problem was first formulated in terms of the activity coefficient—a useful but rather abstract function of no ultimate value. Yet we have seen that knowledge of activity coefficient behavior allows us to make the calculations to obtain the desired result. The information flow consisted of vapor pressure data in the form of Antoine constants and an expression of activity coefficients as functions of composition and temperature. The latter as supplied to us was in the form of Wilson parameters at three temperatures; however, we should appreciate the fact that someone else (Hirata et al.) had for our convenience previously determined these parameters from experimental *Pxy* data taken by yet another group of workers.

Because sufficient data were available, we may characterize calculations such as shown in Ex. 10-1 as the consummate application of the thermodynamic principles of phase equilibrium. This, of course, would be the desired way to proceed; however, there seldom are sufficient data available to allow us this luxury. Instead we are apt to find fragmentary data or no data at all. If there are no data available for the specific system under study, then there is nothing which can be done with thermodynamics alone.[6] However, thermodynamics can be useful in obtaining estimates from fragmentary data. Often for preliminary computations phase equilibrium data of high accuracy are not required, and such estimates are satisfactory.

The Equation-of-State Approach. For systems amenable to this approach (e.g., light hydrocarbons and other simple nonpolar species), a knowledge of k_{12} as a function of temperature would provide the consummate correlation. Fig. 10-5 shows the temperature dependence of k_{12} for a few systems and implies that failure to account for this temperature dependence could result in considerable error.

We will now consider phase equilibrium calculations based on data of less than the consummate quantity or quality.

---

[6] Reasonable estimates of activity coefficients may often be obtained from theories based on molecular models. See Sec. 11-8.

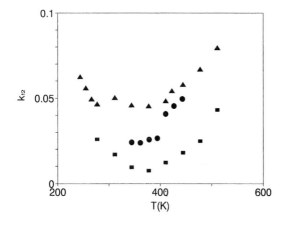

Figure 10-5 Temperature dependence of k_{12}. Methane–decane, triangles: ethane–decane, squares; propane–pentane circles. Data from reference B9 in Table 10-8.

10-2 CONSTANT-PRESSURE VLE DATA

Many systems are studied under the condition of constant pressure, and for the design of separation processes this is the type of data required. If the data are available at the desired pressure, then we would use them directly without recourse to thermodynamics; however, if this is not the case, we will want to use the available information to predict the necessary data at the required conditions. Usually, this will be done by correlating the activity coefficients. With Txy data there can be considerable temperature variation over the composition range, and it may not be possible to separate the effects of composition and temperature on the activity coefficient. The Gibbs-Duhem equation and its integrated forms [Eqs. (9-56)–(9-58)] apply strictly to isothermal conditions, and we may not be able to achieve good representation of activity coefficients with these equations. If the temperature range is small, this problem is usually ignored; however, if it is not, care should be taken in using such data for correlative purposes.

For large temperature variations it is possible to use an activity coefficient equation with prescribed temperature dependence and find the best parameters from non-isothermal VLE data. The parameters in the Wilson equation have a temperature dependence prescribed by Eqs. (9-58d) and (9-58e) and as liquid molar volumes are readily available, the parameters a_{12} and a_{21} can be determined by fitting non-isothermal VLE data. The program WEQ-ISOP.EXE on the accompanying CD-ROM determines a_{12} and a_{21} by means of a pattern search technique that minimizes the objective function OF**Q** using non-isothermal $xyTP$ data, v_1, v_2, and Antoine parameters as input.

EXAMPLE 10-2

Vapor-liquid equilibrium data for the n-hexane–chlorobenzene system at a constant pressure of 759.8 mmHg are shown below. Use this data and the program WEQ-ISOP.EXE to determine the Wilson parameters a_{12} and a_{21}, then use these parameters to estimate vapor-liquid equilibrium data for this system at a constant temperature of 65°C. The Antoine parameters, molecular weights, and liquid densities at 25°C are

		density		Antoine Parameters	
	mol. wt.	g/cc	A	B	C
n-hexane	86.17	0.659	6.8776	1171.53	224.366
chlorobenzene	112.56	1.107	7.17294	1549.2	229.26

TABLE 10-2

VAPOR-LIQUID EQUILIBRIUM DATA[a,b]

x_1	y_1	$t(°C)$	$P(mmHg)$
0.018	0.118	127.56	759.8
0.049	0.282	121.06	
0.081	0.406	115.66	
0.109	0.491	111.53	
0.146	0.527	106.62	
0.200	0.666	101.04	
0.309	0.769	92.7	
0.419	0.835	86.84	
0.516	0.872	82.66	
0.593	0.896	80.13	
0.644	0.912	78.31	
0.737	0.934	75.70	
0.793	0.960	74.14	
0.847	0.965	72.72	

[a] System: n-hexane (1)–chlorobenzene (2) at 759.8 mmHg.
[b] Source: reference B2 in Table 10-8.

Solution 10-2

Because the ratio of liquid molar volumes is required in Eqs. (9-58d) and (9-58e), any units can be used for volume as long as consistency is observed. Likewise, it is expected that the ratio of volumes will be much less sensitive to temperature than the volumes themselves which are only weakly temperature dependent. Thus, the volumes will be evaluated at 25°C using $v = $ M.W./density. For components 1 and 2 we have

$$v_1 = \frac{86.17}{0.659} = 130.8 \text{ cc / gmol}$$

$$v_2 = \frac{112.56}{1.107} = 101.7 \text{ cc / gmol}$$

These volumes along with the Antoine parameters and the $xyTP$ data of Table 10-2 were used with the program WEQ-ISOP.EXE to determine the Wilson parameters a_{12} and a_{21}. Table 10-3 shows the results of that calculation.

TABLE 10-3

RESULTS OF WILSON PARAMETER SEARCH (EX. 10-2)[a, b, c]

Start			Results			Average Values	
G_{12}	G_{21}	Precision	a_{12}/R	a_{21}/R	OFQ	G_{12}	G_{21}
1	1	0.01	264.0	52.14	0.00680	0.377	1.11
1	0.1	0.01	52.60	197.1	0.00857	0.673	0.749
0.67	0.75	0.001	265.9	51.24	0.00681	0.375	1.12
0.1	1	0.01	267.4	50.51	0.00680	0.374	1.12
0.37	1.12	0.001	265.8	51.30	0.00680		

[a] System: n-hexane (1)–chlorobenzene (2) at 759.8 mmHg.
[b] Source: reference B2 Table 10-8.
[c] WEQ-ISOP.EXE operating parameters:
  Initial step size: 2 (default)
  Scaling factor for step-size reduction 0.75 (default)
  Scaling factor for step-size increment: 0.75 (default)

The starting point for the search is specified in terms of G's instead of a's because G's are better behaved and more familiar. In addition to a_{12} and a_{21}, the program WEQ-ISOP.EXE also calculates values of G_{12} and G_{21} evaluated at the average system temperature in order to provide guidance for selection of future starting points. Table 10-3 shows that with the exception of the second run, the results show little variance despite the vastly different starting points. As the third run could be considered a continuation of the second run and the fifth a continuation of the fourth, it is seen that the results are tightly clustered. The value of $\Delta \bar{y}$ was found to be 0.0238 for these runs. From this result we may conclude that a reasonable, but not excellent, fit of the $xyTP$ data was obtained using the Wilson equation with its prescribed temperature dependence.

Now that we have v_1, v_2, a_{12}, and a_{21}, we can evaluate the Wilson parameters G_{12} and G_{21} at 65°C (338 K) using Eqs.(9-58d) and (9-58e)

$$G_{12} = \left(\frac{101.7}{130.8} \right) \exp\left(\frac{-265.8}{338} \right) = 0.3542$$

$$G_{21} = \left(\frac{130.8}{101.7} \right) \exp\left(\frac{-51.30}{338} \right) = 1.1050$$

These parameters can be used to calculate activity coefficients needed to determine partial pressures. The program WILSONEQ.EXE on the accompanying CD-ROM can be used to calculate activity coefficients. Once γ_1 and γ_2 are known, the partial pressures are found from Eq. (9-63a) using vapor pressures obtained from the Antoine equations. These vapor pressures at 65°C are 675.6 and 81.01 mmHg respectively for components 1 and 2. For $x_1 = 0.485$ and using the above G's, the Wilson equation gives

$$\gamma_1 = 1.181; \quad \gamma_2 = 1.189$$

The partial pressures are

$$p_1 = 675.6(0.485)(1.181) = 387.0 \text{ mmHg}$$

$$p_2 = 81.01(1 - 0.485)(1.189) = 49.6 \text{ mmHg}$$

yielding a total pressure of

$$P = 387.0 + 49.6 = 436.6 \text{ mmHg}$$

and

$$y_1 = \frac{387.0}{436.6} = 0.886$$

In this manner y_1 and P were calculated for other values of x_1 and are compared with experimentally determined vapor-liquid equilibrium data for this system at 65°C. A comparison of calculated and experimental values is shown in Fig. 10-6 where it is seen that for hexane mol fractions greater than 0.2 the agreement is quite good. A value of 0.0086 for $\Delta \bar{y}$ was obtained.

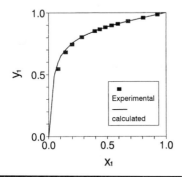

Figure 10-6
Comparison of experimental and calculated VLE data for the *n*-hexane–chloro-benzene system at 65°C. Data from reference B2 inTable 8.

While the data reduction process illustrated in Example 10-2 proved satisfactory, there is, in general, more uncertainty associated with using non-isothermal VLE data than there is with using isothermal VLE data. In fitting isothermal VLE data, one is faced with the question of which of several thermodynamically consistent equations would best represent activity coefficients in the system under consideration. With non-isothermal VLE data, the uncertainty is compounded because even though the Wilson equation is thermodynamically sanctioned and might be an appropriate equation to fit isothermal VLE data for our system, its prescription for temperature dependence enjoys no such sanction and might not be appropriate. Although one should always be aware of these uncertainties, particularly with large extrapolations, constant-pressure VLE data represent a major data source that should not be ignored. Ultimately, we do have measures of the goodness of fit (e.g., OFQ and $\Delta \bar{y}$) that allow us to evaluate the reliability of our data reduction efforts.

Isobaric VLE data can also be used to determine k_{12} in the Peng-Robinson equation by using the program PRKFIT.EXE. The program is written for isothermal VLE data, but can be used with single data points to determine k_{12} as a function of temperature as in Ex. 10-3.

Example 10-3

For the system CO_2–n-C_4H_{10} use *xyt* data at a pressure of 31,000 mmHg with the program PRKFIT.EXE to determine k_{12} for each data point. Plot these values of k_{12} versus temperature and use the value at 71.1°C to calculate the *xyP* data at this temperature with the program PRVLE.EXE. Compare these calculated values with experimental values.

Solution 10-3[7, 8]

The values of k_{12} obtained from each data point are plotted versus the temperatures of those data points on Fig. 10-7 where one sees that there is a systematic dependence of k_{12} on temperature. At 71.1°C k_{12} was estimated to be 0.10. This value was used with the program PRVLE.EXE to calculate y_1 and P at various values of x_1 at the temperature 71.1°C. These calculated values are compared with experimental values in Fig. 10-8 where it is observed that the agreement is excellent.

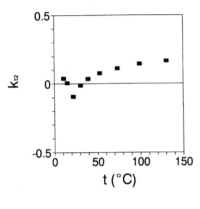

Figure 10-7 Temperature dependence of k_{12} for the $CO_2-n-C_4H_{10}$ system.

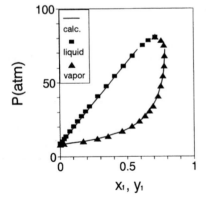

Figure 10-8 Comparison of experimental and calculated VLE data for the $CO_2-n-C_4H_{10}$ system at 71.1°C. Data from reference B9 inTable 10-8.

---

[7] Isobaric and isothermal data from reference B-2 in Table 10-8.
[8] Intermediate results are shown on the spreadsheet EX10-3.WQ1.

10-3 TOTAL PRESSURE DATA

When it is necessary to determine equilibrium data, thermodynamics can sometimes be used to facilitate the experimentation. A case in point is the use of isothermal Px data to determine parameters in an activity coefficient equation. This type of data can be taken in such a way that analysis of liquid and vapor samples is unnecessary, thereby greatly simplifying the experimentation. The parameters in a chosen activity coefficient equation are obtained by first writing the following expression for the total pressure:

$$P = x_1 P_1^{\circ} \gamma_1 + x_2 P_2^{\circ} \gamma_2 \tag{10-1}$$

Because $x_2 = 1 - x_1$ and γ_1 and γ_2 can be represented as specified functions of x_1 and x_2 [Eqs. (9-56)–(9-58)], Eq. (10-1) is a relation involving P and x_1 which contains two adjustable parameters (A's, B's, or G's). The task is then to take a set of isothermal Px data and determine the parameters which produce the best fit of Eq. (10-1). For this purpose the following objective function is minimized:

$$\text{OFP} = \sum_{i=1}^{n} (P^{\text{cal}} - P^{\text{exp}})_i^2$$

where P^{cal} is calculated from Eq. (10-1) at a value of x_1 for which P^{exp} has been measured. The summation is taken over all data points. The values of the parameters which minimize the objective function are found by computation techniques such as used by Hirata et al. and mentioned previously. These parameters are the desired information and can be used to calculate values of y via Eq. (9-63a).

A disadvantage of this technique is that we have no direct measure of how well activity coefficients are being represented by the chosen activity coefficient equation. However, if the minimized value of the objective function is small or if a plot of P^{cal} vs. x agrees well with a plot of P^{exp} vs. x, we need not be concerned about this. If these comparisons are not satisfactory, we may wish to choose another activity coefficient equation and repeat the process.

EXAMPLE 10-4

For the methanol (1)–carbon tetrachloride (2) system of Ex. 9-9, use only the xP data with the program WEQ-ISOT.EXE (on the accompanying CD-ROM) to determine parameters in the Wilson equation.

Solution 10-4

The WEQ-ISOT.EXE program uses the simplex pattern-search technique to determine the Wilson parameter set (G_{12}, G_{21}) which minimizes the objective function OFP when applied to a set of isothermal xP data. To ensure that a global minimum has been reached, the program was executed for four different starting points, yielding the results shown in Table 10-4. The small difference in both the parameter sets and the value of OFP strongly suggests that the global minimum has been reached. For the parameter set of run 1, values of the calculated total pressure and the pressure deviation are shown in Table 10-5, where it is observed that an excellent fit of the xP data was obtained. From this we may conclude that the Wilson equation with this parameter set does an excellent job of representing activity coefficients in this system.

TABLE 10-4
RESULTS OF WILSON PARAMETER SEARCH (EX. 10-4)[a, b, c]

Run	Starting Point		Result		
No.	G_{12}	G_{21}	G_{12}	G_{21}	OFP
1	0.01	0.01	0.04156	0.3369	6.914
2	0.05	0.3	0.04188	0.3361	6.931
3	0.1	0.004	0.04175	0.3364	6.918
4	0.5	0.5	0.04166	0.3371	6.915

[a] System: methanol (1)–carbon tetrachloride (2) at 20°C.
[b] Source: reference B8 in Table 10-8.
[c] WEQ-ISOT.EXE operating parameters:
    Scaling factor for step-size reduction: 0.1
    Scaling factor for step-size increment: 0.1
    Precision: 0.001
    Initial step size: 0.1

For this system we have determined Wilson parameter sets by two techniques: in Ex. 9-9 using xyP data and the objective function OFQ and here using xP data and the objective function OFP. In each case we obtained the best possible parameter set, but we note that the two sets are significantly different and are prompted to ask: Why did this happen and which is the better set?

TABLE 10-5
COMPARISON OF CALCULATED AND EXPERIMENTAL
TOTAL PRESSURES (EX. 10-4)[a, b]

x_1	P^{exp}	P^{cal}	$P^{cal} - P^{exp}$
0.020	136.6	135.5	−1.1
0.040	146.4	147.5	+1.1
0.100	154.8	156.1	+1.3
0.200	158.0	158.7	+0.7
0.300	159.5	159.8	+0.3
0.600	159.9	159.3	−0.6
0.700	157.5	157.1	−0.4
0.800	152.0	150.8	−1.2
0.900	135.5	135.6	+0.1
0.960	117.2	117.0	−0.2

[a] System: methanol (1)–carbon tetrachloride (2) at 20°C, pressures in mmHg.
[b] Source: reference B8 in Table 10-8.

If there were no experimental uncertainties and the Wilson equation perfectly represented activity coefficients in this system, we would expect to obtain the same parameter set by both techniques. However, because these conditions will never be met exactly, different objective functions will yield different parameter sets due to built-in weighting factors.

We may evaluate the efficacy of the parameter sets by comparing values of $\Delta \bar{y}$: 0.014 for the OFQ set and 0.007 for the OFP set. While the parameter set obtained from fitting xP data with OFP is clearly superior, it would be a mistake to generalize from a single example. The recommended use of OFQ was based on results for approximately 800 systems[7] and, of course, is statistically based.

10-4 AZEOTROPES

Azeotropes are often referred to as constant boiling mixtures because they represent the condition where vapor and liquid phases have identical compositions, and therefore continued boiling or distillation of the mixture produces no composition change. Azeotropes always occur at a maximum or minimum in either the Px or Tx curves and therefore are sometimes classified as maximum boiling or minimum boiling azeotropes depending on which behavior is exhibited on a constant-pressure diagram. Referring to Fig. 10-2, it is observed that system b forms a maximum boiling azeotrope and that system d forms a minimum boiling azeo-

---

[7] Hirata et al., *op. cit.*

trope, while systems a and c do not form azeotropes. Observe also that the liquid and vapor curves touch at the azeotropic point.

Although there are certain situations in which the occurrence of an azeotrope can be used to advantage in a separation process, in general the presence of an azeotrope causes difficulty when a separation of components by distillation is desired. In systems which do not form azeotropes, it is possible by distillation to separate any mixture into its pure components. However, when an azeotrope exists, a mixture can only be separated into a pure component and the azeotrope. As the knowledge of the existence of azeotropes is crucial to the design of distillation columns, much effort has been expended in determining and compiling this information.[8]

The azeotropic data are simply a single vapor-liquid equilibrium data point for the special case where $x_i = y_i$ and therefore can be used to determine two parameters in an activity coefficient equation. This is the very minimum amount of data required, and therefore there is no way to tell how well the chosen activity coefficient equation actually represents the activity coefficients.

EXAMPLE 10-5

At $t = 64.3°C$ and $P = 760$ mmHg the system methanol–methyl ethyl ketone forms an azeotrope containing 84.2 mol % of methanol. Use these data to determine the parameters in the Wilson equation, and use that equation to calculate the Txy data for this system at $P = 760$ mmHg. The constants in the Antoine equation are as follows:

	A	B	C
Methanol	7.87863	1474.110	230.0
Methyl ethyl ketone	6.97421	1209.600	216.0

Solution 10-5

At this temperature the vapor pressures calculated from the Antoine equation are

$$P_1^° = 736.94 \text{ mmHg}$$

$$P_2^° = 455.86 \text{ mmHg}$$

---

[8] A two-volume monograph is available: L. H. Horsley, *Azeotropic Data*, Advances in Chemistry Series, No. 6, American Chemical Society, Washington, D C, 1952, and L. H. Horsley, *Azeotropic Data—II*, Advances in Chemistry Series, No. 35, American Chemical Society, Washington, DC, 1962.

and the activity coefficients at $x_1 = 0.842$ are

$$\gamma_1 = \frac{760(0.842)}{736.94(0.842)} = 1.031$$

$$\gamma_2 = \frac{760(0.158)}{455.86(0.158)} = 1.667$$

This information could, in principle, be substituted into Eqs. (9-58a) and (9-58b) so that the parameters G_{12} and G_{21} could be determined. This would have presented no problem if the Margules or van Laar equations had been chosen; however, because of its algebraic complexity, the Wilson equation is difficult to manipulate. For a single data point the simultaneous solution of Eqs. (9-58a) and (9-58b) results in

$$\left\{ \frac{\gamma_2 x_2}{1 - \left[\frac{G_{12}x_1}{(x_1 + G_{12}x_2)} \right] + \left(\frac{x_1}{x_2} \right) \ln \left[\gamma_1(x_1 + G_{12}x_2) \right]} \right\}^{x_2/x_1} [\gamma_1(x_1 + G_{12}x_2)] = 1$$

Substitution of $x_1 = 0.842$, $x_2 = 0.158$, $\gamma_1 = 1.031$ and $y_2 = 1.667$ into this equation allows the determination of G_{12}. We obtain $G_{12} = 1.0818$. Substitution of these values into Eq. (9-58b) yields $G_{21} = 0.3778$.

The normal boiling points of methanol and methyl ethyl ketone are 64.7 and 79.6°C, respectively, and the azeotrope temperature is 64.3°C; therefore, the diagram should span the temperature range 64.3–79.6°C. Because this is not a large temperature range and because we have no way of knowing how well the Wilson equation with these parameters describes the system, a temperature correction of the parameters seems unjustified. We will use the values of G_{12} and G_{21} just determined and vapor pressures obtained from the Antoine equation to calculate the data points at $P = 760$ mmHg in the manner shown in Ex. 10-1. A comparison of the calculated values with experimental data is shown in Table 10-6 where the agreement is seen to be reasonably good. As expected, the best results are in the region of the azeotrope where the parameters were evaluated.

TABLE 10-6

COMPARISON OF CALCULATED AND EXPERIMENTAL

VAPOR-LIQUID EQUILIBRIUM DATA (EX. 10-5)[a, b]

	y_1		$t(°C)$	
x_1	Exptl.	Calc.	Exptl.	Calc.
0.076	0.193	0.185	75.3	75.6
0.197	0.377	0.377	70.7	71.3
0.356	0.528	0.536	67.5	67.8
0.498	0.622	0.637	65.9	66.0
0.622	0.695	0.711	65.1	65.0
0.747	0.777	0.782	64.4	64.3
0.829	0.832	0.833	64.3	64.3
0.936	0.926	0.921	64.4	64.4

[a] System: methanol (1)–methyl ethyl ketone (2) at 760 mmHg.
[b] Source: reference B8 in Table 10-8.

Because azeotropes are important in the design of distillation systems, a study of the reason for their existence is worthwhile. To gain this insight, consider the isothermal behavior of a hypothetical binary system composed of A and B with vapor pressures $P_A^°$ and $P_B^°$. The system is hypothetical because we will suppose that we have the ability to alter the activity coefficients in this system by some unspecified means. We will want to examine the behavior of the system as we systematically increase the activity coefficients and will represent this graphically in Fig. 10-9 where pressure is plotted vs. liquid mol fraction.[11] The dashed line represents ideal liquid-phase behavior, curve 1 represents moderate positive deviations from ideal behavior, and curve 2 represents larger positive deviations. We see that as the deviations (activity coefficients) are increased a maximum will appear in the total pressure curve. This is the condition that applies at an azeotrope. Thus, the existence of an azeotrope (maximum) depends on the difference in volatility (vapor pressures) and the magnitude of deviations from ideal-solution behavior. When the volatility difference is small, only small deviations are needed to produce a maximum and thus form an azeotrope, while large volatility differences require large deviations. It should be noted that no matter how small the volatility difference, an ideal system (linear total pressure curve) cannot exhibit an azeotrope. Identical conclusions would have been obtained if we had considered negative, rather than positive, deviations. Here we would be looking for a minimum in the total pressure curve instead of a maximum.

---

[11] An interactive version of Fig. 10-9 can be found on the spreadsheet FG10-9.WQ1(.XLS) in the COMPUTE directory of the accompanying CD-ROM.

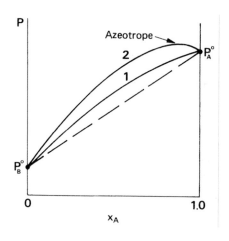

Figure 10-9 Total pressure in the hypothetical system A–B.

When faced with the prospect of an azeotrope in a system we wish to fractionate, it is often desirable to know if the azeotrope composition can be shifted by operating at a different pressure. If sufficient data afforded us the luxury of the consummate treatment, we would merely calculate the Txy diagram at various pressures and make the observation directly. As this is rarely the case, we will consider instead the estimation based on the minimum amount of data—a single azeotrope data point. We begin by writing Eq. (9-63a) for components A and B at the azeotropic point:

$$P = P_A^\circ \gamma_A$$

$$P = P_B^\circ \gamma_B$$

Eliminating P gives

$$\frac{P_A^\circ}{P_B^\circ} = \frac{\gamma_B}{\gamma_A} \tag{10-2}$$

This is the condition which applies for an azeotrope. It is useful in estimating the effect of pressure (which is manifested as a corresponding temperature change) on the azeotrope composition.

The vapor pressures depend only on temperature, while the activity coefficients depend on composition as well as temperature. We will not have enough information about the system to justify estimating the temperature dependence and therefore hope that by working with the ratio of activity coefficients this effect will be minimized. The right-hand side of Eq. (10-2) is assumed to depend only on composition and the left-hand side only on temperature. This allows a simple treatment which can be illustrated by Fig. 10-10 where a plot of P_A° / P_B° vs. T is juxtaposed to a plot of γ_B / γ_A vs. x_A. From vapor pressure data the P_A° / P_B° curve can be established; here, for the sake of example, this ratio is shown to increase with increasing temperature. The γ_A / γ_B curve will be calculated from the activity coefficient equation for which parameters were evaluated from the original azeotropic data point. For the sake of this example, positive deviations are shown. Activity coefficients will then resemble Fig. 9-14, and it is expected that γ_B / γ_A will increase as x_A increases.

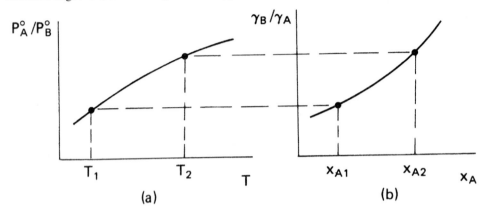

Figure 10-10   Variation of azeotrope composition with temperature.

The ordinate scales of Figs. 10-10a and 10-10b are equal; therefore, the condition for an azeotrope as stated by Eq. (10-2) can be represented by identical ordinate values as shown by one of the dashed horizontal lines which connect the P_A° / P_B° vs. T curve with the γ_B / γ_A vs. x_A curve. Two such connecting lines are shown, one corresponding to T_1 where the azeotrope composition is x_{A1} and the other corresponding to T_2 with azeotrope composition x_{A2}. For this particular example it is seen that the mol fraction of A in the azeotrope increases with increasing temperature. At any azeotrope condition characterized by x_A and T the corresponding pressure can be estimated from the knowledge of P_A°, P_B°, γ_A, and γ_B that we already possess.

10-5 THERMODYNAMIC CONSISTENCY TESTS

In Chap. 9 we saw that the Gibbs-Duhem equation provides a necessary condition that must be obeyed by the components of a solution. This equation forms the basis for testing the thermodynamic consistency of vapor-liquid equilibrium data. The test is usually made in terms of activity coefficients calculated from experimental vapor-liquid equilibrium data. However, before specific methods of testing are considered, we should recall that the Gibbs-Duhem equation was derived for and applies to a single multicomponent phase. Therefore, when considering the dependence of activity coefficients on liquid phase composition as specified by Eq. (9-48) or (9-49), it is legitimate to do so at a constant temperature and pressure because a system comprised of a single multicomponent phase possesses sufficient degrees of freedom. If activity coefficients could be determined from measurements on the liquid phase alone, there would be no problem in imposing the conditions of constant temperature and pressure. However, as we have seen in Chap. 9, the experimental determination of activity coefficients requires vapor-liquid equilibrium data, and a two-phase system possesses insufficient degrees of freedom to permit the fixing of both temperature and pressure while changing liquid composition. We will usually be concerned with applying a consistency test to a system at constant temperature where the pressure changes with liquid composition or a system at a constant pressure where the temperature changes with liquid composition. Isothermal systems usually present no problems because we have previously remarked that activity coefficients are insensitive to pressure. However, because the activity coefficients are sensitive to temperature, isobaric systems exhibiting large temperature ranges require special treatment. Here we will consider only isothermal systems or isobaric systems exhibiting only narrow temperature ranges for which the conditions of constant temperature and pressure are reasonably approximated.

If the activity coefficients in a system can be well fitted by an algebraic equation which is itself consistent with the Gibbs-Duhem equation, it is reasonable to state that the data are thermodynamically consistent. For example, the system treated in Ex. 10-1 was known to be well represented by the Wilson equation, and because of this, there is no need to worry about the consistency of the data. However, the lack of fit of the activity coefficients to a particular equation cannot be interpreted as a lack of consistency because that particular equation is only one of many which are able to satisfy the Gibbs-Duhem equation. If we were unable to reasonably fit a set of activity coefficient data to any of the commonly used activity coefficient equations, we may suspect inconsistency and desire to use a more definite test such as will be treated next.

We have seen that for a binary system the Gibbs-Duhem equation expressed in terms of activity coefficients is

$$x_1 \frac{d \ln \gamma_1}{dx_1} + x_2 \frac{d \ln \gamma_2}{dx_1} = 0 \tag{9-49}$$

or

$$\frac{x_1}{\gamma_1} \frac{d\gamma_1}{dx_1} + \frac{x_2}{\gamma_2} \frac{d\gamma_2}{dx_1} = 0 \tag{9-49a}$$

Either of these equations can be used to test consistency if experimentally determined values of γ or $\ln \gamma$ are plotted vs. liquid mol fraction. An illustration of the execution of this test may be found in Fig. 10-11, where smoothed curves representing experimental data points (not shown) are displayed for the system comprised of components 1 and 2. At a selected liquid composition tangents to each curve are drawn as shown. These graphically determined derivatives are substituted into Eq. (9-49) or (9-49a), and reasonable satisfaction of these equations certifies the data as thermodynamically consistent. This is a test of consistency at a particular composition, and for a complete test it is expected that several liquid compositions would be so examined.

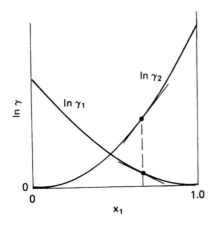

Figure 10-11 Thermodynamic consistency test with activity coefficients.

Examination of Eqs. (9-49) and (9-49a) allows us to make some general observations concerning the behavior of activity coefficients. First, consider the region in which $x_1 \rightarrow 1.0$, where $x_2 \rightarrow 0$ and γ_1 is known to approach unity, and observe that because $d\gamma_2 / dx_1$ must be finite, it is required that $d\gamma_1 / dx_1 \rightarrow 0$. Thus, in the region where the mol fraction of a component approaches unity, the activity coefficient of that component must approach unity with zero slope. Second, observe that the derivatives of the two components must be opposite in sign except for the case where they are both zero. The latter case refers to the existence of a maximum or a minimum and requires that at a composition where the activity coefficient of one component exhibits a maximum the activity coefficient of the other component must exhibit a minimum. Such extrema usually occur near the extremes of composition.

It is possible to test the consistency of vapor-liquid equilibrium data without calculating the activity coefficients. This can be done by using an alternative statement of the Gibbs-Duhem equation which we have previously derived:

$$\sum x_i \, d \ln \hat{f}_i = 0 \tag{9-47}$$

For a binary system at low pressure we write

$$x_1 \frac{d \ln p_1}{dx_1} + x_2 \frac{d \ln p_2}{dx_1} = 0 \tag{9-47a}$$

or

$$\frac{x_1}{p_1} \frac{dp_1}{dx_1} + \frac{x_2}{p_2} \frac{dp_2}{dx_1} = 0 \tag{9-47b}$$

The similarity between these equations and Eqs. (9-49) and (9-49a) is immediately observed. This test is analogous to that previously discussed and can be performed graphically with a plot of partial pressure vs. liquid mol fraction such as those shown in Figs. 9-13 and 9-15. An example of this test is shown in Fig. 10-12 for the chloroform–ethanol system at 45°C with the pertinent information tabulated in Table 10-7.

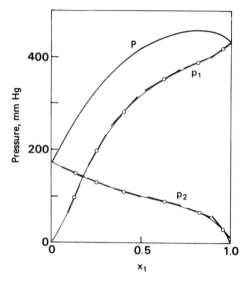

Figure 10-12 Thermodynamic consistency test with partial pressures. System: chloroform (1)–ethanol (2) at 45°C. [Adapted from Hála, J. Pick, V. Fried, and O. Vilím, *Vapor-Liquid Equilibrium*, 2nd ed., 1967, with permission of Pergamon Press Ltd., Elmsford, NY.]

TABLE 10-7
CONSISTENCY TEST OF THE CHLOROFORM (1)–ETHANOL (2) SYSTEM AT 45°C

x_1	p_1	p_2	$\dfrac{dp_1}{dx_1}$	$\dfrac{dp_2}{dx_1}$	$\dfrac{x_1}{p_1}\dfrac{dp_1}{dx_1}$	$\dfrac{x_2}{p_2}\dfrac{dp_2}{dx_1}$
0.1260	99.30	150.62	792	−176	1.01	−1.02
0.2569	199.60	130.02	724	−140	0.932	−0.800
0.4015	279.50	112.01	416	−118	0.598	−0.631
0.6283	349.00	89.89	229	−96	0.412	−0.397
0.8206	388.60	66.96	176	−153	0.372	−0.410
0.9557	418.00	30.49	290	−472	0.663	−0.686

It is seen from the table that, except for one point, the absolute values of

$$\frac{x_1}{p_1}\frac{dp_1}{dx_1}$$

and

$$\frac{x_2}{p_2}\frac{dp_2}{dx_1}$$

differ by less than 10%. Considering the uncertainty inherent in graphical differentiation, the agreement can be considered satisfactory.

Instead of the point-by-point test of consistency we have just considered, it is usually more convenient to test all the data together. To develop this test, we begin with the previously defined function Q,

$$Q = x_1 \ln \gamma_1 + x_2 \ln \gamma_2$$

and differentiate with respect to x_1 at constant T and P:

$$\frac{dQ}{dx_1} = x_1 \frac{d \ln \gamma_1}{dx_1} + \ln \gamma_1 + x_2 \frac{d \ln \gamma_2}{dx_1} + \ln \gamma_2 \frac{dx_2}{dx_1}$$

From Eq. (9-49) it is seen that the first and third terms on the right-hand side add to zero. Because

$$\frac{dx_2}{dx_1} = -1$$

we can write

$$dQ = \ln \frac{\gamma_1}{\gamma_2} dx_1$$

This equation is then integrated between the limits $x_1 = 0$ and $x_1 = 1$. The value of Q at each of these limits is zero, and because Q is a state function, we may write

$$0 = \int_0^1 \ln \frac{\gamma_1}{\gamma_2} dx_1 \tag{10-3}$$

Equation (10-3) is used to test consistency by preparing a plot of $\ln \gamma_1 / \gamma_2$ vs. x_1 such as shown in Fig. 10-13. The condition for consistency is that the positive area A should equal the negative area B. This test, usually called the Redlich-Kister test, is very popular even though it has been shown to have shortcomings.[10]

At this point the question naturally arises as to what degree of conformance is expected in these consistency tests and what are the consequences of the failure of a set of data to meet this degree of conformance. There can be no definite answers which will apply in all situations, and therefore the decision to accept or reject data will depend on the use to which the data are intended and any insight which might be available as to the source of the error. For example, a systematic error in temperature measurement would lead to spurious values

---

[10] H. C. Van Ness, S. M. Byer, and R. E. Gibbs, *AIChEJ.*, *19*, 238 (1973).

of P°'s and could cause activity coefficients to be in error even though values of x and y were correct. Such data could test out inconsistently even though the xy data would be useful if our intended purpose were the design of a distillation column. If a temperature error were suspected, we might find that the data were consistent in terms of the partial

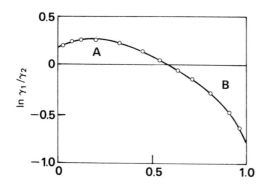

Figure 10-13 Area thermodynamic test. System: chloroform (1)–ethanol (2) at 45°C. [Adapted from E. Hála, J. Pick, V. Fried, and O. Vilím, *Vapor-Liquid Equilibrium*, 2nd ed., 1967, with permission of Pergamon Press Ltd., Elmsford, NY.]

pressure test but inconsistent in terms of activity coefficient tests. Conversely, because the ratio γ_1 / γ_2 does not involve the pressure, data containing a pressure error might satisfy the area test but fail the partial pressure test. For the many ramifications of thermodynamic consistency testing, the article by Van Ness et al.[11] is recommended.

10-6 MULTICOMPONENT VAPOR-LIQUID EQUILIBRIUM

All the activity coefficient equations we have already studied can be written for multicomponent mixtures, and there has been considerable effort directed toward the prediction of ternary equilibrium data from data for the three constituent binary systems. A fairly comprehensive study based on 19 ternary systems[12] has demonstrated that the Wilson equation is fairly effective for this. For the activity coefficient of component k in a multicomponent mixture containing m components the Wilson equation is

$$\ln \gamma_k = 1 - \ln \sum_{j=1}^{m} x_j G_{kj} - \sum_{i=1}^{m} \frac{x_i G_{ik}}{\sum_{j=1}^{m} x_j G_{ij}} \qquad (10\text{-}4)$$

The parameters G's are those which apply to binary mixtures ($G_{ii} = 1$), and thus it should be possible to calculate activity coefficients in the 1-2-3 ternary system from G's determined

[11] Van Ness et al., *ibid.*

[12] M. J. Holmes and M. Van Winkle, *Ind. Eng. Chem.*, 62, 21 (1970).

for the 1-2, 1-3, and 2-3 binary systems. These activity coefficients can be used with Eq. (9-63a) to calculate the ternary equilibrium data. In the study equilibrium data calculated in this manner were then compared to experimental values for the 19 systems. For the total of 262 data points examined, 92% of the predicted vapor mol fractions were within 0.020 mol fraction of the experimental data, and 85% of the predicted temperatures were within 1.0°C of the experimental data. Thus, without having to determine any experimental ternary data, it is possible to obtain reasonably good ternary vapor-liquid equilibrium data from an experimental study of the three constituent binary systems. The program WILSONEQ.EXE on the accompanying CD-ROM calculates activity coefficients for binary systems and for multicomponent systems of up to six components from Eq. (10-4).

Example 10-6

VLE data at 25°C are available for the three binary systems and the ternary system comprised of cyclohexane(1), n-heptane(2), and toluene(3).[13] Use these data to demonstrate the technique of estimating ternary VLE from data available for the constituent binary systems. Use the Wilson equation.

Solution 10-6

The VLE data were reported as partial pressures from which P's and y's were calculated and the resulting xyP data[14] were used in the program WEQ-ISOT.EXE to determine the following parameters in the Wilson equation by a procedure similar to that of Ex. 9-9.

System	Wilson Parameters	$\Delta \bar{y}$
1–2	$G_{12} = 0.8101;\ G_{21} = 1.0934$	0.0012
1–3	$G_{13} = 0.6487;\ G_{31} = 0.8871$	0.0046
2–3	$G_{23} = 0.8044;\ G_{32} = 0.7736$	0.0044

Wilson Parameters: 1 = cyclohexane; 2 = n-heptane; 3 = toluene

The rather small values of $\Delta \bar{y}$ indicate that a good fit was achieved in all three systems.

These parameters were used in the program WILSONEQ.EXE to calculate activity coefficients in the ternary system. The ternary activity coefficients were then used with vapor pressure data to calculate the to-

---

[13] Source: T. Katayama, E. K. Sung, and E. N. Lightfoot, *AIChEJ, 11*, 924 (1965).

[14] Original and calculated data are on the spreadsheet EX10-6.WQ1 in the BACKUP directory of the accompanying CD-ROM.

tal pressure and the vapor mol fractions. A sample calculation is shown here.[15]

At 25°C the vapor pressures in mmHg are:

$$P_1^\circ = 97.58; \; P_2^\circ = 45.72; \; P_3^\circ = 28.45$$

At a liquid composition of $x_1 = 0.333$, $x_2 = 0.333$ the activity coefficients are:

$$\gamma_1 = 1.085; \; \gamma_2 = 1.072; \; \gamma_3 = 1.224$$

and the partial pressures are:

$$p_1 = 0.333(97.58)(1.085) = 35.26 \text{ mmHg}$$

$$p_2 = 0.333(45.72)(1.072) = 16.32 \text{ mmHg}$$

$$p_3 = 0.333(28.45)(1.224) = 11.63 \text{ mmHg}$$

yielding a total pressure of

$$P = 35.26 + 16.32 + 11.63 = 63.21 \text{ mmHg}$$

and vapor mol fractions of

$$y_1 = \frac{35.26}{63.21} = 0.558$$

$$y_2 = \frac{16.32}{63.21} = 0.258$$

Values of y_1 and y_2 calculated in this manner are compared to experimental values in Fig. 10-14 where it is seen that the agreement is quite good. The values of $\Delta \bar{y}$ are 0.0049 and 0.0026 for components 1 and 2 respectively based on the total of 16 ternary data points. These values are seen to be comparable to those obtained in fitting the binary systems.

---

[15] Intermediate values and final results are available on the spreadsheet EX10-6.WQ1.

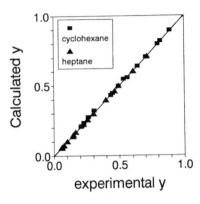

Figure 10-14 Comparison of experimental and calculated VLE data for the cyclohexane–*n*-heptane–toluene system at 25°C.

When equilibrium data are needed for a particular ternary system, we will probably not be so fortunate as to find data available for the three constituent binaries and will therefore have to perform an experimental determination. Here the thermodynamic approach that we have previously applied to binary systems can be employed with enormous savings in experimental effort. For a ternary system there will be six Wilson parameters to be determined, and for each data point an activity coefficient for each of the three components can be calculated. Thus, a minimum of two data points will allow the writing of six equations of the form of Eq. (10-4) which will be sufficient to determine the six parameters. In making an experimental study, we will obviously wish to determine more than the minimum amount of data in order that we might assess the ability of the chosen activity coefficient equation[16] to represent the activity coefficients with sufficient accuracy. However, a little reflection should convince you that this is a far smaller quantity of data than would be needed if the thermodynamic approach were not available and we were forced to rely on some type of graphical presentation of the data.

10-7 PHASE BEHAVIOR IN PARTIALLY MISCIBLE SYSTEMS

There are many familiar examples of partially miscible liquids—notably Italian salad dressing, which consists of an oil phase and an aqueous phase and is applied to salad in the form of an emulsion. The water phase contributes the vinegar flavor and sharpness, while the purpose of the oil is, no doubt, to dissolve those desirable flavors from the secret combination of herbs and spices so that they may be properly blended and subsequently distributed throughout the entire salad by means of the tossing process. Other useful applications of partial liquid miscibility are found in the chemical, petroleum, and related process industries through the application of the separation technique of liquid-liquid extraction. Because the analysis

---

[16] The Margules and van Laar equations can also be written in multicomponent form; however, because of its proven superiority, only the Wilson equation has been listed.

and design of such processes requires an understanding of the principles of liquid-liquid phase equilibrium, we will now develop this subject using the thermodynamic approach. No new functions or concepts are needed for this development.

While the application of the extraction process involves the use of a solvent to preferentially dissolve one of two or more components from another liquid phase and thereby represents a system of three or more components, here most of our attention will be devoted to binary systems. If an understanding of the behavior of binary systems is developed, the extension to ternary systems should come easily. The thermodynamic approach to liquid-liquid equilibrium will be best appreciated after you have become familiar with the qualitative behavior of these systems.

The phase behavior of a partially miscible binary system is depicted in Fig. 10-15, a constant-pressure diagram, and Fig. 10-16, a constant-temperature diagram. When two liquid phases and a vapor phase are in equilibrium, the phase rule specifies only one degree of freedom. Having specified the pressure, Fig. 10-15 shows that there is a single temperature $T_{3\pi}$ and a single set of compositions. The points C and D represent liquid compositions, and E represents the vapor composition. Similarly, at the temperature specified for Fig. 10-16 there is a single pressure $P_{3\pi}$ and a single set of compositions C, D, and E.

At temperatures less than $T_{3\pi}$ a three-phase system would exert a pressure less than the pressure specified for Fig. 10-15, and therefore the vapor phase cannot exist at this specified pressure. Below the temperature $T_{3\pi}$ Fig. 10-15 is divided into three regions labeled L', L'', and $L' + L''$. In the regions L' and L'' only a single liquid phase can exist, and in the region $L' + L''$ two liquid phases exist. The region L' represents solutions which are rich in component 2, and the region L'' represents solutions rich in component 1. Mol fractions in these regions are designated by single-prime and double-prime superscripts, respectively. At the temperature T_a the maximum solubility of component 1 in component 2 is shown by point Q, and the maximum solubility of component 2 in component 1 is shown by point R. The mol fractions of these two saturated liquid phases are designated x'_{1s} and x''_{1s}, respectively. In the region L' homogeneous solutions are possible over the composition range $0 < x'_1 \leq x'_{1s}$, and in region L'' the range is $x''_{1s} \leq x''_1 < 1.0$. The curves GC and HD show the temperature dependence of the mutual solubilities. The region $L' + L''$ represents a mixture of the two saturated liquid phases. In this region mol fraction does not have its usual meaning but is regarded as the overall composition of the two-phase mixture, which, of course, varies with the proportion of the two phases.

Figure 10-16 shows that at pressures higher than $P_{3\pi}$ the vapor phase cannot exist. Above $P_{3\pi}$ there are regions L', L'', and $L' + L''$ which correspond to the same regions of Fig. 10-15. One difference, however, should be noted: The boundaries on Fig. 10-16 are almost vertical lines indicating negligible change in mutual solubility with pressure. Because of this insensitivity to pressure, mutual solubility data are often plotted vs. temperature in the manner of Fig. 10-17. No particular pressure is specified for this type of diagram; all that is required is that at a particular temperature the pressure be at least as high as the three-phase equilibrium pressure $P_{3\pi}$.

Above $T_{3\pi}$ in Fig. 10-15 and below $P_{3\pi}$ in Fig. 10-16 lie the vapor-liquid envelopes. They are interpreted in the same manner as previously encountered vapor-liquid phase diagrams.

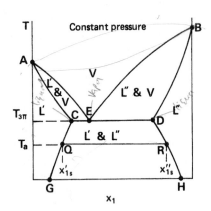

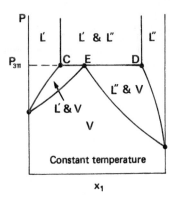

Figures 10-15 and **10-16** Constant-pressure and constant-temperature diagrams for a partially miscible binary system.

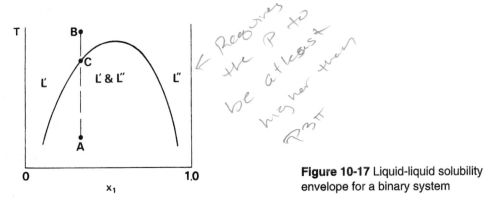

Figure 10-17 Liquid-liquid solubility envelope for a binary system

Experimental binary equilibrium data may be acquired in either of several ways: (1) in the same way vapor-liquid equilibrium data are taken but in equipment especially designed for the presence of two liquid phases; (2) by simply determining by a titration procedure the amount of component 1 which will dissolve in component 2, x'_{1s}, and vice versa, x''_{1s}; and (3) by the *cloud point* technique. The last method is best illustrated by referring to Fig. 10-17, which shows only liquid-liquid equilibrium. Known amounts of the two liquids (corresponding to point A) are introduced into a tube which is subsequently sealed and heated to a temperature where there is certain to be only a single liquid phase (point B). the solution is well mixed and then subjected to a slow cooling process until it becomes cloudy and phase separation occurs (point C). The temperature at the cloud point is recorded, and thus a single point on the solubility envelope is determined. The procedure is repeated for other mixtures until sufficient data are available to construct the envelope. There is quite a bit of this type of data available.[17]

The reason for partial liquid miscibility can be understood if we again consider the hypothetical AB system where we have supposed that we possess the ability to vary the activity coefficients at will. The system will be at a constant temperature, and we will assume that the following simple equations describe the behavior of the activity coefficients:[18]

$$\ln \gamma_A = Bx_B^2 \tag{10-5a}$$

$$\ln \gamma_B = Bx_A^2 \tag{10-5b}$$

We will consider only positive deviations from the ideal-solution model and observe that parameter B alone determines the extent of these deviations. Let us now consider the Gibbs free energy change on forming a solution from n_A mol of pure liquid A and n_B mol of pure liquid B. We may write

$$\Delta G = G - n_A \mu_A - n_B \mu_B \tag{10-6}$$

where G is the Gibbs free energy of the solution and by Eq. (9-43) can be written as

$$G = n_A \hat{\mu}_A + n_B \hat{\mu}_B \tag{10-7}$$

Combining Eqs. (10-6) and (10-7) gives

$$\Delta G = n_A (\hat{\mu}_A - \mu_A) + n_B (\hat{\mu}_B - \mu_B) \tag{10-8}$$

If Eq. (10-8) is divided by $n_A + n_B$, it will contain only intensive properties:

$$\frac{\Delta G}{n_A + n_B} = \Delta g = x_A (\hat{\mu}_A - \mu_A) + x_B (\hat{\mu}_B - \mu_B) \tag{10-9}$$

From Eq. (9-38) we have seen that the differences in chemical potential are expressible in terms of mol fractions and activity coefficients. Use of Eq. (9-38) reduces Eq. (10-9) to

$$\frac{\Delta g}{RT} = x_A \ln x_A + x_B \ln x_B + x_A \ln \gamma_A + x_B \ln \gamma_B \tag{10-10}$$

When Eqs. (10-5a) and (10-5b) are used, the result is

$$\frac{\Delta g}{RT} = x_A \ln x_A + x_B \ln x_B + x_A (Bx_B^2) + x_B (Bx_A^2) \tag{10-11}$$

---

[17] A. W. Francis, *Liquid-Liquid Equilibriums*, Wiley-Interscience, New York, 1963, and A. W. Francis, *Critical Solution Temperatures*, Advances in Chemistry Series, No. 31, American Chemical Society, Washington, DC, 1961.

[18] These equations result from either the Margules or van Laar equations when the parameters are equal.

It is seen that Eq. (10-11) relates the Gibbs free energy of mixing to liquid mol fraction and the parameter B which measures the extent of deviations from ideal solutions. A value of zero for B represents ideal solutions, and thus the first two terms on the right-hand side of Eq. (10-11) are identified with the ideal Gibbs free energy of mixing. This can be stated more generally as

$$\Delta g^i = RT \sum x_i \ln x_i \tag{10-12}$$

A quantity called the excess Gibbs free energy of mixing is defined by

$$\Delta g = \Delta g^i + \Delta g^e \tag{10-13}$$

From Eqs. (10-10), (10-12), and (10-13) we write the excess Gibbs free energy of mixing as

$$\Delta g^e = RT \sum x_i \ln \gamma_i \tag{10-14}$$

Thus, the last two terms in Eq. (10-11) are identified with Δg^e.

In Fig. 10-18 $\Delta g / RT$ as determined by Eq. (10-11) is plotted vs. x_A for various values of the parameter B increasing in the order $0, B', B'', B'''$. The dashed portion of curve B''' as calculated by Eq. (10-11) represents a single liquid phase; however, it is to be noted that points on this curve possess a higher free energy than points on the line CD. This line represents mixtures of the two liquid phases having compositions corresponding to the points C and D. Because the Gibbs free energy will decrease when the system moves from a point on the dashed curve to a point on CD, this process is spontaneous, and separation into two liquid phases results. Partial liquid miscibility is therefore seen to be caused by extremely large positive deviations;[19] negative deviations do not produce limited miscibility.

By using stability criteria developed in Chap. 7, it can be shown that a portion of the curve labeled B''' in Fig. 10-18 represents an unstable system. We begin by differentiating Eq. (10-9) with respect to x_A at constant temperature and pressure and obtain

$$\frac{\partial \Delta g}{\partial x_A} = (\hat{\mu}_A - \mu_A) - (\hat{\mu}_B - \mu_B) + x_A \frac{\partial \hat{\mu}_A}{\partial x_A} + x_B \frac{\partial \hat{\mu}_B}{\partial x_A}$$

---

[19] With Eq. (10-5) partial miscibility occurs with $B > 2.0$. $B = 2$ corresponds to activity coefficients as high as 7.4. Activity coefficients in the range of hundreds or thousands are not uncommon in partially miscible systems.

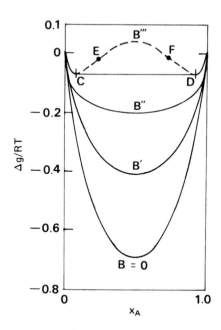

Figure 10-18 Free energy of mixing by Eq. (10-11). [Adapted from J. Hildebrand and R. Scott, *The Solubility of Nonelectrolytes*, copyright © 1950 by Reinhold Publishing. Reprinted by permission of Dover Publications Inc., New York.]

The Gibbs-Duhem equation requires the last two terms to add to zero, and the remaining terms are again differentiated with respect to x_A:

$$\frac{\partial^2 \Delta g}{\partial x_A^2} = \frac{\partial \hat{\mu}_A}{\partial x_A} - \frac{\partial \hat{\mu}_B}{\partial x_A}$$

Because $dx_A = -dx_B$, we write

$$\frac{\partial^2 \Delta g}{\partial x_A^2} = \frac{\partial \hat{\mu}_A}{\partial x_A} + \frac{\partial \hat{\mu}_B}{\partial x_B}$$

The stability criterion developed in Sec. 7-5 requires that each derivative on the right-hand side be positive, and we may therefore write the stability criterion as

$$\frac{\partial^2 \Delta g}{\partial x_A^2} > 0$$

Between the points E and F in Fig. 10-18 this second derivative is seen to be negative, and therefore a single liquid phase is not stable.[20]

While we are still considering the hypothetical AB system, let us examine its isothermal Pxy behavior as the positive deviations are increased. Displayed in Fig. 10-19 is a series of five Pxy diagrams corresponding to the complete range of positive deviations beginning in part a with ideal behavior $(B = 0)$ and ending in part e with total immiscibility $(B \to \infty)$. As deviations are increased, the system is seen to pass from an ideal system with no azeotrope to a homogeneous azeotrope (part b) to partial miscibility and a heterogeneous azeotrope (part c) to more limited miscibility and finally to complete immiscibility (part e). Thus, the type of phase behavior should provide a clue as to the order of magnitude of the deviations from ideal-solution behavior.

For the purposes of illustrating the influence of the activity coefficients on the behavior of our hypothetical system, we chose the simplest algebraic form; however, it should be recognized that the enormous range of activity coefficient values found in partially miscible systems requires expressions of greater complexity for satisfactory representation. Attempts to use the Margules and van Laar equations [Eqs. (9-56) and (9-57)] with partially miscible systems have met with little success. The Wilson equation, which serves so well for vapor-liquid equilibrium calculations, does not produce the type of dashed curve shown in Fig. 10-18 and therefore is unsuitable for partially miscible systems. Recently, new activity coefficient equations have been developed through the guidance of molecular theory which are more capable of describing the behavior of activity coefficients in partially miscible systems with only two adjustable parameters.[21]

10-8 LIQUID-LIQUID EQUILIBRIUM

In a binary system containing two liquid phases and a vapor phase at equilibrium we may write

$$\hat{f}_1' = \hat{f}_1'' = \hat{f}_1^G \tag{10-15a}$$

$$\hat{f}_2' = \hat{f}_2'' = \hat{f}_2^G \tag{10-15b}$$

where the primed and double-primed superscripts refer to the liquid phases. Substitutions of the low-pressure versions of Eqs. (9-25) and (9-35) into the preceding conditions of equilibrium results in

---

[20] An interactive version of Fig. 10-18 can be found on the spreadsheet FG10-18.WQ1(.XLS) in the COMPUTE directory of the accompanying CD-ROM.

[21] They are the NRTL equation [*AIChEJ.*, *14*, 135 (1968)] and the UNIQUAC equation [*AIChEJ.*,*21*, 116 (1975) and *Ind. Eng. Chem. Process Des. Dev.*, *17*, 561 (1978)].

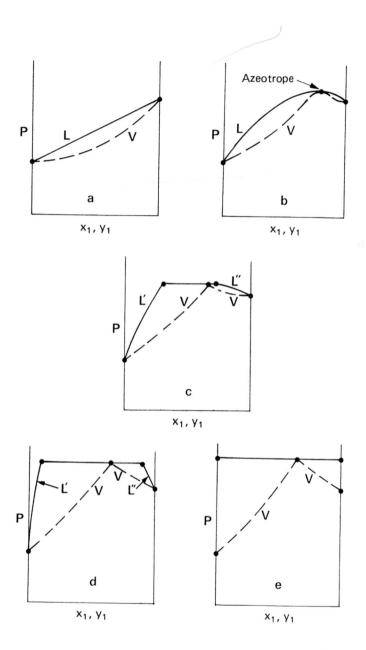

Figure 10-19 Constant-temperature diagrams for increasing positive deviations.

$$x'_{1s}\gamma'_{1s}P_1^\circ = x''_{1s}\gamma''_{1s}P_1^\circ = Py_1 \tag{10-16a}$$

$$x'_{2s}\gamma'_{2s}P_2^\circ = x''_{2s}\gamma''_{2s}P_2^\circ = Py_2 \tag{10-16b}$$

where the subscript s refers to a saturated liquid phase. Without this subscript, x'_1 refers to the unsaturated primed liquid phase where $0 < x'_1 \le x'_{1s}$, and x''_1 refers to the unsaturated double-primed liquid phase where $x''_{1s} \le x''_1 < 1.0$. In these two regions liquid mol fraction has meaning, but there is no liquid corresponding to $x'_{1s} < x_1 < x''_{1s}$.

While Eqs. (10-15) and (10-16) are similar to those which we have applied to vapor-liquid equilibrium, calculations involving liquid-liquid equilibrium are much more complex, because in addition to the two liquid phases there can be a vapor phase present—a total of three phases which according to the phase rule leaves only one degree of freedom for a binary system. Thus, if the temperature is fixed, the system is defined, and all the terms in Eq. (10-16) are fixed even though we may not know their values. Suppose that we have an activity coefficient equation which we have reason to believe represents activity coefficients in this system and desire to use it to calculate the equilibrium values of x'_{1s}, x''_{1s}, y_1, and P at a specified temperature. Such a calculation is possible but is quite involved. It can be visualized functionally if we rewrite Eqs. (10-16) to show that the γs are known functions of liquid mol fraction:

$$x'_{1s}\gamma_1(x'_{1s}) = x''_{1s}\gamma_1(x''_{1s}) \tag{10-16c}$$

$$(1 - x'_{1s})\gamma_2(x'_{1s}) = (1 - x''_{1s})\gamma_2(x''_{1s}) \tag{10-16d}$$

These two equations contain the two unknowns x'_{1s} and x''_{1s} and in principle can be solved. After they have been obtained, they can be used in Eqs. (10-16a) and (10-16b) to calculate y_1 and P.

Occasionally, liquid-liquid equilibrium data are used to provide information useful for vapor-liquid equilibrium calculations. In this case Eqs. (10-16) are used to evaluate or estimate activity coefficients. Two methods are available for accomplishing this depending on the extent of mutual solubility. When appreciable mutual solubility exists, a knowledge of x'_{1s} and x''_{1s} may be used to evaluate two parameters in a chosen activity coefficient equation. This is essentially the inverse of the previously discussed calculation and again can be visualized functionally by restating Eqs. (10-16):

$$x'_{1s}\gamma_1(x'_{1s}, A, B) = x''_{1s}\gamma_1(x''_{1s}, A, B) \tag{10-16e}$$

$$(1 - x'_{1s})\gamma_2 \ (x'_{1s}, \ A, B) = (1 - x''_{1s})\gamma_2 \ (x''_{1s}, A, B) \tag{10-16f}$$

Because we have chosen the activity coefficient equation, the functional forms for γ_1 and γ_2 are known, and the only unknowns in these equations are the parameters A and B. They can be solved for and used to calculate activity coefficients for subsequent vapor-liquid equilibrium calculations. An obvious disadvantage to this method is that there is no way of knowing how well the chosen equation represents the activity coefficients.[22]

When the degree of mutual solubility is small, it is assumed that Raoult's law applies to the component in high concentration and that Henry's law applies to the component in low concentration. Thus, we could write

$$\gamma''_{1s} \rightarrow 1.0$$

$$\gamma'_{2s} \rightarrow 1.0$$

which allows Eqs. (10-16) to be written as

$$x'_{1s}\gamma'_{1s} = x''_{1s} \tag{10-16g}$$

$$(1 - x'_{1s}) = (1 - x''_{1s})\gamma''_{2s} \tag{10-16h}$$

This allows γ'_{1s} and γ''_{2s} to be determined from the known values of x'_{1s} and x''_{1s}. These activity coefficients correspond to dilute solutions of each of the two components where Henry's law is assumed valid, and for these solutions we write

$$p_1 = P_1^\circ \gamma'_1 x'_1 = k_1 x'_1 \tag{10-16i}$$

$$p_2 = P_2^\circ \gamma''_2 x''_2 = k_2 x''_2 \tag{10-16j}$$

where $k_1 = P_1^\circ \gamma'_{1s}$ and $k_2 = P_2^\circ \gamma''_{2s}$. For Henry's law to be followed, we see that it is necessary for γ'_1 to be constant at γ'_{1s} over the range $0 < x'_1 \leq x'_{1s}$ and that γ''_2 should have the value γ''_{2s} over the range $0 < x''_2 \leq x''_{2s}$. When these concentration ranges are small, the assumptions seem reasonable.

---

[22] For a detailed evaluation of the deficiencies of the method, see D. S. Joy and B. G. Kyle, *AIChEJ.*, *15*, 298 (1969).

10-9 TERNARY LIQUID-LIQUID EQUILIBRIUM

Ternary liquid-liquid equilibria are usually represented on triangular diagrams such as shown in Figs. 10-20 and 10-21. The system displayed in Fig. 10-20 has only one partially miscible binary pair and is referred to as a type I system, while Fig. 10-21 shows a type II system which has two partially miscible binary pairs. In these figures the solubility envelope shows the region in which two liquid phases exist, and tie lines are used to indicate compositions of each phase which are in equilibrium. This type of data is usually acquired by equilibrating a mixture at the desired temperature and withdrawing samples of each liquid phase for analysis. Usually pressure measurements and vapor samples are not taken. The compositions of the two liquid phases in equilibrium constitute a tie line.

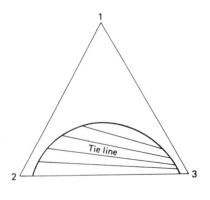

Figure 10-20 Typical type I ternary liquid-liquid system.

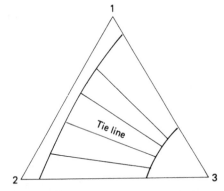

Figure 10-21 Typical type II ternary liquid-liquid system.

Following the successful attempts to calculate ternary vapor-liquid equilibrium data from only binary data, the corresponding calculation of ternary liquid-liquid equilibrium from only binary data has been attempted. Fewer systems have been studied, and results are not nearly as striking as for the corresponding vapor-liquid calculation, although generally positive results have been obtained. There is little doubt that acceptable results can be calculated if high-quality binary data are available for parameter determination in a reliable activity coefficient equation. For liquid-liquid equilibrium the calculation procedure is more involved, and because of the difficulty in analytically representing the widely varying activity coefficients in these types of systems, the accuracy is often a problem.

The procedure for calculating ternary liquid-liquid equilibrium from an activity coefficient equation we believe to be reliable is based on solving the following set of equilibrium statements:

$$x'_{1s}\gamma_1(x'_{1s}, x'_{2s}) = x''_{1s}\gamma_1(x''_{1s}, x''_{2s})$$

$$x'_{2s}\gamma_2(x'_{1s}, x'_{2s}) = x''_{2s}\gamma_2(x''_{1s}, x''_{2s})$$

$$(1 - x'_{1s} - x'_{2s})\gamma_3(x'_{1s}, x'_{2s}) = (1 - x''_{1s} - x''_{2s})\gamma_3(x''_{1s}, x''_{2s})$$

The mol fraction of component 3 has been eliminated from these equations. It is assumed that parameters in the activity coefficient equation are available, and thus the functional dependence of activity coefficient on liquid mol fractions is known. Values of x'_{1s}, x'_{2s} and x''_{1s}, x''_{2s} which satisfy these equations are at the ends of a tie line. Because there are four of these variables and only three equations, the system is undefined. This is in accordance with the phase rule which specifies that a three-component, three-phase (two liquids and a vapor) system at a specified temperature has only one degree of freedom. Specification of a single liquid mol fraction will then allow the remaining mol fractions to be determined. Because of their complexity, these calculations are usually computerized. The literature should be consulted for the computational technique and details.[23]

The inverse procedure of determining activity coefficient parameters from ternary liquid-liquid equilibrium data has also been attempted.[24] It should always be possible, at least in principle, to accomplish this if the chosen equation is capable of representing the activity coefficients. To visualize this procedure, we write the equations of equilibrium slightly differently:

$$x'_{1s}\gamma_1(x'_{1s}, x'_{2s}, x'_{3s}, A's) = x''_{1s}\gamma_1(x''_{1s}, x''_{2s}, x''_{3s}, A's)$$

$$x'_{2s}\gamma_2(x'_{1s}, x'_{2s}, x'_{3s}, A's) = x''_{2s}\gamma_2(x''_{1s}, x''_{2s}, x''_{3s}, A's)$$

$$x'_{3s}\gamma_3(x'_{1s}, x'_{2s}, x'_{3s}, A's) = x''_{3s}\gamma_3(x''_{1s}, x''_{2s}, x''_{3s}, A's)$$

This set of three equations can be written for each tie line where the saturated mol fractions are known. For the chosen activity coefficient equation the functional dependence of the γ's is known, and therefore only the parameters (A's) are unknown. If six parameters are re-

[23] For example, see H. Renon and J. M. Prausnitz, *AIChEJ.*, *14*, 135 (1968), or D. S. Joy and B. G. Kyle, *AIChEJ.*, *15*, 298 (1969).

[24] See D. S. Joy and B. G. Kyle, *Ind. Eng. Chem. Process Des. Dev.*, *9*, 244 (1970), and G. Varhegyi and C. H. Eon, *Ind. Eng. Chem. Fundam.*, *16*, 182 (1977).

quired, a minimum of two tie lines are needed; however, several tie lines can be used so that the best set of parameters can be determined. The following objective function has been used:

$$OF = \sum_{j=1}^{m} \sum_{i=1}^{3} (x'_{ij}\gamma'_{ij} - x''_{ij}\gamma''_{ij})^2$$

where $i = 1,2,3$ represents the components and j refers to tie lines of which there are m. The parameters which minimize this objective function can be found by various techniques such as a pattern-search technique. Although this method has been tested on only a few systems, reasonable results were obtained. Thus, ternary liquid-liquid equilibrium data are a potential source of thermodynamically useful information.

10-10 ESTIMATES FROM FRAGMENTARY DATA

In contrast to the consummate thermodynamic treatment, we have seen that in the case of azeotropes fragmentary data can be used for making estimates suitable for preliminary design studies. Here some examples will be presented which further illustrate the use of fragmentary data for estimation purposes. Central to each application is the estimation of activity coefficients at the desired conditions from data available at other conditions.

EXAMPLE 10-7

Equilibrium data are needed for the design of an absorber to remove acetone vapor from an air stream using water as a solvent. Use available vapor-liquid equilibrium data for the acetone–water system at 25°C to calculate mol fractions (m.f.) of acetone in air which are in equilibrium with mol fractions of acetone in water at various pressures and at 25°C. The equilibrium data[25] for the acetone–water system are tabulated as follows:

x_A	0.0194	0.0289	0.0449	0.0556	0.0939
y_A	0.5234	0.6212	0.7168	0.7591	0.8351
P (mmHg)	50.1	61.8	81.3	91.9	126.1

At 25°C the vapor pressures are

---

[25]Source: reference B2 in Table 10-8.

EXAMPLE 10-7 CON'T

$$P_A^\circ = 230.05 \, \text{mmHg}$$

$$P_W^\circ = 23.76 \, \text{mmHg}$$

Solution 10-7

From the preceding $x_A y_A P$ data, activity coefficients for acetone are calculated by

$$\gamma_A = \frac{P y_A}{230.05 x_A}$$

They are as follows:

x_A	0.0194	0.0289	0.0449	0.0556	0.0939
γ_A	5.88	5.77	5.64	5.45	4.88

At equilibrium in the air–water–acetone system up to moderate pressures, for the acetone we may write

$$P y_A = 230.05 \gamma_A x_A$$

We will assume that air is essentially insoluble in acetone–water solutions, and therefore the liquid phase behavior is described adequately by the preceding relation between x_A and γ_A. Thus, the equilibrium data needed for the absorber design can be calculated from the preceding equation and the table of x_A and γ_A values at the temperature of 25°C and any pressure P. For $P = 760$ mmHg the following data were calculated:

y_A, m.f. acetone in air	0.0345	0.0505	0.0767	0.0917	0.1387
x_A, m.f. acetone in water	0.0194	0.0289	0.0449	0.0556	0.0939

If high pressures are to be considered, we may want to replace the total pressure and the vapor pressure with the appropriate fugacities and will assume activity coefficients insensitive to pressure. Also, if the mol fraction of water vapor in the gas phase were desired, it could be calculated in the same manner.

EXAMPLE 10-8

At 25°C the solubility of ethane in *n*-heptanol at a pressure of 1 atm is x_E = 0.0159. Assuming that the activity coefficients in this system can be represented by

$$\ln \gamma_E = B(1 - x_E)^2$$

estimate the solubility of ethane in *n*-heptanol at 25°C and 20 atm. K values for ethane are as follows:

At 25°C and 1 atm: K_E = 27.0
At 25°C and 20 atm: K_E = 1.62

n-heptanol may be considered nonvolatile.

Solution 10-8

We begin by applying Eq. (9-50) to ethane:

$$f_E^G y_E = f_E^L x_E \gamma_E$$

or

$$y_E = \frac{f_E^L}{f_E^G} x_E \gamma_E = K_E x_E \gamma_E$$

Because *n*-heptanol is considered nonvolatile, $y_E = 1.0$, and from the solubility measurement at $P = 1$ atm we write

$$1 = 27.0(0.0159)\gamma_E$$

or

$$\gamma_E = 2.33$$

This is the very minimum amount of data which can be used—one data point will allow the determination of one parameter. (This form of activity coefficient equation results when both parameters are equal in either the Margules or van Laar equations.) Substitution of the data point $\gamma_E = 2.33$ at $x_E = 0.0159$ results in $B = 0.859$.

We can now estimate γ_E other values of x_E by

$$\ln \gamma_E = 0.859(1 - x_E)^2$$

At 20 atm we again write

$$y_E = 1.0 = K_E x_E \gamma_E$$

With $K_E = 1.62$, on taking logarithms, we have

$$0 = \ln 1.62 + \ln x_E + 0.859(1 - x_E)^2$$

On solving the preceding equation for x_E, we obtain

$$x_E = 0.50$$

Because we have used the bare minimum of data, we do not know how well our chosen equation represents activity coefficients in this system; therefore, we can only regard the preceding calculated solubility as an estimate.

EXAMPLE 10-9

A hydrocarbon has a vapor pressure of 2 atm at 20°C. The solubility of water in this liquid hydrocarbon at 20°C has been measured and is $x_w = 0.00021$. We wish to dry this hydrocarbon to a very low water content and wish to know if distillation would be a feasible process and, if so, to estimate the necessary equilibrium data.

Solution 10-9

This is almost a completely immiscible system, and we can therefore expect that the solubility of liquid hydrocarbon in liquid water is negligible. We let primes refer to the hydrocarbon phase and double primes refer to the water phase and for two liquid phases and a vapor phase in equilibrium write

$$P_H^\circ x'_{Hs} \gamma'_{Hs} = P_H^\circ x''_{Hs} \gamma''_{Hs} = p_H$$

$$P_W^\circ x'_{Ws} \gamma'_{Ws} = P_W^\circ x''_{Ws} \gamma''_{Ws} = p_W$$

where

$$x'_{Hs} = 1 - 0.00021 \doteq 1.0$$

$$\gamma'_{Hs} = 1.0$$

$$x''_{Ws} \doteq 1.0$$

$$\gamma''_{Ws} = 1.0$$

Thus, we see that

$$p_H = P_H^\circ$$

$$p_W = P_W^\circ$$

as we would expect for an immiscible system. This gives a vapor mol fraction of

$$y_W = \frac{p_W}{p_W + p_H} = \frac{P_W^\circ}{P_W^\circ + P_H^\circ}$$

At $20°C$, $P_W^\circ = 17.5\,\text{mmHg}$ and we find $y_W = 0.0114$. We note that the mol fraction of water in the vapor is many times larger than the mol fraction of water in the liquid hydrocarbon, and therefore distillation appears to be feasible for drying the hydrocarbon.

For the estimation of equilibrium data for the hydrocarbon liquid containing less than the saturation water content, we assume that Henry's law applies to the water in this phase:

$$p_W = k_W x_W'$$

where k_w is evaluated from the saturation condition:

$$p_W = P_W^\circ = k_W x_{W_s}'$$

$$17.5 = k_W(0.00021)$$

$$k_W = 8.33(10^4)$$

Now

$$p_W = 8.33(10^4) x_W'$$

Because this phase is practically pure hydrocarbon, it will be reasonable to state that

$$p_H = 2(760) = 1520 \text{ mmHg}$$

for all $x_W' < x_{W_s}' = 0.00021$. The vapor mol fraction of water is

$$y_W = \frac{8.33(10^4) x_W'}{8.33(10^4) x_W' + 1520}$$

Values of y_w evaluated at various values of x'_w are tabulated along with the relative volatility of water to hydrocarbon α_{WH}, which is a measure of the ease of separation by distillation. The relative volatility is defined as

$$\alpha_{WH} = \frac{y_W/x_W}{y_H/x_H}$$

The greater this ratio deviates from unity, the better the separation. The value of 55 for this system indicates that drying by distillation would be very easy to accomplish.

x'_w	(10^{-4})	(10^{-5})	(10^{-6})
y_W	$5.45(10^{-3})$	$5.5(10^{-4})$	$5.5(10^{-5})$
α_{WH}	55	55	55

In this example we were only interested in equilibrium between the liquid hydrocarbon phase and the vapor phase but found it convenient to visualize the partial pressures for the situation where a saturated aqueous phase was present. It should be obvious that this should present no problem as the equilibrium which exists between the saturated liquid hydrocarbon phase and the vapor phase would be unaffected by the presence or absence of the saturated aqueous phase.

EXAMPLE 10-10

Use the following vapor-liquid equilibrium data for the two binary systems dioxane–benzene and dioxane–water at 25°C to estimate the distribution of dioxane between liquid benzene and liquid water at 25°C.

BENZENE–DIOXANE AT 25°C[a]

Mol fraction of benzene		Pressure (mmHg)
In liquid	In vapor	
0	0	35.5
0.194	0.398	49.0
0.311	0.555	58.7
0.522	0.741	72.0
0.666	0.824	80.7
0.788	0.894	88.5
1.0	1.0	94.4

[a] Source: reference B2 in Table 10-8.

EXAMPLE 10-10 CON'T

WATER–DIOXANE AT 25°C[a]

Mol fraction of water		Pressure (mmHg)
In liquid	In vapor	
0.050	0.1545	42.7
0.100	0.241	45.2
0.240	0.310	47.5
0.450	0.350	47.7
0.720	0.405	45.5
0.770	0.440	43.9
0.890	0.560	37.5
0.960	0.770	30.0

[a] Source: reference B2 in Table 10-8.

Solution 10-10

Liquid water and liquid benzene can be considered immiscible, and the ternary system will be visualized as a liquid phase containing only dioxane and water in equilibrium with a liquid phase containing only dioxane and benzene. This may be a reasonable approximation at low dioxane concentrations and will allow us to write an equilibrium statement for dioxane:

$$\hat{f}'_D = \hat{f}''_D$$

At these low pressures we will replace fugacities with partial pressures. The vapor-liquid equilibrium data allow us to determine the partial pressure of dioxane over each solution as a function of liquid mol fraction of dioxane. The partial pressure of dioxane is plotted vs. the liquid mol fraction of dioxane in Fig. 10-22 for each of the binary systems. Mol fractions of dioxane in equilibrium in the ternary liquid-liquid system are obtained from Fig. 10-22 by reading off values of dioxane mol fraction in each binary system corresponding to a given partial pressure. An example of this procedure is shown in the figure where x'_D and x''_D are the compositions so determined. A comparison of the calculated data with experimentally determined liquid-liquid equilibrium data is shown in Fig. 10-23 where, as expected, the agreement is found to be good at low dioxane concentrations.

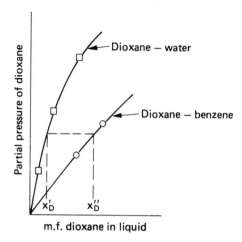

Figure 10-22 Partial pressures of dioxane for the dioxane–water and dioxane–benzene binary systems. [Data from J. C. Chu, S. L. Wang, S. L. Levy, and R. Paul, *Vapor-Liquid Equilibrium Data*, J. W. Edwards Publisher, inc., Ann Arbor, MI, 1956.]

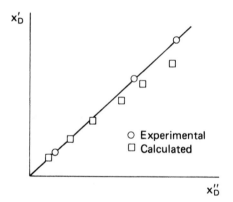

Figure 10-23 Comparison of calculated and experimental distribution data for the dioxane–water–benzene system. [Data from R. J. Berndt and C. C. Lynch, *J. Am. Chem. Soc., 66,* 282 (1944).]

10-11 RECAPITULATION

It is now time to reflect on the topics of this chapter and attempt to answer the previously raised question concerning the usefulness of the thermodynamic approach to phase equilibrium. Foremost, in doing this, it is hoped that you will be able to say that through the systematic thermodynamic approach you have obtained insight into the reasons for the observed diversity of phase behavior. For example, the reasons for the existence of azeotropes or two liquid phases can be understood in terms of vapor pressures and activity coefficients.

The treatment of every topic in this chapter has hinged on our ability to obtain activity coefficients from experimental data and to use them for calculating equilibria under different conditions. Thus, the thermodynamic approach is seen to be a systematic means of deriving the most use from experimentally determined information. This can range from what we have called the consummate treatment to estimates from fragmentary data. Sources of various types of phase equilibrium data are given in Table 10-8.

We have also seen how the thermodynamic approach can be used to facilitate the experimental data-taking process. In the case of total pressure measurements we saw that the actual experimentation was made easier, and in the case of multicomponent vapor-liquid equilibrium we saw that considerably less data would be required for a description of the system.

Finally, we have seen that thermodynamics provides a means of evaluating vapor-liquid equilibrium data through various consistency tests.

TABLE 10-8
SOURCES OF PHASE EQUILIBRIUM DATA

A. *General*

1. International Critical Tables.

2. J. Timmermans, *Physico-Chemical Constants of Binary Systems*, Wiley Interscience, New York, 1960.

3. J. Wisniak, *Phase Diagrams, A Literature Sourcebook*, Elsevier Scientific Publication Co., Amsterdam, 1981. A comprehensive bibliography of reported systems containing organic and inorganic substances.

B. *Vapor-Liquid Equilibria*

1. J. C. Chu, R. J. Getty, L. F. Brennecke, and R. Paul, *Distillation Equilibrium Data*, Reinhold Publishing Co., New York, 1950. Contains tabulated data.

2. J. C. Chu, S. L. Wang, S. L. Levy, and R. Paul, *Vapor-Liquid Equilibrium Data*, J. W. Edwards, Publisher, Ann Arbor, MI, 1956. Contains tabulated data.

3. J. Gmehling and U. Onken, *Vapor-Liquid Equilibrium Data Collection*, DECHEMA, Frankfort. A projected multivolume series (first issue 1978) containing data and van Laar, Margules, Wilson, NRTL, and UNIQUAC parameters. A comprehensive collection of systems.

4. E. Hála, I. Wichterle, J. Polak, and T. Boublik, *Vapor-Liquid Equilibrium Data at Normal Pressures*, Pergamon Press Ltd., Oxford, 1968. Contains data and van Laar and Margules parameters for many binary and several ternary and quaternary systems.

5. I. Wichterle, J. Linek, and E. Hála, *Vapor-Liquid Equilibrium Data Bibliography*, Elsevier Scientific Publication Co., Amsterdam, 1973. With periodic supplements; No. IV, 1985. A periodically updated bibliography of reported systems.

6. L. H. Horsley, *Azeotropic Data*, Advances in Chemistry Series, No. 6, American Chemical Society, Washington, DC, 1952. An azeotropic data compilation.

7. L. H. Horsley, *Azeotropic Data—II*, Advances in Chemistry Series, No. 35, American Chemical Society, Washington, DC, 1962. An updated azeotropic data compilation.

8. M. Hirata, S. Ohe, and K. Nagahama, *Computer-Aided Data Book of Vapor-Liquid Equilibria*, Kodansha/Elsevier, Tokyo/New York, 1975. Contains tabulated data and Wilson parameters for 800 binary systems.

9. S. Ohe, *Vapor-Liquid Equilibrium Data at High Pressure*, Kodansha/Elsevier, Tokyo/Amsterdam, 1990. Contains Peng-Robinson interaction parameters and graphical expressions of the data fit for 700 binary systems.

C. *Liquid-Liquid Equilibria*

1. A. W. Francis, *Liquid-Liquid Equilibriums*, Wiley Interscience, New York, 1963. A bibliography of reported systems.

2. A. W. Francis, *Critical Solution Temperatures*, Advances in Chemistry Series, No. 31, American Chemical Society, Washington, DC, 1961. Contains binary solubility data.

3. J. Wisniak and A. Tamir, *Liquid-Liquid Equilibrium and Extraction: A Literature Source Book*, Elsevier Scientific Publication Co., Amsterdam, 1981. A bibliography of binary and ternary systems.

D. *Liquid-Liquid and Solid-Liquid Equilibria*

1. H. Stephen and T. Stephen, *Solubilities of Inorganic and Organic Compounds*, Macmillan, New York, 1963. Contains various solubility data.

2. W. F. Linke, *Solubilities of Inorganic and Metal Organic Compounds*, Van Nostrand, Princeton, NJ, 1958. Contains various solubility data.

E. *Data Bases*

The annual Software Directory accompanying the January issue of *Chemical Engineering Progress* contains brief descriptions of commercially available data bases and data estimation programs. The directory is scheduled for inclusion on *AIChE*'s Web site (www.aiche.org).

PROBLEMS

10-1. Use data found in Ex. 10-7 to calculate the following for the air–water–acetone system:

(a) the vapor mol fraction of acetone in equilibrium with a liquid containing 0.0939 mol fraction acetone at 25°C and a pressure of 2 atm.

(b) the vapor mol fraction of water in equilibrium with the liquid described in part (a).

10-2. Use the data of Ex. 10-7 to estimate for the air–water–acetone system the vapor mol fractions of water and acetone in equilibrium with a liquid containing 50 mol % acetone at 25°C and 2 atm.

10-3. The following table of VLE data from an old company report has just been faxed to your office, but, unfortunately, it contains several illegible entries. Your boss has asked you if you can estimate the values of the missing entries. Fill in the blanks as best you can and justify your work.

SYSTEM A–B AT 80°C

x_A	y_A	P (mmHg)
0	0	700
—	0.427	1042
0.500	0.533	1118
0.750	—	—
1.0	1.0	800

10-4. At 35°C a liquid solution containing 40.5 mol % ethanol and 59.5 mol % methylcyclohexane exerts a pressure of 152.4 mmHg. The vapor phase composition is 54.7 mol % ethanol and 45.3 mol % methylcyclohexane. The vapor pressures of the pure compounds at 35°C are as follows:

Ethanol: 103.1 mmHg
Methylcyclohexane: 73.6 mmHg

Estimate the mol fraction of ethanol in the vapor phase in equilibrium with a liquid phase containing 60 mol % ethanol at 35°C.

10-5. At 150°F a liquid solution containing 80 mol % A and 20 mol % B is in equilibrium at 760 mmHg with a vapor containing 84.3 mol % A and 15.7 mol % B. Estimate the pressure and the vapor composition in equilibrium with a liquid containing 50% A and 50% B at 150°F. At this temperature the vapor pressures are

$$P_A^\circ = 800 \text{ mmHg}; \quad P_B^\circ = 600 \text{ mmHg}$$

10-6. Industrial ethanol is produced by the catalytic hydration of ethylene carried out at high pressure. Only about 5% conversion is obtained in the reactor, and the ethanol product is removed from the ethylene by absorption in water. To design the absorber which operates at 500 psia and a temperature close to 25°C, we need equilibrium data relating mol fraction of ethanol in the gas phase to mol fraction of ethanol in the liquid phase. In the literature we find experimental Pxy data for the ethanol–water system at temperatures of 45 and 55°C. Show how these data could be used to calculate the desired xy data assuming that ethylene solubility in the liquid phase can be ignored. Give a detailed account of the calculation procedure.

10-7. Vapor-liquid equilibrium data for a binary mixture of two organic compounds are needed over a wide range of temperature at relatively low pressures so that separational equipment can be designed. No data are available in the literature, and you are asked to specify what type of data should be taken in the laboratory and how you intend to use the laboratory data to make the necessary calculations. Bear in mind that laboratory data are expensive to determine as well as time consuming, and do not request more data than are necessary.

10-8. A form of the virial equation can be used to describe gas phase behavior up to moderate pressures. For gas mixtures,

$$Z = 1 + \frac{BP}{RT}$$

where

$$B = y_1^2 B_{11} + 2 y_1 y_2 B_{12} + y_2^2 B_{22}$$

or

$$B = y_1 B_{11} + y_2 B_{22} + y_1 y_2 \delta_{12}$$

with

$$\delta_{12} = 2 B_{12} - B_{11} - B_{22}$$

Show that up to moderate pressures the fundamental equation of vapor-liquid equilibrium [Eq. (9-63)] may be written as

$$\ln \gamma_1 = \ln \frac{y_1 P}{x_1 P_1^\circ} + \frac{B_{11}(P - P_1^\circ) + P\delta_{12} y_2^2}{RT}$$

$$\ln \gamma_2 = \ln \frac{y_2 P}{x_2 P_2^\circ} + \frac{B_{22}(P - P_2^\circ) + P\delta_{12} y_1^2}{RT}$$

10-9. Use the Wilson parameters determined in Ex. 9-9 and the prescribed temperature dependence of Eq. (9-58) to calculate for the methanol–carbon tetrachloride system:

(a) the Txy data for a pressure of 760 mmHg.
(b) the azeotrope composition when the pressure is 2 atm.

The liquid densities and Antoine constants are as follows.

	$\rho(g/cm^3)$	A	B	C
Methanol	0.792	8.08097	1582.271	239.726
Carbon tetrachloride	1.595	6.87926	1212.021	226.409

10-10. Use the data of Ex. 9-8 to determine Wilson parameters for the ethanol–benzene system at 40°C via the program WEQ-ISOT.EXE. Further, use these parameters and the prescribed temperature dependence of Eq. (9-58) to calculate for this system:

(a) the Txy data at a pressure of 900 mmHg.
(b) the azeotrope composition at a pressure of 2000 mmHg.

The liquid densities and Antoine constants are as follows:

	$\rho(g/cm^3)$	A	B	C
Ethanol	0.789	8.11220	1592.864	226.184
Benzene	0.879	6.89272	1203.531	219.888

10-11. At 60°C compounds A and B each have vapor pressures of 800 mmHg. At 60°C the system A–B forms an azeotrope containing 50 mol % A and exerts a pressure of 1000 mmHg.

(a) Calculate the equilibrium pressure and vapor composition over a liquid solution containing 25 mol % A.
(b) If A and B have equal latent heats of vaporization, how do you expect the azeotrope composition to respond to an increase in temperature?

10-12. At 80°C compounds A and B each have vapor pressures of 700 mmHg. At 80°C this system forms an azeotrope containing 50 mol % A and exerts a pressure of 960 mmHg. Calculate the equilibrium pressure and vapor composition at 80°C over a liquid solution containing 25 mol % A.

10-13. Using only the data available in Ex. 10-7, make the best possible determination as to the existence of an azeotrope in the acetone–water system at 25°C.

10-14. Use any information available in Ex. 10-1 and make the best estimate of the azeotrope composition and pressure in the chloroform–n-hexane system at 55°C.

10-15. Use any information in Ex. 10-5 to calculate the temperature and pressure when the mol fraction of methanol in the azeotrope is 0.88.

10-16. An azeotrope occurs in the a–b system but the only information that is known is the temperature (366 K) and pressure (1 atm.). The vapor pressures of a and b at 366 K are 700 and 600 mmHg respectively. Calculate the azeotrope composition at 366 K. Use the Margules equation with $A_{12} = A_{21}$.

10-17. In the system A–B, activity coefficients can be expressed by

$$\ln \gamma_A = 0.5x_B^2$$

$$\ln \gamma_B = 0.5 x_A^2$$

The vapor pressures of A and B at 80°C are

$$P_A^\circ = 900 \text{ mmHg}; \quad P_B^\circ = 600 \text{ mmHg}$$

Is there an azeotrope in this system at 80°C, and if so, what is the azeotrope pressure and composition?

10-18. At 1 atm the system ethylacetate (A)–ethanol (B) exhibits an azeotrope ($x_A = 0.539$, t = 71.8°C). Determine whether the mol fraction of ethylacetate in the azeotrope increases or decreases as the pressure is increased. The vapor pressures are as follows:

$t(^\circ C)$	P_A° (atm)	P_B° (atm)
70	596	543
80	833	813
90	1130	1188

10-19. At 80°C compounds A and B each have a vapor pressure of 800 mmHg. At this temperature the A–B system forms an azeotrope containing 50 mol % A at a pressure of 1050 mmHg. Compound A has a heat of vaporization of 8000 cal/gmol and B has a heat of vaporization of 10,000 cal/gmol. Estimate the azeotrope composition and pressure at 60°C.

10-20. The following vapor-liquid equilibrium data indicate that the benzene (1)–hexa-fluorobenzene (2) system exhibits two azeotropes [*Nature, 212* (Oct. 15, 1966), p. 283]. We are naturally concerned as to whether this is thermodynamically permissible. Based on the information at hand what is your verdict—are two azeotropes thermodynamically permissible?

x_1	y_1	$P(kPa)$		x_1	y_1	$P(kPa)$
0	0	73.408		0.5267	0.5035	72.451
0.0938	0.0991	74.168		0.6014	0.5826	71.966
0.1847	0.1831	74.318		0.7852	0.7834	71.294
0.2740	0.2624	74.110		0.8959	0.8995	71.374
0.3468	0.3447	73.627		1	1	71.851
0.4539	0.4299	73.002				

10-21. We need vapor-liquid equilibrium data for a particular binary system (known here as the A–B system). No data can be found in the literature; however, in checking the azeotropic data compilation, we find that the system has been studied at 1 atm but that no azeotrope exists. Some members of our design group claim that the nonexistence of an azeotrope implies an ideal system, but others disagree. What do you think? If your answer is no, would it be possible to estimate the error which might be involved in approximating the system at 1 atm as an ideal system? The following vapor pressure data are available:

$t(°C)$	$P_A^°$ (atm)	$P_B^°$ (atm)
80	1.00	0.704
100	1.42	1.00

10-22. In a binary system where activity coefficients can be represented by the simple equations

$$\ln \gamma_1 = Bx_2^2$$

$$\ln \gamma_2 = Bx_1^2$$

show that an azeotrope occurs when

$$|B| \geq \left|\ln \frac{P_1^°}{P_2^°}\right|$$

Also show that for

$$\left|\ln \frac{P_1^°}{P_2^°}\right| < |B|; \quad \text{and} \quad B < 2$$

the azeotrope composition is given by

$$x_1 = \frac{1}{2}\left(1 + \frac{1}{B} \ln \frac{P_1^°}{P_2^°}\right)$$

10-23. The following vapor-liquid equilibrium data at a constant pressure of 760 mmHg were uncovered in an old company research report:

SYSTEM A–B		
x	y	y'
0.2	0.50	0.38
0.4	0.72	0.61
0.6	0.86	0.78
0.8	0.94	0.90

No temperature data are given. x and y are the mol fractions of the more volatile component in the liquid and vapor phase, respectively, and y' is the vapor mol fraction the more volatile component would have if the system behaved ideally. Can these data be analyzed for thermodynamic consistency? If so, are they thermodynamically consistent?

10-24. Show that for a binary system thermodynamic consistency requires on a plot of partial pressure vs. liquid mol fraction that the ratio of the slope of the partial pressure curve to the slope of the line joining the partial pressure to the origin ($x_i = 0$, $p_i = 0$) must be the same for each component. Further, show that if the tangent to one partial pressure curve passes through the origin the same must hold for the other component.

10-25. The following vapor-liquid equilibrium data are available for the A–B system at a constant pressure of 760 mmHg. Also tabulated are the vapor pressures of A and B.

x_A	y_A	$t(°C)$	P_A^o mmHg	P_B^o mmHg
0.116	0.239	105	1569	654
0.244	0.439	100	1369	563
0.390	0.612	95	1192	484
0.559	0.761	90	1034	413
0.755	0.887	85	893	351

(a) Test the data for thermodynamic consistency.
(b) Calculate the vapor composition and total pressure when the liquid mol fraction is 0.50 and the temperature is 100°C.

10-26. Test the following data for thermodynamic consistency:

(a) ethanol–water at 25°C; see Prob. 14-4.
(b) benzene–dioxane at 25°C; see Ex. 10-10.
(c) water–dioxane at 25°C; see Ex. 10-10.

10-27. Test the following isothermal formaldehyde–water VLE data for thermodynamic consistency. The vapor pressures of formaldehyde and water are 1596 and 28.57 kPa respectively.

SYSTEM: FORMALDEHYDE (1)–WATER (2) AT 68°C[a]

x_1	y_1	P (kPa)
0.076	0.155	405.3
0.106	0.188	414.4
0.137	0.226	423.1
0.139	0.221	423.7
0.171	0.264	432.9
0.227	0.307	443.1
0.245	0.334	442.7
0.304	0.375	449.1

[a] Source: M. Albert, I Hahnenstein, H. Hasse, and G. Maurer, *A.IChEJ.*, *42*, 1741 (1996).

10-28. Solution theory predicts that certain thermodynamic properties should depend on volume fraction of the liquid phase rather than mol fraction. It has therefore been proposed that an expression analogous to Raoult's law might have a broader range of application than Raoult's law. For a binary solution this law would be written as

$$p_1 = P_1^\circ \left(\frac{x_1 v_1}{x_1 v_1 + x_2 v_2} \right)$$

$$p_2 = P_2^\circ \left(\frac{x_2 v_2}{x_1 v_1 + x_2 v_2} \right)$$

Comment on the validity of this proposed model.

10-29. Equations (9-58d) and (9-58e) that prescribe the temperature dependence of the Wilson parameters are not valid for negative values of the parameters. Are negative values permitted by the Gibbs-Duhem equation?

10-30. It has been stated [W. H. Prahl, *Chem. Eng.*, *118* (Oct. 22, 1979)] that binary vapor-liquid equilibrium data can be well fitted by relating the relative volatility α to the liquid mol fraction by the following expression:

$$\ln \alpha = a + bx_1 + cx_1^z$$

where x_1 is the liquid mol fraction of the more volatile component ($P_1^\circ > P_2^\circ$) in a binary mixture and a, b, c, and z are empirically determined parameters. The relative volatility is defined as

$$\alpha = \frac{y_1/x_1}{y_2/x_2}$$

but can also be expressed as

$$\alpha = \frac{P_1^\circ \gamma_1}{P_2^\circ \gamma_2}$$

Here y is vapor mol fraction, P° is vapor pressure, and γ is the activity coefficient. In fitting this equation (α vs. x) to isothermal vapor-liquid equilibrium data, does the Gibbs-Duhem equation allow all parameters (a, b, c, and z) to be independently determined? Explain.

10-31. Figure 10-23 shows that a plot of x_D' vs. x_D'' is linear. Show that this linearity is expected for the distribution of a solute (in this case, dioxane) between two essentially immiscible solvents (in this case, water and benzene) in the region of dilute solutions.

10-32. A stream contains 30 mol % toluene, 40 mol % ethylbenzene, and 30 mol % water. Assuming that mixtures of ethylbenzene and toluene obey Raoult's law and that the hydrocarbons are completely immiscible in water, calculate the temperature and compositions at the

    (a) bubble point.
    (b) dew point.

The total pressure is 1 atm.

10-33. Mutual liquid solubilities in the $A–B$ system are expected to be quite low (no more than a few mol %), but rather than measure them directly we have determined vapor dew points and used this information to find the vapor composition($y_A = 0.602$) and pressure (976 mmHg) when two liquid phases and a vapor phase are in equilibrium at a temperature of 350 K.

    (a) Use this information and the following vapor pressures to determine the composition of the two liquid phases at 350 K.
    (b) Calculate the vapor composition and the total pressure in equilibrium with a liquid containing 1 mol % A.

$$P_A^\circ = 600 \text{ mmHg}; \quad P_B^\circ = 400 \text{ mmHg}$$

10-34. At a temperature of 320 K the system $A–B$ exhibits partial liquid miscibility. At this temperature it is known that the saturated A-rich phase contains 73 mol % A and is in equilibrium with a vapor containing 53.3 mol % A at an equilibrium pressure of 1282 mmHg. The vapor pressures of A and B are 800 and 700 mmHg respectively. Calculate:

    (a) the mol % of A in the saturated B-rich phase in equilibrium with the vapor and the saturated A-rich phase.
    (b) the pressure and vapor phase mol % of A in equilibrium with a liquid containing 85 mol % A.

Use the Margules equation with $A_{12} = A_{21}$.

10-35. At 70°C the system $a–b$ exhibits partial liquid miscibility with $x_a' = 0.3$ and $x_a'' = 0.7$. The vapor pressures of a and b are 600 and 500 mmHg respectively. Calculate:

    (1) the vapor m.f. of a and the total pressure when two liquid phases and a vapor phase are in equilibrium.
    (2) the vapor m.f. of a and the total pressure when a vapor phase is in equilibrium with a liquid phase containing an a m.f. of 0.1.

Sketch the phase diagram for this system at a temperature of 70°C labeling the L', L'', V, $L' + V$, and $L'' + V$ areas.

Use the Margules equation with $A_{12} = A_{21}$.

10-36. The following set of activity coefficients was obtained from the UNIFAC estimation method for the system n-heptane–acetonitrile at 298 K. One of the members of your group has remarked that they are quite large—so large, in fact, that there may be partial liquid miscibility. As the junior member of the group, you are given the task of testing for partial miscibility. What is your verdict?

ACTIVITY COEFFICIENTS VIA UNIFAC

SYSTEM: n-HEPTANE (1)–ACETONITRILE (2) AT 298 K

x_1	γ_1	γ_2
0	32.93	1
0.1	13.75	1.045
0.2	7.159	1.172
0.3	4.324	1.385
0.4	2.901	1.716
0.5	2.108	2.228
0.6	1.632	3.046
0.7	1.333	4.433
0.8	1.146	6.990
0.9	1.038	12.313
1	1	25.523

10-37. For the benzene–water–dioxane system at 25°C it is possible for two saturated liquid phases and a vapor phase to coexist at equilibrium. Using information available in Ex. 10-10, estimate the equilibrium pressure and vapor-phase composition in the three-phase system when the water-rich phase contains 4 mol % dioxane.

10-38. At 100°C nitrobenzene and water are only partially miscible with saturated liquid compositions at 0.147 mol % nitrobenzene and 91.7 mol % nitrobenzene. At this temperature the vapor pressure of nitrobenzene is 21 mmHg.

> (a) Estimate the total pressure and the vapor composition when two liquid phases and a vapor phase are in equilibrium.
> (b) To design a stripping column to remove nitrobenzene from dilute aqueous solutions, it is necessary to have vapor-liquid equilibrium data. Estimate the vapor composition in equilibrium with an aqueous solution containing 0.0100 mol % nitrobenzene at a temperature of 100°C. Also estimate the total pressure.

10-39. At 25°C a binary system containing components A and B is in a state of liquid-liquid-vapor equilibrium. The compositions of the saturated liquid phases are

$$x'_{As} = 0.02 \qquad x'_{Bs} = 0.98$$

$$x''_{As} = 0.98 \qquad x''_{Bs} = 0.02$$

The vapor pressures at 25°C are

$$P_A^\circ = 0.1 \text{ atm}; \qquad P_B^\circ = 1.0 \text{ atm}$$

Making reasonable assumptions (state them explicitly and give your justification), determine good estimates for the following:

(a) The vapor composition and total pressure for this three-phase system at equilibrium.

(b) The vapor composition and total pressure in equilibrium with $x_A' = 0.01$.

(c) At 25°C a vapor containing 20% A and 80% B initially at 0.1 atm is compressed iso-themally. Find the dew point pressure and liquid composition.

10-40. For a partially miscible system with slight mutual solubility, show that the assumption of Henry's law leads to a Gibbs free energy of mixing curve with two minima similar to that shown in Fig. 10-18. For a symmetric system ($x_{1s}' = x_{2s}''$) show that common tangency occurs at x_{1s}' and x_{2s}''.

10-41. At 20°C water (1) and methylethyl ketone (2) are only partially miscible in the liquid phase with

$$x_{1s}' = 0.3637; \qquad x_{1s}'' = 0.9150$$

$$x_{2s}' = 0.6363; \qquad x_{2s}'' = 0.0850$$

Use this information to estimate the possibility of an azeotrope in this system at atmospheric pressure. Use the van Laar equation with the following temperature dependence:

$$B_{12} = \frac{b_{12}}{T}; \qquad B_{21} = \frac{b_{21}}{T}$$

where b_{12} and b_{21} are constants. The vapor pressure of methylethyl ketone is given by

$$\log P_2^\circ = 7.06356 - \frac{1261.339}{t + 221.969}$$

10-42. Use the following VLE data for the acetone–water system at 760 mmHg to obtain Wilson parameters via the program WEQ-ISOP.EXE. Further, use these parameters to estimate acetone activity coefficients at 25°C and at the liquid mol fractions found in Ex. 10-7. Compare your estimated results with the experimental values and comment on the agreement.

VAPOR - LIQUID EQUILIBRIUM DATA

SYSTEM: ACETONE (1)–WATER (2) AT 760 mmHg[a]

x_1	y_1	$t(°C)$	P(mmHg)
0.010	0.335	87.8	760
0.023	0.462	83.0	
0.041	0.585	76.5	
0.120	0.756	66.2	
0.264	0.802	61.8	
0.300	0.809	61.1	
0.444	0.832	60.0	
0.506	0.837	59.7	
0.538	0.840	59.5	
0.609	0.847	58.9	
0.661	0.860	58.5	
0.793	0.900	57.4	
0.850	0.918	57.1	

[a] Source: reference B2 in Table 10-8.

10-43. Use the following VLE data for the methanol–water system at 760 mmHg to obtain Wilson parameters via the program WEQ-ISOP.EXE. Use this information to estimate equilibrium data needed to design a column for absorbing methanol from a methanol–air mixture with water as the solvent. Specifically, at 25°C and at pressures of 1 and 2 atm. prepare plots of the mol fraction of methanol in the gas phase vs. mol fraction of methanol in the liquid phase. Also prepare similar plots for water.

VAPOR - LIQUID EQUILIBRIUM DATA [a]

SYSTEM: METHANOL (1)–WATER (2) AT 760 mmHg

x_1	y_1	$t(°C)$	P(mmHg)
0.0531	0.2834	92.9	760
0.0767	0.4001	90.3	
0.0926	0.4353	88.9	
0.1315	0.5455	85.0	
0.2083	0.6273	81.6	
0.2818	0.6775	78.0	
0.3333	0.6918	76.7	
0.4620	0.7756	73.8	

	VAPOR - LIQUID EQUILIBRIUM DATA CON'T		
0.5292	0.7971	72.7	
0.5937	0.8183	71.3	
0.6849	0.8492	70.0	
0.8562	0.8962	68.0	
0.8741	0.9194	66.9	

[a] Source: reference B2 Table 10-8.

10-44. The system acetone–methanol forms an azeotrope as can be seen from the tabulated VLE data. Will the addition of water to this system break the azeotrope? VLE data at 760 mmHg for the three binary systems formed from acetone, methanol, and water are given below and in Problems 10-42 and 10-43. Because a ternary azeotrope would occur at the lowest temperature, we can expect it to lie close to the binary azeotrope which has the lowest temperature of all the binaries. You can, therefore, limit the region of your search to liquid compositions in the range of, but less than, 80 mol % acetone. To simplify the calculations, determine whether there is a ternary azeotrope at 60°C.

VAPOR - LIQUID EQUILIBRIUM DATA

SYSTEM: ACETONE (1)–METHANOL (2) AT 760 mmHg[a]

x_1	y_1	$t(°C)$	$P(mmHg)$
0.050	0.108	63.0	760
0.100	0.196	61.6	
0.200	0.335	59.5	
0.300	0.432	58.1	
0.400	0.514	56.9	
0.500	0.588	56.2	
0.600	0.655	55.8	
0.700	0.726	55.5	
0.800	0.800	55.4	
0.900	0.885	55.6	
0.950	0.941	55.8	

[a] Source: reference B2 in Table 10-8.

10-45. Use the program WEQ-ISOP.EXE and the experimental xyt data for the chloroform–hexane system at 760 mmHg (found in Ex. 10-1) to determine the Wilson parameters a_{12} and a_{21}. From these parameters calculate values of G_{12} and G_{21} at temperatures of 35, 45, and 55°C and compare with the values presented in Ex. 10-1.

10-46. Use any information found in Ex. 10-2 and calculate the xyt data for the hexane–chlorobenzene system at a constant pressure of 500 mmHg. Prepare a xyt plot for the system similar to those shown in Fig. 10-2.

10-47. Use the program PRVLE.EXE and k_{12}'s determined from Fig. 10-5 to determine Henry's law constants for methane in decane over the range 255 to 400 K. Repeat the calculations using an average value of 0.05 for k_{12}. Compare results on a plot of logarithm of Henry's law constant vs. reciprocal of absolute temperature.

10-48. Use the program PRKFIT.EXE and the following Pxy data to determine k_{12} for the $CH_4(1)$–$CO_2(2)$ system at 250 K then use the program PRVLE.EXE with this k_{12} to calculate the VLE data and prepare a Pxy plot in the style of Fig. 9-24 to show the fit.

x_1	y_1	P (atm)
0	0	12.85
0.024	0.134	15.10
0.089	0.365	22.00
0.097	0.383	22.50
0.196	0.540	32.50
0.320	0.643	45.00
0.426	0.673	55.20
0.546	0.673	65.7

Source: J. Davalos, W. R. Anderson, R. E. Phelps, and A. J. Kidnay, J. *Chem. Eng. Data, 21*, 81 (1976).

10-49. Use the value of k_{12} determined in Prob. 10-48 along with the value of 0.0184 for the CH_4–C_2H_6 system[26] and 0.1355 for the CO_2–C_2H_6 system[27] to calculate equilibrium ratios, K's, in the ternary system at 250 K. Compare your results with the following reported values.

SYSTEM: $CH_4(1)$–$C_2H_6(2)$–$CO_2(3)$

x_1	x_2	P (atm)	K_1	K_2	K_3
0.0352	0.7898	21.00	7.102	0.7850	0.7429
0.0950	0.8097	25.00	3.436	0.6502	1.545
0.1351	0.7350	30.00	2.976	0.6165	1.1138
0.0520	0.2495	30.00	5.058	0.9098	0.7198

Source: J. Davalos, W. R. Anderson, R. E. Phelps, and A. J. Kidnay, *J. Chem. Eng. Data, 21*, 81 (1976).

10-50. The CO_2–C_2H_6 system exhibits an azeotrope at 250 K where the pressure is 21.0 atm and the composition is 67 mol % CO_2. Use this data point and the program PRKFIT.EXE to determine k_{12} then use this value with the program PRVLE.EXE to calculate the VLE data. Compare your calculated data graphically with the experimental data found in the spreadsheet FG9-24A.WQ1.

---

[26] Source: reference B-9 in Table 10-8.
[27] See Fig. 9-24a.

Additional Topics in Phase Equilibrium

11-1 PARTIAL MOLAR PROPERTIES

When a substance becomes part of a mixture, it loses its identity; however, it still contributes to the mixture properties because we know that these depend on the relative amounts of its various components. In applying thermodynamics to mixtures, it is necessary to have some means of identifying the contribution of each component to the thermodynamic properties of the mixture. A convenient way of dealing with this problem is to define a partial molar property which is identified with a particular component in a mixture and measures the contribution of that component to the mixture property. Because it is easiest to measure and to visualize, we will begin with the volume of the mixture V, an extensive property which at a specified temperature and pressure will depend on the number of mols of the constituent components. The partial molar volume of component i, $\overline{V}_i$ in a mixture is defined as

$$\overline{V}_i = \left(\frac{\partial V}{\partial n_i} \right)_{T,P,n_j} \tag{11-1}$$

We have seen this derivative before [Eqs. (8-3) and (9-23)] and can visualize it in physical terms as the incremental change in mixture volume which occurs when a small quantity of component i is added at constant temperature and pressure. The amount of component i added is small enough that no detectable change in composition occurs. While only positive values are possible for V, it is possible and not uncommon to find negative values for the partial molar volume, $\overline{V}_i$. This would result when the addition of component i resulted in a decrease in mixture volume.[1]

In general, the partial molar property assigned to component i, $\overline{M}_i$, can be determined from any corresponding extensive mixture property M in the same manner:

$$\overline{M}_i = \left(\frac{\partial M}{\partial n_i} \right)_{T,P,n_j} \tag{11-2}$$

Just as we know that M depends on T, P, and composition, we can expect that the intensive property $\overline{M}_i$ will depend on these same variables.

To determine how these partial molar properties are related to the mixture property, we express the mixture property as a function of temperature, pressure, and the mol numbers of the components:

$$M = M(T,P,n_1,n_2,\ldots,n_C)$$

and write

$$dM = \left(\frac{\partial M}{\partial T} \right)_{P,n_i} dT + \left(\frac{\partial M}{\partial P} \right)_{T,n_i} dP + \sum \left(\frac{\partial M}{\partial n_i} \right)_{T,P,n_j} dn_i \tag{11-3}$$

Integration of this equation at constant T and P and for the condition where the size of the phase is increased but its composition remains constant gives

$$\Delta M = \sum \overline{M}_i \, \Delta n_i$$

With the same procedure used to obtain Eq. (9-43), we obtain

$$M = \sum \overline{M}_i n_i \tag{11-4}$$

and observe that Eq. (9-43) is a specific statement of this more general equation. We further note that the chemical potential $\hat{\mu}_i$ is identical with the partial molar Gibbs free energy

---

[1] For an example of this behavior, see J. M. Prausnitz, *Chem. Eng. Sci.*, *18*, 613 (1963).

$\overline{G}_i$. The rather restricted conditions under which Eq. (11-4) was obtained from Eq. (11-3) in no way restricts its usefulness. This equation relates the partial molar properties of the constituent components to the mixture property and applies at any given temperature, pressure, and composition.

The partial molar properties are related in the same way that the extensive properties are related. This can be seen by writing

$$G = U + PV - TS$$

and differentiating with respect to n_i at constant T, P, and n_j:

$$\left(\frac{\partial G}{\partial n_i}\right)_{T,P,n_j} = \left(\frac{\partial U}{\partial n_i}\right)_{T,P,n_j} + P\left(\frac{\partial V}{\partial n_i}\right)_{T,P,n_j} - T\left(\frac{\partial S}{\partial n_i}\right)_{T,P,n_j}$$

Applying Eq. (11-2), we see that this can be written as

$$\overline{G}_i = \overline{U}_i + P\overline{V}_i - T\overline{S}_i$$

In a similar manner it can be shown that

$$\overline{H}_i = \overline{U}_i + P\overline{V}_i$$

and

$$\overline{A}_i = \overline{U}_i - T\overline{S}_i$$

Returning to Eqs. (8-3) and (8-4), we see that the temperature and pressure dependence of the chemical potential are expressible in terms of partial molar properties:

$$\left(\frac{\partial \hat{\mu}_i}{\partial P}\right)_{T,n_i} = \overline{V}_i \tag{11-5}$$

$$\left(\frac{\partial \hat{\mu}_i}{\partial T}\right)_{P,n_i} = -\overline{S}_i \tag{11-6}$$

It is more convenient to express the temperature dependence in terms of $\overline{H}_i$. This can be done by writing

$$\hat{\mu}_i = \overline{G}_i = \overline{H}_i - T\overline{S}_i$$

Substituting Eq. (11-6) for $\overline{S}_i$ gives

$$\hat{\mu}_i = \overline{H}_i + T\left(\frac{\partial \hat{\mu}_i}{\partial T}\right)_{P,n_i}$$

which can be manipulated to give

$$\left(\frac{\partial \hat{\mu}_i / T}{\partial T} \right)_{P, n_i} = -\frac{\overline{H}_i}{T^2} \tag{11-7}$$

For pure substances Eqs. (11-5)–(11-7) are valid when written in terms of pure-component properties.[2]

An equation relating the changes in partial molar properties analogous to the Gibbs-Duhem equation will now be derived. We begin by differentiating Eq. (11-4):

$$dM = \sum \overline{M}_i \, dn_i + \sum n_i \, d\overline{M}_i$$

Equating this with Eq. (11-3) reveals that

$$\sum n_i \, d\overline{M}_i = \left(\frac{\partial M}{\partial T} \right)_{P, n_i} dT + \left(\frac{\partial M}{\partial P} \right)_{T, \, n_i} dP \tag{11-8}$$

This equation is quite general, and the Gibbs-Duhem equation can be seen as a specific statement of it. At constant temperature and pressure it becomes

$$\left[\sum n_i \, d\overline{M}_i = 0 \right]_{T, P} \tag{11-9}$$

11-2 EXPERIMENTAL DETERMINATION OF MIXTURE AND PARTIAL MOLAR PROPERTIES

We have seen that all thermodynamic properties must be either measured directly or calculated from experimentally determined data. Here we will be concerned with the type of data needed and the procedures used to calculate property changes on mixing and partial molar properties. Of the various mixture properties only the volume can be determined absolutely; therefore, for a general treatment it becomes necessary to work in terms of the change in a property which results when a solution is formed from pure components. This property change can be written as

$$\Delta M = M - \sum n_i m_i$$

where m_i is the pure-component intensive property. When Eq. (11-4) is used to express the mixture property, we have

---

[2] See Sec. 8-2.

$$\Delta M = \sum n_i (\overline{M}_i - m_i) = \sum n_i \Delta \overline{M}_i \tag{11-10}$$

Thus, $\Delta \overline{M}_i$ is the partial molar property change and is the difference between the partial molar property of component i in solution and the property of pure component i. For $\Delta \overline{M}_i$ and ΔM to have meaning, it is necessary to specify the state of the pure unmixed components. Usually, all the pure components can exist in the same phase, temperature, and pressure as the solution. However, there will be instances where some components cannot exist in a pure state in the same phase as the solution,[3] and therefore care must be taken to specify the state corresponding to the property m_i. Here, unless specified otherwise, m_i will represent the pure component property at the same conditions as the solution.

In Sec. 10-7 we have already considered the Gibbs free energy change on forming a mixture from pure components. We saw that

$$\Delta G = \sum n_i (\hat{\mu}_i - \mu_i) \tag{10-8}$$

which can be written as

$$\Delta G = \sum n_i (\overline{G}_i - g_i) = \sum n_i \Delta \overline{G}_i \tag{11-11}$$

Further, we saw that the Gibbs free energy change on forming a liquid solution from pure components is[4]

$$\Delta G = RT \sum n_i \ln x_i \gamma_i \tag{11-12}$$

Per mol of solution this becomes

$$\Delta g = RT \sum x_i \ln x_i \gamma_i \tag{11-12a}$$

Comparison of Eqs. (11-11) and (11-12) shows that

$$\Delta \overline{G}_i = RT \ln x_i \gamma_i \tag{11-13}$$

Equations (11-12) and (11-13) allow the calculation of Δg and $\Delta \overline{G}_i$ from experimentally determined activity coefficients. Evaluation of other partial molar property changes is usually accomplished from experimentally determined property changes: ΔV or ΔH.

When the property change on mixing ΔM has been determined for several solution compositions,[5] a graphical procedure may be used to calculate $\Delta \overline{M}_1$ and $\Delta \overline{M}_2$ for a binary system. The values of Δm, ΔM based on a mol of solution, are plotted vs. mol fraction in Fig.

---

[3] For example, a component which is a gas above its critical temperature can exist in a liquid solution but cannot exist as a pure liquid.
[4] The product $x_i \gamma_i$ is equal to the activity $\hat{a}_i$. This function is useful in the treatment of heterogeneous chemical equilibrium and is discussed in Secs. 12-2 and 12-9.
[5] In Sec. 6-5 we saw how Δh or Δh_s could be experimentally determined.

11-1 as the curve passing through the data points. At any composition x_1 represented by the point G a tangent is drawn to the curve and extended to its intersection with $x_1 = 0$ and $x_1 = 1$. It can be shown that the intercept at $x_1 = 1$, the line segment DC, is $\Delta \overline{M}_1$ and that the intercept at $x_1 = 0$, the line segment FE, is $\Delta \overline{M}_2$.

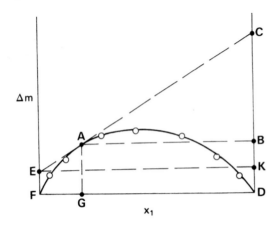

Figure 11-1 Illustration of the slope-intercept method.

We start by writing

$$\Delta M = n_1 \Delta \overline{M}_1 + n_2 \Delta \overline{M}_2$$

and divide by $n_1 + n_2$ to obtain

$$\frac{\Delta M}{n_1 + n_2} = \Delta m = x_1 \Delta \overline{M}_1 + x_2 \Delta \overline{M}_2 \tag{11-14}$$

We now differentiate Δm with respect to x_1:

$$\frac{d \Delta m}{dx_1} = \Delta \overline{M}_1 - \Delta \overline{M}_2 + x_1 \frac{d \Delta \overline{M}_1}{dx_1} + x_2 \frac{d \Delta \overline{M}_2}{dx_1}$$

In accordance with Eq. (11-10) the last two terms on the right-hand side can be written as

$$x_1 \frac{d\overline{M}_1}{dx_1} - x_1 \frac{dm_1}{dx_1} + x_2 \frac{d\overline{M}_2}{dx_1} - x_2 \frac{dm_2}{dx_1}$$

As m_1 and m_2 are pure component properties, their derivatives are zero. Comparison with Eq. (11-9) shows the sum of the remaining two terms to be zero and allows the slope of the Δm vs. x_1 curve to be written as

$$\frac{d \Delta m}{dx_1} = \Delta \overline{M}_1 - \Delta \overline{M}_2 \tag{11-15}$$

Eliminating $\Delta \overline{M}_2$ from Eqs. (11-14) and (11-15) gives

$$\Delta \overline{M}_1 = \Delta m + (1 - x_1)\frac{d\,\Delta m}{dx_1} \qquad (11\text{-}16\text{a})$$

This equation can be represented graphically in Fig. 11-1, where Δm is represented by AG or BD, $1 - x_1$ is represented by GD or AB, and the slope $d\,\Delta m\,/\,dx_1$ is CB/AB or KC. Graphically, Eq. (11-16a) can be written as

$$\Delta \overline{M}_1 = BD + AB\left(\frac{CB}{AB}\right) = BD + CB = CD$$

From Eq. (11-15) we can write

$$\Delta \overline{M}_2 = \Delta \overline{M}_1 - \frac{d\,\Delta m}{dx_1} \qquad (11\text{-}16\text{b})$$

or

$$\Delta \overline{M}_2 = CD - \frac{KC}{1} = KD = EF$$

Thus, for a binary system a simple graphical technique can be used to determine partial molar properties from property changes on mixing. Other methods of calculating partial molar properties are available in the literature.[6]

We have already seen that one of the advantages of thermodynamics is that it allows properties of one type to be calculated from other properties. A specific application is the calculation of heat of mixing from free energies of mixing, which are in turn evaluated from vapor-liquid equilibrium data. To show this, we begin with Eq. (9-38),

$$\hat{\mu}_i^L - \mu_i^L = RT \ln x_i + RT \ln \gamma_i \qquad (9\text{-}38)$$

divide by T, and differentiate with respect to T at constant pressure and composition:

$$\left(\frac{\partial \hat{\mu}_i^L/T}{\partial T}\right)_{P,x} - \left(\frac{\partial \mu_i^L/T}{\partial T}\right)_P = R\left(\frac{\partial \ln \gamma_i}{\partial T}\right)_{P,x}$$

on applying Eq. (11-7) we obtain

$$\left(\frac{\partial \ln \gamma_i}{\partial T}\right)_{P,x} = \left(\frac{h_i^L - \overline{H}_i}{RT^2}\right) = -\frac{\Delta \overline{H}_i}{RT^2} \qquad (11\text{-}17)$$

[6] G. N. Lewis and M. Randall, *Thermodynamics*, 2nd ed., revised by K. S. Pitzer and L. Brewer, McGraw-Hill, New York, 1961, Chap. 17; and B. F. Dodge, *Chemical Engineering Thermodynamics*, McGraw-Hill, New York, 1944, Chap. IV.

Thus, it is seen that if the activity coefficients in a system are known at several temperatures, partial molar enthalpies may be obtained from their temperature derivative, and the heat of mixing may be subsequently determined from Eq. (11-10). This procedure is used frequently to calculate heats of mixing via vapor-liquid equilibrium measurements. Equation (11-17) also proves useful for interpolating and extrapolating activity coefficients with respect to temperature. If it is assumed that $\Delta \overline{H}_i$ is constant over the temperature range and the usually minor effect of pressure can be ignored, a plot of $\ln \gamma_i$ vs. $1/T$ should be linear when γ_i is evaluated at a constant liquid-phase composition.

EXAMPLE 11-1

Demonstrate for a binary system that it is possible to calculate vapor-liquid equilibrium data from isothermal heat of mixing data and pure-component vapor pressures.

Solution 11-1[7]

This technique is based on the use of an activity coefficient equation in which the parameters have a prescribed temperature dependence. We use the Wilson equation and begin by writing for the heat of mixing

$$\Delta h = x_1 \Delta \overline{H}_1 + x_2 \Delta \overline{H}_2$$

By using Eq. (11-17), this becomes

$$\Delta h = -RT^2 \left[x_1 \left(\frac{\partial \ln \gamma_1}{\partial T} \right)_{P,x} + x_2 \left(\frac{\partial \ln \gamma_2}{\partial T} \right)_{P,x} \right] \qquad (11\text{-}18)$$

When the Wilson equation [Eqs. (9-58a) and (9-58b)] with its parameter temperature dependence [Eqs. (9-58d) and (9-58e)] is used in Eq. (11-18), the result after some manipulation is

$$\Delta h = x_1 a_{12} \left(\frac{x_2 G_{12}}{x_1 + G_{12} x_2} \right) + x_2 a_{21} \left(\frac{x_1 G_{21}}{x_2 + x_1 G_{21}} \right) \qquad (11\text{-}19)$$

The G's are dependent on the a's, and therefore Eq. (11-19) contains only two adjustable parameters which can be determined by fitting it to a set of Δh vs. x_2 data as shown in Fig. 11-2a. Parameter estimation methods similar to those described in Sec. 9-11 would be used for this task. Once the parameters a_{12} and a_{21} have been obtained, Wilson parameters may be evaluated at the temperature of the calculation and may

---

[7] This example is based on the work of R. W. Hanks, A. C. Gupta, and J. J. Christensen, *Ind. Eng. Chem. Fundam.*, *10*, 504 (1971).

be estimated at other temperatures from Eqs. (9-58d) and (9-58e). Activity coefficients are then known as a function of composition (and temperature), and the vapor-liquid equilibrium data may be calculated. Vapor-liquid equilibrium data calculated via this method at the temperature of the heat of mixing data are compared with experimentally determined values in Fig. 11-2b.

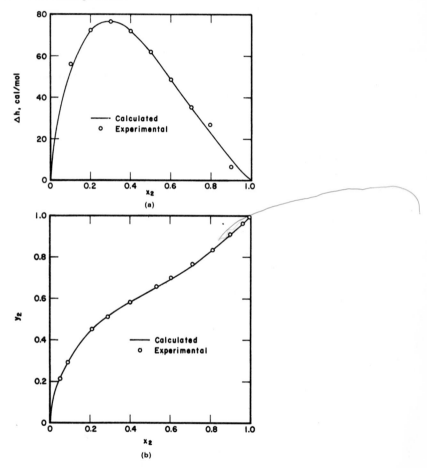

Figure 11-2 Comparison of calculated and experimental data for the carbon tetrachloride (1)–acetone(2) system at 45°C. Solid curves calculated from the Wilson equation. [Reprinted by permission from R. W. Hanks, A. C. Gupta, and J. J. Christensen, *Ind. Eng. Chem. Fundam.*, *10*, 504 (1971), copyright 1971 by American Chemical Society, Washington, DC.]

The agreement is seen to be excellent. Because these results depend on the validity of the prescribed temperature dependence of the parameters, it is reasonable to expect acceptable results at other temperatures.

11-3 MIXTURE PROPERTIES FOR IDEAL SOLUTIONS

Because we have seen that at low to moderate pressures most gas mixtures conform reasonably well to the ideal-solution model, we can expect that mixture and mixing properties based on that model will be directly useful. While we observed that liquid solutions rarely conform to this model, these properties will nevertheless be useful because of the insight provided into the mixing process and because they may serve as benchmarks when evaluating the behavior of nonconforming systems.

In Chap. 10 we identified the ideal Gibbs free energy change on mixing and found for the formation of a solution from pure components

$$\Delta g^i = RT \sum x_i \ln x_i \tag{10-12}$$

or

$$\Delta G^i = RT \sum n_i \ln x_i \tag{10-12a}$$

While these equations were originally derived for liquid solutions, they also apply to ideal gaseous solutions. Comparison of Eq. (10-12a) with Eq. (11-10) reveals that

$$\Delta \overline{G}_i = \overline{G}_i - g_i = RT \ln x_i$$

where g_i is the molar Gibbs free energy of pure component i. Recalling that $\overline{G}_i = \hat{\mu}_i$ and $g_i = \mu_i$, for the ideal solution we write

$$\hat{\mu}_i = \mu_i + RT \ln x_i \tag{11-20}$$

Differentiation of this equation with respect to pressure at constant temperature and composition yields

$$\left(\frac{\partial \hat{\mu}_i}{\partial P} \right)_{T,x} = \left(\frac{\partial \mu_i}{\partial P} \right)_T$$

From Eq. (11-5) we may write

$$\overline{V}_i = v_i \tag{11-21}$$

When Eq. (11-20) is divided by T and then differentiated with respect to temperature at constant pressure and composition, we obtain

$$\left(\frac{\partial \hat{\mu}_i/T}{\partial T}\right)_{P,x} = \left(\frac{\partial \mu_i/T}{\partial T}\right)_P$$

and from Eq. (11-7) we see that

$$\overline{H}_i = h_i \tag{11-22}$$

Thus, it is observed that there is no change in enthalpy or volume on forming an ideal solution, and

$$\Delta H^i = 0 \tag{11-23}$$

$$\Delta V^i = 0 \tag{11-24}$$

For an isothermal mixing process we write

$$\Delta G = \Delta H - T \Delta S$$

and through Eqs. (10-12a) and (11-23) obtain

$$\Delta S^i = -R \sum n_i \ln x_i \tag{11-25}$$

$$\Delta s^i = -R \sum x_i \ln x_i \tag{11-25a}$$

From Eq. (11-25) it is also seen that

$$\overline{S}_i - s_i = \Delta \overline{S}_i = -R \ln x_i \tag{11-26}$$

For ideal solutions, liquid or gaseous, these properties are summarized in Table 11-1.

TABLE 11-1 PROPERTIES OF IDEAL SOLUTIONS[a]

$\overline{V}_i - v_i = 0$	$\Delta V^i = 0$
$\overline{H}_i - h_i = 0$	$\Delta H^i = 0$
$\overline{G}_i - g_i = RT \ln x_i$	$\Delta G^i = RT \sum n_i \ln x_i$
$\hat{\mu}_i - \mu_i = RT \ln x_i$	$\Delta g^i = RT \sum x_i \ln x_i$
$\overline{S}_i - s_i = -R \ln x_i$	$\Delta S^i = -R \sum n_i \ln x_i$
	$\Delta s^i = -R \sum x_i \ln x_i$

[a]Property changes are written for liquid solutions. For gaseous solutions, substitute gaseous mol fraction y_i for liquid mol fraction x_i.

11-4 ACTIVITY COEFFICIENTS BASED ON HENRY'S LAW

Up to this point the activity coefficient we have been dealing with has been viewed as a correction for the ideal-liquid solution model. We have seen that although few systems conform to this model, ideal behavior is approached by all components as their liquid mol fractions approach unity. This, of course, applies to those components which are able to exist as pure liquids at the conditions of the system. For most systems this condition is met, but when considering solutions of solids or gases above their critical temperatures where the terms f_i^L in Eq. (9-63) and P_i° in Eq. (9-63a) have no physical significance, it is convenient to work in terms of Henry's law and define another activity coefficient as a correction to it. This activity coefficient γ_i^* is defined by

$$\hat{f}_i^L = k_i x_i \gamma_i^* \qquad (11\text{-}27a)$$

or at low pressures

$$p_i = k_i x_i \gamma_i^* \qquad (11\text{-}27b)$$

where k_i is the Henry's law constant for component i. Unlike f_i^L in the ideal solution model which depends only on component i and the temperature and pressure, k_i depends on the component and specific system as well as temperature and pressure but is not dependent on composition. Because we have seen that Henry's law is approached at low concentration, we write

$$\lim_{x_i \to 0} \gamma_i^* = 1 \qquad (11\text{-}28)$$

As before with γ, we see that γ^* is defined in terms of thermodynamic properties and is therefore a thermodynamic property whose temperature, pressure, and composition dependence we will attempt to discern. First, we should attempt to acquire a physical understanding of γ^*, and for this we turn to Fig. 11-3. In this figure the fugacity of component 1 in the liquid phase is plotted vs. liquid mol fraction of component 1 as the solid curve. When component 1 is a solid or a gas above its critical temperature, liquid solutions cannot exist over the entire range of composition, as indicated in the figure. Also shown in the figure is a dashed line representing Henry's law. As expected, it is seen that this line coincides with the fugacity curve as x_1 approaches zero. At a liquid mol fraction corresponding to the vertical line ABC the segment AB corresponds to the Henry's law fugacity, while the segment AC is the actual fugacity. According to our definition, Eq. (11-27a), it is seen that γ_i^* is the ratio AC/AB.

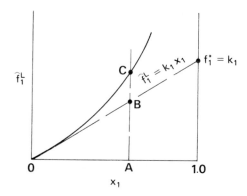

Figure 11-3 Fugacity of a nonliquid component vs. solution liquid mol fraction.

In determining the temperature and pressure dependence of γ^*, it is advantageous to visualize the Henry's law constant k_i as a fugacity. It has the same units as fugacity, and from Fig. 11-3 we see that it may be visualized as the fugacity of pure component i predicted from an extrapolation of Henry's law to $x_i = 1.0$. While this is obviously a hypothetical state, it is nevertheless convenient to visualize k_i as a pure component fugacity and replace it by f_i^* in Eq. (11-27a). We will then write Eq. (11-27) in logarithmic form and apply Eq. (9-7) in integrated form to obtain

$$\hat{\mu}_i^L - \mu_i^* = RT \ln \frac{\hat{f}_i^L}{f_i^*} = RT \ln x_i + RT \ln \gamma_i^* \tag{11-29}$$

Differentiation of Eq. (11-29) with respect to pressure at constant temperature and composition yields[8]

$$\left(\frac{\partial \hat{\mu}_i^L}{\partial P} \right)_{T,x} - \left(\frac{\partial \mu_i^*}{\partial P} \right)_T = RT \left(\frac{\partial \ln \gamma_i^*}{\partial P} \right)_{T,x}$$

Application of Eq. (11-5) results in

$$\overline{V}_i - v_i^* = RT \left(\frac{\partial \ln \gamma_i^*}{\partial P} \right)_{T,x} \tag{11-30}$$

Dividing Eq. (11-29) by T and differentiating with respect to T at constant pressure and composition yields

---

[8] Note that μ_i^* is identified as a pure-component property in a hypothetical state.

$$\left(\frac{\partial \hat{\mu}_i^L/T}{\partial T}\right)_{P,x} - \left(\frac{\partial \mu_i^*/T}{\partial T}\right)_P = R\left(\frac{\partial \ln \gamma_i^*}{\partial T}\right)_{P,x}$$

Application of Eq. (11-7) results in

$$\frac{-\overline{H}_i}{T^2} + \frac{h_i^*}{T^2} = R\left(\frac{\partial \ln \gamma_i^*}{\partial T}\right)_{P,x} \tag{11-31}$$

We will now attempt to attach physical significance to the quantities v_i^* and h_i^* in Eqs. (11-30) and (11-31) which are identified with a pure component in a hypothetical system which obeys Henry's law from $0 < x_i \leq 1$. For this hypothetical system we write

$$\hat{f}_i^L = k_i x_i = f_i^* x_i$$

In terms of chemical potentials this becomes

$$\hat{\mu}_i^L - \mu_i^* = RT \ln x_i \tag{11-32}$$

Using the procedure for deriving Eqs. (11-30) and (11-31), we obtain the corresponding equations for the hypothetical system:

$$\overline{V}_i = v_i^*$$

and

$$\overline{H}_i = h_i^*$$

v_i^* and h_i^* are identified with a pure substance and are thus independent of composition. Therefore, $\overline{H}_i$ and $\overline{V}_i$ are constant over the range $0 < x_i \leq 1$. It is more natural to view v_i^* and h_i^* as properties of a real system obeying Henry's law and identify them with composition range $x_i \to 0$ where Henry's law is actually followed. For this reason we will replace them with $\overline{V}_i^\infty$ and $\overline{H}_i^\infty$ which are limiting values as x_i approaches zero. We now rewrite Eqs. (11-30) and (11-31):

$$\left(\frac{\partial \ln \gamma_i^*}{\partial P}\right)_{T,x} = \frac{\overline{V}_i - \overline{V}_i^\infty}{RT} = \frac{\Delta \overline{V}_i - \Delta \overline{V}_i^\infty}{RT} \tag{11-30a}$$

and

$$\left(\frac{\partial \ln \gamma_i^*}{\partial T}\right)_{P,x} = \frac{-(\overline{H}_i - \overline{H}_i^\infty)}{RT^2} = -\frac{\Delta \overline{H}_i - \Delta \overline{H}_i^\infty}{RT^2} \tag{11-31a}$$

The quantities $\Delta \overline{V}_i^{\infty}$ and $\Delta \overline{H}_i^{\infty}$ are referred to as infinite dilution properties and are partial molar properties which can be evaluated from experimental data. Figure 11-4 shows the evaluation of $\Delta \overline{H}_1^{\infty}$ and $\Delta \overline{H}_2^{\infty}$ from Δh vs. x_1 data by means of the slope-intercept method. While it was necessary to resort temporarily to the use of a hypothetical state, the defining equation for γ_i^* and the equations stating its dependence on temperature and pressure are expressed in terms of properties which can actually be measured and therefore are physically significant.

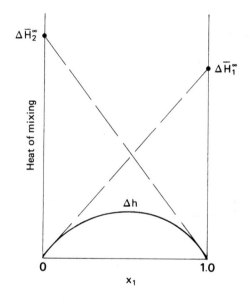

Figure 11-4 Determination of partial property changes at infinite dilution.

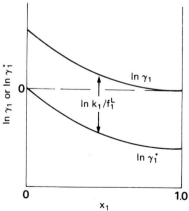

Figure 11-5 Relationship between γ_i and γ_i^*.

Because we have defined two different activity coefficients, it will be instructive to determine how they are related. For this let us consider a binary system composed of components which can exist as pure liquids and for which either activity coefficient could be used. At a definite liquid mol fraction, component 1 has a definite fugacity regardless of whether we use Eq. (9-35) or (11-27) to express it. Equating these expressions gives

$$\hat{f}_1^L = f_1^L x_1 \gamma_1 = k_1 x_1 \gamma_1^*$$

which simplifies to

$$f_1^L \gamma_1 = k_1 \gamma_1^*$$

and we see that for a given component the ratio of these activity coefficients will always be constant and that their logarithms will differ by a constant. Therefore, as shown in Fig. 11-5, a plot of $\ln \gamma_1^*$ vs. x_1 will have the same shape as a plot of $\ln \gamma_1$ and will simply be displaced by the amount $\ln (k_1 / f_1^L)$. Because $d \ln \gamma_1^* = d \ln \gamma_1$ the Gibbs-Duhem equation applies when the activity coefficients of one or all of the components are expressed in terms of Henry's law.

11-5 THE SOLUBILITY OF GASES IN LIQUIDS

The solubility of gases in liquids is a special case of vapor-liquid equilibrium where the gas cannot exist as a pure liquid at the conditions of the solution and has a low solubility in the liquid. When the gas solubility is less than a few mol percent, it is generally assumed that Henry's law is valid for the dissolved gas $(\gamma^* = 1.0)$, and at equilibrium we can then write

$$\hat{\mu}_1^G = \hat{\mu}_1^L$$

where

$$\hat{\mu}_1^G = \mu_1^\circ + RT \ln \hat{f}_1^G \qquad (9\text{-}6)$$

and

$$\hat{\mu}_1^L = \mu_1^* + RT \ln x_1 \qquad (11\text{-}32)$$

which leads to

$$\ln k_1 = \ln \frac{\hat{f}_1^G}{x_1} = \frac{\mu_1^* - \mu_1^\circ}{RT} \qquad (11\text{-}33)$$

Differentiating Eq. (11-33) with respect to pressure at constant temperature yields

$$\left(\frac{\partial \ln k_1}{\partial P}\right)_T = \frac{1}{RT}\left[\left(\frac{\partial \mu_1^*}{\partial P}\right)_T - \left(\frac{\partial \mu_1^\circ}{\partial P}\right)_T\right] \tag{11-34}$$

From the defining equation μ_i° is identified as a function only of temperature. Because Eq. (9-6) must be valid for all compositions, it must also hold at $y_1 = 1$, and μ_1° can be identified as the chemical potential of pure gaseous component 1 when its fugacity is equal to 1 atm. Thus, the second term on the right-hand side of Eq. (11-34) is zero. We have previously encountered the first term and may write

$$\left(\frac{\partial \ln k_1}{\partial P}\right)_T = \frac{v_1^*}{RT} = \frac{\overline{V}_1^\infty}{RT} \tag{11-35}$$

where $\overline{V}_1^\infty$ is the partial molar volume of the dissolved gas at infinite dilution.

When Eq. (11-33) is differentiated with respect to temperature at constant pressure, we obtain

$$\left(\frac{\partial \ln k_1}{\partial T}\right)_P = \frac{1}{R}\left(\frac{\partial \mu_1^*/T}{\partial T}\right)_P - \frac{1}{R}\left(\frac{\partial \mu_1^\circ/T}{\partial T}\right)_P \tag{11-36}$$

We have already seen that

$$\left(\frac{\partial \mu_1^*/T}{\partial T}\right)_P = -\frac{h_1^*}{T^2} = -\frac{\overline{H}_1^\infty}{T^2}$$

and because μ_1° refers to pure gaseous component 1,

$$\left(\frac{\partial \mu_1^\circ/T}{\partial T}\right)_P = -\frac{h_1^G}{T^2}$$

Equation (11-36) becomes

$$\left(\frac{\partial \ln k_1}{\partial T}\right)_P = \frac{h_1^G - \overline{H}_1^\infty}{RT^2} = -\frac{\Delta \overline{H}_1^\infty}{RT^2} \tag{11-37}$$

where h_1^G is the enthalpy of the pure gas and $\overline{H}_1^\infty$ is the partial molar enthalpy of the dissolved gas at infinite dilution. As we have seen, the partial molar enthalpy change at infinite dilution $\Delta \overline{H}_1^\infty$ can be determined experimentally.

A useful equation relating gas solubility to pressure can be obtained by the isothermal integration of Eq. (11-35) with the assumption that $\overline{V}_1^\infty$ is independent of pressure. The result, referred to as the Krichevsky-Kasarnovsky equation, is

$$\ln k_1 = \ln k_1^0 + \frac{\overline{V}_1^\infty (P - P_0)}{RT} \tag{11-38}$$

In this equation P_0 is a reference pressure and k_1^0 is the Henry's law constant at that pressure. The equation predicts that an isothermal plot of $\ln k_1$ vs. P should be linear. General conformance to this equation has been observed as evidenced by the data for the solubility of nitrogen in water shown in Fig. 11-6. The equation is used mainly for correlation, interpolation, and extrapolation of gas solubility data, although the slope of the isotherms yields $\overline{V}_1^\infty$ which is generally in good agreement with values determined from density measurements.

Guidance for the correlation, interpolation, and extrapolation of gas solubility with respect to temperature may be obtained from a rearrangement of Eq. (11-37):

$$\left(\frac{\partial \ln k_1}{\partial 1/T} \right)_P = \frac{\overline{H}_1^\infty - h_1^G}{R} = \frac{\Delta \overline{H}_1^\infty}{R} \tag{11-37a}$$

This equation resembles the Clausius-Clapeyron equation and indicates that if $\Delta \overline{H}_1^\infty$ is constant over a temperature range a plot of $\ln k_1$ vs. $1/T$ will be linear. An example of such a correlation is shown in Fig. 11-7 where the Henry's law constant for hydrogen in various solvents is plotted vs. $1/T$ on semilog coordinates. Reasonably good linearity is observed.

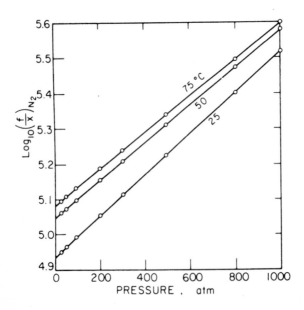

Figure 11-6 The solubility of nitrogen in water. [From J. M. Prausnitz, *Molecular Thermodynamics of Fluid - Phase Equilibria*, 1969, p. 358. Reprinted by permission of Prentice-Hall, Inc., Englewood Cliffs, NJ.]

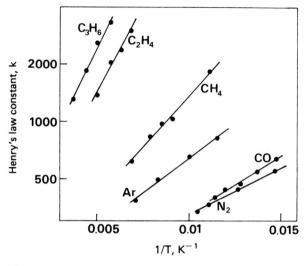

Figure 11-7 Henry's law constants for hydrogen in various cryogenic solvents. [Data from M. Orentlicher and J. M. Prausnitz, *Chem. Eng. Sci., 19,* 775 (1964).]

Conventional wisdom has it that the solubility of a gas always decreases with an increase in temperature, although the opposite behavior is often observed. An understanding of this phenomenon can be obtained through Eq. (11-37), which can be written in terms of entropy changes as

$$\left(\frac{\partial \ln k_1}{\partial T} \right)_P = \frac{h_1^G - \overline{H}_1^\infty}{RT^2} = \frac{s_1^G - \overline{S}_1^\infty}{RT} \tag{11-37b}$$

The entropy change can be expressed as

$$s_1^G - \overline{S}_1^\infty = (s_1^G - s_1^L) - (\overline{S}_1^\infty - s_1^L)$$

where s_1^L is the entropy of pure liquefied gas at the temperature of the solution. Although this may be a hypothetical state, we may still visualize the first term as the entropy of vaporization, which we know will always be positive. The second term is the partial molar entropy of mixing. Again a hypothetical state is involved, but this property change can be estimated by the assumption of an ideal solution as $-R \ln x_1$ as per Eq. (11-26). The entropy change in Eq. (11-37b) may now be approximated by

$$s_1^G - \overline{S}_1^\infty = (s_1^G - s_1^L) + R \ln x_1$$

It is seen that the sum of two terms of opposite sign determines the sign of the temperature dependence of the Henry's law constant. When the solubility is small, the second term will overpower the first, and the Henry's law constant will decrease with increasing temperature. The converse will hold for gases with appreciable solubility. Because at a given pressure there is an inverse relationship between Henry's law constant and solubility, we may state that, with increasing temperature, gases with low solubility will exhibit increased solubility while the solubility of moderately soluble gases decreases.

11-6 SOLID-LIQUID EQUILIBRIA

Phase behavior in a binary system is best visualized on a plot of temperature vs. mol fraction. Constant pressure is implied, but because condensed-phase properties are insensitive to pressure, its value is often not stated. Figure 11-8 shows a Tx diagram for the system A-B, where the melting points of pure A and B are represented by the points A and B, respectively. Two single-phase regions exist: liquid solution above the curve AEB and solid solution to the left of curve ACF. All other regions represent conditions where two phases coexist: solid solution and liquid in ACE, liquid and pure solid B in BED, and solid solution and pure solid B in $FCED$ ($x_B = 1$). A system such as this is called a eutectic system, and the temperature corresponding to CED is the eutectic temperature.

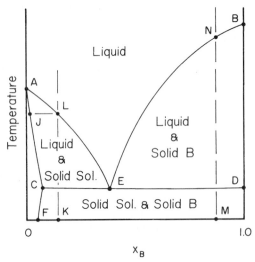

Figure 11-8 Solid-liquid phase diagram for an eutectic system.

When a liquid corresponding to the composition MN is cooled, pure solid B forms at the temperature N. As the temperature is decreased, pure solid B continues to form, and consequently the solution becomes richer in component A and moves along the curve BE toward E. When a liquid of composition KL is cooled, a solid phase forms at the temperature corresponding to L. This solid phase is a solution of mol fraction J and is in equilibrium with a liquid solution of mol fraction L. On continued cooling, the liquid composition moves along the curve AE toward E and the solid composition moves along the curve AC toward C. Thus, all liquids on cooling move toward the point E, the eutectic point, where three phases coexist: liquid of composition E, solid solution of composition C, and pure solid B.

For equilibrium between a saturated solid phase of mol fraction X_{is} and a saturated liquid phase of mol fraction x_{is} we write

$$\hat{\mu}_i^S = \hat{\mu}_i^L$$

Applying Eq. (9-38) to each phase and rearranging yields

$$RT \ln \frac{x_{is} \gamma_{is}}{X_{is} \Gamma_{is}} = \mu_i^S - \mu_i^L \qquad (11\text{-}39)$$

where Γ_i is the activity coefficient of component i in the solid phase and is defined in a manner analogous to its liquid-phase counterpart, γ_i The chemical potentials refer to pure solid and pure liquid component i. This difference can be calculated via the computational path shown in Fig. 11-9

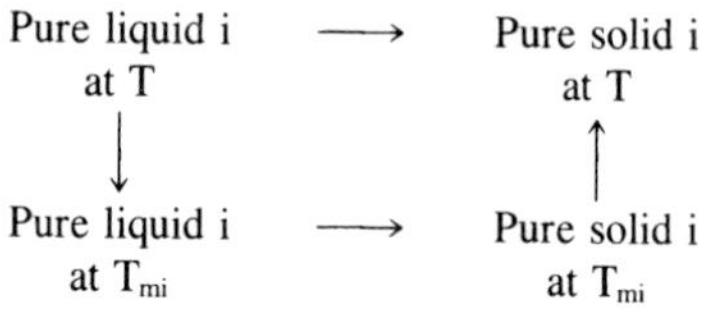

$$
\begin{array}{ccc}
\text{Pure liquid } i & \longrightarrow & \text{Pure solid } i \\
\text{at T} & & \text{at T} \\
\downarrow & & \uparrow \\
\text{Pure liquid } i & \longrightarrow & \text{Pure solid } i \\
\text{at T}_{mi} & & \text{at T}_{mi}
\end{array}
$$

Figure 11-9 Computational path for $\mu_i^S - \mu_i^L$ and $s_i^L - s_i^S$.

and is

$$\mu_i^S - \mu_i^L = -\int_T^{T_{mi}} s_i^L \, dT + 0 - \int_{T_{mi}}^T s_i^S \, dT$$

which can be written

$$\mu_i^S - \mu_i^L = \int_{T_{mi}}^T (s_i^L - s_i^S) \, dT \qquad (11\text{-}40)$$

where T_{mi} is the melting point of component i. Also, from the computational path of Fig. 11-9 we write

$$s_i^S - s_i^L = \int_T^{T_{mi}} \frac{c_{Pi}^L}{T} \, dT - \frac{L_{mi}}{T_{mi}} + \int_{T_{mi}}^T \frac{c_{Pi}^S}{T} \, dT$$

which we restate as

$$s_i^L - s_i^S = \frac{L_{mi}}{T_{mi}} + \int_{T_{mi}}^T \frac{(c_P^L - c_P^S)_i}{T} \, dT \qquad (11\text{-}41)$$

where L_{mi} is the latent heat of melting of component i. When combined, Eqs. (11-39), (11-40), and (11-41) represent the most general formulation of solid-liquid phase equilibrium. Heat capacities of liquids below their freezing points or of solids above their melting points are required for implementing these equations but are seldom available. Often, however, the heat capacities are available at the melting temperature, and the assumption is made that $\Delta c_P = c_P^L - c_P^S$ is independent of temperature. With this assumption Eqs. (11-39)–(11-41) lead to

$$RT \ln \frac{x_{is}\gamma_{is}}{X_{is}\Gamma_{is}} = \frac{L_{mi}(T - T_{mi})}{T_{mi}} + \Delta c_{Pi}(T_{mi} - T) + T\,\Delta c_{Pi} \ln \frac{T}{T_{mi}} \qquad (11\text{-}42)$$

The right-hand side of this equation is seen to depend on the properties of pure component i and is a function only of temperature.

On surveying solid-liquid phase diagrams, one finds considerable variation and, often, formidable complexity. Still, there are only two basic situations: The solid phase is either pure or a solid solution. In Sec. 10-7 we saw that liquid miscibility was determined by the magnitude of the activity coefficients, a manifestation of the intermolecular forces (see Sec. 11-8). While this is an important factor, spatial considerations are more important for solid solubility. In order to have solid solutions, it is necessary that the species be able to fit into a structured crystal lattice. This is likely only for species that are compact and closely sized; therefore, complete solid miscibility is rarely found. Figure 11-10 shows four systems displaying varying degrees of solid miscibility. There is no solid miscibility in system (a) and at the eutectic temperature pure solid A, pure solid B, and a liquid of composition E are in equilibrium. Slight solid miscibility is exhibited in system (b), where at the eutectic temperature solid solutions of composition C and D are in equilibrium with a liquid of composition E. System (c) shows complete solid miscibility. Peritectic behavior is exhibited by system (d); there is partial solid miscibility and three solutions, C, D, and E are in equilibrium at the peritectic temperature. In the eutectic state the liquid composition, E, is bracketed by the compositions of the two solid phases, C and D. In the peritectic state E lies outside the range CD.

Pure Solid Phase.   For equilibrium between a liquid solution and a pure solid, we merely set $X_{is}\Gamma_{is}$ in Eq. (11-42) equal to unity.

$$RT \ln x_{is}\gamma_{is} = \frac{L_{mi}(T - T_{mi})}{T_{mi}} + \Delta c_{Pi}(T_{mi} - T) + T\,\Delta c_{Pi} \ln \frac{T}{T_{mi}} \qquad (11\text{-}43)$$

The simplest application of this equation will occur when ideal-solution behavior is found. This, we know, occurs in mixtures of isomers and homologs and also for any component in which mol fraction approaches unity.

For dilute solutions, Eq. (11-43) can be simplified when applied to the solvent, component 1. With x_{1s} approaching unity, we may set γ_{1s} to unity, and because T will be close to T_{m1}, we may drop the Δc_P terms. Further, we define the freezing point depression, θ, as $T_{m1} - T$ and replace x_{1s} by $1 - x_2$ to obtain

$$\ln(1 - x_2) = -\frac{L_{m1}\theta}{RT_{m1}^2}$$

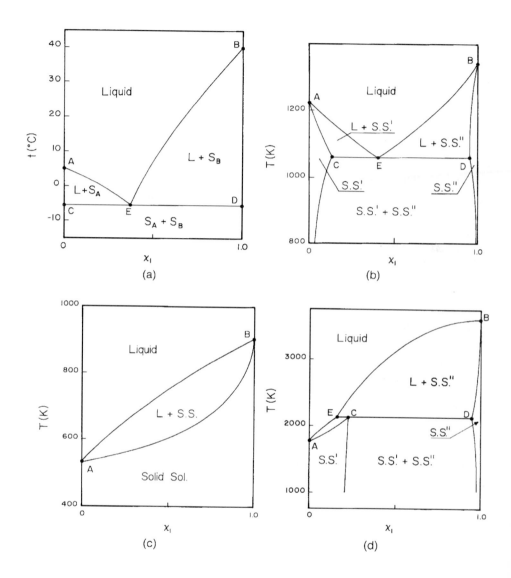

Figure 11-10 Solid-liquid diagrams: (a) phenol (1)–benzene (2), (b) copper (1)–silver (2), (c) antimony (1)–bismuth (2), (d) tungsten (1)–palladium (2).

For small x_2 the left-hand side may be approximated by $-x_2$, and we obtain

$$x_2 = \frac{L_{m1}\theta}{RT_{m1}^2} \tag{11-43a}$$

Equation (11-43a) is useful for determining the purity of a known substance or for obtaining the molecular weight of an unknown solute. The former application is straightforward. In the latter application the freezing point depression of a selected solvent is experimentally determined for a solution containing a small but known weight fraction of the solute in the solvent. The solute mol fraction obtained from Eq. (11-43a) and its known weight fraction allow the determination of the solute molecular weight. It is to be remembered that Eq. (11-43a) gives the impurity mol fraction in terms of the solvent properties L_{m1} and T_{m1}.

EXAMPLE 11-2

The temperature at which pure ice is in equilibrium with air-saturated water at 1 atm is 0.0100°C lower than the triple point where ice, liquid, and vapor are all in equilibrium at a pressure of 4.6 mmHg. How much of this freezing point depression can be attributed to the pressure increase and how much can be attributed to the solubility of air in liquid water? Following are the data involved:

    Density of ice: 0.917 g/cm^3
    Density of liquid water: 1.000
    Latent heat of fusion: 79.6 cal/g
    Henry's law for air dissolved in liquid water
    $P = 4.32(10^4)x$ with P in atmospheres

Solution 11-2

The effect of pressure on the freezing point of pure water was determined in Ex. 8-1 to be –0.00746 K. The freezing point of water is further lowered by the presence of dissolved air at a mol fraction of

$$x = \frac{1}{4.32(10^4)} = 2.31(10^{-5})$$

This amounts to

$$\theta = \frac{RT_m^2 x}{L_m}$$

$$= \frac{1.987(273)^2(2.31)(10^{-5})}{79.6(18)} = 0.00239 \; K$$

The total freezing point lowering is calculated to be

$$0.00746 + 0.00239 = 0.00985 \; K$$

in very good agreement with the experimental value of 0.0100 K.

A phase diagram such as that shown in Fig. 11-10(a) is nothing more than the intersection of two freezing curves: AE for component A, and BE for component B. In general, information concerning activity coefficients is necessary to implement the calculations with Eq. (11-43); however, for ideal solutions only the physical properties of the pure components are required (L_m, Δc_P, and T_m). The calculations are easily extended to multicomponent ideal solutions.

EXAMPLE 11-3

How much pure *p*-xylene can be obtained from a liquid mixture of 50% *p*-xylene, 25% *o*-xylene, and 25% *m*-xylene by a crystallization process? What is the minimum temperature to be used? The following data are given:

	L_m (cal/g mol)	T_m (K)
Ortho	3250	248
Meta	2765	226
Para	4090	286

Solution 11-3

Setting Δc_P to zero in Eq. (11-43) and using the preceding values of L_m and T_m provides an equation of the freezing-point curve for each component:

$$\ln x_{os} = 6.595 - \frac{1636}{T} \qquad (o)$$

$$\ln x_{ms} = 6.157 - \frac{1391}{T} \qquad (m)$$

$$\ln x_{ps} = 7.197 - \frac{2058}{T} \qquad (p)$$

The initial freezing temperature may be found from the solution composition:

$$x_o = 0.25$$

$$x_m = 0.25$$

$$x_p = 0.50$$

Using $x_{os} = 0.25$ in Eq. (o) gives $T = 205$ K; similarly, we obtain $T = 184$ K for $x_{ms} = 0.25$ and $T = 261$ K for $x_{ps} = 0.50$ and observe that pure p-xylene solid will form first at 261 K. As the temperature is lowered, pure p-xylene will continue to form while the solution becomes more concentrated with respect to the other isomers. This will continue until one of the other isomers reaches its saturation temperature, and it will form a pure solid. Because we have no way of separating the crystals of the pure isomers, this becomes the minimum temperature to be used for obtaining crystals of pure p-xylene. Until this temperature is reached, it is permissible to use a material balance to determine all x's in terms of the recovery of pure p-xylene crystals. We take as a basis 100 mol of original solution and let P equal the mols of pure p-xylene crystals recovered. At any point in the process, until another isomer begins to freeze out, we have

50–P mol of p-xylene
25 mol of m-xylene
25 mol of o-xylene

Liquid mol fractions may now be calculated for any value of P and inserted into equations (o), (m), and (p) as before to obtain the saturation temperatures. As long as the saturation temperatures of o-xylene and m-xylene are below that of p-xylene, we are above the minimum temperature. This can be found by a trial and error procedure to be 221 K when $P = 43.5$ mol of p-xylene crystals, which represents a recovery of 87% of the p-xylene in the original solution.

Because these isomers have normal boiling points within a span of a few degrees, this is an attractive separation technique and is used commercially to produce p-xylene.

In dealing with nonideal systems more information is required, although it has been demonstrated[9] that when a reliable activity coefficient equation is available, good estimates of solubility can be obtained for systems of nonelectrolytes. Alternatively, Eq. (11-43) may be used with experimental solubility data to obtain activity coefficients which may subsequently be used to estimate equilibria under other conditions.

EXAMPLE 11-4

At 20°C the solubility of solid naphthalene in hexane is 0.090 mol/mol of solution. Use this information and the properties of the pure components to estimate the following for this system:
(a) the mol fraction of naphthalene in the vapor phase in equilibrium with a saturated solution of naphthalene at 20°C.
(b) the solubility of naphthalene in hexane at 40°C.
(c) the mol fraction of naphthalene in the vapor phase in equilibrium with a liquid containing 5.0 mol % naphthalene and 95 mol % hexane at 20°C.
The following data are given:

Heat of melting of naphthalene: 4610 cal/g mol
Melting point of naphthalene: 80.2°C
Vapor pressure at 20°C:
naphthalene: 0.054 mmHg
hexane: 122.5 mmHg

Solution 11-4

(a) Equation (11-43) with $\Delta c_p = 0$ is used to calculate γ_{1s}.

$$\ln (0.090\gamma_{1s}) = \frac{4610}{1.987}\left(\frac{1}{353.2} - \frac{1}{293}\right)$$

$$\gamma_{1s} = 2.89$$

Because γ is known at a single x, we can evaluate the parameter in the simple activity coefficient equation:[10]

$$RT \ln \gamma_1 = B(1-x_1)^2$$

With this data point we obtain

[9] J. G. Gmehling, T. F. Anderson, and J. M. Prausnitz, *Ind. Eng. Chem. Fundam.*, 17, 269 (1978).
[10] The temperature dependence of this equation is suggested by regular solution theory (see Sec. 11-8).

$$1.987(293) \ln 2.89 = B(1 - 0.090)^2$$

$$B = 745$$

The activity coefficient of the hexane in this saturated solution may now be calculated:

$$RT \ln \gamma_2 = B(1 - x_2)^2$$

$$\gamma_2 = 1.01$$

To calculate the naphthalene mol fraction in the vapor phase, we apply the condition of equilibrium to naphthalene:

$$f_1^S = \hat{f}_1^L = \hat{f}_1^G$$

Replacing fugacities with pressures, we see that the partial pressure of naphthalene is equal to the vapor pressure of pure solid naphthalene:

$$p_1 = 0.054 \text{ mmHg}$$

The partial pressure of hexane is

$$p_2 = x_2 P_2^\circ \gamma_2 = 0.91(122.5)(1.01) = 112.5 \text{ mmHg}$$

which results in

$$P = 0.054 + 112.5 = 112.554 \text{ mmHg}$$

and

$$y_1 = \frac{0.054}{112.554} = 0.00048$$

(b) To estimate the solubility of naphthalene at 40°C, we apply Eq. (11-43):

$$\ln \gamma_{1s} + \ln x_{1s} = \frac{4610}{1.987}\left(\frac{1}{353.2} - \frac{1}{313}\right)$$

Using the previously determined value of B we write

$$\frac{745}{1.987(313)}(1 - x_{1s})^2 + \ln x_{1s} = -0.8437$$

which yields

$$x_{1s} = 0.20$$

This compares well with the experimental value of 0.22 reported by Sunier [*J. Phys. Chem.*, *34*, 2582 (1930)].

(c) The activity coefficient we have used for naphthalene is defined in terms of Eq. (9-35) or its low-pressure form, Eq. (9-51), which specify a pure liquid fugacity or vapor pressure. This presented no problem in part (b), but the vapor pressure of pure subcooled liquid naphthalene is required to calculate the partial pressure of naphthalene via Eq. (9-51). While this is a hypothetical state for naphthalene, a value for this vapor pressure may be obtained from a statement of equilibrium applied to naphthalene in the solid-liquid-vapor system:

$$p_1 = 0.054 = P_1^{\circ} x_{1s} \gamma_{1s}$$

$$P_1^{\circ} = \frac{0.054}{0.09(2.89)} = 0.208 \text{ mmHg}$$

At 20°C the partial pressure of naphthalene is

$$p_1 = 0.208 x_1 \gamma_1$$

At $x_1 = 0.05$ we calculate γ_1 and γ_2:

$$1.987(293) \ln \gamma_1 = 745(1 - 0.05)^2$$

$$\gamma_1 = 3.18$$

$$1.987(293) \ln \gamma_2 = 745(0.05)^2$$

$$\gamma_2 = 1.00$$

The partial pressures are

$$p_1 = 0.208(0.05)(3.18) = 0.033 \text{ mmHg}$$

$$p_2 = 122.5(0.95)(1.00) = 116.3 \text{ mmHg}$$

$$P = 116.333 \text{ mmHg}$$

and

$$y_1 = \frac{0.033}{116.333} = 0.00028$$

Solid Solutions.   For solid solutions Eq. (11-42) is written for each component.

$$RT \ln \frac{x_{1s}\gamma_{1s}}{X_{1s}\Gamma_{1s}} = \frac{L_{m1}(T-T_{m1})}{T_{m1}} + \Delta c_{P1}(T_{m1}-T) + T\,\Delta c_{P1}\ln\frac{T}{T_{m1}} \qquad (11\text{-}42a)$$

$$RT \ln \frac{x_{2s}\gamma_{2s}}{X_{2s}\Gamma_{2s}} = \frac{L_{m2}(T-T_{m2})}{T_{m2}} + \Delta c_{P2}(T_{m2}-T) + T\,\Delta c_{P2}\ln\frac{T}{T_{m2}} \qquad (11\text{-}42b)$$

The simplest application of these equations occurs for ideal solutions in both phases. This situation gives rise to lens-shaped phase diagrams similar to that of Fig. 11-10(c). At a given temperature the equations yield values of x_{1s}/X_{1s} and x_{2s}/X_{2s} which are necessary and sufficient to determine the individual mol fractions.

For nonideal solutions, the use of these equations will require activity coefficient equations for both phases. The Γ's and γ's are defined similarly and are both governed by the Gibbs-Duhem equation; therefore, the activity coefficient equations should be similar. In general, for any given system the Γ's are expected to be larger than the γ's because, in addition to intermolecular effects, there can also be effects arising from lattice distortion or strain.[11] Although any activity coefficient equation used for γ's could be used in Eq. (11-42), only those with a specified temperature dependence need be considered. A frequently used equation is

$$RT \ln \Gamma_1 = \frac{A}{[1+(AX_1/BX_2)]^2} \qquad (11\text{-}44a)$$

$$RT \ln \Gamma_2 = \frac{B}{[1+(BX_2/AX_1)]^2} \qquad (11\text{-}44b)$$

This is an empirical version of the regular solution theory, Eq. (11-55), and is essentially a van Laar equation, Eq. (9-57), with temperature dependence. The equation can be further simplified by setting A equal to B:

$$RT \ln \Gamma_1 = AX_2^2 \qquad (11\text{-}44c)$$

$$RT \ln \Gamma_2 = AX_1^2 \qquad (11\text{-}44d)$$

Using this one-parameter equation for each phase, Lupis[12] has shown the dramatic effect of these parameters upon the shape and complexity of binary phase diagrams. These same "regular-solution" equations were used by Kaufman and Bernstein[13] to simulate binary

[11] Factors affecting Γ's are discussed by J. H. Hildebrand and R. L. Scott, *Solubility of Nonelectrolytes*, 3rd ed., Van Nostrand Reinhold, New York, 1950.

[12] C. H. P. Lupis, *Chemical Thermodynamics of Materials*, North-Holland, NY, 1983, Chap. 8.

[13] L. Kaufman and H. Bernstein, *Computer Calculation of Phase Diagrams*, Academic Press, New York, 1970.

phase diagrams for many systems, often with good results. For the reverse calculation, a procedure for the determination of parameters in activity coefficient equations from binary phase diagrams has been developed and successfully applied.[14]

EXAMPLE 11-5[15]

The phase diagram for the aluminum-silicon system[16] resembles Fig. 11-8. The melting temperatures for aluminum and silicon are 933.6 K and 1687 K, respectively. At the eutectic temperature of 850 K three phases are in equilibrium: pure solid silicon, a liquid containing 12.2 atom % silicon, and a solid solution containing 1.5 atom % silicon (corresponding, respectively, to points D, E, and C in Fig. 11-8).

This system is important to the semiconductor industry because aluminum is sometimes used for current-carrying connections in integrated circuits on silicon wafers. If during subsequent processing the wafer is heated to a high temperature, the diffusion of silicon into aluminum may occur. The driving force for this diffusion will depend on the silicon atom fraction of the saturated aluminum-silicon solid solution. Estimate this atom fraction as a function of temperature.

Solution 11-5

At the eutectic temperature, 850 K, the eutectic liquid is in equilibrium with a solid solution containing 1.5 atom % silicon and with pure solid silicon. Thus we write for silicon, component 1

$$\hat{\mu}_1^L = \hat{\mu}_1^{S.S.} \quad \text{and} \quad \hat{\mu}_1^L = \mu_1^S$$

which leads to

$$\mu_1^S = \hat{\mu}_1^{S.S.}$$

By analogy with liquid solutions, we use Eq. (9-38) for the solid solution and write

$$\mu_1^S = \mu_1 + RT \ln X_{1s}\Gamma_{1s}$$

Because μ_1 is the chemical potential of pure solid silicon, we have

$$\ln X_{1s}\Gamma_{1s} = 0 \tag{a}$$

---

[14] R. Hiskes and W. A. Tiller, *Mater. Sci. Eng.*, 2, 320 (1968); 4, 163 (1969).

[15] Suggested by Prof. T. J. Anderson, University of Florida.

[16] J. L. Murray and A. J. McAlister, *Bull. Alloy Phase Diagrams*, 5 (1), 74 (1984).

$$\Gamma_{1s} = \frac{1}{X_{1s}} = \frac{1}{0.015} = 66.7$$

Using the equation

$$RT \ln \Gamma_1 = A(1 - X_1)^2 \tag{b}$$

we determine A

$$A = \frac{1.987(850) \ln 66.7}{(1 - 0.015)^2} = 7312 \frac{\text{cal}}{\text{g atom}}$$

The desired relationship between X_{1s} and T is obtained by combining Eqs. (a) and (b):

$$RT \ln X_{1s} + 7312(1 - X_{1s})^2 = 0$$

The following values have been computed

X_{1s}	0.015	0.010	0.005	0.002
$T(K)$	850	783	688	590

EXAMPLE 11-6

Potassium and cesium each crystallize on a body-centered-cubic lattice and form solid solutions over the entire composition range. On the phase diagram,[17] Fig. 11-11, the solidus and liquidus lines touch at 50.5 atom % potassium, where a minimum temperature of –38.05°C is exhibited. This is called a congruent melting point and is similar to an azeotrope in a vapor-liquid system. Use this data point and the following pure-component data[18] to calculate the solid-liquid phase diagram for this system.

---

[17] C. W. Bale and A. D. Pelton, *Bull. Alloy Phase Diagrams, 4*(4), 379 (1983).
[18] Source: references 1D and 10 from Table 12-1.

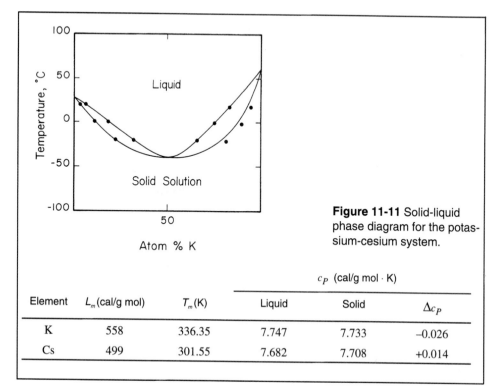

Figure 11-11 Solid-liquid phase diagram for the potassium-cesium system.

Element	L_m (cal/g mol)	T_m (K)	c_P (cal/g mol · K)		
			Liquid	Solid	Δc_P
K	558	336.35	7.747	7.733	−0.026
Cs	499	301.55	7.682	7.708	+0.014

Solution 11-6

With the supplied data Eq. (11-42) is applied to each component

$$\ln \frac{x_{1s}\gamma_{1s}}{X_{1s}\Gamma_{1s}} = 0.9242 - \frac{285.23}{T} - 0.0131 \ln T \qquad \text{(a)}$$

$$\ln \frac{x_{2s}\gamma_{2s}}{X_{2s}\Gamma_{2s}} = 0.7856 - \frac{249.01}{T} + 0.00705 \ln T \qquad \text{(b)}$$

For the congruent melting point at $T = 235.10$ with $x_{1s} = X_{1s}$ and $x_{2s} = X_{2s}$, we have

$$\ln \frac{\gamma_{1s}}{\Gamma_{1s}} = -0.3606$$

$$\ln \frac{\gamma_{2s}}{\Gamma_{2s}} = -0.2350$$

With these two equations it is possible to determine the value of two parameters. Our first attempt would be to use a one-parameter activity coefficient equation for each phase, such as

$$RT \ln \gamma_1 = B(1-x_1)^2 \qquad\qquad RT \ln \gamma_2 = Bx_1^2$$

and

$$RT \ln \Gamma_1 = C(1-X_1)^2 \qquad\qquad RT \ln \Gamma_2 = CX_1^2$$

This results in

$$B(1-x_1)^2 - C(1-X_1)^2 = -0.3606RT$$

$$Bx_1^2 - CX_1^2 = -0.2350RT$$

Because $x_1 = X_1$, the two equations are seen to be inconsistent. Even if they were consistent, we could obtain only the difference between the two parameters. In any event, we see that these two simple equations cannot be used. We note, however, that these equations show that C is larger than B, as would be expected. This suggests that the system could be modeled by assuming ideal-liquid solutions and using Eqs. (11-44a) and (11-44b) for activity coefficients in the solid solution:

$$RT \ln \Gamma_1 = \frac{A}{[1+(AX_1/BX_2)]^2}$$

$$RT \ln \Gamma_2 = \frac{B}{[1+(BX_2/AX_1)]^2}$$

Setting γ_{1s} and γ_{2s} equal to unity results in

$$\ln \Gamma_{1s} = 0.3606$$

$$\ln \Gamma_{2s} = 0.2350$$

which on substitution into the van Laar equations with $X_1 = 0.505$ results in

$$A = 453 \text{ cal}/\text{g atom} \cdot \text{K}$$

$$B = 719 \text{ cal}/\text{g atom} \cdot \text{K}$$

The equations for phase equilibrium are now

$$\ln x_{1s} = \ln X_{1s} + \frac{453/RT}{[1 + 0.630(X_1/X_2)]^2} + 0.9242 - \frac{285.23}{T} - 0.0131 \ln T \quad (c)$$

$$\ln x_{2s} = \ln X_{2s} + \frac{719/RT}{[1 + 1.59(X_2/X_1)]^2} + 0.7856 - \frac{249.10}{T} + 0.00705 \ln T \quad (d)$$

At a given temperature, selected values of X_{1s} are used to calculate x_{1s} and x_{2s} from Eqs. (c) and (d), respectively. (*Note:* $X_2 = 1 - X_1$.)When the calculated x_{1s} and x_{2s} sum to unity, the equations are satisfied and we have an X_{1s}, x_{1s} pair for the phase diagram. The calculation details for $t = 0°C$ are as follows.

X_{1s}	x_{1s} via Eq. (c)	x_{2s} via Eq. (d)	$x_{1s} + x_{2s}$
0.05	0.0901	0.872	0.962
0.10	0.171	0.830	1.001
0.15	0.243	0.790	1.033
0.80	0.705	0.362	1.067
0.85	0.729	0.309	1.038
0.90	0.756	0.239	0.995

By graphical interpolation we obtain

$$X_{1s} = 0.099 \qquad\qquad X_{1s} = 0.894$$

and

$$x_{1s} = 0.170 \qquad\qquad x_{1s} = 0.752$$

X_{1s}, x_{1s} pairs have been calculated at three temperatures and are plotted in Fig. 11-11, where it is seen that reasonable agreement with the experimentally determined phase diagram is obtained despite the crudeness of our model.

11-7 SOLID-SUPERCRITICAL FLUID EQUILIBRIUM

Supercritical fluids are becoming well-established as solvents[19] because of their high diffusivity, low viscosity, and a widely varying solute solubility with pressure. Here, we are interested in the effect of pressure on the isothermal solubility of a solid in a gas as we move
from moderate pressures to pressures in the supercritical region. We begin with the equilibrium condition

$$\mu_1^S = \hat{\mu}_1^F$$

where the superscript S refers to the solid which is assumed to be pure and F refers to the
fluid phase which is a binary mixture. For the fluid phase the chemical potential can be expressed by Eq. (9-6). This leads to

$$\mu_1^S = \mu_1^\circ + RT \ln \hat{f}_1$$

Using Eqs. (9-9) and (9-24) and rearranging we obtain

$$RT \ln y_1 = \mu_1^S - \mu_1^\circ - RT \ln P - RT \ln \hat{\phi}_1$$

The chemical potentials are pure-component properties: μ_1^S refers to the solid at the system
pressure and μ_1° refers to the gas at a fugacity of one atm. Their difference can be determined
from the computational path shown in Fig. 11-12.

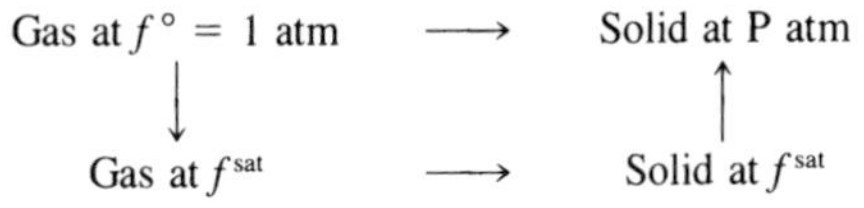

Gas at $f^\circ = 1$ atm	$\longrightarrow$	Solid at P atm
$\downarrow$		$\uparrow$
Gas at f^{sat}	$\longrightarrow$	Solid at f^{sat}

Figure 11-12 Computational path
for evaluating $\mu_1^S - \mu_1^\circ$.

From this three-step path we obtain

$$\mu_1^S - \mu_1^\circ = RT \ln \frac{f_1^{\text{sat}}}{1} + 0 + \int_{P_1^{\text{sat}}}^{P} v_1^S \, dP$$

which on substitution into the previous equation, followed by rearrangement, yields

$$y_1 = \frac{f_1^{\text{sat}} \exp[v_1^S (P - P_1^{\text{sat}})/RT]}{P \hat{\phi}_1} \tag{11-45}$$

From this equation it is seen that $\hat{\phi}_1$ is the factor which produces the dramatic pressure dependence of solubility. This fluid-phase fugacity coefficient could be calculated from Eq. (9-
62) if T_c, P_c, and ω for both components were known as well as the binary interaction pa-

---

[19] M. A. McHugh and V. J. Krukonis, *Supercritical Fluid Extraction*, 2nd ed., Butterworth-Heinemann, Boston,
1994.

rameter k_{12}. If this information is available, the program PRSGE.EXE on the CD-ROM accompanying this text can be used to execute the calculations via Eq. (11-45).

When T_c, P_c, and ω are available for both components, another program, KFIT-SGE.EXE on the accompanying CD-ROM, can be used to fit a value of the Peng-Robinson binary interaction parameter k_{12} to a set of isothermal $y_1 P$ data. The optimum value of k_{12} is determined by minimizing an objective function. Two objective functions are used:

$$\textbf{OF1} = \sum (y_1^{exp} - y_1^{cal})^2$$

$$\textbf{OF2} = \sum (\log y_1^{exp} - \log y_1^{cal})^2$$

in each case the summation is taken over all data points. As T, P, and y_1 are specified for each data point, each objective function depends only on k_{12}. Because of the wide range of y_1 values and concomitant range of deviations, **OF1** virtually ignores the smaller y_1 values that make a minute contribution to the sum. In order to accommodate this wide range of y_1 values, a logarithmic scale is usually used to represent $y_1 P$ data as is shown in Fig. 11-13. The objective function **OF2** sums the squares of the distances between the points and the fitted curve when a logarithmic scale is used for y_1 and therefore gives equal weight to each data point. Figure 11-13 shows the fit produced from each objective function to a set of experimental data[20] where it is seen that neither objective function yields a k_{12} capable of closely fitting the data over the entire range of y_1. This confirms the reported observation[21] that van der Waals–type equations of state do not fit well in the low-pressure range.

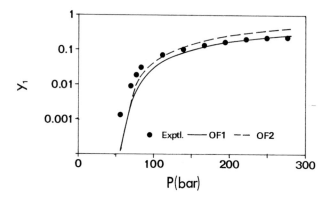

Figure 11-13 Fit of experimental data to Eq. (11-45) with k_{12} as a parameter using the program KFIT-SGE.EXE. System: Phenanthrene—Ethylene at 25°C. [K.P. Johnston and C.A. Eckert, *A.I.Ch.E.J.*, 27, 773 (1981).]

At low pressure Eq. (11-45) reduces to

---

[20] The data for construction of this figure is available on the spreadsheet FG11-13.WQ1.
[21] K. P. Johnston and C. A. Eckert, *AIChE.J.*, 27, 773 (1981).

$$y_1 = \frac{P_1^{sat}}{P}$$

and it is seen that as the pressure is lowered y_1 increases, becoming unity at P_1^{sat}. This low-pressure branch of the solubility curve would not be discernible on Fig. 11-13; however, it is obvious that the solubility curve would exhibit a minimum if extended to low pressures. Equation (11-45) together with Eq. (9-62) predicts a low-pressure minimum as well as a high-pressure maximum,[22] which has been observed in a few systems studied under sufficiently high pressure. For a critical appraisal of modeling supercritical fluid mixtures see the review by Johnston et al.[23]

11-8 PREDICTION OF SOLUTION BEHAVIOR

We have seen that information concerning solution behavior can be organized in terms of the activity coefficient and that some type of experimental data is usually required. Here we wish to consider ways in which theories based on molecular structure or properties can be useful in obtaining qualitative and quantitative estimates of activity coefficients. While these theories deal with a thermodynamic property, the activity coefficient, they are of extrathermodynamic origin and therefore are simply theories and do not possess the certitude accorded purely thermodynamic laws and relationships. Nevertheless, they are a valuable adjunct to thermodynamics in the treatment of phase equilibrium.

Qualitative Classification of Solutions.   A scheme based on hydrogen bonding behavior[24] is useful for qualitative estimates of solution behavior. As shown in Table 11-2, substances can be placed into five classes depending on their ability to form hydrogen bonds. Positive or negative deviations from ideal solution behavior are determined by the extent of hydrogen bonding in the mixture. This behavior is summarized in Table 11-3 where it is observed that the great preponderance of systems exhibit positive deviations.

TABLE 11-2

CLASSIFICATIONS OF SUBSTANCES ACCORDING TO HYDROGEN BONDING CAPABILITIES

CLASS I. Substances capable of forming three-dimensional networks of strong hydrogen bonds, e.g., water, glycol, glycerol, amino alcohols, hydroxylamine, hydroxy acids, polyphenols, amides, etc. Compounds such as nitromethane and acetonitrile also form three-dimensional networks of hydrogen bonds, but the bonds are much weaker than those involving OH and NH groups. Therefore, these types of compounds are placed in class II.

---

[22] R. T. Kurnik and R. C. Reid, *AIChE J., 27*, 861 (1981).

[23] K. P. Johnston, D. G. Peck, and S. Kim, *Ind. Eng. Chem. Res., 28*, 1115 (1989).

[24] R. H. Ewell, J. M. Harrison, and L. Berg, *Ind. Eng. Chem., 36*, 871 (1944).

TABLE 11-2 Con't

CLASS II. Other molecules containing both active hydrogen atoms and donor atoms (oxygen, nitrogen, and fluorine), e.g., alcohols, acids, phenols, primary and secondary amines, oxides, nitro compounds with α-hydrogen atoms, ammonia, hydrazine, hydrogen fluoride, hydrogen cyanide, etc.

CLASS III. Molecules containing donor atoms but no active hydrogen atoms, e.g., ethers, ketones, aldehydes, esters, tertiary amines (including pyridine type), nitro compounds and nitriles without α-hydrogen atoms, etc.

CLASS IV. Molecules containing active hydrogen atoms but no donor atoms. These are molecules having two or three halogen atoms on the same carbon as a hydrogen atom or one halogen on the same carbon atom and one or more halogen atoms on adjacent carbon atoms, e.g., $CHCl_3$, CH_2Cl_2, CH_3CHCl_2, etc.

CLASS V. All other liquids (i.e., liquids having no hydrogen bond-forming capabilities), e.g., hydrocarbons, carbon disulfide, sulfides, mercaptans, halohydrocarbons not in class IV, nonmetallic elements such as iodine, phosphorus, sulfur, etc.

TABLE 11-3
QUALITATIVE BEHAVIOR OF SOLUTIONS

Classes	Type of deviations	Hydrogen bonding behavior
I + V and II + V	Always + deviations I + V frequently limited solubility	H bonds broken only
III + IV	Always – deviations	H bonds formed only
I + IV and II + IV	Always + deviations I + IV frequently limited solubility	H bonds both broken and formed, but dissociation of class I or II liquid is more important effect
I + I, I + II, I + III, II + II, and II + III	Usually + deviations	H bonds both broken and formed
III + III, III + V, IV + IV, V + V, and IV + V	Always + deviations or ideal	No H bonds involved`

The Regular Solution Theory. By definition a *regular solution* has an ideal entropy of mixing (cf. Table 11-1), although its heat of mixing may be nonzero, and in that sense it may behave nonideally. Its excess entropy of mixing[25] Δs^e is zero, and deviations from ideal-solution behavior are consequently attributed only to the heat of mixing:

$$\Delta g^e = \Delta h$$

The theory of regular solutions was first advanced by Scatchard[26] and later refined and expanded by Hildebrand[27] and is often identified as the Scatchard-Hildebrand theory.

Scatchard regarded the internal energy of a liquid as consisting of two parts: the energy under perfect gas conditions and the *cohesive energy*. The first type is assumed unchanged in the mixing process and therefore does not appear in the expression for the energy of mixing. The cohesive energy u is essentially the interaction energy between molecules and is approximated by the energy of liquefaction. Scatchard further assumed no volume change on mixing and wrote the following expression for the cohesive energy of a binary mixture:

$$-u_{mixture} = \frac{c_{11}v_1^2 x_1^2 + 2c_{12}v_1 v_2 x_1 x_2 + c_{22}v_2^2 x_2^2}{v_1 x_1 + v_2 x_2} \tag{11-46}$$

where the c's are *cohesive energy densities* or the energy per unit volume. For pure liquids c is defined by

$$c_{ii} = -\frac{u_i}{v_i} \tag{11-47}$$

Equation (11-46) is much easier to visualize in terms of the cohesive energy density of the mixture. Thus, dividing both sides by $x_1 v_1 + x_2 v_2$ and transforming to volume fractions, we obtain

$$c_{mixture} = c_{11}\varphi_1^2 + 2c_{12}\varphi_1\varphi_2 + c_{22}\varphi_2^2 \tag{11-48}$$

where

$$\varphi_i = \frac{x_i v_i}{\sum x_i v_i} \tag{11-49}$$

The significance of Eq. (11-48) is easily grasped if the terms φ_1^2, $2\varphi_1\varphi_2$, and φ_2^2 are regarded as the probabilities of 1-1, 1-2, and 2-2 pairs, respectively, or the fraction of the total volume attributable to these interactions.

The energy change on forming 1 mol of mixture is

---

[25] All excess thermodynamic property changes on mixing are defined as is Δg^e, Eq. (10-13).
[26] G. Scatchard, *Chem. Rev., 8*, 321 (1931).
[27] J. H. Hildebrand and R. L. Scott, *Solubility of Nonelectrolytes*, 3rd ed., Van Nostrand Reinhold, New York, 1950.

$$\Delta u = u_{mixture} - x_1 u_1 - x_2 u_2 \tag{11-50}$$

Utilizing Eqs. (11-46) and (11-47) and simplifying, we obtain

$$\Delta u = (x_1 v_1 + x_2 v_2)\varphi_1 \varphi_2 (c_{11} + c_{22} - 2c_{12}) \tag{11-51}$$

By assuming that $c_{12} = \sqrt{c_{11} c_{22}}$, Scatchard then obtained[28]

$$\Delta u = (x_1 v_1 + x_2 v_2)\varphi_1 \varphi_2 (c_{11}^{1/2} - c_{22}^{1/2})^2 \tag{11-52}$$

The square root of the cohesive energy density is termed the solubility parameter and is denoted by δ. Solubility parameters are usually evaluated from the energy of vaporization and the liquid molar volume:

$$\delta_i = \sqrt{\frac{\Delta u_{vi}}{v_i}} \tag{11-53}$$

The Scatchard-Hildebrand energy of mixing is normally written in terms of the solubility parameters:

$$\Delta u = (x_1 v_1 + x_2 v_2)(\delta_1 - \delta_2)^2 \varphi_1 \varphi_2 \tag{11-52a}$$

Because the volume change was assumed zero, Eqs. (11-52) and (11-52a) also express the enthalpy change on mixing.

The excess entropy change on mixing has been assumed zero, a step which Scatchard considered "daring" and remarked that efforts at justification were "more entertaining than convincing." This step, however, allows us to write

$$\Delta g^e = (x_1 v_1 + x_2 v_2)(\delta_1 - \delta_2)^2 \varphi_1 \varphi_2 \tag{11-54}$$

Expressions for the activity coefficients follow from the application of Eqs. (10-14), (11-10), and (11-16):

$$RT \ln \gamma_1 = v_1 (\delta_1 - \delta_2)^2 \varphi_2^2 \tag{11-55a}$$

$$RT \ln \gamma_2 = v_2 (\delta_1 - \delta_2)^2 \varphi_1^2 \tag{11-55b}$$

Extension of the derivation to multicomponent mixtures is straightforward and for the kth component yields

$$RT \ln \gamma_k = v_k \left(\sum_i (\delta_k - \delta_i)\varphi_i \right)^2 \tag{11-56}$$

---

[28] Molecular theory indicates that the geometric mean is appropriate only for molecules with spherically symmetric force fields.

For convenience this expression may be transformed to

$$RT \ln \gamma_k = v_k (\delta_k - \overline{\delta})^2 \tag{11-57}$$

where $\overline{\delta}$ is regarded as the average solubility parameter of the entire mixture and is defined as

$$\overline{\delta} = \sum \delta_i \varphi_i \tag{11-58}$$

    It is apparent that the regular solution theory predicts only endothermic heat effects and hence positive deviations from ideal-solution behavior. These deviations arise from a difference in cohesive energy densities or solubility parameters of the constituents and may be estimated from only pure component properties. We also observe that the volume fraction is the proper means of representing composition in regular solutions.

    The nature of the assumptions and approximations involved in the development of the regular solution theory leads one to expect that its applicability is limited to mixtures composed of molecules with spherically symmetric force fields. While this is the case, nevertheless, when combined with some empiricism, the theory has been successfully applied to systems of more complex molecules and is often useful in providing insight into the behavior of solutions. In their excellent monograph Hildebrand and Scott have used this theory to bring unity and understanding to various types of solubility phenomena.

    The Chao-Seader Correlation. In 1961 Chao and Seader[29] published a correlation based on the regular solution theory and the principle of corresponding states which allowed vapor-liquid equilibrium in light hydrocarbon systems to be predicted from only pure-component parameters. While the work was restricted to a narrow class of compounds and numerous refinements and improvements to it have since appeared in the literature,[30] nevertheless, as the first generalized and unified approach it represented a milestone in the quest of predicting mixture behavior from pure-component parameters.

    The approach was based on correlating the individual terms in the following expression for the equilibrium ratio:[31]

$$K_i \equiv \frac{y_i}{x_i} = \frac{\phi_i^L \gamma_i}{\hat{\phi}_i} \tag{11-59}$$

This equation derives from Eq. (9-63) and the following definition for ϕ_i^L the liquid-phase fugacity coefficient:

---

[29] K. C. Chao and J. D. Seader, *AIChEJ.,* **7**, 598 (1961).
[30] See, for example, J. M. Prausnitz and P. L. Chueh, *Computer Calculations for High-Pressure Vapor-Liquid Equilibria*, Prentice-Hall, Englewood Cliffs, NJ, 1968.
[31] In Sec. 9-7 K_i was defined as

$$f_i^L / f_i^G$$

and was shown to equal y_i/x_i for an ideal solution. The following definition is more general but leads to a K which can be composition dependent.

$$f_i^L = \phi_i^L P \qquad (11\text{-}60)$$

As a pure-component property ϕ_i^L depends only on temperature and pressure. This property was correlated with the acentric factor version of the corresponding states principle (cf. Sec. 3-4), and the result is an analytical, although cumbersome, expression relating ϕ_i^L to reduced temperature and pressure.

The Redlich-Kwong equation of state was used to calculate the vapor-phase fugacity coefficient $\hat{\phi}_i$ utilizing an appropriate form of Eq. (9-61). Because the two parameters in this equation are expressible in terms of critical properties (cf. Sec. 3-2), only these pure component properties are necessary for the determination of $\hat{\phi}_i$. Again, an analytical expression relating $\hat{\phi}_i$ to temperature, pressure, and vapor mol fractions was obtained.

The liquid-phase activity coefficients γ_i were calculated from Eqs. (11-57) and (11-58). The parameters needed to implement this calculation, δ and v, while strictly functions of temperature, were regarded as constants and were tabulated for hydrogen and many familiar hydrocarbons. Thus, an analytical expression relating γ to temperature and liquid-phase composition is available.

The method is suitable for computer implementation as the three components of K_i are expressed analytically. Predictions based on this method are in reasonable agreement with experimental results for light hydrocarbon systems.

UNIFAC. For a long time the idea that solution behavior is expressible not by inter-actions between molecules but by interactions between the functional groups comprising the molecules has been appealing. This concept has recently come to fruition in the development of the UNIFAC method for organizing and extending experimentally gained information about solution behavior in the form of activity coefficients.[32] In this method the activity coefficient is expressible in terms of two types of contributions, combinatorial and residual:

$$\ln \gamma_i \;=\; \underset{\text{combinatorial}}{\ln \gamma_i^c} \;+\; \underset{\text{residual}}{\ln \gamma_i^R}$$

The combinatorial part is determined by the size and shape of the molecules, while the residual part arises from their interaction energies. This separation of causative factors led to the development of the UNIQUAC equation[33] which can successfully represent activity coefficients in a wide range of solution environments.

The UNIQUAC prescription for $\ln \gamma_i^c$ is used for UNIFAC:

$$\ln \gamma_i^c = \ln \frac{\phi_i}{x_i} + \frac{z}{2} q_i \ln \frac{\theta_i}{\phi_i} + l_i - \frac{\phi_i}{x_i} \sum_j x_j l_j \qquad (11\text{-}61)$$

[32] A. Fredenslund, R. L. Jones, and J. M. Prausnitz, *AIChEJ.*, *21*, 1086 (1975); A. Fredenslund, J. Gmehling, and P. Rasmussen, *Vapor-Liquid Equilibria Using UNIFAC*, Elsevier, Amsterdam, 1977.

[33] D. S. Abrams and J. M. Prausnitz, *AIChEJ.*, *21*, 116 (1975). UNIQUAC stands for UNIversal QUAsi Chemical, and UNIFAC stands for UNIquac Functional-Group Activity Coefficients.

where

$$l_i = \frac{z}{2}(r_i - q_i) - (r_i - 1) \qquad \theta_i = \frac{q_i x_i}{\sum_j q_j x_j} \qquad \phi_i = \frac{r_i x_i}{\sum_j r_j x_j}$$

The molecular volume parameter r_i and area parameter q_i are determined by summing group contributions:

$$r_i = \sum_k v_k^i R_k \qquad \text{and} \qquad q_i = \sum_k v_k^i Q_k$$

where v_k^i is the number of groups of type k in molecule i and R_k and Q_k are the volume and area parameters for that group. Group parameters R_k and Q_k are evaluated from atomic and molecular structure data.

The functionality of the UNIFAC residual contribution is similar to that of UNIQUAC except that it is expressed in terms of the group area parameters Q_k and binary group interaction energy parameters a_{kj}. The residual contribution contains the temperature dependency.

The approach can be demonstrated with the ethanol–hexane system. Expressed in terms of their constituent groups, these compounds are

Ethanol: CH_3–CH_2–OH
Hexane: CH_3–CH_2–CH_2–CH_2–CH_2–CH_3

Thus, in this binary system there are three types of groups, CH_3, CH_2, and OH—the same as would be found in any multicomponent aliphatic alcohol-aliphatic hydrocarbon system or aliphatic alcohol-aliphatic alcohol system.

UNIFAC is a data reduction and organization principle in which experimental data are used to determine the binary parameter a_{kj}, which can in turn be used to estimate activity coefficients for systems which have not been studied experimentally but which contain the same functional groups. The beauty of the method derives from the dramatic reduction in the number of systems to be treated. Instead of all possible mixtures of compounds, it is only necessary to consider the much smaller number of constituent groups and their interactions.[34]

The original publication of the UNIFAC method was accompanied by R_k and Q_k for 18 functional groups and the interaction energy parameters a_{kj} for a large fraction of the possible $18^2/2$ pairs of groups. Additional parameters are determined as the necessary equilibrium data become available and have been published periodically. Presently the number of groups stands at 40 with interaction energy parameters available for approximately half of the pos-

---

[34] While this may be somewhat reminiscent of the simplification achieved through the use of heats of formation (cf. Sec. 6-7), it must be remembered that unlike that thermodynamically rigorous data reduction procedure, UNIFAC is based on a particular theory and therefore has no general thermodynamic validity. Nevertheless, for estimates which will usually satisfy the requirements of engineering accuracy, the method finds considerable use.

sible group pairs.[35] Because of its universality, UNIFAC is widely used in data bases and process simulation programs.

Calculations using UNIFAC can be executed with the program UNIFAC.EXE on the CD-ROM accompanying this text. The UNIFAC parameters are contained in the files UNIFAC.RQ and UNIFAC.AIJ and the user need only identify the molecules and the groups (by number) and supply the v_k^i's. The user must therefore parse the molecules into their constituent groups. Toward this end the program displays a table listing the groups, their numbers, and an example parsed molecule for each group. The table can be re-displayed as often as necessary. This table can be printed from the file UNIFAC.TBL if desired for study before running the program. Example 11-7 illustrates the parsing procedure.

EXAMPLE 11-7

Use the UNIFAC method and the program UNIFAC.EXE to estimate activity coefficients in the *n*-hexane–chlorobenzene system at 65°C. Use these activity coefficients to calculate VLE data for the system and compare with experimental values.

Solution 11-7

The first step in applying the UNIFAC method is to parse the molecules into their constituent groups. A tabulation of these groups can be viewed as often as necessary by the user of the program. As shown above, *n*-hexane contains 2 CH_3 groups (group #1) and 4 CH_2 groups (group #2). The table shows that chlorobenzene contains 5 aromatic CH groups (group #10) and one aromatic CCl group (group #54). After giving the user unlimited access to the table, the program asks for the number of different groups in the system (in this case, 4). The program then asks for the group numbers[36] (in this case, 1, 2, 10, and 54) followed by requests for the number of each group in each molecule (in this case, 2, 4, 0, 0 for *n*-hexane and 0, 0, 5, 1 for chlorobenzene). The user then chooses calculation option 3 and enters the experimental values of *x*. The results obtained from UNIFAC.EXE are shown in Table 11-4 along with calculated and experimental values of y_1. The procedure for the calculation of y_1 is illustrated in Ex. 9-8.

Examination of Table 11-4 shows that good agreement was obtained between experimental and calculated y_1's; this is underscored by the relatively low value of 0.0059 obtained for $\Delta \bar{y}$.

---

[35] J. Gmehling, P. Rasmussen, and A. Fredenslund, *Ind. Eng. Chem. Process Des. Dev.*, **21**, 188 (1982).
[36] Group numbers need not be entered in numerical order.

TABLE 11-4

COMPARISON OF EXPERIMENTAL VLE DATA WITH CALCULATED DATA
USING UNIFAC[a,b]

| x_1 | UNIFAC Results | | | Experimental |
	γ_1	γ_2	y_1	y_1
0.083	1.667	1.004	0.556	0.544
0.144	1.558	1.013	0.683	0.679
0.201	1.469	1.025	0.750	0.744
0.284	1.358	1.052	0.810	0.803
0.394	1.240	1.101	0.859	0.852
0.438	1.202	1.127	0.874	0.866
0.485	1.165	1.157	0.888	0.882
0.540	1.128	1.196	0.902	0.896
0.591	1.099	1.238	0.915	0.910
0.679	1.059	1.321	0.934	0.929
0.806	1.021	1.469	0.960	0.957
0.927	1.003	1.645	0.985	0.984

^{a} System: n-hexane (1)–chlorobenzene (2) at 65°C.
^{b} Source: reference B2 Table 10-8.

PROBLEMS

11-1. In some binary systems activity coefficients defined by Eq. (9-35) can be represented by the following simple expressions:

$$\ln \gamma_1 = bx_2^2 \qquad \ln \gamma_2 = bx_1^2$$

Assuming that these expressions are valid, derive expressions relating the activity coefficient based on Henry's law, γ^*, to composition.

11-2. Calculate ($\overline{H}^\infty - h^G$) for HCl using the data from Fig. 6-5. Also calculate this quantity from the following solubility data for HCl in water (J. H. Perry, *Chemical Engineers' Handbook*, 3rd ed., McGraw-Hill, New York, 1950, p. 167):

Weight of HCl,	p_{HCl} (mmHg)	
100 weights of water	20°C	30°C
2.04	0.000044	0.000151
4.17	0.00024	0.00077
8.70	0.00178	0.00515
13.64	0.00880	0.0234
19.05	0.0428	0.106
25.0	0.205	0.48
31.6	1.00	2.17
38.9	4.90	9.90
47.0	23.5	44.5
56.3	105.5	188
66.7	399	627

Note: To obtain an accurate value for k, plot on an expanded scale near $x_{HCl} \to 0$. Also use these data to

(a) calculate γ^*_{HCl} for the various values of x_{HCl} given. Make a plot of $\ln \gamma^*_{HCl}$ vs. x_{HCl}.
(b) test the suitability of the Margules or van Laar equation for representing γ^*_{HCl}.
(c) calculate p_{water} for various values of x_{HCl}.

11-3. In studying the HCl–H_2O system, we have determined Henry's law constants and $\Delta \overline{H}^\infty$ for HCl in the temperature range 20–30°C. Use this information to estimate vapor mol fractions and total pressure for the system containing only HCl and water at 40°C when $x_{HCl} = 0.10$. List any assumptions or approximations that you consider affected your answer. The following data are given:

Henry's law constant for HCl at 20°C: $k = 0.00334$ mmHg.
Henry's law constant for HCl at 30°C: $k = 0.00957$ mmHg.

$(\overline{H}^\infty - h^G) = -17{,}850$ cal/g mol.

Parameter in the equation $\ln \gamma^*_{HCl} = A(1 - x_w^2)$:

$$A = 30.5 \text{ at } 20°C; A = 29.0 \text{ at } 30°C$$

Vapor pressure of water at 40°C = 55.3 mmHg.

11-4. We have studied the HCl–H_2O system and have determined the Henry's law constant and activity coefficients at 20°C. Use this information to determine whether an azeotrope is possible at this temperature. Henry's law constant at 20°C is 0.00334 mmHg. Activity coefficients for HCl can be represented by

$$\ln \gamma^*_{HCl} = 30.5(1 - x_w^2)$$

The vapor pressure of water at 20°C is 17.5 mmHg.

11-5. Because of dissociation, Henry's law is not obeyed by HCl in aqueous solution, and the limiting slope of a partial pressure vs. liquid mol fraction curve approaches zero. (See Sec. 12-11.) This state of affairs has been ignored in the formulation of Probs. (11-2)–(11-4), and it is desired to determine whether such an application of the methods of thermodynamics is legitimate. Note that in plotting the partial pressure data supplied in Prob. 11-2 it is possible to obtain a nonzero limiting slope even though the curve is not linear in this region. Present a reasoned argument to show whether the methods of thermodynamics can be applied properly to this system if dissociation is ignored.

11-6. Prove the following pressure dependence of the Henry's law constant:

$$\left(\frac{\partial \ln k_2}{\partial P} \right)_T = \frac{\overline{V}_2^{\infty}}{RT}$$

where $\overline{V}_2^{\infty}$ is the partial molar volume of the solute at infinite dilution.

11-7. The quantity $(\partial \mu_1 / \partial x_1)_{T, P}$ can be obtained from measurements performed on a liquid solution subjected to a centrifugal field. Show how we could use information in the form of $(\partial \mu_1 / \partial x_1)_{T, P}$ vs. x_1 to calculate vapor-liquid equilibrium data for a binary system at the temperature of the measurements.

11-8. It has been proposed to measure differential heats of mixing for the lithium bromide–calcium chloride–water system at 298 K. These differential heats of mixing will be determined by adding a small quantity of solute or water to a large quantity of solution and measuring the heat effect. Because the mol ratio of lithium bromide to calcium chloride will be kept constant at 2 to 1, we would like to treat the solute as a single component and therefore consider this to be a pseudo-binary system. We would also like to base the results on a kilogram of solution and use weight fractions instead of mol fractions in presenting the data. Will such data be thermodynamically significant and could they be tested for thermodynamic consistency? Specifically, is it permissible to use mass units instead of mol units and to lump the two solutes as a single solute? Could such data be used to determine integral heats of solution? It is visualized that we would acquire differential heats of mixing for water and the pseudo-solute at various solution compositions.

11-9. The following vapor-liquid equilibrium data have been obtained for the A–B system at temperatures of 25 and 50°C

	25°C		50°C	
x_A	y_A	P (mmHg)	y_A	P(mmHg)
0	0	450	0	1230
0.2	0.364	608.6	0.358	1634
0.4	0.426	627.9	0.440	1701
0.6	0.449	624.3	0.483	1692
0.8	0.514	585.9	0.567	1580
1.0	1.0	350	1.0	1050

Use these data to prepare plots of:

(a) the Gibbs free energy change on mixing vs. x_1.
(b) the enthalpy change on mixing vs. x_1.
(c) the entropy change on mixing vs. x_1.

11-10. At 322 K the following mixing properties have been reported for a binary system at a liquid mol fraction of 0.5

$$\Delta h = -920 \text{ cal / g mol}; \quad \Delta s = -2.132 \text{ cal / g mol} \cdot K$$

Is $\Delta s < 0$ possible for a solution formed at constant T and P? Should the data be questioned?

11-11. Show how the Pxy data listed in Prob. 14-4 and the heat of mixing data of Fig. 6-4 could be used to estimate Txy data for the ethanol–water system at 760 mmHg. To demonstrate the method, perform a sample calculation for $x = 0.5$.

11-12. Partial pressures of water over a solution containing 70 mol % water and 30 mol % glycerol have been measured at 47.5 and 100.3°C and are 50 and 500 mmHg respectively. At these temperatures the vapor pressures of water are 81.6 and 768.2 mmHg respectively. Make the best possible estimate of the partial pressure of water over this solution at a temperature of 25°C where the vapor pressure of water is 23.8 mmHg. Provide thermodynamic justification for your result.

11-13. The partial molar heat of mixing of water in a solution of water and a nonvolatile substance has been determined from the temperature dependence of water partial pressures. These data can be represented by the simple algebraic expression

$$\Delta \overline{H}_w = b(1 - x_w)^2$$

Is it possible to determine the heat of mixing, Δh, from this information? If so, determine the expression for Δh.

11-14.   In Sec. 10-4 it is stated that an azeotrope occurs at a maximum or minimum in the total pressure curve or in the temperature vs. composition curve. Generalize this statement by showing that for a two-phase binary system the condition

$$\left(\frac{\partial P}{\partial x}\right)_T = 0$$

or

$$\left(\frac{\partial T}{\partial x}\right)_P = 0$$

occurs when $x = y$. *Hint*: Start by writing the conditions of equilibrium for each component and the Gibbs-Duhem equation for each phase.

11-15.   We have defined activity coefficients by the following equation:

$$\hat{f}_i^L = f_i^L x_i \gamma_i$$

where f_i^L is the fugacity of pure liquid i at the T and P of the system—the equilibrium T and P for the two-phase system. We may also define an activity coefficient γ_i' by

$$\hat{f}_i^L = f_i^{\circ} x_i \gamma_i'$$

where f_i° is the fugacity of pure liquid i at the temperature of the system and a reference pressure P_0. Normally, the effect of pressure on activity coefficients is ignored. However, if we are considering constant-temperature, vapor-liquid equilibrium, there can sometimes be a large pressure variation with liquid composition. We desire to correct these activity coefficients for the effect of pressure and will do so by adjusting them to a reference pressure. Derive equations to be used for correcting γ_i and γ_i' to the reference pressure. Can this reference pressure be different from that which defines f_i°? Identify and discuss any similarities or differences in these equations.

11-16.   It is desired to extrapolate some data for solubility of liquid water in a liquid hydrocarbon, and thermodynamic guidance is sought. Making reasonable assumptions, derive a thermodynamically realistic equation, giving the temperature dependence of the solubility of one liquid in another in the range where the solubility is very low.

11-17.   The following method of determining activity coefficients in binary systems has been proposed [G. Arich, I. Kikic, and P. Alessi, *Chem. Eng. Sci.*, *30*, 187 (1975)]: Binary liquid solutions of components 1 and 2 are equilibrated with a solvent in which the components are only slightly soluble and which dissolves only slightly in the individual components or the binary solution. The mol fractions in the binary solution x_1 and x_2 are known from its preparation, and the ratio of mol fractions of 1 and 2 in the solvent phase x_1^s/x_2^s is measured. Also the solubility of the pure components in the solvent x_1^{so} and x_2^{so} is measured. By performing these measurements for several binary solution concentrations at a constant temperature, it is claimed that the excess free energy of mixing Δg^e can be obtained from

$$\Delta g^e = RT \int_0^{x_1} \ln \left[\left(\frac{x_2^{so}}{x_1^{so}} \right) \left(\frac{x_1^s}{x_2^s} \right) \left(\frac{x_2}{x_1} \right) \right] dx_1$$

Verify this relation for Δg^e, and identify all necessary assumptions. Also show how activity coefficients may be obtained.

11-18. Reported solubility data for nitrogen in liquid ammonia at 0°C are listed below [R. Wiebe and V. L. Gaddy, *J. Am. Chem. Soc., 59,* 1984 (1937)]. Use these data to estimate the solubility of nitrogen in liquid ammonia at 0°C and 1000 atm, and compare with the reported value of $s = 29.69$.

P (atm)	100	200	400	600
s (cc of N_2(STP)/g NH_3)	7.90	13.73	20.76	24.95

11-19. Compounds A and B are believed to form a simple eutectic system (no solid solutions), and the solid-liquid phase diagram has been calculated on the basis of ideal-solution behavior. Someone has pointed out, however, that this system is known to form minimum-boiling azeotropes (the system exhibits a maximum in the isothermal total pressure vs. liquid-composition curve), and therefore the solid-liquid phase diagram calculated on the assumption of ideal-solution behavior would be in error. Do you agree? If so, would you expect this calculated eutectic temperature to be too high or too low? Justify your answers.

11-20. *p*-Dichlorobenzene and *p*-dibromobenzene are solids at 50°C. The sublimation pressures at this temperature are

$C_6H_4Cl_2$: 7.48 mmHg
$C_6H_4Br_2$: 0.56 mmHg

The two solids are intimately mixed in various ratios, and the equilibrium vapor pressures and vapor-phase compositions are determined. From the following data, determine

   (a) the nature of the solid phase (i.e., are the compounds immiscible, partially miscible, or completely miscible in the solid phase?).
   (b) whether the data can be tested for thermodynamic consistency.
   (c) if the answer to part (b) is yes, test the thermodynamic consistency of these data.

The following data were obtained by charging various known mixtures of the two compounds intimately mixed into an equilibrium vessel. The vapor space is small so that it is assumed that the overall solid composition is the same as charged. The total pressure over the solid is measured, and the vapor phase is analyzed.

Mols of $C_6H_4Cl_2$ / mol of $C_6H_4Br_2$	Mol fraction $C_6H_4Cl_2$ in vapor	Total pressure (mmHg)
0.055	0.518	1.10
0.247	0.823	2.55
0.699	0.919	4.27
1.000	0.936	4.81
1.25	0.954	5.40
2.22	0.962	5.82
9.63	0.987	6.93

11-21.   Use the data of Prob. 11-20 and the following data to estimate the solid-liquid phase diagram for that system.

	melting point	heat of melting
p-dichlorobenzene	325.7 K	4366 cal/g mol
p-dibromobenzene	360 K	4836 cal/g mol

List the assumptions necessary for your estimate and comment on their soundness.

11-22.   Substances A and B have melting points of 400 K and 390 K respectively and are believed to form a eutectic system with no solid solutions. At a temperature of 370 K it is known that pure solid A is in equilibrium with a liquid solution containing 60 mol % A and 40 mol % B. Use this information to estimate the mol % of B in a liquid solution in equilibrium with pure solid B at 370 K. The heats of melting of A and B are respectively 3000 and 2800 cal/gmol.

11-23.   For vapor-liquid equilibrium in the A–B system at 420 K, use any information from Prob. 11-22 to calculate the vapor mol fraction of A and the total pressure when the liquid mol fraction of A is 0.500. The vapor pressures of A and B at 420 K are 10 and 8 mmHg respectively.

11-24.   At 20°C the solubility of solid naphthalene in carbon tetrachloride is 0.205 mols per mol of solution. Estimate the solubility of naphthalene in this solvent at 50°C. Naphthalene data can be found in Ex. 11-4. The freezing point of carbon tetrachloride is –23°C.

11-25.   Estimate the freezing point of an aqueous solution of 10 mol % HCl. Use any information provided in Prob. 11-3. The heat of fusion of water is 143 Btu/lb at its freezing point. Take the solid phase to be pure water.

11-26.   From glycerol–water solutions both components crystallize as pure solids and hence a simple eutectic system is observed. Use the eutectic point of $x_G = 0.280$ and $T = 228.7$ K along with the following pure component data to calculate the solid-liquid phase diagram for this system and compare with the experimental freezing points.

	T_m K	ΔH_m cal/g mol
water	273.2	1436
glycerol	291.2	4421

FREEZING POINTS OF GLYCEROL–WATER SOLUTIONS

x_G	T (K)
0.0213	271.6
0.0466	268.4
0.0774	263.7
0.1154	257.7
0.1646	251.2
0.2269	239.6
0.2800	228.7
0.3134	235.4
0.4390	254.0
0.6378	271.6

11-27. For the solubility of solid benzene in liquid propane it is observed that a plot of $\ln x_B$ vs. $1/T$ is linear in the region where x_B approaches zero. This is attributed to conformance to Henry's law. Is such linearity expected? Is it possible to obtain a heat effect from the data listed below? If so, identify and determine it. No solid solutions are formed.

x_B	0.0543	0.0274	0.0128	0.00736	0.00548
Temperature (K)	205	190	175	165	160

Data from W. L. Chen, K. D. Luks, and J. P. Kohn, *J. Chem. Eng. Data, 26*, 310 (1981).

11-28. Methane freezes at 90.5 K and has a heat of fusion of 232 cal/g mol. Estimate the solubility of methane in liquid nitrogen at 78 K assuming an ideal solution is formed. What assumptions did you make in your calculations?

11-29. The solubilities of stannic iodide in carbon disulfide at several temperatures are given below. Estimate the melting point and heat of fusion of stannic iodide in cal/g mol.

t (°C)	10.0	25.0	40.0
s (g/100 g solution)	49.01	58.53	67.56

11-30.   Bronsted measured the solubility of rhombic and monoclinic sulfur at 25.3°C in various solvents (benzene, diethyl ether, ethyl bromide, ethanol). In all cases the solubility of monoclinic sulfur was 1.28 times that of rhombic sulfur. In these solvents sulfur exists exclusively as S_8 molecules, as has been proved, for example, by the freezing point depression. Find the free energy change in cal/g mol when monoclinic S is transformed into rhombic S, and deduce which form is the more stable at 25.3°C.

11-31.   When crystallization is employed to separate components in a binary mixture where no solid solubility occurs, one obtains pure crystals of component A and a liquid solution close to the eutectic composition. The question has arisen as to whether the eutectic composition might be shifted by increasing the pressure, thereby allowing pure crystals of A and pure crystals of B to be produced in a two-step process (one step at atmospheric pressure and the other step at an elevated pressure). Outline in detail how you would calculate the shift in eutectic composition with pressure. Derive all necessary equations, list all assumptions, and specify the physical property data necessary to carry out the calculations. Assume the liquid is an ideal solution.

11-32.   Two liquids, A and B, are immiscible. These liquids are allowed to equilibrate with a solid (component 1) so that at equilibrium there are four phases: a vapor, pure solid 1, a saturated solution of 1 in A, and a saturated solution of 1 in B. The system was studied at 0 and 20°C and the following saturated mol fractions of component 1 were found

t (°C)	A-rich solution	B-rich solution
0	0.0221	0.0173
20	0.0287	0.0251

Assume that the solutions are dilute enough that Henry's law is followed and calculate the resulting solution concentrations when 1 mol of A, 1 mol of B, and 0.02 mol of component 1 come to equilibrium at 40°C and a pressure high enough to suppress the vapor phase. The melting point of component 1 is 80°C.

11-33.   Show that the knowledge of the eutectic point for nonideal solid-liquid equilibria (no solid solutions) is analogous to knowledge of an azeotrope composition for a nonideal vapor-liquid system. Write the necessary equations, and describe step-by-step how you would use the eutectic point to calculate the solid-liquid diagram. List any major assumptions or approximations required.

11-34.   Dopants often have extremely low solubility in silicon. Show that the following data for the solubility of copper in silicon can be well fitted to a straight line when log of concentration is plotted versus reciprocal of absolute temperature. Provide thermodynamic justification for this correlation.

t (°C)	640	720	820	890	1060	1170
x (atoms/cm^3)	$3(10^{15})$	10^{16}	$5(10^{16})$	10^{17}	$5(10^{17})$	10^{18}

11-35.   Compare and contrast the equations for liquid-solid solution equilibrium with those for constant-pressure vapor-liquid equilibrium.

11-36. Estimate the eutectic temperature and composition in the hexane–naphthalene system. Use any information from Ex. 11-4. For hexane $T_m = 179$ K and $L_m = 3156$ cal/g mol.

11-37. Figure 11-11 shows a solid-liquid phase diagram with a minimum congruent melting point. This diagram resembles that of a minimum-boiling azeotrope which is characteristic of systems exhibiting positive deviations from ideal-solution behavior. Is there a similar generalization which can be made for the solid-liquid system? Maximum-boiling azeotropes are much less common than minimum-boiling azeotropes. Would you expect maximum congruent melting points likewise to be rarer than minimum congruent melting points? Explain.

11-38. Use any information from Ex. 11-5 and the following data to calculate the solid-liquid phase diagram for the aluminum-silicon system.[37]

	T_m (K)	L_m (kJ/kmol)	c_P^L (kJ/kmol $\cdot$ K)	c_P^S (kJ/kmol $\cdot$ K)
Al	933	10,790	31.96	33.72
Si	1685	50,550	25.25	28.33

11-39. The solubility of pyridine hydrobromide in chloroform is reported to be 18 weight % at 0°C and 10 weight % at 60°C. Apparently no solid solutions are formed. Is this behavior thermodynamically consistent? Explain.

11-40. For the system carbon dioxide–ethylene at –80°F, use the programs PREOS.EXE and PRVLE.EXE to determine the pressure and liquid and vapor compositions when equilibrium exists between vapor, liquid, and pure solid carbon dioxide. Consult *Perry's Handbook* for the necessary data.

11-41. Use the UNIFAC method and the program UNIFAC.EXE to estimate VLE data for the following systems and compare with the experimental results.

    (a)   ethanol–benzene at 40°C (Ex. 9-8)
    (b)   methanol–carbon tetrachloride at 20°C (Ex. 9-9)
    (c)   hexane–chlorobenzene at 65°C (Ex. 10-2)
    (d)   heptane–toluene at 25°C (Ex. 10-6)
    (e)   cyclohexane–heptane–toluene at 25°C (Ex. 10-6)

11-42. Activity coefficients for the system *n*-heptane–acetonitrile estimated at 25°C by the UNIFAC method are quite large and suggest partial liquid miscibility. Use the program UNIFAC.EXE to estimate activity coefficients for this system at higher temperatures and estimate the temperature above which there is complete miscibility.

---

[37] Suggested by Prof. T. J. Anderson, University of Florida.

11-43. A convenient way of removing the vapor of a condensable substance from a noncondensable gas is by compression followed by cooling. If the condensable substance forms a solid phase, we can expect the vapor mol fraction of the condensable component to show a minimum with respect to pressure at constant temperature. This would be the optimal pressure for the purification process. Use the virial equation, Eq. (3-3), truncated after the second term to determine $\hat{\phi}_1$ Eq. (9-26), and show that at the minimum

$$\ln y_1 = 1 + \ln\left(\frac{-2B_{12}P^{\text{sat}}}{RT}\right)$$

$$P = -\frac{RT}{2B_{12}}\left(1 - \frac{B_{22}}{2B_{12}}\right)$$

Note that B_{12} and B_{22} have negative values.

11-44. Show that when the fugacity coefficient in Eq. 11-45 is determined by the truncated virial equation (Eqs. 9-28) and a few simplifying assumptions are made, a linear relationship between $\ln y_1 P$ vs. P should result. Test this relationship with the following data for the phenanthrene(1)–ethylene(2) system at 70°C.[38] and comment on the results.

P (atm)	$y_1 \times 10^3$
69.0	0.0531
96.2	0.207
123.4	0.714
150.7	1.73
177.9	2.92
205.1	4.18
232.3	6.42
273.1	8.90

11-45. The isothermal solubility of a solid in a gas phase varies widely with pressure and sometimes exhibits extrema. At temperatures above the critical temperature of the gas, minima have been observed and experimental measurements suggest that a maximum may occur in the superfluid region (very high pressures).

Derive a rigorous thermodynamic equation relating mol fraction of the solid solute in the fluid phase to pressure at a constant temperature and indicate the conditions under which extrema are possible. Ignore any solubility of the gas in the solid.

---

[38] Data of K. P. Johnston and C. A. Eckert, *AIChEJ.*, 27, 773 (1981).

11-46. A gas stream of 1% CO_2 and 99% N_2 at 100 atm is to be cooled at constant pressure to a temperature of 178 K, and someone has raised the concern that solid CO_2 may condense and eventually plug the heat-exchanger tubes. You are to assess the possibility of this occurrence. For this system, in the PR equation $k_{12} = -0.02$.

11-47. Show that the van Laar equations can be regarded as empericized versions of Eqs. (11-55) of the regular solution theory.

11-48. Use the UNIFAC method to estimate the solid-liquid phase diagram for the phenol–benzene system. Assume that the solid phases are pure and use the following data.

Component	t_m °C	L_m cal/g
benzene	5.5	30.1
phenol	42	29.0

Compare your results with the following experimental data[39]

m.f. Phenol	t °C	
0.2	−1	
0.35	−7	eutectic
0.4	−3	
0.6	12	
0.8	28	

---

[39] As estimated from graph in J. G. Gmehling, T. F. Anderson, and J. M. Prausnitz, *Ind. Eng. Chem. Fundam., 17,* 269 (1978).

Chemical Equilibrium

The equilibrium state has been defined in terms of time invariance and reproducibility, and up to now we have studied systems which reach equilibrium within a reasonable time span. With such systems there is no problem in defining an equilibrium state. The same cannot always be said of equilibrium with respect to chemical transformations. Rates of chemical reaction are strongly dependent on temperature but may be extremely slow or even imperceptible until high temperatures are reached. A well-known example of this is a mixture of hydrogen and oxygen gas. We know that a spark or an appropriate catalyst will cause a violent reaction, but we also know that in the absence of a catalyst such a mixture shows no evidence of reaction even when heated to moderately high temperatures. We can only say that a gaseous mixture of hydrogen and oxygen has a tendency to move toward an equilibrium state[1] where a certain fraction of the reactants has been converted to water. While the uncatalyzed system of hydrogen and oxygen would be

---

[1] Recall that the movement of a system toward an equilibrium state is always spontaneous.

reproducible and appear to be time-invariant, it would not be in a state of ultimate equilibrium but could be conceived as moving toward that equilibrium state at an imperceptibly slow rate.[2] Thus, from a practical standpoint, unless we are dealing with reactions at elevated temperatures, we are seldom interested in the state of ultimate equilibrium. Rather we are concerned about specific reactions, which by virtue of temperature level or catalysis, proceed at reasonable rates and can be observed to approach an equilibrium state which is for practical purposes reproducible and time-invariant. This may be referred to as a restricted equilibrium state.

To illustrate the concept of a restricted equilibrium state, consider the several possible reactions involving hydrogen and carbon monoxide:

$$CO + 3H_2 = CH_4 + H_2O \quad \text{methanation}$$
$$CO + 2H_2 = CH_3OH \quad \text{methanol synthesis}$$
$$xCO + yH_2 = \text{hydrocarbons} \quad \text{Fischer-Tropsch synthesis}$$
$$x'CO + y'H_2 = C,\ CO_2,\ H_2O,\ CH_4,\ \text{etc.} \quad \text{ultimate equilibrium}$$

With the proper selection of temperature, pressure, and catalyst it is possible to selectively promote either of the first three reactions and obtain the desired products. However, we must also recognize that other reactions with undesirable products such as H_2O, CO_2, and solid carbon are possible and that in a state of ultimate equilibrium these species, because of their great stability, will predominate. Thus, while a state of restricted equilibrium may be established with respect to either of the first three reactions, it should be recognized that the system is not in a state of ultimate equilibrium. In fact, from the standpoint of obtaining products of economic value, a state of ultimate equilibrium is usually undesirable.

The reason that a single reaction out of many possible may be promoted is that an effective catalyst accelerates tremendously the rate of the desired reaction. Because the rates of the competing reactions are extremely slow, a condition of restricted equilibrium with respect to the desired reaction may be approached. Thus, it is appropriate to consider equilibrium with respect to a specific chemical reaction.

A catalyst is a substance which may participate in an intermediate step of a reaction mechanism but is regenerated and therefore is not consumed in the reaction. While the nature of the chemical species present in the equilibrium system is seen to depend on the choice of catalyst, the catalyst affects only the rate and not the equilibrium state of the promoted reaction. Thus, two different methanol synthesis catalyst formulations may produce different reaction rates, but neither can alter the maximum conversion to methanol. Therefore, for chemical equilibrium calculations the identity of the catalyst is not required as long as we know the reaction which is promoted. While the identity of the catalyst is unnecessary for chemical equilibrium calculations, the vital role played by catalysis in the petrochemical and organic chemical industry should not be ignored. Many industrially important chemicals

---

[2] Although we have used time invariance as a definition of equilibrium, it must be recognized that time is not a thermodynamic variable and that thermodynamics has nothing to say about rates or the time required to reach equilibrium.

are produced via catalytic reactions because the many possible uncatalyzed reactions would lead to the undesirable state of ultimate equilibrium with products of little value.

The focus of this chapter will be on specific chemical reactions. Emphasis will be placed on determining the extent of reaction at equilibrium and the effect of temperature, pressure, and reactant proportions on equilibrium conversion. Often the design of reactor systems depends strongly on equilibrium considerations. While equilibrium may be closely approached in only a few reactors, a knowledge of this conversion is valuable because it represents the best that can be expected.

12-1 GENERALIZED STOICHIOMETRY

Because we desire thermodynamic results of maximum generality, it will be convenient to generalize the stoichiometry of a reaction. The reaction equation may be written as

$$\sum v_i A_i = 0 \qquad (12\text{-}1)$$

where the A_i's represent chemical species and the v_i's are the stoichiometric coefficients which are taken positive for products and negative for reactants. The change in the number of mols of each species due to the chemical reaction is related by

$$\frac{dn_1}{v_1} = \frac{dn_2}{v_2} = \ldots = \frac{dn_i}{v_i} = d\alpha \qquad (12\text{-}2)$$

All the ratios are equal and are equated to a change in α, which will be called the *extent of reaction*. Thus, a single variable α can be used to express the extent of reaction, and stoichiometric calculations can be performed systematically. As an example, consider the methanation reaction

$$CO + 3H_2 = CH_4 + H_2O$$

We have

$$
\begin{aligned}
A_1 &= CO & v_1 &= -1 \\
A_2 &= H_2 & v_2 &= -3 \\
A_3 &= CH_4 & v_3 &= 1 \\
A_4 &= H_2O & v_4 &= 1
\end{aligned}
$$

From Eq. (12-2) we may write

$$\frac{dn_1}{-1} = \frac{dn_2}{-3} = \frac{dn_3}{1} = \frac{dn_4}{1} = d\alpha$$

The number of mols of each species present after any extent of reaction can be determined by integration of the four differential equations with the appropriate boundary conditions:

$$\int_{n_{01}}^{n_1} dn_1 = -\int_0^{\alpha} d\alpha$$

$$\int_{n_{02}}^{n_2} dn_2 = -3\int_0^{\alpha} d\alpha$$

$$\int_{n_{03}}^{n_3} dn_3 = \int_0^{\alpha} d\alpha$$

$$\int_{n_{04}}^{n_4} dn_4 = \int_0^{\alpha} d\alpha$$

The resulting expressions are

$$n_1 = n_{01} - \alpha$$

$$n_2 = n_{02} - 3\alpha$$

$$n_3 = n_{03} + \alpha$$

$$n_4 = n_{04} + \alpha$$

These equations express the number of mols of a species in the reacting mixture in terms of the number of mols of the species originally present and the extent of reaction. If desired, mol fractions of each species could be calculated and expressed in terms of the n_o's and α. From the preceding example it is seen that the number of mols of any species present at any extent of reaction can be generalized to

$$n_i = n_{0i} + v_i \alpha \tag{12-3}$$

Examination of Eq. (12-3) shows the product $v_i \alpha$ to have the units of mols. If, as is customary, the v_i's are considered dimensionless, then α will have the units of mols and can be considered the *mols of reaction*. Note that in general α can be greater than unity but could be normalized to a fractional extent of reaction, or conversion, when Eq. (12-3) is divided by $n_0 / |v|$ of the limiting reactant.[3]

---

[3] The limiting reactant has the smallest value of $n_{0i} / |v_i|$.

12-2 THE CONDITION OF EQUILIBRIUM FOR A CHEMICAL REACTION

For a process involving a composition change and occurring in a closed system, we apply the fundamental equation

$$dG = V\,dP - S\,dT + \sum \hat{\mu}_i\,dn_i \qquad (7\text{-}20)$$

When the composition change is the result of a single chemical reaction such as represented by Eq. (12-1), we may write

$$dn_i = v_i\,d\alpha \qquad (12\text{-}4)$$

Substitution of Eq. (12-4) into Eq. (7-20) yields

$$dG = V\,dP - S\,dT + \left(\sum v_i \hat{\mu}_i\right) d\alpha \qquad (12\text{-}5)$$

Inspection of this differential equation reveals that

$$\sum v_i \hat{\mu}_i = \left(\frac{\partial G}{\partial \alpha}\right)_{T,P}$$

In Sec. 7-3 we saw that the condition $dG_{T,P} = 0$ applies to variations constrained to constant temperature and pressure which originate from an equilibrium state. Applying this condition to Eq. (12-5) results in

$$\left(\frac{\partial G}{\partial \alpha}\right)_{T,P} = 0$$

or

$$\sum v_i \hat{\mu}_i = 0 \qquad (12\text{-}6)$$

Equation (12-6) is the condition of equilibrium for a chemical reaction, but in this form it has little utility. A more useful relation may be obtained by expressing the chemical potentials in terms of fugacities. For this purpose we recall the defining equation for the fugacity of a substance in a mixture:

$$d\hat{\mu}_i = RT\,d\ln \hat{f}_i \qquad (9\text{-}7)$$

When this equation is integrated between a standard state, designated by the superscript °, and the equilibrium state, we obtain

$$\hat{\mu}_i = \mu_i^\circ + RT \ln \frac{\hat{f}_i}{f_i^\circ} \tag{12-7}$$

This ratio of fugacities appears quite often, and it is convenient to define it as a new thermodynamic function, the activity $\hat{a}_i$:

$$\hat{a}_i = \frac{\hat{f}_i}{f_i^\circ} \tag{12-8}$$

While the fugacity $\hat{f}_i$ depends only on the state of the system, the numerical value of the activity will depend as well on the choice of standard state. Therefore, when dealing with the activity, it is necessary that the standard state be specified and well defined. Fortunately, in applying the activity in calculations involving chemical equilibrium, the standard state is well defined once the state of aggregation of the substance is specified.

Equation (12-8) defines the activity of a substance in a solution. For a pure substance the activity is

$$a_i = \frac{f_i}{f_i^\circ} \tag{12-9}$$

By writing Eq. (12-7) as

$$\hat{\mu}_i = \mu_i^\circ + RT \ln \hat{a}_i \tag{12-10}$$

and substituting this expression into Eq. (12-6), the condition of equilibrium becomes

$$\sum v_i \mu_i^\circ + RT \sum v_i \ln \hat{a}_i = 0$$

or

$$\frac{\sum v_i \mu_i^\circ}{RT} = -\sum \ln \hat{a}_i^{v_i}$$

or

$$\frac{\sum v_i \mu_i^\circ}{RT} = -\ln \Pi \, \hat{a}_i^{v_i}$$

where Π signifies the product over all species. This product is called the equilibrium constant and is given the symbol K:

$$K = \Pi \, \hat{a}_i^{v_i} \tag{12-11}$$

Thus, the condition of equilibrium now becomes

$$\sum v_i \mu_i^\circ = -RT \ln K \tag{12-12}$$

Recalling that μ_i° is the chemical potential of component i when in its standard state and that chemical potentials are identical with Gibbs free energies, the left-hand side of Eq. (12-12) is called the standard Gibbs free energy change of reaction and is designated ΔG°. We now write Eq. (12-12) as

$$\Delta G^\circ = -RT \ln K \tag{12-13}$$

While this equation was derived from consideration of a variation constrained to constant temperature and pressure, it is completely general because it characterizes the original equilibrium state and not the variation away from that state. The manner in which the system actually arrived at this equilibrium state is irrelevant, and all that is necessary is to know that Eq. (12-13) determines the equilibrium compositions for a reaction at specified temperature, pressure, and initial mol numbers.

When more than one chemical reaction is necessary to describe chemical equilibrium in a system, the condition of equilibrium is obtained by recognizing that for each reaction expressed in the form of Eq. (12-1) an α defined in terms of Eq. (12-2) measures the extent of reaction. For such a system Eq. (12-5) can be written at constant temperature and pressure:

$$dG = \sum_{j=1}^{k} \left(\sum_i v_{ij} \hat{\mu}_{ij} \right) d\alpha_j$$

where the subscript i identifies a species participating in the jth reaction.[4] At equilibrium G exhibits a minimum, and all partial derivatives must equal zero,

$$\left(\frac{\partial G}{\partial \alpha_j} \right)_{T,P} = 0$$

which reduces to

$$\sum v_{ij} \hat{\mu}_{ij} = 0$$

for each reaction. Therefore, Eq. (12-13) applies to each reaction, and at equilibrium the equilibrium constant for each reaction must be satisfied.

---

[4] There are k reactions, and the summation in parentheses is performed over all participants in the jth reaction.

12-3 STANDARD STATES AND $\Delta G°$

The original condition of equilibrium for a chemical reaction $\Sigma v_i \hat{\mu}_i = 0$ or $\Delta G = 0$ is not directly useful because the absolute value of a chemical potential can never be known, nor can it be expressed in terms of a single fugacity. However, use can be made of this condition by incorporating an equilibrium step in a computational path between two other states. This is illustrated on the path diagram shown in Fig. 12-1 for carrying out the reaction

$$aA(g) + bB(g) = cC(g) + dD(g)$$

isothermally. For convenience all species are assumed to be gaseous, as indicated. Any initial and final state could be chosen, but for the sake of convenience standard states are used. The Gibbs free energy change for the reaction occurring between reactants in their standard states and yielding products in their standard states is called the standard free energy change and is designated $\Delta G°$. As we saw in Sec. 6-7, the use of a standard property change for a reaction allows us to identify and isolate effects due only to chemical reaction. The fact that such a property change usually represents a hypothetical process was seen to present no problems so long as a consistent computational path was devised for its use.

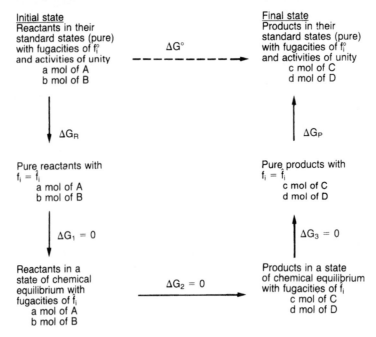

Figure 12-1 Computational path for an isothermal chemical reaction.

Because the Gibbs free energy is a state property, its change, in this case ΔG°, may be evaluated via any convenient path such as shown in Fig. 12-1. The execution of the reaction via this computational path proceeds as follows:[5]

Step 1. The pure reactant gases in their standard states with $f_i = f_i^\circ$ are expanded or compressed isothermally to a fugacity equal to their fugacities in the reacting system at equilibrium. The Gibbs free energy change for this process ΔG_R, from Eq. (8-7) and integration of Eq. (9-2), is

$$\Delta G_R = aRT \ln \frac{\hat{f}_A}{f_A^\circ} + bRT \ln \frac{\hat{f}_B}{f_B^\circ}$$

or

$$\Delta G_R = aRT \ln \hat{a}_A + bRT \ln \hat{a}_B$$

or

$$\Delta G_R = RT \ln \hat{a}_A^a + RT \ln \hat{a}_B^b$$

Step 2. The pure reactant gases are introduced into the reacting system through membranes permeable only to a single species. Because the fugacities on either side of the membrane are equal, there is no Gibbs free energy change for this step.

Step 3. The introduction of differential quantities of the reactants slightly displaces the reacting system from equilibrium, and the reaction proceeds to the right in order to restore the system to the equilibrium state. Because the system is conceived to never be more than infinitesimally displaced from equilibrium, the Gibbs free energy change is zero according to our condition of equilibrium.

Step 4. The small quantities of products produced in the reaction step are removed from the reacting system as pure gases at $f_i = \hat{f}_i$. Membranes permeable only to a single species allow this step. As in step 2, $\Delta G = 0$.

Step 5. The pure product gases are compressed or expanded isothermally from $\hat{f}_i$ to f_i°. As in step 1, we have

$$\Delta G_P = cRT \ln \frac{f_C^\circ}{\hat{f}_C} + dRT \ln \frac{f_D^\circ}{\hat{f}_D}$$

or

$$\Delta G_P = -RT \ln \hat{a}_C^c - RT \ln \hat{a}_D^d$$

---

[5] Each of the five steps is conceived as being reversible, and the apparatus for such a process is referred to as a van't Hoff equilibrium box.

For the five-step path from the initial to final state, both of which are standard states, we have

$$\Delta G° = \Delta G_R + 0 + 0 + 0 + \Delta G_P$$

$$= -RT(\ln \hat{a}_C^c + \ln \hat{a}_D^d - \ln \hat{a}_A^a - \ln \hat{a}_B^b)$$

or

$$\Delta G° = -RT \ln \frac{\hat{a}_C^c \hat{a}_D^d}{\hat{a}_A^a \hat{a}_B^b} = -RT \ln K$$

Thus, we see that Eq. (12-13) relates the equilibrium constant K involving the equilibrium-state activities (or fugacities) to $\Delta G°$, the Gibbs free energy change representing the conversion of reactants from their standard states to products in their standard states. Any initial and final state could be used to obtain a working relation similar to Eq. (12-13), and therefore the specification of standard states is arbitrary. However, for the sake of convenience in using and tabulating thermochemical data, the following standard states are universally accepted:

> *Gases.*   The standard state is the pure gas at the reaction temperature and at a pressure such that the fugacity is 1 atm. Setting $f° = 1$ has the advantage of making the activity numerically equal to the fugacity.[6]
> *Liquids and solids.*   The standard state is the pure substance at the reaction temperature and at a pressure of 1 atm.

Because the standard state is defined in terms of a specified pressure, the properties of a substance in its standard state do not depend on pressure and are fixed as soon as the state of aggregation and temperature are specified. Therefore, $\Delta G°$, the difference of such properties, is independent of pressure and depends only on temperature. Thus, from Eq. (12-13) the equilibrium constant K is seen to be a function only of temperature.

Because thermochemical data are usually tabulated only at 298 K, it is sometimes erroneously assumed that this is the only temperature which specifies the standard state. The fallacy in this interpretation may be discerned by recalling that the path diagram of Fig. 12-1 represents an isothermal process from initial to final standard state at the equilibrium temperature which was not specified. Methods of evaluating $\Delta G°$ or employing Eq. (12-13) at temperatures other than 298 K are presented in the next section.

We have used $\Delta G°$ to represent the term $\Sigma \ v_i \mu_i°$ and now need to consider how it is to be evaluated. Because the absolute value of a chemical potential can never be known, a direct calculation cannot be made. However, because the change in a state property is desired, any convenient path may be devised for the calculation. As in Sec. 6-7 we choose the path shown in Fig. 12-2 where reactants are first dissociated into their elements,

---

[6] See Prob. 12-65 for the effect of using S. I. units in place of atm.

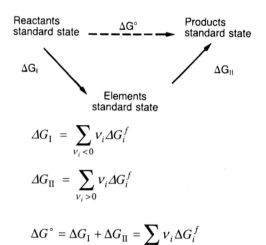

$$\Delta G_{\mathrm{I}} = \sum_{v_i < 0} v_i \Delta G_i^f$$

$$\Delta G_{\mathrm{II}} = \sum_{v_i > 0} v_i \Delta G_i^f$$

$$\Delta G^\circ = \Delta G_{\mathrm{I}} + \Delta G_{\mathrm{II}} = \sum v_i \Delta G_i^f$$

Figure 12-2 Path for evaluating ΔG°.

which are subsequently recombined to form the products. To implement this approach, we define the free energy of formation ΔG_i^f of a compound A_i as the Gibbs free energy change for the reaction forming 1 mol of this compound in its standard state from the elements in their standard states. From Fig. 12-2 it is seen that

$$\Delta G^\circ = \sum v_i \Delta G_i^f \qquad (12\text{-}14)$$

It should be recalled that $v_i > 0$ for products and that $v_i < 0$ for reactants. By definition $\Delta G_i^f = 0$ for an element[7] which takes part in the reaction. This calculation procedure is illustrated for the shift reaction:

$$CO(g) + H_2O(g) = CO_2(g) + H_2(g)$$

We write the following formation reactions:

---

[7] An element may exist in different forms; for example, hydrogen could be H_2 or H. The free energies of formation of all hydrogen-containing compounds are based on the more common form H_2, and for this form of the element $\Delta G_{H_2}^f = 0$.

The free energy of formation of atomic hydrogen H is not zero but is ΔG° for the following reaction:

$$\tfrac{1}{2} H_2 = H$$

$$CO(g) = C(c) + \frac{1}{2}O_2(g) \qquad \Delta G° = -\Delta G_{CO}^f$$

$$H_2O(g) = H_2(g) + \frac{1}{2}O_2(g) \qquad \Delta G° = -\Delta G_{H_2O}^f$$

$$C(c) + O_2(g) = CO_2(g) \qquad \Delta G° = \Delta G_{CO_2}^f$$

$$CO(g) + H_2O = CO_2(g) + H_2(g)$$

The sum of these three reactions is the shift reaction, and the sum of the $\Delta G°$'s is $\Delta G°$ for the shift reaction:

$$\Delta G° = \Delta G_{CO_2}^f - \Delta G_{CO}^f - \Delta G_{H_2O}^f$$

This, of course, is the result which would be obtained from Eq. (12-14) with

$$v_{CO} = -1 \qquad v_{H_2O} = -1 \qquad v_{CO_2} = 1 \qquad v_{H_2} = 1 \qquad \text{and} \quad \Delta G_{H_2}^f = 0$$

12-4 TEMPERATURE DEPENDENCE OF THE EQUILIBRIUM CONSTANT

It has been shown that $\Delta G°$ is a function only of temperature. We now wish to determine that dependence and begin by writing Eq. (12-13) as

$$\frac{\Delta G°}{RT} = \sum \frac{v_i \mu_i°}{RT} = -\ln K$$

and differentiating with respect to temperature:[8]

$$\frac{d(\Delta G°/RT)}{dT} = \frac{1}{R}\sum v_i \frac{d\mu_i°/T}{dT} = -\frac{d \ln K}{dT}$$

Because $\mu_i°$ is the chemical potential of component i in the standard state, the derivative, according to Eq. (11-7), is expressed in terms of the enthalpy of component i in the standard state,

---

[8] Note that because $\Delta G°$ and the $\mu°$'s depend only on temperature, the ordinary, rather than a partial, derivative is written.

$$\frac{d\mu_i^{\circ}/T}{dT} = -\frac{h_i^{\circ}}{T^2}$$

and we have

$$\frac{d(\Delta G^{\circ}/RT)}{dT} = -\frac{1}{RT^2}\sum v_i h_i^{\circ} = -\frac{d \ln K}{dT}$$

The quantity $\sum v_i h_i^{\circ}$ is the difference between standard state enthalpies of products and reactants and is the standard enthalpy change for the reaction ΔH° which we have already seen in Chap. 6. For the temperature dependence of K we now write

$$\frac{d \ln K}{dT} = \frac{\Delta H^{\circ}}{RT^2} \tag{12-15}$$

The sign of the derivative is seen to depend only on ΔH°; therefore, for exothermic reactions, $\Delta H^{\circ} < 0$, K decreases with increasing T, while for endothermic reactions, $\Delta H^{\circ} > 0$, K increases with increasing T. Again, because of the way in which standard states are defined, ΔH°, like ΔG°, depends only on temperature. In Chap. 6 it was shown that ΔH° at any temperature T could be obtained through a three-step process of bringing the reactants from T to 298 K, carrying out the reaction at 298 K, and returning products to T. In terms of the general stoichiometric notation this becomes

$$\Delta H^{\circ} = \Delta H_{298}^{\circ} + \int_{298}^{T}\left(\sum v_i c_{P_i}^{\circ}\right)dT \tag{12-16}$$

When enthalpies of formation are available for evaluation of ΔH_{298} and heat capacities are known over the temperature interval 298 to T, Eq. (12-16) can be integrated to obtain an algebraic expression relating ΔH° to temperature. Substitution of this relation for ΔH° would allow Eq. (12-15) to be integrated, and an algebraic relation between the equilibrium constant and temperature could be obtained. As an illustration, consider the special case where the heat capacities of all reaction participants are available as the following power series in T:

$$c_{P_i}^{\circ} = a_i + b_i T + c_i T^2$$

The standard enthalpy change ΔH° can be written as

$$\Delta H^{\circ} = \Delta H_0 + \left(\sum v_i a_i\right)T + \left(\frac{\sum v_i b_i}{2}\right)T^2 + \left(\frac{\sum v_i c_i}{3}\right)T^3 \tag{12-17}$$

where ΔH_0 is a constant of integration. Substitution of this expression into Eq. (12-15) and subsequent integration leads to

$$\ln K = -\frac{\Delta H_0}{RT} + \left(\frac{\sum v_i a_i}{R}\right) \ln T + \left(\frac{\sum v_i b_i}{2R}\right) T + \left(\frac{\sum v_i c_i}{6R}\right) T^2 + I \qquad (12\text{-}18)$$

where I is another constant of integration which can be determined from the value of the equilibrium constant at a single temperature.

When the necessary heat capacity data are unavailable or when a quick estimate is desired, Eq. (12-15) may be integrated on the assumption that $\Delta H°$ is constant over the temperature interval. The result is

$$\ln \frac{K}{K_1} = \frac{\Delta H°}{R}\left(\frac{1}{T_1} - \frac{1}{T}\right) \qquad (12\text{-}19)$$

This equation suggests that a plot of $\ln K$ vs. $1/T$ will be linear. Over a small temperature interval this relation should yield good estimates for most reactions.

12-5 EXPERIMENTAL DETERMINATION OF THERMOCHEMICAL DATA

It should be remembered that thermodynamic properties must be evaluated from experimental data and that thermodynamics may be regarded as a tool for utilizing experimental information. Here we will be concerned with the experimental determination of thermochemical data. However, a little thought will reveal that the thermodynamic approach has resulted in a tremendous savings in experimental effort. Instead of requiring values of $\Delta H°$ and $\Delta G°$ for every reaction to be considered, it has been shown that only ΔG^f and ΔH^f for the compounds involved in the reactions are needed. The values of ΔG^f and ΔH^f have to be obtained from studies of specific reactions which yield values of $\Delta G°$ and $\Delta H°$. Through a systematic study of a network of reactions it is possible to determine values of ΔG^f or ΔH^f for various compounds.[9] The following example involving the species CH_4, CO, CO_2, and H_2O should illustrate how this can be accomplished. Suppose that the following reactions have been studied:

$$H_2(g) + \frac{1}{2}O_2(g) = H_2O(g) \qquad (a)$$

$$C(c) + \frac{1}{2}O_2(g) = CO(g) \qquad (b)$$

---

[9] Besides the direct measurement of the equilibrium dissociation of water at elevated temperatures, G. N. Lewis and M. Randall (*Thermodynamics and the Free Energy of Chemical Substances*, McGraw-Hill, New York, 1923) demonstrated the calculation of ΔG^f of water via three other reaction networks involving, respectively, compounds of silver, mercury, and chlorine.

$$CO(g) + \frac{1}{2}O_2(g) = CO_2(g) \tag{c}$$

$$CO(g) + 3H_2(g) = CH_4(g) + H_2O(g) \tag{d}$$

Information on reactions (a) and (b) leads directly to ΔG^f (or ΔH^f) for $H_2O(g)$ and $CO(g)$. Knowing this and ΔG° for reaction (c) permits the determination of $\Delta G^f_{CO_2}$ (or $\Delta H^f_{CO_2}$) from the relations

$$\Delta G^\circ = \Delta G^f_{CO_2} - \Delta G^f_{CO}$$

$$\Delta H^\circ = \Delta H^f_{CO_2} - \Delta H^f_{CO}$$

Similarly, ΔG^f_{CO}, $\Delta G^f_{H_2O}$ (or ΔH^f_{CO}, $\Delta H^f_{H_2O}$), and ΔG° (or ΔH°) for reaction (d) leads to the evaluation of $\Delta G^f_{CH_4}$ (or $\Delta H^f_{CH_4}$).

Two methods for determining free energies of formation from experimental data are available: a second law method and a third law method.

Second law method. This method depends on the sampling of a system in which a chemical reaction is in a state of equilibrium. The subsequent analysis of the samples provides the equilibrium compositions which are used to calculate the equilibrium constant. When the reaction is studied for various reactant ratios and the results lead to the same equilibrium constant, the experimenter is assured that equilibrium was indeed attained.

This method is usually employed at high temperatures where the reaction rates are sufficient for equilibrium to be established in a reasonable time. Through the use of Eq. (12-18) and the known value of K at the higher temperature, K's, and subsequently ΔG°'s, may be obtained at other temperatures. This requires heat capacities for all the species participating in the reaction. Alternatively, sufficient K vs. T data may be used to empirically determine the parameters in an equation of the form of (12-18) which is then used for extrapolation to lower temperatures.

Third law method. The third law allows an absolute value of the entropy to be determined. Hence, from the calorimetric measurements described in Chapter 4 absolute values of the entropies in the standard state S_i° may be calculated. From these the standard entropy change for the reaction ΔS° may be evaluated from

$$\Delta S^\circ = \sum v_i S_i^\circ \tag{12-20}$$

When ΔH° is known (usually from combustion calorimetry), ΔG° may be calculated from

$$\Delta G^\circ = \Delta H^\circ - T\Delta S^\circ$$

Values of S_i° may also be obtained from spectroscopic data using the methods of statistical and quantum mechanics. Details of this method may be found elsewhere.[10]

Whenever reliable data have been used to calculate ΔG° for specific reactions by each of these two methods, the agreement has always been good. This, of course, constitutes a test of the laws of thermodynamics, especially the third law.

Whichever method is used to determine ΔG^f, the data must be exceptionally accurate in order that the error in the equilibrium constant be tolerable because the relationship between K and ΔG° is exponential and because ΔG° is an algebraic sum, $\Sigma\, v_i \Delta G_i^f$, which is often much smaller than individual members of the sum. For example, it has been shown[11] that for the dehydrogenation of isopropanol to acetone

$$CH_3CHOHCH_3(g) = CH_3COCH_3(g) + H_2(g)$$

an error of 1% in the heat of combustion of isopropanol produces a 3000-fold error in the equilibrium constant at 298 K.

Thermochemical data sufficient to work the problems in this text are tabulated in Tables D-1 and D-2 and in Figure A-1 of the Appendix. Sources of thermochemical data are given in Table 12-1. In the absence of published thermochemical data, methods based on molecular structure[12] may be used for estimates; however, as we have seen, extremely accurate data are needed for realistic equilibrium calculations.

TABLE 12-1
SOURCES OF THERMOCHEMICAL DATA

1. National Bureau of Standards

    A. NBS Cicular 500, *Selected Values of Chemical Thermodynamic Properties*, 1952. ΔH_0^f, ΔH_{298}^f, ΔG_{298}^f, log K_{298}^f, S°, and c_P° are tabulated for many compounds, radicals, and ions.

    B. NBS Technical Notes 270–3, 4, 5, 6, *Selected Values of Chemical Thermodynamic Properties*, 1968, 1969, 1970, and 1971. These are revisions of data in Circular 500.

    C. NBS Circular 461, *Selected Values of Properties of Hydrocarbons*, 1947. ΔH^{vap}, ΔS^{vap}, ΔH_{298}^c, ΔH_{298}^f, ΔG_{298}^f, ΔS_{298}°, $(H^\circ - H_0^\circ)\,/\,T$, and $(G^\circ - H_0^\circ)\,/\,T$ are tabulated.

    D. *JANAF Thermochemical Tables*, 2nd ed., National Standards Reference Data Service, NBS, *37*, 1971, and also supplements in *J. Phy. Chem. Ref. Data, 3,* 311 (1974) and *4,* 1 (1975). c_P°, S°,

---

[10] F. Rossini, *Chemical Thermodynamics*, Wiley, New York, 1950, Chap. 19; D. R. Stull, E. F. Westrum, Jr., and G. C. Sinke, *The Chemical Thermodynamics of Organic Compounds*, Wiley, New York, 1969, Chap. 4; G. N. Lewis, M. Randall, K. S. Pitzer, and L. Brewer, *Thermodynamics*, McGraw-Hill, New York, 1961, Chap. 27.

[11] D. R. Stull et al., *op. cit.*, Chap. 7.

[12] R. C. Reid, J. M. Prausnitz, and T. K. Sherwood, *Properties of Gases and Liquids*, Mc Graw-Hill, New York, 1977.

TABLE 12-1 con't

$(G° - \Delta H°_{298})/T$, $H° - H°_{298}$, ΔH^f, ΔG^f, and log K^f are tabulated.

   E. *The NBS Tables of Chemical Thermodynamic Properties. Selected Values for Inorganic and C_1 and C_2 Organic Substances in SI Units, J. Phys. Chem. Ref. Data, 11*, Supplement 2, 1982. ΔH^f_0, ΔH^f_{298}, ΔG^f_{298}, $S°$, and $c°_p$ are tabulated.

2. American Petroleum Institute, Project 44, *Selected Values of Properties of Hydrocarbons and Related Compounds*, Texas A. & M. University, College Station, a continuation of NBS Circular 461 (an ongoing project).

3. D. R. Stull, E. F. Westrum, Jr., and G. C. Sinke, *The Chemical Thermodynamics of Organic Compounds*, Wiley, New York, 1969, $c°_P$, $S°$, $(G° - H°_{298})/T$, $H° - H°_{298}$, ΔH^f, ΔG^f, and log K^f are tabulated.

4. I. Barin and O. Knacke, *Thermodynamical Properties of Inorganic Substances*, Springer-Verlag, Berlin, 1973, supplement 1977, $c°_P$, $S°$, $H°$, $G°$, and $G°/T$ vs. temperature are tabulated.

5. M. K. Karapet'yants and M. L. Karapet'yants, *Thermodyamic Constants of Inorganic and Organic Compounds*, Ann-Arbor-Humphrey Science Pub., Ann Arbor, 1970. $c°_P$, ΔH^f_{298}, ΔG^f_{298}, and $\Delta S°_{298}$ are tablulated.

6. B. J. McBride, S. Heimel, J. G. Ehlers, and S. Gordon, *Thermodynamic Properties to 6000°K for 210 Substances Involving the First 18 Elements*, NASA SP-3001, Washington, DC, 1963. The same properties as in the JANAF tables are tabulated.

7. *Physical and Thermodynamic Properties of Elements and Compounds*, United Catalysts Inc., Louisville, KY, 1969, $c°_P$, $H° - H°_0 - \Delta H^f_0$, and log K^f are tabulated.

8. K. A. Kobe and associates, *Thermochemistry of Petrochemicals*, Reprint No. 44, Bureau of Engineering Research, University of Texas, Austin, $c°_P$, ΔH^f_{298}, and ΔG^f_{298} are tabulated.

9. *Journal of Physical and Chemical Reference Data*, published quarterly by the American Chemical Society and the American Institute of Physics for the National Bureau of Standards, publishes newly determined reference data.

10. L. V. Gurvich, I. V. Veyts, and C. B. Alcock, eds., *Thermodynamic Properties of Individual Substances*, 4th ed., Hemisphere Publications Corp., New York, 1989. ΔH^f_0 and ΔH^f_{298}, along with $c°_p$, $S°$, $(G° - H°)/T$, $H° - H°_0$, and log K^f vs. temperature are tabulated.

11. *Data Bases*. The annual Software Directory accompanying the January issue of *Chemical Engineering Progress* contains brief descriptions of commercially available data bases and data estimation programs. The directory is scheduled for inclusion on *AIChE*'s Web site (www.aiche.org).

EXAMPLE 12-1

Develop a spreadsheet using Eqs. (12-15) and (12-16) to calculate ΔH^{f} $(H_{T}^{\circ} - H_{298}^{\circ} + \Delta H_{298}^{f})$, and log K^{f} at various temperatures for gaseous substances comprised of the elements carbon, hydrogen, oxygen, nitrogen, chlorine, and sulfur.

Solution 12-1

For maximum generality the compound is written as $C_{\alpha}H_{\beta}O_{\gamma}N_{\delta}Cl_{\varepsilon}S_{\eta}$ where the Greek letters are formula coefficients which are either zeros or small integers. The formation reaction can be written as

$$\alpha C(c) + \frac{\beta}{2}H_{2}(g) + \frac{\gamma}{2}O_{2}(g) + \frac{\delta}{2}N_{2}(g) + \frac{\varepsilon}{2}Cl_{2}(g) + \eta S(c) = C_{\alpha}H_{\beta}O_{\gamma}N_{\delta}Cl_{\varepsilon}S_{\eta}(g)$$

The property changes for this reaction are referred to the crystalline state for sulfur and will be designated ΔH^{f} and log K^{f} (these correspond to values found in App. D). For use at higher temperatures it is more convenient to refer these properties to the gaseous state for sulfur, the $S_{2}(g)$ state. These properties will be designated $\Delta H^{f'}$ and log $K^{f'}$ and are obtained by adding to the original formation reaction the following reaction

$$\frac{\eta}{2}S_{2}(g) = \eta S(c)$$

$$\Delta H_{298}^{\circ} = -129.1 \text{ MJ / kmol}; \qquad \Delta G_{298}^{\circ} = -80.07 \text{ MJ / kmol}$$

to obtain the desired reaction

$$\alpha C(c) + \frac{\beta}{2}H_{2}(g) + \frac{\gamma}{2}O_{2}(g) + \frac{\delta}{2}N_{2}(g) + \frac{\varepsilon}{2}Cl_{2}(g) + \frac{\eta}{2}S_{2}(g) = C_{\alpha}H_{\beta}O_{\gamma}N_{\delta}Cl_{\varepsilon}S_{\eta}(g)$$

The desired property changes at 298 K are

$$\Delta H_{298}^{f'} = \Delta H_{298}^{f} + \eta \left(\frac{-129.1}{2}\right); \qquad \Delta G_{298}^{f'} = \Delta G_{298}^{f} + \eta \left(\frac{-80.07}{2}\right)$$

The heat of formation of species i at any temperature is obtained by integration of Eq. (12-16) with

$$\sum v_{i} c_{Pi}^{\circ} = c_{Pi} - \alpha c_{PC} - \frac{\beta}{2}c_{PH_{2}} - \frac{\gamma}{2}c_{PO_{2}} - \frac{\delta}{2}c_{PN_{2}} - \frac{\varepsilon}{2}c_{PCl_{2}} - \frac{\eta}{2}c_{PS_{2}}$$

where the heat capacities are of the form in App. B except for that of carbon which takes the form[13]

$$c_P^\circ = 2.673 + 0.002617T - \frac{116,900}{T^2}$$

Having $\Delta H^{f\prime}$ as a function of temperature allows integration of Eq. (12-15) yielding log $K^{f\prime}$ as a function of temperature.

For any component i the total enthalpy, $(H_T^\circ - H_{298}^\circ + \Delta H_{298}^f)$, can be calculated from

$$(H_T^\circ - H_{298}^\circ + H_{298}^f)_i = (\Delta H_{298}^f)_i + \int_{298}^{T} c_{P_i}^\circ dT$$

The spreadsheet LOGKF(T).WQ1(or .XLS) on the CD-ROM accompanying this text calculates $\Delta H^{f\prime}$, $(H^\circ - H_{298}^\circ + \Delta H_{298}^f)$, and log $K^{f\prime}$ for any substance comprised of the elements C, H, O, N, Cl, and S at temperatures ranging from 298 K to 1500 K.[14] Algebraic details may be discerned from the appropriate spreadsheet cells. For several substances the necessary data are stored in named blocks and can be easily accessed.

EXAMPLE 12-2

Calculate ΔG_{298}^f for gaseous water based on the second law method.

Solution 12-2

In their landmark work published in 1923, Lewis and Randall[15] presented the following calculations based on data then available in the literature. Four independent investigations of the water dissociation reaction provided values of the equilibrium constant or ΔG° at temperatures above 1300 K. The heat of formation of gaseous water at 298 K, available from calorimetric combustion data, was −57,820 cal. The gaseous heat capacities were as follows:

---

[13] K. K. Kelley, *U.S. Bureau Mines Bull.*, 371 (1934).

[14] The upper limit for most of the heat capacity equations of Table B-1 is 1500 K. Be sure to check this upper limit when using the spreadsheet.

[15] G. N. Lewis and M. Randall, *Thermodynamics and the Free Energy of Chemical Substances*, McGraw-Hill, New York, 1923.

Water: $c_P = 8.81 - 1.9(10^{-3})T + 2.22(10^{-6})T^2$

Hydrogen: $c_P = 6.50 + 0.90(10^{-3})T$

Oxygen: $c_P = 6.50 + 1.0(10^{-3})T$

Evaluation of the parameters in Eqs. (12-17) and (12-18) proceeds as follows: From the heat capacities,

$$\sum v_i a_i = 8.81 - \frac{1}{2}(6.50) - 6.50 = -0.94$$

$$\sum v_i b_i = [-1.9 - \frac{1}{2}(1.0) - 0.90](10^{-3}) = -3.30(10^{-3})$$

$$\sum v_i c_i = 2.22(10^{-6}) - \frac{1}{2}(0) - 0 = 2.22(10^{-6})$$

From ΔH_{298}^f and Eq. (12-17),

$$\Delta H_0 = \Delta H^f - \left(\sum v_i a_i\right)T - \left(\frac{\sum v_i b_i}{2}\right)T^2 - \left(\frac{\sum v_i c_i}{3}\right)T^3$$

At 298 K this becomes

$$\Delta H_0 = -57,820 + 0.94(298.1) + \frac{3.30(10^{-3})}{2}(298.1)^2 - \frac{2.22(10^{-6})}{3}(298.1)^3$$

with

$$\Delta H_0 = -57,410 \text{ cal}$$

Combining Eqs. (12-18) and (12-13) gives

$$\Delta G^\circ = \Delta H_0 - \left(\sum v_i a_i\right)T \ln T - \left(\frac{\sum v_i b_i}{2}\right)T^2 - \left(\frac{\sum v_i c_i}{6}\right)T^3 + I'T$$

which on substituting the previously determined parameters becomes

$$\Delta G^\circ = -57,410 + 0.94T \ln T + 1.65(10^{-3})T^2 - 0.37(10^{-6})T^3 + I'T$$

When determined from the four sets of chemical equilibrium data, the constant I' ranged from 3.55 to 3.81. When a weighted mean value of 3.66 is used, the resulting equation is

$$\Delta G^{\circ} = -57,410 + 0.94T \ln T + 1.65(10^{-3})T^2 - 0.37(10^{-6})T^3 + 3.66T$$

At 298 K a value of –54,590 cal is obtained for the free energy of formation of gaseous water.

EXAMPLE 12-3

Calculate ΔG_{298}^f for liquid water from ΔG_{298}^f for gaseous water.

Solution 12-3

The ΔG^f calculated in Ex. 12-2 for $H_2O(g)$ refers to the formation of gaseous water at a pressure of 1 atm.[16] At 298 K the vapor pressure of water is 23.756 mmHg, and it is recognized that pure water vapor cannot exist at a pressure of 1 atm. This standard state for $H_2O(g)$, while well defined, is nevertheless hypothetical. However, this should cause little concern when it is recalled from Sec. 12-3 that for practical application of the chemical equilibrium condition, specification of an initial unreacted state and a final reacted state was necessary, although the conditions of these states could be arbitrary. The selection of conditions which correspond to a hypothetical state is therefore legitimate and will not affect calculations involving the equilibrium constant as long as consistency is maintained. Because ideal-gas behavior may be assumed, all thermodynamic property changes can be evaluated.

The Gibbs free energy change for the process

$$H_2O(g, \ 1 \ atm) = H_2O(l, \ 1 \ atm)$$

is needed and is obtained through the following sequence of steps:

$$H_2O(g, \ 1 atm) = H_2O(g, \ P^{\circ}) = H_2O(l, \ P^{\circ}) = H_2O(l, \ 1 \ atm)$$

$$\text{step 1} \qquad\qquad \text{step 2} \qquad\qquad \text{step 3}$$

Step 1 involves an isothermal change of pressure for an ideal gas, and

---

[16] More precisely, at a pressure such that the fugacity is 1 atm. At a presure this low most gases closely approach ideal-gas behavior, and the difference between this pressure and 1 atm is not significant and is usually neglected.

$$\Delta G_1 = \int_1^{P^\circ} v \, dP = RT \ln P^\circ = RT \ln \frac{23.756}{760} = -2053 \text{ cal}$$

Step 2 occurs at conditions of equilibrium between liquid and vapor phases, and

$$\Delta G_2 = 0$$

Step 3 involves an isothermal change in pressure on the liquid which is assumed incompressible, and

$$\Delta G_3 = \int_{P^\circ}^1 v \, dP = v(1 - P^\circ)$$

$$= 18\left(\frac{760 - 23.756}{760}\right) = 17.4 \text{ cc} \cdot \text{atm} = 0.42 \text{ cal}$$

From these values and the ΔG^f of gaseous water, the ΔG^f of liquid water is found by adding the following equations:

$H_2(g) + \frac{1}{2}O_2(g) = H_2O(g, 1 \text{ atm})$	$\Delta G^f =$	$-54,590$ cal
$H_2O(g, 1 \text{ atm}) = H_2O(g, P^\circ)$	$\Delta G_1 =$	$-2,053$ cal
$H_2O(g, P^\circ) = H_2O(1, P^\circ)$	$\Delta G_2 =$	0 cal
$H_2O(1, P^\circ) = H_2O(1, 1 \text{ atm})$	$\Delta G_3 =$	0.42 cal
$H_2(g) + \frac{1}{2}O_2(g) = H_2O(l)$	$\Delta G^f =$	$-56,643$ cal

12-6 OTHER FREE ENERGY FUNCTIONS

Besides ΔG^f values, there are other ways of presenting and utilizing free energy data. This involves use of the functions

$$\ln K^f \qquad \frac{G_T^\circ - H_0^\circ}{T} \quad \text{and} \quad \frac{G_T^\circ - H_{298}^\circ}{T}$$

ln K^f or log K^f.   For the reaction in which the compound i is formed from its elements, the standard Gibbs free energy change ΔG° has been designated ΔG_i^f, the free energy of formation of the compound. For this reaction Eq. (12-13) is written as

$$\Delta G_i^f = -RT \ln K_i^f \tag{12-13}$$

where K_i^f is the equilibrium constant for the reaction written for 1 mol of compound i and is called the equilibrium constant of formation of compound i. When combined with Eq. (12-14), this relation yields

$$\Delta G^\circ = \sum v_i \, \Delta G_i^f = -RT \sum v_i \ln K_i^f$$

Use of Eq. (12-13),

$$\Delta G^\circ = -RT \ln K \qquad (12\text{-}13)$$

Kf is eq. of the formation of the compounds

allows us to write

$$\ln K = \sum v_i \ln K_i^f \qquad (12\text{-}21)$$

which can also be written in terms of base 10 logarithms. Thus, it is seen that ΔG_i^f and $\ln K_i^f$ (or $\log K_i^f$) provide the same information.

The Gibbs Energy Functions. Often one encounters tables of the Gibbs energy functions $(G_T^\circ - H_0^\circ)/T$ and $(G_T^\circ - H_{298}^\circ)/T$ at various temperatures. Because only standard state properties are involved, these functions depend only on temperature, and their slight temperature dependence makes them desirable for tabular interpolation. The Gibbs energy function referred to 0 K may be calculated for gases from spectroscopic data using the methods of statistical and quantum mechanics. This accounts for the origin of the function, but its application is not restricted to gases or use of spectroscopic data. The Gibbs energy function may also be referred to 298 K. These functions may be written as

$$\left(\frac{G_T^\circ - H_0^\circ}{T} \right) = \frac{H_T^\circ - H_0^\circ}{T} - S_T^\circ \qquad (12\text{-}22)$$

$$\left(\frac{G_T^\circ - H_{298}^\circ}{T} \right) = \frac{H_T^\circ - H_{298}^\circ}{T} - S_T^\circ \qquad (12\text{-}23)$$

and are related by

$$\left(\frac{G_T^\circ - H_{298}^\circ}{T} \right) = \frac{G_T^\circ - H_0^\circ}{T} - \frac{H_{298}^\circ - H_0^\circ}{T} \qquad (12\text{-}24)$$

In terms of the Gibbs energy functions the standard Gibbs free energy change for a reaction is

$$\frac{\Delta G^\circ}{T} = \sum_i v_i \left(\frac{G_T^\circ - H_0^\circ}{T} \right)_i + \frac{\sum_i v_i \Delta H_{0i}^f}{T} \tag{12-25}$$

and

$$\frac{\Delta G^\circ}{T} = \sum_i v_i \left(\frac{G_T^\circ - H_{298}^\circ}{T} \right)_i + \frac{\sum_i v_i \Delta H_{298i}^f}{T} \tag{12-26}$$

Because heats of formation are more readily available at 298 K, $(G_T^\circ - H_{298}^\circ)/T$ is the more convenient function to use.

12-7 HOMOGENEOUS GAS-PHASE REACTIONS

When all the participants in a reaction are gaseous, activities become numerically equal to fugacities,[17] and the equilibrium constant can be written as

$$K = \Pi \, \hat{f}_i^{\,v_i} \qquad \text{when gaseous} \tag{12-27a}$$

In terms of fugacity coefficients [Eq. (9-24)] we have

$$K = \Pi \, (p_i \hat{\phi}_i)^{v_i} \tag{12-27b}$$

For low to moderate pressures $\hat{\phi} \cong 1$, and

$$K = \Pi \, p_i^{\,v_i} = \Pi \, (Py_i)^{\,v_i} \tag{12-28a}$$

$$= P^{\sum_i v_i} \Pi \, y_i^{\,v_i} \tag{12-28b}$$

Equations (12-28) express the equilibrium constant in terms of partial pressures or vapor mol fractions at equilibrium. We have seen that K depends only on temperature, but Eq. (12-28b) shows that the equilibrium mol fraction product $\Pi \, y_i^{\,v_i}$ can also depend on pressure. If $\sum v_i > 0$, which corresponds to the mols of products exceeding the mols of reactants, then $\Pi \, y_i^{\,v_i}$ will decrease with increasing pressure. A decrease in $\Pi \, y_i^{\,v_i}$ corresponds to a decrease in extent of reaction. Conversely, if there is a decrease in the number of mols as the reaction

---

[17] This is true only when fugacities (or pressures) are expressed in atmospheres. Care must be taken to use only this unit of pressure.

proceeds, then $\Sigma v_i < 0$, and the extent of reaction (and $\Pi y_i^{v_i}$) increases with increasing pressure.

EXAMPLE 12-4

Investigate the effect of temperature, pressure, and reactant ratio on the production of methanol via the reaction

$$CO(g) + 2H_2(g) = CH_3OH(g)$$

The Gibbs energy functions are as follows:[18]

	$(G° - H_{298}°)/T$ (cal / K)		
T(K)	CH_3OH	CO	H_2
500	−58.61	−48.10	−31.997
600	−59.67	−48.68	−32.575
700	−60.81	−49.27	−33.156

The heats of formation at 298 K are

$CH_3OH(g)$: $\Delta H_{298}^f = -48.08$ kcal

$CO(g)$: $\Delta H_{298}^f = -26.42$ kcal

Solution 12-4

At 500 K we find $\Delta G° / T$ via Eq. (12-26):

$$-R \ln K = \frac{\Delta G°}{T} = [-58.61 - (-48.10) - 2(-31.997)]$$

$$+[-48.08 - (-26.42)]\frac{1000}{500}$$

$$K = 6.00(10^{-3})$$

Similarly,

At $T = 600$ K, $K = 1.13(10^{-4})$.

At $T = 700$ K, $K = 6.17(10^{-6})$.

For the generalized stoichiometry we specify

---

[18] Source: reference 3 in Table 12-1.

CH_3OH: $v = 1$, $n_0 = 0$

CO: $v = -1$, $n_0 = 1$

H_2: $v = -2$, $n_0 = h$

$\sum v_i = -2$

Therefore,

$$n_{CH_3OH} = \alpha \qquad y_{CH_3OH} = \frac{\alpha}{h + 1 - 2\alpha}$$

$$n_{CO} = 1 - \alpha \qquad y_{CO} = \frac{1 - \alpha}{h + 1 - 2\alpha}$$

$$n_{H_2} = h - 2\alpha \qquad y_{H_2} = \frac{h - 2\alpha}{h + 1 - 2\alpha}$$

$$n_{total} = h + 1 - 2\alpha$$

and

$$K = P^{-2} \frac{\alpha/(h + 1 - 2\alpha)}{[(1 - \alpha)(h - 2\alpha)^2]/(h + 1 - 2\alpha)^3} \qquad (12\text{-}29)$$

To determine the effect of temperature, we solve Eq. (12-29) for the case of $P = 1$ atm and $h = 2$:

$$K = \frac{\alpha(3 - 2\alpha)^2}{(1 - \alpha)(2 - 2\alpha)^2} \qquad (12\text{-}30)$$

Because K is quite small, α will be also; therefore, a good approximation will be to set α equal to zero except where it stands alone. Thus,

$$K \doteq \frac{3^2 \alpha}{1 \times 2^2}$$

At 500 K,

$$K = 6.0(10^{-3}) \quad \text{and} \quad \alpha \doteq 2.67(10^{-3})$$

On substituting this value of α back into Eq. (12-30), the value of 0.00603 is obtained for K, which is of sufficient accuracy for our calculations. The equilibrium extents of reaction α for the three temperatures are calculated to be

$T(K)$	500	600	700
α	$2.67(10^{-3})$	$5.02(10^{-5})$	$2.74(10^{-6})$

These values are quite small but increase with decreasing temperature. Low temperatures are desirable from equilibrium considerations; however, kinetic considerations dictate that for a reasonable reaction rate the temperature should be above 550 K. Because $\Sigma \nu_i < 0$, the extent of reaction may be increased by increasing the pressure. At a pressure of 100 atm and for $h = 2$ the following values of α and y_{CH_3OH} are obtained from Eq. (12-29):

$T(K)$	α	y_{CH_3OH}
500	0.815	0.595
600	0.280	0.115
700	0.0263	0.009

These values were calculated on the assumption that the gas phase behaved ideally. For a better estimate Eq. (12-27b) should be used with the ideal gaseous solution model. For this model,

$$\hat{\phi}_i = \phi_i$$

and

$$K = P^{\Sigma \nu_i} \Pi \phi_i^{\nu_i} \Pi y_i^{\nu_i} \tag{12-31}$$

Values of ϕ_i estimated from Fig. 9-7 are used to calculate $\Pi \phi_i^{\nu_i}$:

At 500 K, $\Pi \phi_i^{\nu_i} = \phi_{CH_3OH}/\phi_{CO}\phi_{H_2}^2 = 0.530/[0.960(1.08)^2] = 0.473.$

At 600 K, $\Pi \phi_i^{\nu_i} = 0.677.$

At 700 K, $\Pi \phi_i^{\nu_i} = 0.742.$

The values of α and y_{CH_3OH} corrected for gas-phase nonideality are calculated via Eq. (12-31):

$T(K)$	α	y_{CH_3OH}
500	0.86	0.672
600	0.35	0.152
700	0.035	0.012

In comparing these values with the previous values, it is seen that gas-phase nonideality is significant at 100 atm. The effect of pressure on conversion for this reaction is manifested in two ways: mainly through the $P^{\Sigma v_i}$ term but not insignificantly through the $\Pi \, \phi_i^{y_i}$ term.

At 600 K and 100 atm the effect of reactant ratio is studied by using values of 1 and 3 for h with nonideal-gas behavior accounted for by the term $\Pi \, \phi_i^{y_i}$. Calculated values of α and y_{CH_3OH} are as follows:

h	α	y_{CH_3OH}
1	0.194	0.120
2	0.350	0.152
3	0.436	0.139

It is seen that the maximum mol fraction of methanol at equilibrium occurs at the stoichiometric ratio. Increasing the ratio of hydrogen to carbon monoxide above this value does increase the extent of reaction, but the unreacted hydrogen dilutes the methanol and lowers its equilibrium mol fraction.

EXAMPLE 12-5

For the catalytic dehydrogenation of 1-butene to 1,3-butadiene,

$$C_4H_8(g) = C_4H_6(g) + H_2(g)$$

carried out at 900 K and 1 atm and with a ratio of 10 mol of steam per mol of butene, determine the extent of reaction at equilibrium. Also determine the extent of reaction in the absence of steam.

Solution 12-5

For this reaction $\Sigma \, v_i > 0$, and conversion is improved by reducing the pressure. The pressure of the reacting system can be reduced rather effectively by the addition of a diluent gas such as steam which can be easily separated from the reaction participants by condensation. In addition the presence of the steam allows the temperature to be controlled more easily and also helps to minimize the deposition of carbon on the catalyst.

The following thermochemical data are available:[19]

---

[19] From reference 2 in Table 12-1.

	$\dfrac{G^{\circ}_{900} - H^{\circ}_{298}}{T}$ (kcal / K)	ΔH^{f}_{298} (kcal)
C_4H_6	−80.35	26.33
C_4H_8	−88.03	−0.03
H_2	−34.762	——

From Eq. (12-26),

$$-R \ln K = \frac{\Delta G^{\circ}}{T} = -80.35 + (-34.762) - (-88.03)$$

$$+ \frac{26.33 - (-0.03)}{900} \cdot 1000$$

$$K = 0.329$$

In the generalized stoichiometric notation we have

C_4H_6: $v = 1, n_0 = 0$

H_2: $v = 1, n_0 = 0$

C_4H_8: $v = -1, n_0 = 1$

and

$$n_{C_4H_6} = n_{H_2} = \alpha$$

$$n_{C_4H_8} = 1 - \alpha$$

$$n_{\text{steam}} = 10$$

$$\overline{n_{\text{total}} = 11 + \alpha}$$

$$y_{C_4H_6} = y_{H_2} = \frac{\alpha}{11 + \alpha}$$

$$y_{C_4H_8} = \frac{1 - \alpha}{11 + \alpha}$$

Equation (12-28b) becomes

$$0.329 = 1 \frac{[\alpha/(11+\alpha)]^2}{(1-\alpha)/(11+\alpha)} = \frac{\alpha^2}{(1-\alpha)(11+\alpha)}$$

Solving gives

$$\alpha = 0.825$$

In the absence of steam Eq. (12-28b) becomes

$$0.329 = 1 - \frac{[\alpha/(1+\alpha)]^2}{(1-\alpha)/(1+\alpha)} = \frac{\alpha^2}{1-\alpha^2}$$

with $\alpha = 0.498$.

Thus, it is noted that considerable improvement in the extent of reaction is realized when the partial pressures of the reacting species are
lowered by the addition of a diluent gas. In many cases this will be a
more attractive course of action than carrying out the reaction at reduced
pressure. The major drawback to either reducing the pressure or adding
a diluent gas is that both of these actions cause a decrease in the rate of
reaction.

12-8 HETEROGENEOUS CHEMICAL EQUILIBRIUM

Here we are concerned with equilibrium in a system containing a gas phase and one or more
pure condensed phases. Each pure condensed phase has a definite vapor pressure, although it
may be immeasurably small. Therefore, we may refer to its partial pressure, or better its fugacity, in the gas phase. When the heterogeneous system is in equilibrium, there will be
equilibrium with respect to the chemical reaction in the gas phase and also phase equilibrium
between components in the gas phase and their pure condensed phases. When a pure condensed phase is present, the partial pressure (or fugacity) of that component in the gas phase
will equal the vapor pressure (or saturation fugacity) of the pure condensed phase. Thus,
when a pure condensed phase is present at equilibrium, the partial pressure (or fugacity) of
that component will have its maximum value. Conversely, when the partial pressure (or fugacity) of that component is less than this maximum value, the pure condensed phase cannot
exist.

Our inability to measure infinitesimally small vapor pressures, however, is not a hin-
drance in making chemical equilibrium calculations because the equilibrium constant is defined in terms of activities. Consider the activity of a pure condensed phase, say a solid,

$$a = \frac{f}{f^{\circ}}$$

where f is the fugacity of the pure solid in the equilibrium system and f° is the fugacity of pure solid at 1 atm pressure. These two fugacities refer to states which differ only in pressure; therefore, by Eq. (9-11) we write

$$RT \int_{f^\circ}^{f} d \ln f = \int_{1}^{P} v \, dP$$

Condensed phases are normally assumed incompressible, and on integration we obtain

$$RT \ln \frac{f}{f^\circ} = RT \ln a = v(P - 1) \tag{12-32a}$$

or

$$a = \exp \left[\frac{v(P-1)}{RT} \right] \tag{12-32b}$$

Except for very high pressures the argument of the exponential is small, and the activity is close to unity. In other words, from the standpoint of the pure condensed phase the equilibrium state is not significantly different from the standard state. Considerable simplification of the equilibrium calculations results at low to moderate pressures because the activities of all pure condensed phases present may be set equal to unity. This, of course, does not apply to condensed phases which are not pure because the fugacity depends on composition as well as pressure.

Heterogeneous reactions can be made to go to completion in the sense that a condensed phase will disappear. Consider the well-known reaction involving the dissociation of calcium carbonate to lime with the evolution of carbon dioxide:

$$CaCO_3(c) = CaO(c) + CO_2(g)$$

The equilibrium constant for this reaction is written as

$$K = \frac{a_{CaO} a_{CO_2}}{a_{CaCO_3}}$$

Because the activity of CO_2 is referred to the state of $f^\circ = 1$ atm, it is numerically equal to $\hat{f}_{CO_2}$. When both solid phases are present and the pressure is not high, their activities are close to unity, and we write

$$K = \hat{f}_{CO_2}$$

At low pressure this becomes

$$K = p_{CO_2}$$

Thus, when both $CaCO_3(c)$ and $CaO(c)$ are present at equilibrium, there is a single partial pressure of CO_2. Because K depends only on temperature, so does this equilibrium pressure. The system will always try to maintain this equilibrium pressure. If the partial pressure is reduced below the equilibrium value, $CaCO_3$ will decompose in an effort to maintain equilibrium, and a continued maintenance of a lower partial pressure will result in the disappearance of $CaCO_3$. Conversely, if a higher partial pressure is imposed, CaO will react with CO_2 in an attempt to reestablish equilibrium and will continue to do so until solid CaO disappears.

EXAMPLE 12-6

Determine the stability of calcium carbonate at ambient conditions. Also determine the temperature to which it must be heated at atmospheric pressure by direct contact with combustion gas containing 15% CO_2 in order to produce lime.

Solution 12-6

The following data are available at 298 K.[20]

	ΔG^f (cal/g mol)	ΔH^f (cal/g mol)
$CaCO_3(c)$	−269,780	−288,450
$CaO(c)$	−144,400	−151,900
$CO_2(g)$	−94,208	−94,051

and for the reaction

$$CaCO_3(c) = CaO(c) + CO_2(g)$$

from Eqs. (12-13) and (12-14) we have

$$-1.987(298) \ln K = -144,440 + (-94,258) - (-269,780)$$

$$K = 1.5(10^{-23})$$

or

$$p_{CO_2} = 1.5(10^{-23}) \text{ atm}$$

We see that at 298 K $CaCO_3(c)$ and $CaO(c)$ will coexist when the partial pressure of CO_2 is $1.5(10^{-23})$ atm. Because the partial pressure of CO_2 in the atmosphere is of the order of $3(10^{-4})$ atm, the reaction is driven to the left, and there is no tendency of $CaCO_3(c)$ to dissociate.

---

[20] Source: reference 1 in Table 12-1.

To produce lime in the presence of a carbon dioxide partial pressure of 0.15 atm, the temperature must be raised until

$$K = p_{CO_2} > 0.15$$

Knowledge of K as a function of T is needed so that T corresponding to $K = 0.15$ may be found. Equation (12-15) will be used with the following heat capacity data.[21] For the equation

$$c_P^\circ = A + B(10^{-3})T + C(10^5)T^{-2}$$

the constants are

	A	B	C
CaO(c)	11.86	1.08	−1.66
CaCO$_3$(c)	24.98	5.24	−6.20
CO$_2$(g)	10.55	2.16	−2.04

Using Eqs. (6-17) and (12-16), we obtain

$$\Delta H^\circ = 42{,}499 + \int_{298}^{T} \left(\sum v_i c_{P_i}^\circ \right) dT \qquad (12\text{-}16)$$

where

$$\sum v_i c_{P_i}^\circ = \sum v_i A_i + 10^{-3} \sum v_i B_i T + 10^5 \sum v_i C_i T^{-2}$$

$$= - \ 2.57 - 2.00(10^{-3})T + 2.5(10^5)T^{-2}$$

Substituting for $\sum v_i c_{P_i}^\circ$ gives

$$\Delta H^\circ = 42{,}499 - 2.57 \int_{298}^{T} dT - 2.00(10^{-3}) \int_{298}^{T} T \, dT + 2.5(10^5) \int_{298}^{T} T^{-2} \, dT$$

which results in

$$\Delta H^\circ = 44{,}193 - 2.57T - (10^{-3})T^2 - \frac{2.5(10^5)}{T}$$

Substituting for ΔH° in Eq. (12-15) gives

---

[21] Source: reference 4 in Table 12-1.

$$R \ln K - R \ln 1.5(10^{-23})$$

$$= 44,193 \int_{298}^{T} \frac{dT}{T^2} - 2.57 \int_{298}^{T} \frac{dT}{T} - 10^{-3} \int_{298}^{T} dT - 2.5(10^5) \int_{298}^{T} \frac{dT}{T^3}$$

$$R \ln K = 57.411 - \frac{44,193}{T} - 2.57 \ln T - 10^{-3} T + \frac{1.25(10^5)}{T^2}$$

From this equation at $K = 0.15$ we find by trial and error that $T = 1042$ K.

At temperatures higher than 1042 K calcium carbonate should dissociate when the partial pressure of CO_2 is 0.15 atm.

EXAMPLE 12-7

Mixtures of CO and CO_2 are to be processed at temperatures between 900 and 1000 K, and it is desired to know under what conditions solid carbon might deposit according to the reaction

$$CO_2(g) + C(c) = 2CO(g)$$

For this reaction the equilibrium constants are

$T(K)$	K
900	0.178
1000	1.58

Solution 12-7

For this reaction K is written as

$$K = \frac{\hat{f}_{CO}^2}{\hat{f}_{CO_2} a_C}$$

We will be working at low pressure so that

$$K a_C = \frac{p_{CO}^2}{p_{CO_2}} = \frac{P y_{CO}^2}{y_{CO_2}}$$

for the binary mixture $y_{CO} + y_{CO_2} = 1$, and

$$\frac{y_{CO}^2}{1 - y_{CO}} = \frac{Ka_C}{P}$$

When solid carbon is present at equilibrium, its activity is unity, and if solid carbon is not present, its activity is less than unity. Therefore, we want the condition $a_C < 1$ which results in

$$\frac{y_{CO}^2}{1 - y_{CO}} < \frac{K}{P}$$

The maximum permissible y_{co} is found by equating $y_{co}^2 / (1 - y_{co})$ to K/P; gases containing less CO should not deposit carbon. Over the range 900–1000 K and 1–10 atm the maximum permissible CO percentage is calculated to be

	y_{co}	
$T(K)$	$P = 1$ atm	$P = 10$ atm
900	0.342	0.125
1000	0.695	0.326

It is observed that at any given pressure the maximum permissible y_{co} increases with increasing temperature. At a given temperature it decreases with increasing presure.

EXAMPLE 12-8

The following reaction has been suggested as part of a thermochemical hydrogen production cycle:

$$6Fe(OH)_3(c) = 2Fe_3O_4(c) + 9H_2O(g) + \frac{1}{2}O_2(g)$$

At 1500 K with $\Delta G^\circ = -783$ KJ the reaction should proceed spontaneously; however, it does not.[22] Show that this behavior can be explained on the basis that the reaction represents the sum of the following reactions:

a: $6Fe(OH)_3(c) = 3Fe_2O_3(c) + 9H_2O(g)$ $\Delta G^\circ = -815$ kJ

b: $3Fe_2O_3(c) = 2Fe_3O_4(c) + \frac{1}{2}O_2(g)$ $\Delta G^\circ = 31.96$ kJ

---

[22] C. E. Bamberger, J. Braunstein, and D. M. Richardson, *J. Chem. Educ.*, 55, 561 (1978).

Solution 12-8

For the two reactions the equilibrium constants are as follows:

$$\text{Reaction } a\text{: } K_a = 2.4(10^{28})$$

$$\text{Reaction } b\text{: } K_b = 0.0711$$

For reaction a the equilibrium partial pressure of water is

$$K_a = 2.4(10^{28}) = p_{H_2O}^9$$

$$p_{H_2O} = 1400 \text{ atm}$$

For reaction b the equilibrium partial pressure of oxygen is

$$K_b = 0.0711 = p_{O_2}^{1/2}$$

$$p_{O_2} = 0.00505 \text{ atm}$$

For water partial pressures less than 1400 atm $Fe_2O_3(c)$ will form from the decomposition of $Fe(OH)_3(c)$; hence there is no problem with this reaction. However, the partial pressure of oxygen must be kept less than 0.00505 atm in order to decompose Fe_2O_3 to Fe_3O_4. To keep the oxygen partial pressure this low would require that either a vacuum or a large flow of inert gas be maintained.

This problem illustrates the potential difficulties which can arise through the superficial application of thermodynamics. It has been repeatedly emphasized that thermodynamics is a tool for utilizing and extending knowledge; however, some knowledge and a good understanding of the specific system under study are necessary for the intelligent application of thermodynamics. In this specific application knowledge of the existence of Fe_2O_3, a species not included in the original reaction, was required.

12-9 REACTIONS IN SOLUTION

When a reaction occurs in a solution, there may or may not be a vapor phase depending on the pressure. If vapor and liquid phases coexist, equilibrium will be established with respect to the reaction in both phases, and phase equilibrium will be established with respect to components between the liquid and vapor. For the vapor phase the equilibrium constant is written in terms of fugacities or partial pressures and for the liquid phase in terms of activities. Therefore, the numerical value of K (and $\Delta G°$) will be different. Liquid phase compositions at equilibrium may be obtained by finding the equilibrium partial pressures for the vapor phase reaction with the subsequent determination of liquid compositions using the methods previously developed for phase equilibrium. Or they may be determined from the equilibrium constant for the reaction in the liquid phase and a known relationship between activities and liquid compositions. In any event, the behavior of the liquid phase must be known. As we have seen, the liquid phase behavior is characterized by activity coefficients which are related to activities. This relationship is determined by writing the defining equations for the activity in a liquid solution as

$$\hat{a}_i = \frac{\hat{f}_i^L}{f_i°} \tag{12-8}$$

and for the activity coefficient as

$$\hat{f}_i^L = f_i^L x_i \gamma_i \tag{9-35}$$

Elimination of $\hat{f}_i^L$ gives

$$\hat{a}_i = \frac{f_i^L x_i \gamma_i}{f_i°}$$

f_i^L is the fugacity of pure liquid i at the pressure of the system and $f_i°$ is the fugacity of pure liquid i at a pressure of 1 atm. As long as the system pressure is not too high, this ratio is essentially unity, and we write

$$\hat{a}_i = x_i \gamma_i \tag{12-33}$$

For a reaction occurring in the liquid phase the equilibrium constant becomes

$$K = \Pi \hat{a}_i^{\nu_i} = \Pi x_i^{\nu_i} \Pi \gamma_i^{\nu_i}$$

The data needed to determine activity coefficients are usually unavailable, and except for situations where solutions are dilute and Raoult's and Henry's laws may be applied or where solutions behave ideally, exact equilibrium compositions cannot be determined. Usually, except for some dissociation, isomerization, or polymerization reactions, the assumption of ideal-solution behavior for a reacting system is highly questionable.

EXAMPLE 12-9

Butane exists as either of two isomers: normal butane and iso butane. (a) At 300°F and 50 psia, calculate the equilibrium composition of butane gas. (b) The temperature is kept at 300°F, while the pressure is increased until a liquid phase appears. What is the pressure, and

---

EXAMPLE 12-9 CON'T

what is the composition of the liquid phase? Can the pressure be further increased? (c) An inert nonvolatile solvent which forms ideal solutions with butane is added so that in the liquid phase there are 2 mol of solvent for each mol of butanes. Calculate the equilibrium mol fractions of the butanes. If the temperature remains at 300°F, what can be said about the allowable pressures in the system?

Solution 12-9

(a) For the reaction

$$n - C_4H_{10}(g) = i - C_4H_{10}(g)$$

the equilibrium constant at 300°F is

$$K = 1.096 = \frac{Py_i \, \hat{\phi}_i}{Py_n \, \hat{\phi}_n}$$

At a pressure of 50 psia the gas will be assumed ideal, and we write

$$K = 1.096 = \frac{y_i}{y_n}$$

For the binary mixture $y_i + y_n = 1$, and we find that

$$y_i = 0.523$$

$$y_n = 0.477$$

for the equilibrium vapor compositions.

(b) Isothermally increasing the pressure will eventually cause condensation with vapor and liquid phases in equilibrium. For this hydrocarbon system phase equilibrium may be expressed in terms of equilibrium ratios:

$$y_i = K_i x_i$$

$$y_n = K_n x_n$$

These K's are functions of temperature and pressure; therefore at 300°F a trial and error solution with assumed pressures is necessary. When a pressure is assumed, K's are evaluated and x's calculated from the known y's. The correct pressure is found when the sum of the calculated x's adds to unity. From Fig. 9-10 at 600 psia we find that $K_n = 0.89$ and $K_i = 1.12$:

$$x_n = \frac{0.477}{0.89} = 0.536$$

$$x_i = \frac{0.523}{1.12} = \frac{0.467}{1.003}$$

This checks very well, and hence the assumed pressure of 600 psia is correct.

For this system of isomers ideal-solution behavior is expected, and for γ's of unity it is seen from Eq. (12-33) that activities are equal to mol fractions. We may then evaluate the equilibrium constant for the reaction in the liquid phase:

$$K_L = \frac{\hat{a}_i}{\hat{a}_n} = \frac{x_i}{x_n}$$

$$= \frac{0.467}{0.536} = 0.871$$

The effect of pressure on equilibrium in the gas-phase reaction is manifested only through the ratio ϕ_i / ϕ_n, which we have set equal to unity and which would be expected to change only slightly over the pressure range 50–600 psia. As the pressure is increased from 50 psia, we expect to observe no change in the gas phase composition, and when the pressure of 600 psia is reached, the liquid phase will form. At this temperature liquid and vapor of these fixed compositions can coexist only at 600 psia, and a further increase in the pressure would eliminate the vapor phase. This can be seen from the fact that the fixed vapor and liquid compositions determine the value of the equilibrium ratios, which can be realized only at a single pressure.

(c) Because the solvent is nonvolatile, equilibrium vapor compositions will be unaffected by its addition. Equilibrium with respect to the reaction will be established in the liquid phase so that the equilibrium constant is satisfied:

$$K_L = 0.871 = \frac{x_i}{x_n}$$

The ratio of isomers remains the same, but the mol fractions change in conformance with our specification that

$$x_i + x_n = \frac{1}{3}$$

This results in

$$x_i = 0.155 \qquad x_n = 0.178$$

Again, at this temperature the y's and x's are fixed and hence K_i and K_n. The two phases will coexist at only one pressure. For

$$K_i = \frac{0.523}{0.155} = 3.37$$

and

$$K_n = \frac{0.477}{0.178} = 2.68$$

the pressure is approximately 140 psia.

EXAMPLE 12-10

During the Arab oil embargo of 1973 scarcities spawned "black markets" in petroleum products. In the Midwest some of the black market propane was occasionally sold to rural residents in anhydrous ammonia cylinders. This prompted a local radio station to warn of the dangers of cyanide gas formation. Make a thermodynamic evaluation of the risk of this occurrence.

Solution 12-10

We will first explore possible reactions at ambient temperature. Also, because propane is used mainly as a fuel, we should consider the stability of hydrogen cyanide at combustion temperatures in the presence of excess oxygen. The following equilibrium constants of formation will be useful in our analysis.[23]

	$\log K^f$ at 77°F	$\log K^f$ at 2200°F
HCN(g)	−21.038	−2.878
NH_3(g)	2.831	−4.182
C_3H_8(g)	4.12	−12.18
CH_4(g)	8.8985	−2.5438
C_2H_6(g)	5.76	−7.52

---

[23] Source: reference 7 in Table 12-1.

	$\log K^f$ at 77°F	$\log K^f$ at 2200°F
$C_3H_6(g)$	−10.988	−9.511
$C_5H_{12}(g)$	1.466	−21.442
$CO_2(g)$	69.0915	14.0072
$H_2O(g)$	40.0470	5.8574

At ambient temperature there are many possible reactions to consider with ammonia and propane as reactants. Because $\log K^f$ for both reactants is positive and $\log K^f$ for HCN is a large negative number, the equilibrium constant for any possible reaction will be extremely low unless products with large positive $\log K^f$'s are formed. Definite trends exist in the variation of $\log K^f$ with carbon number within a given family of compounds. For saturated and unsaturated hydrocarbons $\log K^f$ decreases with increasing carbon number, and therefore a reaction in which methane is a product will be the most favorable. Consider the following set of possible reactions and their equilibrium constants:

$$C_3H_8(g) + NH_3(g) = HCN(g) + C_2H_6(g) + 2H_2(g) \qquad K = 5.9(10^{-23})$$

$$C_3H_8(g) + NH_3(g) = HCN(g) + 2CH_4(g) + H_2(g) \qquad K = 6.4(10^{-11})$$

$$C_3H_8(g) + 3NH_3(g) = 3HCN(g) + 7H_2(g) \qquad K = 1.9(10^{-76})$$

$$2C_3H_8(g) + NH_3(g) = HCN(g) + C_5H_{12}(g) + 3H_2(g) \qquad K = 2.3(10^{-31})$$

Although all equilibrium constants are extremely small, the reaction producing methane is the most favorable. Normally, with equilibrium constants this low such reactions would be dismissed; however, because of the extreme toxicity of hydrogen cyanide, the calculations will be carried further. For the reaction producing methane the equilibrium constant is written as

$$K = 6.4(10^{-11}) = \frac{p_{HCN}\, p_{CH_4}^2\, p_{H_2}}{p_{C_3H_8}\, p_{NH_3}}$$

The cylinder will contain both liquid and gaseous propane. If it is assumed that the products are only slightly soluble in the liquid propane, then they exist only in the gas phase, and their partial pressures are related through the stoichiometry of the reaction

$$p_{H_2} = p_{HCN}$$

$$p_{CH_4} = 2p_{HCN}$$

The equilibrium constant now becomes

$$K = 6.4(10^{-11}) = \frac{4\,p_{HCN}^4}{p_{C_3H_8}\,p_{NH_3}}$$

or

$$p_{HCN} = \left[\frac{6.4(10^{-11})}{4}\,p_{C_3H_8}\,p_{NH_3}\right]^{1/4}$$

We now need reasonable estimates for the partial pressures of propane and ammonia. For propane we will use the vapor pressure, 8.8 atm. This reduces to

$$p_{HCN} = 3.4(10^{-3})\,p_{NH_3}^{1/4}$$

It hardly seems likely that the partial pressure of ammonia would exceed 1 atm, and therefore the maximum partial pressure of hydrogen cyanide would be $3.4(10^{-3})$ atm. With these assumed partial pressures the vapor mol fraction of hydrogen cyanide would be $3.5(10^{-4})$ or 350 parts per million. Thus, we may conclude that reactions between propane and ammonia would produce very little hydrogen cyanide.

It should be noted that unsaturated hydrocarbons which might be present in propane as impurities would be more reactive. For example, consider propylene, C_3H_6, and the reaction

$$C_3H_6(g) + NH_3(g) = HCN(g) + 2CH_4(g) \qquad K = 8.2(10^5)$$

The large equilibrium constant indicates that this reaction would go virtually to completion, and the amount of hydrogen cyanide formed would depend on the limiting reactant. All unsaturated hydrocarbons would behave similarly and the maximum mols of hydrogen cyanide possible would equal the mols of unsaturated hydrocarbons originally present with the propane. Because of their extremely small K's unsaturated hydrocarbons probably do not occur in natural petroleum sources, but they are formed in the various processes to which petroleum is subjected in the refinery. Because of their reactivity, they are valuable as feedstocks for various petrochemicals and are therefore separated as completely as possible from propane sold as bottled gas. The total amount of unsaturated hydrocarbons in bottled propane is not expected to exceed 1%; therefore, if sufficient ammonia were present, we could expect that the maximum percentage of hydrogen cyanide would also be 1%.

In the application of thermodynamics we can definitely say that reactions with extremely small equilibrium constants will not proceed to a measurable extent. However, about reactions with large equilibrium constants we can only say that a tendency to react exists. We can calculate the maximum amount of hydrogen cyanide which would be formed, but we know nothing about the rate at which the reaction would proceed.

At temperatures prevailing in a combustion zone, most reactions occur fast enough that equilibrium is approached. When an excess of oxygen is present, we consider the following reaction:

$$HCN(g) + 1\frac{1}{4}O_2(g) = \frac{1}{2}H_2O(g) + CO_2(g) + \frac{1}{2}N_2(g)$$

at 2200°F,

$$K = 6.5(10^{19})$$

With an equilibrium constant this large the combustion gas would be expected to contain only infinitesimal quantities of hydrogen cyanide. Any hydrogen cyanide formed in the cylinder from the reaction of ammonia with unsaturated hydrocarbons should be completely oxidized in a combustion zone.

The only possible danger would arise from leakage from the cylinder into an enclosed space. Exposure to the level of 135 parts per million for a period of 30 minutes is lethal.[24]

EXAMPLE 12-11[25]

Storage of highly volatile and reactive liquids often presents some unusual and interesting problems pertaining to plant safety. One particular problem is the possibility of explosion of an acetaldehyde storage tank in which the acetaldehyde has been contaminated with acid, thus causing the acetaldehyde to trimerize to paraldehyde (acid catalyzes this reaction). The exothermic trimerization reaction raises the temperature, which, in turn, causes a rapid increase in pressure due to the high vapor pressure of acetaldehyde. Hence, prime importance is given to the following question: How large will be the pressure increase, and is there an upper limit to this pressure? The following data are available: At equilibrium a liquid solution of acetaldehyde and paraldehyde contains 60.6 mol % acetaldehyde at 50.5°C and 15.3 mol % acetaldehyde at 15.5°C.

---

[24] *Hazardous Chemical Data*, No. CG-446-2, U.S. Department of Transportation, Washington, DC, Jan. 1974.
[25] Based on the work of I. Pliskin, *Can. J. Chem. Eng., 45,* 327 (1967).

PHYSICAL PROPERTIES

Acetaldehyde	Paraldehyde
Normal boiling point, 20.2°C	Normal boiling point, 124°C
Vapor pressure at 80°C, 5.3 atm	
Liquid heat capacity 23 cal/g mol °C	Liquid heat capacity (estimated), 61 cal/g mol °C

Solution 12-11

The trimerization reaction may be written as

$$3A(l) = P(l)$$

where A is acetaldehyde and P is paraldehyde. On the reasonable assumption that these species form an ideal solution, the equilibrium constant is written as

$$K = \frac{x_P}{x_A^3}$$

and because of binary mixtures, $x_A + x_P = 1$. From the given data we find the following:

$$\text{At } 50.5°C, K = \frac{0.394}{(0.606)^3} = 1.77.$$

$$\text{At } 15.5°C, K = \frac{0.847}{(0.153)^3} = 236.$$

From Eq. (12-19) the standard heat of reaction is found:

$$\ln \frac{1.77}{236} = \frac{\Delta H°}{1.987}\left(\frac{1}{288.65} - \frac{1}{323.65}\right)$$

$$\Delta H° = -26,000 \text{ cal}$$

K may be evaluated at other temperatures from

$$\ln K = \frac{13,085}{T} - 39.8681$$

Because the reaction is exothermic, the temperature will rise as an adiabatic reaction proceeds; however, at higher temperatures the equilibrium conversion decreases, and there will be an upper limit to the extent of reaction. There will also be an upper limit to the temperature. This limiting condition, the adiabatic reaction temperature, may be found from the solution of a first-law equation relating temperature to conversion and an expression relating equilibrium conversion to temperature. The first law expression is determined by requiring $\Delta H = 0$ for the adiabatic process which begins at 298 K and is pictured as follows:

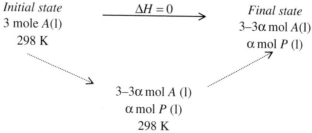

We write

$$\Delta H = 0 = \alpha(-26{,}000) + (3 - 3\alpha)(23)(T - 298) + \alpha(61)(T - 298)$$

In terms of the extent of reaction (or equilibrium conversion) the equilibrium constant is

$$K = \frac{\alpha(3 - 2\alpha)^2}{(3 - 3\alpha)^3}$$

and the known relationship between K and T allows us to determine the equilibrium conversion corresponding to any termperature. This relationship is plotted in Fig. 12-3 along with the first law relationship.

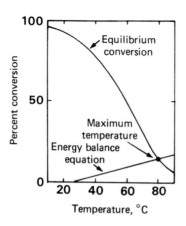

Figure 12-3 Determination of maximum reaction temperature. [Adapted from I. Pliskin, *Can. J. Chem. Eng., 45,* 327 (1967), by permission.]

The maximum temperature, or adiabatic reaction temperature, is found from the intersection of these two curves to be 80° C. This temperature

cannot be exceeded in an adiabatic system. At this tempera-
ture $\alpha = 0.15$, and $x_A = 0.944$. The maximum pressure may be estimated
on the assumption that paraldehyde is nonvolatile. For the ideal solution,

$$P = p_A = x_A P_A^\circ = 0.944(5.3) = 5.0 \text{ atm}$$

Instead of a graphical solution, this problem can be solved with
POLYMATH. The problem is in the POLYMATH library as EX12-11
and the worksheet appears as follows:

POLYMATH EX12-11

Equations:
f(T) = –26000*a + (3 – 3*a)*23*(T – 298) + 61*a*(T – 298)
f(a) = K – a*(3 – 2*a)**2/(3 – 3*a)**3
K = exp(13085/T – 39.8681)

Initial values: $T_0 = 300$, $a_0 = 0.1$

The solution, $T = 352.8$ K and $a = 0.143$ agree with the above values of t
= 80°C and $\alpha = 0.15$.

12-10 REACTIONS IN AQUEOUS SOLUTION

For many reactions which occur in dilute aqueous solution it is advantageous to employ a
special standard state. Because we have seen that the major problem in dealing with reactions in solution is the lack of information concerning the relationship between activity and
composition, a standard state where f° is equal to the Henry's law constant is convenient.
For dilute solutions where Henry's law is obeyed, we write

$$\hat{f}_i^L = k_i x_i \tag{9-55}$$

At low concentrations the mol fraction of a solute is proportional to the molality[26] m, and this
equation may be restated as

$$\hat{f}_i^L = k_i' m_i \tag{12-34}$$

If we set $f_i^\circ = k_i'$ then the activity is

---

[26] Molality = g mol solute/1000 g water.

$$\hat{a}_i = \frac{\hat{f}_i^L}{k_i'}$$ (12-35)

and where Henry's law is obeyed, we have

$$\hat{a}_i = m_i$$ (12-35a)

Thus, with this standard state the activity will approach the molality as the molality approaches zero, and calculations involving the equilibrium constant can be easily carried out.

This standard state may be visualized with the aid of Fig. 12-4, where the fugacity of a component in an aqueous solution is plotted vs. its molality[27] and shown as the solid curve. The linear relationship, Henry's law, which applies at low molality is extrapolated to higher molality and is represented by the dashed line. As indicated in the figure, if a solution were to obey Henry's law all the way to a molality of unity, the fugacity of the solute would be equal to k', the Henry's law constant. The formal description of this standard state is as follows: a molality of unity for a hypothetical solution in which the solute obeys Henry's law over the molality range 0 to 1.[28] This is referred to as the aqueous or hypothetical 1-molal standard state and is denoted by (ao).[29] It is, of course, a hypothetical state, but we have previously seen that so long as properties may be calculated and consistency is observed in the calculations, no problems are encountered.

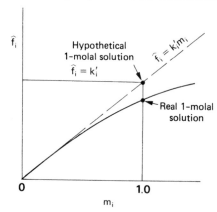

Figure 12-4 Illustration of aqueous standard state.

---

[27] Some type of experimental data (e.g., vapor-liquid equilibrium measurements) is necessary in order to establish such a relationship.

[28] In the unlikely event that the solute actually obeyed Henry's law up to a molality of unity, the standard state of unit molality would be a real state instead of a hypothetical state.

[29] The notation ao has been recently introduced (reference 1-E in Table 12-1) and refers to nondissociating solutes; a complementary notation ai refers to dissociating solutes. These replace the older notation aq, which referred to both types of solutes.

Only for dilute solutions will the activity be equal to the molality of a solute. Where Henry's law is not applicable, activities may be related to molalities through an activity coefficient. We have previously defined an activity coefficient based on Henry's law:

$$\hat{f}_i^{\,L} = k_i x_i \gamma_i^* \qquad\qquad (11\text{-}27a)$$

In terms of molalities we could write this as

$$\hat{f}_i^{\,L} = k_i' m_i \gamma_i^{\square} \qquad\qquad (12\text{-}36)$$

and determine the activity based on the ao standard state via Eq. (12-35):

$$\hat{a}_i = m_i \gamma_i^{\square} \qquad\qquad (12\text{-}37)$$

The activity coefficient $\gamma_i^{\square}$ is proportional to γ_i at low concentrations where x_i and m_i are proportional, but because they are based on different concentration scales, the proportionality does not exist at higher concentrations. Examination of Eqs. (12-37) and (12-35a) shows that $\gamma^{\square}$ approaches unity as the molality approaches zero.

EXAMPLE 12-12

Illustrate the use of the various standard states in dealing with equilibrium calculations for the reaction producing ethyl acetate from ethyl alcohol and acetic acid:

$$C_2H_5OH + CH_3COOH = CH_3COOC_2H_5 + H_2O$$

$$\text{Et} \qquad \text{AA} \qquad\qquad \text{EA} \qquad\qquad \text{W}$$

Solution 12-12

Consider first the case where the reaction occurs in the gas phase at 298 K:

$$\text{Et(g)} + \text{AA(g)} = \text{EA(g)} + \text{W(g)} \qquad\qquad (a)$$

The free energies of formation are as follows:[30]

	ΔG^f (g) (cal)
Ethyl acetate, EA	−78,250
Water, W	−54,640
Acetic acid, AA	−90,030
Ethyl alcohol, Et	−40,220

---

[30] Source: reference 3 in Table 12-1.

And for reaction (a) the standard Gibbs free energy change and equilibrium constant are

$$\Delta G_a^\circ = -78{,}250 - 54{,}640 - (-90{,}030) - (-40{,}220)$$

$$= -2640 \text{ cal}$$

$$K_a = \exp\left[\frac{2640}{1.987(298)}\right] = 86.3$$

For each species the standard state is the pure gas at a fugacity of 1 atm, and for low pressure we write

$$K_a = 86.3 = \frac{p_{EA}p_W}{p_{AA}p_{Et}} = \frac{y_{EA}y_W}{y_{AA}y_{Et}}$$

Considering the reaction of an equimolar mixture of acetic acid, AA, and ethyl alcohol, Et, we have

$$v_{AA} = -1 \qquad v_{Et} = -1 \qquad v_{EA} = 1 \qquad v_W = 1$$
$$n_{0AA} = 1 \qquad n_{0Et} = 1 \qquad n_{0EA} = 0 \qquad n_{0W} = 0$$

$$n_{EA} = n_W = \alpha \qquad y_{EA} = y_W = \frac{\alpha}{2}$$

$$n_{AA} = n_{Et} = 1 - \alpha \qquad y_{AA} = y_{Et} = \frac{1-\alpha}{2}$$

$$n_{\text{total}} = 2$$

$$K_a = 86.3 = \frac{(\alpha/2)^2}{[(1-\alpha)/2]^2} = \left(\frac{\alpha}{1-\alpha}\right)^2$$

$$\alpha = 0.903$$

With this extent of reaction the equilibrium mol fractions are

$$y_{EA} = y_W = 0.4515 \qquad y_{AA} = y_{Et} = 0.0485$$

We will now consider the reaction carried out in the liquid phase and will use for each species the standard state of pure liquid:

$$Et(l) + AA(l) = EA(l) + W(l) \tag{b}$$

In Ex. 12-3 we saw how $\Delta G^f(g)$ and $\Delta G^f(l)$ are related. At 298 K the vapor pressures and $\Delta G^f(l)$ values are as follows:[31]

	$P°$ (atm)	$\Delta G^f(l)$ (cal)
Ethyl acetate, EA	0.126	−79,479
Water, W	0.0314	−56,690
Acetic acid, AA	0.0171	−92,439
Ethyl alcohol, Et	0.0776	−41,733

For reaction (b) we find that

$$\Delta G_b^° = -79{,}479 - 56{,}690 - (-92{,}439) - (-41{,}733)$$

$$= -1997$$

and that

$$K_b = \exp\left[\frac{1997}{1.987(298)}\right]$$

$$= 29.2 = \frac{\hat{a}_{EA}\hat{a}_W}{\hat{a}_{AA}\hat{a}_{Et}}$$

In this liquid solution we may hardly expect ideal solutions and therefore must employ Eq. (12-33) to write

$$K_b = 29.2 = \frac{x_{EA}\,x_W}{x_{AA}\,x_{Et}}\left(\frac{\gamma_{EA}\,\gamma_W}{\gamma_{AA}\,\gamma_{Et}}\right)$$

Without some experimental data for the quaternary system the term in parentheses which characterizes the liquid-phase behavior cannot be evaluated, and we are unable to calculate equilibrium compositions.

It should be noted that the preceding expression could have been obtained from K_a. By visualizing vapor and liquid phases in equilibrium, for each component we write

$$p_i = P_i^° x_i \gamma_i \tag{9-51}$$

and substituting into K_a, we obtain

---

[31] Source: reference 3 in Table 12-1.

$$K_a = \frac{p_{EA}\, p_W}{p_{AA}\, p_{Et}} = \frac{x_{EA}\, x_W}{x_{AA}\, x_{Et}} \left(\frac{\gamma_{EA}\, \gamma_W}{\gamma_{AA}\, \gamma_{Et}} \right) \frac{P^\circ_{EA}\, P^\circ_W}{P^\circ_{AA}\, P^\circ_{Et}}$$

$$K_a \left(\frac{P^\circ_{AA}\, P^\circ_{Et}}{P^\circ_{EA}\, P^\circ_W} \right) = \frac{x_{EA}\, x_W}{x_{AA}\, x_{Et}} \left(\frac{\gamma_{EA}\, \gamma_W}{\gamma_{AA}\, \gamma_{Et}} \right) = K_b$$

$$= \frac{86.3(0.0171)(0.0776)}{0.126(0.0314)} = 29.2$$

We now consider the reaction carried out in dilute aqueous solution and employ the ao standard state for all species except water:

$$Et(ao) + AA(ao) = EA(ao) + W(l) \tag{c}$$

Values of $\Delta G^f(ao)$ are available for ethanol and acetic acid;[32] the value for ethyl acetate may be calculated from $\Delta G^f(g)$ via the following two-step path:

$$f^\circ = 1 \text{ atm} \xrightarrow{\Delta G_1} f = k' \xrightarrow{\Delta G_2} \text{hypothetical solution, } m = 1$$

$$\text{pure gas} \qquad\qquad \text{pure gas}$$

In the first step the pure gas is expanded isothermally from a fugacity of 1 atm to a fugacity equal to the Henry's law constant:

$$\Delta G_1 = RT \ln \frac{k'}{1} = RT \ln k'$$

In the second step 1 mol of the gas is dissolved in an aqueous solution of $m = 1$. The quantity of solution is large enough so that no appreciable composition change occurs. For this equilibrium step $\Delta G_2 = 0$, and we have

$$\Delta G^f(ao) = \Delta G^f(g) + RT \ln k' \tag{12-38}$$

The ethyl acetate–water system exhibits partial liquid miscibility and hence is highly nonideal; however, parameters in an activity coefficient equation have been determined from an experimental study of vapor-liquid and liquid-liquid equilibrium.[33] From these parameters γ^∞ for

[32] Source: reference 1E in Table 12-1.
[33] F. van Zandijcke and L. Verhoeye, *J. Appl. Chem. Biotechnol.* **24**, 709 (1974).

EA is found to be 50.2. The Henry's law constant, defined by Eq. (9-55) and calculated via Eq. (10-16i), is

$$k = 50.2(0.126) = 6.33 \text{ atm}$$

Multiplying by 0.0180 converts this into k', defined by Eq. (12-34):

$$k' = 0.0180(6.33) = 0.115 \text{ atm / unit molality}$$

With this Henry's law constant $\Delta G^f(\text{ao})$ is calculated via Eq. (12-38). For the three organic compounds we have the following.

	ΔG^f (ao) (cal)
Acetic acid, AA	−94,690
Ethyl alcohol, Et	−43,120
Ethyl acetate, EA	−79,526

For reaction (c) we find that

$$\Delta G_c^\circ = -79,526 - 56,690 - (-94,960) - (-43,120)$$

$$= 1594 \text{ cal}$$

$$K_c = 0.0677$$

With these standard states the equilibrium constant is written as

$$K_c = 0.0677 = \frac{m_{EA} x_W}{m_{AA} m_{Et}} \left[\frac{\gamma_{EA}^\square \gamma_W}{\gamma_{AA}^\square \gamma_{Et}^\square} \right]$$

As written, this expression is no more useful than that for K_b; however, for the special case of dilute solutions where solutes obey Henry's law and water obeys Raoult's law, it reduces to

$$0.0677 = \frac{m_{EA} (1)}{m_{AA} m_{Et}} \left[\frac{1(1)}{1(1)} \right]$$

$$= \frac{m_{EA}}{m_{AA} m_{Et}}$$

Let us start with a dilute equimolar mixture of acetic acid and ethanol:

$$n_{0EA} = 0 \qquad n_{0W} = 1000 \qquad n_{0AA} = 1 \qquad n_{0Et} = 1$$

At any extent of reaction we will have

$$n_{EA} = \alpha$$

$$n_{Et} = n_{AA} = 1 - \alpha$$

$$n_W = 1000 + \alpha$$

$$\text{grams of water} = (1000 + \alpha)18 \doteq 18{,}000$$

$$m_{EA} = \frac{\alpha}{18}$$

$$m_{AA} = m_{Et} = \frac{1 - \alpha}{18}$$

The equilibrium constant is now

$$K_c = 0.0677 = \frac{\alpha/18}{(1-\alpha)^2/18^2} = \frac{18\alpha}{(1-\alpha)^2}$$

The equilibrium extent of reaction is

$$\alpha = 0.00376$$

and equilibrium molalities are

$$m_{EA} = \frac{0.00376}{18} = 2.09(10^{-4})$$

$$m_{AA} = m_{Et} = \frac{1 - 0.00376}{18} = 5.55(10^{-2})$$

The α of 0.00376 for the reaction in dilute aqueous solution differs considerably from the α of 0.903 for the gas-phase reaction. There are two reasons for this: liquid-phase solution effects and the presence of a large amount of product (water) which forces the equilibrium to the left.

Even though $\ln K_a > 0$ and $\ln K_c < 0$, there is still consistency between these two equilibrium constants. This can be demonstrated by caculating the equilibrium partial pressures corresponding to the equilibrium composition found for the dilute aqueous reaction. The Henry's law constants required for this caclulation may be obtained through the use of Eq. (12-38). For acetic acid and ethanol these are, respectively, $3.82(10^{-4})$ and $7.46(10^{-3})$ atm/unit molality. The partial pressures are

$$p_W = P_W^\circ = 0.0314 \text{ atm}$$

$$p_{EA} = k'm = 0.115(2.09)(10^{-4}) = 2.40(10^{-5}) \text{ atm}$$

$$p_{Et} = k'm = 0.00746(0.0555) = 4.14(10^{-4}) \text{ atm}$$

$$p_{AA} = k'm = 0.000382(0.0555) = 2.12(10^{-5}) \text{ atm}$$

Substituting these partial pressures into the equilibrium constant expression for reaction (a) gives

$$K_a = \frac{0.0314(2.40)(10^{-5})}{4.14(10^{-4})(2.12)(10^{-5})} = 85.9$$

This simply emphasizes the fact that reaction equilibrium prevails within a phase and that phase equilibrium prevails between phases.

It is to be noted that liquid phase behavior must be known in order to utilize either the liquid or ao standard state. However, since use of the ao standard state implies application to dilute solution, only binary solute–water data are required to execute the calculations.[34] These data are more likely to be available than the multicomponent data required for systems outside this restricted composition range where one is more apt to apply the liquid standard state.

One final note about standard states: They may be chosen for convenience of calculation; however, there must be consistency between those used to determine $\Delta G°$ and the corresponding activities in the equilibrium constant expression.

12-11 ELECTROLYTE SOLUTIONS

Strong Electrolytes.   From our recognition that electrolytes in solution are at least partially—and often completely—dissociated, we would expect the thermodynamic treatment of these solutions to differ significantly from that which we have developed for nonelectrolyte solutions. The difference is immediately apparent when we examine the behavior of dilute aqueous solutions of HCl. As shown in Fig. 12-5, a plot of partial pressure of HCl vs. molality, m, exhibits no linearity as m approaches zero, but rather suggests a limiting slope of zero. This implies a Henry's law constant, or standard state fugacity, of zero, which will not be useful.

The problem of a zero Henry's law constant can be avoided when it is noted that for this system a plot of partial pressure vs. m^2, as shown in Fig. 12-6, becomes linear as m^2 approaches zero. This allows us to write as the limiting relation

$$p_{HCl} = \hat{f}_{HCl} = k_{HCl}\, m_{HCl}^2 \tag{12-39}$$

The explanation for this relation lies in the dissociation of HCl into ionic species according to

$$HCl = H^+ + Cl^-$$

---

[34] While the dilute aqueous reacting system is in reality a multicomponent system, the dilution is sufficient that solute molecules "see" predominantly water molecules, and thus the behavior of each solute closely approaches its behavior in the solute–water binary system.

Invoking Eqs. (12-6) and (12-11) we write for this reaction

$$\mu_{H^+} + \mu_{Cl^-} = \hat{\mu}_{HCl} \tag{12-40a}$$

and

$$K = \frac{a_{H^+} a_{Cl^-}}{\hat{a}_{HCl}} \tag{12-40b}$$

and rearrange the last equation to

$$\hat{a}_{HCl} = \frac{\hat{f}_{HCl}}{f_{HCl}^\circ} = \frac{1}{K} a_{H^+} a_{Cl^-} \tag{12-41}$$

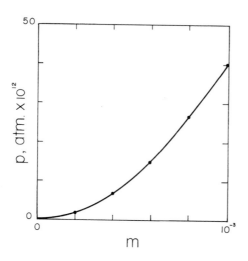

Figure 12-5 Partial pressure HCl vs. molality for aqueous HCl solutions at 25°C. [Constructed from data compiled by G. N. Lewis, M. Randall, K. S. Pitzer, and L. Brewer, *Thermodynamics*, 2nd ed., McGraw-Hill, New York, 1961, Chap. 22.]

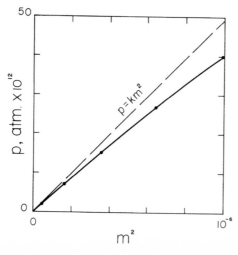

Figure 12-6 Partial pressure of HCl vs. the squre of molality for aqueous HCl solutions at 25°C. [Constructed from data compiled by G. N. Lewis, M. Randall, K. S. Pitzer, and L. Brewer, *Thermodynamics*, 2nd ed., McGraw-Hill, New York, 1961, Chap. 22.]

The numerical value of the equilibrium constant K depends on the standard states chosen for the various activities. For the ionic species the ao standard state is chosen so that

$$\lim_{m \to 0} a_{H^+} = m_{H^+} \tag{12-42a}$$

$$\lim_{m \to 0} a_{Cl^-} = m_{Cl^-} \tag{12-42b}$$

At infinite dilution $(m \to 0)$ there is good evidence to support the assumption of complete dissociation and thus we may state[35]

$$m_{H^+} = m_{Cl^-} = m \tag{12-43}$$

and use this relation along with Eq. (12-42) in Eq. (12-41) to obtain

$$\hat{f}_{HCl} = \frac{f^{\circ}_{HCl}}{K} m^2 \tag{12-44}$$

At low pressure where fugacity may be replaced by partial pressure, Eq. (12-44) predicts a limiting slope of f°_{HCl} / K for Fig. 12-6. A comparison of Eqs. (12-39) and (12-44) shows that

$$k_{HCl} = \frac{f^{\circ}_{HCl}}{K}$$

and we see that K will have the value of unity if we choose f°_{HCl} equal to k_{HCl}. In making this choice we have defined the standard state for "HCl in solution" as a hypothetical 1-molal solution obeying Eq. (12-39) up to a molality of unity, and we see that in the region where Eq. (12-39) is followed,

$$\lim_{m \to 0} \hat{a} = m^2 \tag{12-45}$$

It will be more convenient to work with an activity that approaches the first power of the molality and therefore we define the *mean ion activity*, $a_{\pm}$:

$$a_{\pm} = (a_{H^+} a_{Cl^-})^{1/2} \tag{12-46}$$

Substituting Eq. (12-46) into Eq. (12-40b) with K set equal to unity gives

$$\hat{a} = (a_{\pm})^2 \tag{12-47}$$

which allows us to restate Eq. (12-45):

$$\lim_{m \to 0} (a_{\pm})^2 = m^2 \tag{12-48a}$$

---

[35] m is the molality of the electrolye "as a whole" as would be determined from the preparation of the solution.

or

$$\lim_{m \to 0} a_{\pm} = m \tag{12-48b}$$

Thus, $a_{\pm}$ shows the desired limiting behavior at infinite dilution. At higher molalities it is necessary to relate $a_{\pm}$ to m through the *mean ion activity coefficient*, $\gamma_{\pm}$:

$$a_{\pm} = m\gamma_{\pm} \tag{12-49}$$

which shows the following limiting behavior:

$$\lim_{m \to 0} \gamma_{\pm} = 1 \tag{12-50}$$

Up to this point we have considered a specific system: HCl–water. Because this is one of the simplest possible electrolyte systems, it affords us the best opportunity of visualizing a physical picture and grasping the rationale behind the use of the mean ion activity $a_{\pm}$. The concepts we have developed for this system can now be generalized by considering the electrolyte $C_{v+}A_{v-}$ which dissociates into v_{+} cations of charge z_{+} and v_{-} anions of charge z_{-}:

$$C_{v+}A_{v-} = v_{+}C^{z+} + v_{-}A^{z-}$$

Electrical neutrality is always maintained and requires that

$$v_{+}z_{+} + v_{-}z_{-} = 0 \tag{12-51}$$

The equilibrium constant for dissociation is written

$$K = \frac{a_{+}^{v_{+}} a_{-}^{v_{-}}}{\hat{a}} \tag{12-52}$$

and again with K equal to unity, we obtain

$$\hat{a} = a_{+}^{v_{+}} a_{-}^{v_{-}} \tag{12-53}$$

The mean ion activity is defined as

$$a_{\pm} = \left[a_{+}^{v_{+}} a_{-}^{v_{-}} \right]^{1/v} \tag{12-54}$$

where

$$v = v_{+} + v_{-} \tag{12-55}$$

For the mean ion activity coefficient, $\gamma_{\pm}$, to conform to the limiting condition expressed by Eq. (12-50), it is necessary to define a *mean ion molality*:

$$m_{\pm} = \left[m_{+}^{v_{+}} m_{-}^{v_{-}} \right]^{1/v} \tag{12-56}$$

With $m_\pm$ as the concentration variable, the mean ion activity coefficient, defined by

$$a_\pm = m_\pm \gamma_\pm \tag{12-57}$$

approaches unity at infinite dilution.

Combining Eqs. (12-53) and (12-54) gives

$$\hat{a} = a_\pm^v \tag{12-58}$$

In recalling the aqueous HCl system, we are reminded that $\hat{a}$ is the quantity that determines the partial pressure; hence the phase behavior of "HCl in solution." Thus, we identify $\hat{a}$ as the activity of the "electrolyte in solution" and see that it is a thermodynamically meaningful quantity. As we have seen, however, there is a disadvantage to employing $\hat{a}$; at infinite dilution

$$\lim_{m_\pm \to 0} \hat{a} = m_\pm^v \tag{12-59}$$

The use of $a_\pm = \hat{a}^{1/v}$ removes this difficulty and produces the desired linearity

$$\lim_{m_\pm \to 0} a_\pm = m_\pm \tag{12-60}$$

or

$$\lim_{m_\pm \to 0} \gamma_\pm = 1 \tag{12-61}$$

For this reason electrolyte solutions are treated in terms of mean ion molality, mean ion activity, and mean ion activity coefficient.

The mean ion activity coefficients of several common electrolytes are plotted vs. molality in Fig. 12-7, where it is observed that only in extremely dilute solution does $\gamma_\pm$ closely approach unity. This is because of the long-range nature of coulombic forces operating between ions. These activity coefficients can be determined from several types of measurements. In the case of a volatile substance such as HCl, vapor-liquid equilibrium data could be used. For nonvolatile electrolytes, measurements from electrochemical cells are frequently used.[36] It is also possible to determine solvent activity coefficients from partial pressure measurements with the subsequent use of the Gibbs-Duhem equation to calculate mean ion activity coefficients of the electrolyte solute.[37]

Reliable values of $\gamma_\pm$ must be obtained experimentally, although it is possible to obtain reasonable estimates in very dilute solution from the theoretically derived Debye-Hückel equation. This theory is based on the assumption that deviations from Henry's law are due to

---

[36] See, for example, G. N. Lewis, M. Randall, K. S. Pitzer, and L. Brewer, *Thermodynamics*, 2nd ed., McGraw-Hill, New York, 1961, Chap. 22.

[37] *Ibid.*

electrostatic interaction of the ions and when simplified for dilute solutions gives the activity coefficient of an ionic species as

$$\log \gamma_i^{\square} = -A z_i^2 I^{1/2} \qquad (12\text{-}62)$$

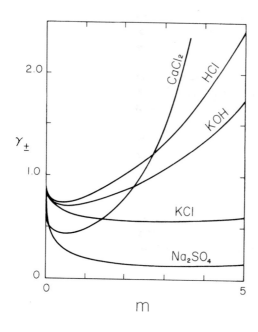

Figure 12-7 Mean ion activity coefficients of selected electrolytes at 25°C. [Constructed from data compiled by R. A. Robinson and R. H. Stokes, *Electrolyte Solutions*, 2nd ed., Butterworth Scientific Publications, London, 1959.]

where I is the ionic strength, defined as

$$I = \frac{1}{2} \sum z_i^2 m_i \qquad (12\text{-}63)$$

and includes contributions from all ionic species present in the solution.

Although it is not possible to experimentally determine $\gamma_i^{\square}$ for an individual ionic species, nevertheless, in certain types of calculations this is a useful quantity. The more commonly used mean molal activity coefficient can be obtained from Eq. (12-62). Utilizing Eq. (12-51) and Eqs. (12-54)–(12-57), one obtains (after some algebraic manipulation)

$$\log \gamma_{\pm} = -A |z_+ z_-| I^{1/2} \qquad (12\text{-}64)$$

The parameter A in Eqs. (12-62) and (12-64) is temperature dependent; at 298 K it has the value 0.5085. The accuracy of this equation improves as I approaches zero; it is seldom used at ionic strengths above 0.01. It should be noted that in Fig. 12-7 the three curves for $|z_+ z_-| = 1$ coalesce as m approaches zero, as do the two curves for $|z_+ z_-| = 2$. This is in accordance with the Debye-Hückel equation.

Several theoretically inspired, empirically employed equations are available for the correlation of mean ion activity coefficients. These equations cover a wide range of molality and apply both to single and mixed electrolyte solutions; they have been collected, described, and evaluated.[38]

EXAMPLE 12-13

Use the following selected data from Figs. 12-5 and 12-6 to determine $\gamma_\pm$ for aqueous HCl. Also, calculate $\gamma_\pm$ from the Debye-Hückel theory.

m (g mol/kg)	0.0005	0.001	0.01
p (atm)	$1.17(10^{-13})$	$4.58(10^{-13})$	$4.03(10^{-11})$

The limiting slope of the isotherm in Fig. 12-6 is

$$\lim_{m^2 \to 0} \frac{dp}{d(m^2)} = k = 4.92(10^{-7}) \frac{\text{atm}}{(\text{g mol}/\text{kg})^2}$$

Solution 12-13

Combining Eqs. (12-53), (12-54), and (12-57) we write

$$\hat{a} = (m_\pm \gamma_\pm)^2$$

where

$$\hat{a} = \frac{\hat{f}}{k}$$

Combining these equations, replacing fugacity with partial pressure, and noting that for this system $m_\pm = m$, we have

$$p = k(m\gamma_\pm)^2$$

and may calculate $\gamma_\pm$ for each of the data points. For this solute $I = m$ and the calculation of $\gamma_\pm$ from Eq. (12-64) is straightforward. The results are summarized in the following table.

---

[38] J. F. Zemaitis, Jr., D. M. Clark, M. Rafal, and N. C. Scrivner, *Handbook of Aqueous Electrolyte Thermodynamics*, American Institute of Chemical Engineers, New York, 1986.

Molality	0.0005	0.001	0.01
$\gamma_\pm$ (experimental)	0.975	0.964	0.905
$\gamma_\pm$ (Eq. 12-64)	0.974	0.964	0.890

Even at these very low molalities $\gamma_\pm$ does not closely approach unity. On the other hand, the Debye-Hückel theory is quite effective in this range.

Note: The data used to construct Figs. 12-5 and 12-6 are not directly measured partial pressures but were calculated from electrochemical measurements which can be carried out in extremely dilute solution. For details of this calculation see Lewis et al.[39]

Weak Electrolytes. For strong electrolytes that lose their identity in solution through dissociation, the activity is defined in terms of the activities of the constituent ions, $a_\pm$. The setting to zero of $\Delta G°$ for the ionization reaction defines the standard state, which is denoted by ai.

In contrast to strong electrolytes, weak electrolytes are only slightly ionized. Because undissociated molecules are the predominant species in solution, the activity of the solute is usually based on its actual molality, as was done with nondissociating solutes where we employed Eqs. (12-34) and (12-35) and identified this standard state by ao. Acetic acid is a good example of a weak electrolyte. It dissociates according to

$$CH_3COOH = H^+ + CH_3COO^-$$

$$HA \quad H^+ \quad A^-$$

Experimental study has established the value of the equilibrium constant for this reaction:[40]

$$K = 1.758(10^{-5}) = \frac{a_{H^+} a_{A^-}}{\hat{a}_{HA}}$$

We have seen that the activity of an ionic species is specified so as to approach the molality at infinite dilution—the ao standard state. Also, the activity of undissociated acetic acid approaches its molality at infinite dilution when the ao standard state is used. Therefore, in dilute solution we may state

$$a_{H^+} = a_{A^-} = \alpha$$

[39] G. N. Lewis et al., *op. cit.*

[40] D. A. MacInnes and T. Shedlovsky, *J. Am. Chem. Soc., 54,* 1429 (1932).

and

$$\hat{a}_{HA} = m - \alpha$$

where α is the extent of ionization and m is the stoichiometric molality of acetic acid. The value of α is seen to depend on m:

$$1.758(10^{-5}) = \frac{\alpha^2}{m - \alpha}$$

More informative is the fraction ionized, α/m. This has been calculated for several values of m:

m	10^{-1}	10^{-3}	10^{-5}	10^{-7}
α/m	0.013	0.124	0.712	0.986

Experience leads us to expect that as a solution is diluted, an undissociated solute would display conformance to Henry's law before the molality reached the 10^{-1} level. Thus, our calculations for acetic acid show that dissociation could be ignored and an extrapolation to zero on a fairly coarse scale of molality should reveal the Henry's law region defined by Eq. (12-34). This is shown in Fig. 12-8. On the other hand, our figures show that in ultra-dilute solutions one could expect partial pressure proportional to m^2—the typical strong electrolyte behavior.

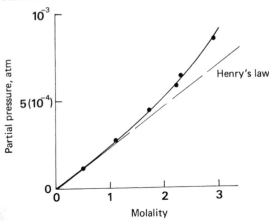

Figure 12-8 Determination of Henry's law constant for acetic acid in water. Constructed from data in reference B-2 in Table 10-6.

Although it is conceivable that a weak electrolyte could be treated on the same basis as a strong electrolyte and be referred to the ai standard state, this would be realistic only for ultra-dilute solutions. We are normally interested in solutions of higher concentration where there is little dissociation and an activity based on the ao standard state at least approximates an actual molality. The activities of all ionic species are defined so that they approach molalities at infinite dilution (the ao standard state) regardless of whether the original solute was a strong or weak electrolyte. For this reason the Debye-Hückel theory can be applied to solutions of both strong and weak electrolytes.

Free Energy Relations. The convenience of mean ion activities based on a hypothetical 1-molal solution has been demonstrated. Thermochemical data referring to this state are identified by the notation ai and can be determined by way of the chemical potential. We begin with Eq. (12-10) and employ Eq. (12-58) to obtain

$$\hat{\mu}_i = \mu_i^{\pm} + v\,RT \ln a_{\pm} \tag{12-65}$$

where $\mu_i^{\pm}$ is the chemical potential of the electrolyte in the ai standard state. This equation allows us to deal with phase equilibrium involving an aqueous solution of the electrolyte. Two cases will be of interest: solid-solution equilibrium and gas-solution equilibrium.

For a pure solid electrolyte in equilibrium with its saturated aqueous solution we state

$$\mu_i^S = \hat{\mu}_i$$

and use Eq. (12-65) to write

$$\mu_i^{\pm} - \mu_i^S = -RT \ln a_{\pm}^v \tag{12-66}$$

The left-hand side is the Gibbs free energy change between two standard states: the ai state and the pure crystalline state. It is easily shown that

$$\mu_i^{\pm} - \mu_i^S = \Delta G^f(\text{ai}) - \Delta G^f(\text{c})$$

EXAMPLE 12-14

The reported value of ΔG^f at 298 K for NaCl(c) is –91,785 cal/g mol.[41] Determine the value of ΔG^f for NaCl(ai) at this temperature.

Solution 12-14

Restatement of Eq. (12-66) as

$$\Delta G^f(\text{ai}) - \Delta G^f(\text{c}) = \mu_i^{\pm} - \mu_i^S = -v\,RT \ln m_{\pm}\gamma_{\pm}$$

shows that the desired quantity can be determined from $m_{\pm}$ and $\gamma_{\pm}$ at saturation. These values are available[42] and are

$$m = m_{\pm} = 6.144 \text{ g mol} / \text{kg}$$

---

[41] Source: reference 1-A in Table 12-1.
[42] J. F. Zemaitis et al., *op. cit.*

and

$$\gamma_{\pm} = 1.005$$

Thus, we have

$$\Delta G^f \text{(ai)} = -91{,}785 - 2(1.987)(298.15) \ln[6.144(1.005)]$$

$$\Delta G^f \text{(ai)} = -93{,}942 \text{ cal} / \text{g mol}$$

$$= -393.3 \text{ MJ} / \text{kmol}$$

Because the left-hand side of Eq. (12-66) depends only on temperature, the argument of the logarithm, $a_{\pm}^{\nu}$, is constant at any temperature. This constant is referred to as the solubility product, S.P., and can be expressed as

$$\text{S.P.} = (m_{\pm}\gamma_{\pm})^{\nu} \qquad (12\text{-}67)$$

For the binary system of electrolyte and water, there is, of course, only a single value of $m_{\pm}$ corresponding to saturation. However, if other electrolytes are present in solution, the value of $\gamma_{\pm}$ and hence the value of $m_{\pm}$ will change. The solubility product concept is used almost exclusively for electrolytes of extremely low solubility and therefore is applied at molalities low enough for the Debye-Hückel theory to be applicable. Equation (12-67) taken together with Eq. (12-64) indicates that the increasing of I by the addition of another electrolyte will lower $\gamma_{\pm}$ and thereby increase $m_{\pm}$. Thus, the solubility of a sparingly soluble electrolyte should increase with the addition of another electrolyte to the solution.

EXAMPLE 12-15

At 25°C the solubility of PbI_2 in water is 0.00166 g mol/kg, in 0.01 molal NaCl solution it increases to 0.00186 g mol/kg, but in 0.01 molal KI solution it decreases to 0.00028 g mol/kg.[43] Is our treatment of electrolyte solutions capable of explaining this effect?

Solution 12-15

From the solubility in water we can calculate the solubility product, S.P.

---

[43] Source: reference D2 in Table 10-6.

$$\frac{1}{3}\log \text{S.P.} = \log m_{\pm} + \log \gamma_{\pm}$$

where

$$m_{+} = m_{\text{Pb}^{++}} = m$$

$$m_{-} = m_{I^{-}} = 2m$$

$$m_{\pm} = [m(2m)^2]^{1/3} = 1.587m$$

$$I = \frac{1}{2}[2^2 m + 2m] = 3m$$

and from Eq. (12-64)

$$\log \gamma_{\pm} = -0.5085(2)(3m)^{1/2}$$

With $m = 0.00166$ we obtain

$$\frac{1}{3}\log \text{S.P.} = -2.651$$

For the other solutions saturated with PbI_2 we can now state

$$\frac{1}{3}\log \text{S.P.} = -2.651 = \log m_{\pm} - 0.5085(2)I^{1/2} \qquad \text{(a)}$$

For the NaCl solution we write

$$m_{\text{Pb}^{++}} = m \quad m_{I^{-}} = 2m \quad m_{\text{Na}^{+}} = m_{\text{Cl}^{-}} = 0.01$$

$$m_{\pm} = [m(2m)^2]^{1/3} = 1.587m$$

$$I = \frac{1}{2}[2^2 m + 2m + 2(0.01)]$$

$$I = 3m + 0.01$$

When substituted into Eq. (a), these values of $m_{\pm}$ and I yield

$$m = 0.00189$$

For the KI solution we write

$$m_{Pb^{++}} = m \qquad m_{I^-} = 2m + 0.01 \qquad m_{K^+} = 0.01$$

$$m_{\pm} = [m(2m + 0.01)^2]^{1/3}$$

$$I = \frac{1}{2}[2^2 m + 2m + 2(0.01)]$$

$$I = 3m + 0.01$$

Using these values of $m_{\pm}$ and I in Eq. (a) gives

$$m = 0.00024$$

For both systems the calculated and experimental values are in good agreement. The large decrease in PbI_2 solubility in the presence of KI is due to the need to include all sources of I^- ions in $m_{\pm}$ and is referred to as the common ion effect.

For a gaseous electrolyte in equilibrium with its aqueous solution we write

$$\hat{\mu}_i^G = \hat{\mu}_i$$

and substitute Eqs. (9-6) and (12-65) to obtain

$$\mu_i^{\pm} - \mu_i^{\circ} = RT \ln \frac{\hat{f}_i}{a_{\pm}^{\nu}} \tag{12-68}$$

The left-hand side of Eq. (12-68) is the Gibbs free energy difference between the ai standard state and the pure-gas standard state ($f^{\circ} = 1$ atm) and, as seen next, can be calculated from measurable quantities.

EXAMPLE 12-16

Table D-2 shows the value of $\Delta G'$ of HCl(g) to be –95,330 kJ/kmol at 298 K. Calculate $\Delta G'$ of HCl(ai) at this temperature.

Solution 12-16

As applied to this system, we write Eq. (12-68)

$$\Delta G^f(\text{ai}) - \Delta G^f(\text{g}) = \mu_i^{\pm} - \mu_i^{\circ} = RT \ln \frac{p_i}{(m_{\pm}\gamma_{\pm})^2}$$

where partial pressure has replaced fugacity. From data of Ex. 12-13 we select a VLE data point

$$m = m_{\pm} = 0.01 \text{ g mol / kg}$$

$$p_i = 4.03(10^{-11}) \text{ atm}$$

where we had found $\gamma_{\pm}$ to be 0.905 and we obtain

$$\Delta G^f(\text{ai}) = -95{,}330 + 8.314(298.15) \ln \frac{4.03(10^{-11})}{[0.01(0.905)]^2}$$

$$\Delta G^f(\text{ai}) = -131{,}333 \text{ kJ / kmol}$$

While it is not possible to separate the effects of the cations and anions in solution, it is possible by a suitable convention to determine a consistent table of thermochemical properties for individual ions. Consider the dissociation of our generalized electrolyte, for which Eq. (12-52) expresses the dissociation equilibrium constant. Because we have set K equal to unity, it follows that $\Delta G^{\circ} = 0$ and from Eq. (12-14) that we may state

$$\Delta G^{\circ} = 0 = \nu_{+}\Delta G_{C^{z+}}^f + \nu_{-}\Delta G_{A^{z-}}^f - \Delta G_{C_{\nu_{+}}A_{\nu_{-}}(\text{ai})}^f$$

The last right-hand term has physical significance and, as we have seen, can be determined. The anion and cation terms have no separate physical significance but can be assigned a numerical value if we adopt the convention that the Gibbs free energy of formation of the hydrogen ion is zero. To illustrate this approach we begin with HCl:

$$\Delta G^{\circ} = 0 = \Delta G_{\text{H}^+}^f + \Delta G_{\text{Cl}^-}^f - \Delta G_{\text{HCl(ai)}}^f$$

$$\Delta G_{\text{Cl}^-}^f = \Delta G_{\text{HCl(ai)}}^f = -131.3 \text{ MJ / kmol}$$

With this value for $\Delta G_{\text{Cl}^-}^f$ we can determine $\Delta G_{\text{Na}^+}^f$ from $\Delta G_{\text{NaCl(ai)}}^f$:

$$\Delta G_{\text{Na}^+}^f = \Delta G_{\text{NaCl(ai)}}^f - \Delta G_{\text{Cl}^-}^f$$

$$\Delta G^f_{Na^+} = -393.3 - (-131.3) = -262.0 \text{ MJ / kmol}$$

Proceeding in like fashion a tabulation of ΔG^f for ions can be developed; Table D-3 contains values for some of the more common ions.

EXAMPLE 12-17

Use the data of Tables D-2 and D-3 to calculate the solubility of PbI_2 in water at 25°C.

Solution 12-17

The value of ΔG^f (ai) needed in Eq. (12-66) is first obtained:

$$\Delta G^f_{PbI_2 \, (ai)} = \Delta G^f_{Pb^{++}} + 2\Delta G^f_{I^-}$$

$$\Delta G^f_{PbI_2 \, (ai)} = -24.3 + 2(-51.71)$$

$$= -127.7$$

With $\Delta G^f_{PbI_2 (c)} = -173.9$ we obtain, via Eq. (12-66),

$$\ln m_{\pm}\gamma_{\pm} = -\frac{[-127.7 - (-173.9)](1000)}{3(8.314)(298.15)}$$

$$m_{\pm}\gamma_{\pm} = 0.00200$$

Now

$$m_+ = m \qquad v_+ = 1 \qquad z_+ = 2$$

$$m_- = 2m \qquad v_- = 2 \qquad z_- = 1$$

and

$$m_{\pm} = [m(2m)^2]^{1/3} = 1.587m$$

$$I = \frac{1}{2}[2^2 m + 1^2 (2m)] = 3m$$

Hence

$$m\gamma_{\pm} = 0.00127$$

In this concentration range the Debye-Hückel theory should provide a reasonable estimate of $\gamma_{\pm}$. Taking logarithms to the base 10 and utilizing Eq. (12-64) results in

$$\log m - 0.5085(2)(3m)^{1/2} = -2.896$$

which is satisfied by $m = 0.00148$. The reported solubility[44] of 0.00166 is only slightly higher, which shows that at this ionic strength ($I = 0.0045$) the Debye-Hückel theory gives a reasonable estimate of $\gamma_{\pm}$.

EXAMPLE 12-18

Use the thermochemical data of Tables D-2 and D-3 to estimate the pH of a solution of CO_2 in water in equilibrium with atmospheric CO_2 at 298 K and a partial pressure of $3.2(10^{-4})$ atm—e.g., a rain drop.

Solution 12-18

Phase equilibrium can be expressed as

$$CO_2(g) = CO_2(ao) \qquad \text{(a)}$$

and chemical equilibrium in the aqueous phase as

$$CO_2(ao) + H_2O(l) = H_2CO_3(ao) \qquad \text{(b)}$$
$$\quad 1 \qquad\quad 2 \qquad\qquad 3$$

$$H_2CO_3(ao) = H^+(ao) + HCO_3^-(ao) \qquad \text{(c)}$$
$$\qquad 3 \qquad\qquad 4 \qquad\quad 5$$

$$HCO_3^-(ao) = H^+(ao) + CO_3^=(ao) \qquad \text{(d)}$$
$$\qquad 5 \qquad\qquad 4 \qquad\quad 6$$

$$H_2O(l) = H^+(ao) + OH^-(ao) \qquad \text{(e)}$$
$$\quad 2 \qquad\quad 4 \qquad\quad 7$$

where the aqueous components have been numbered from 1 to 7.

---

[44] *Ibid.*

From Tables D-2 and D-3, $\Delta G°$ and K are calculated for each reaction:

$$\Delta G_a° = -386.5 - (-394.6) = 8.1 \text{ MJ}$$

$$K_a = 0.0380 = \frac{m_1 \gamma_1^\square}{p}$$

$$\Delta G_b° = -623.8 - (-386.5) - (-237.4) = 0.1 \text{ MJ} \quad [= 0]$$

$$K_b = 1 = \frac{m_3}{m_1 x_2} \left(\frac{\gamma_3^\square}{\gamma_1^\square \gamma_2} \right)$$

$$\Delta G_c° = -587.44 + 0 - (-623.8) = 36.36 \text{ MJ}$$

$$K_c = 4.23(10^{-7}) = \frac{m_4 m_5}{m_3} \left(\frac{\gamma_4^\square \gamma_5^\square}{\gamma_3^\square} \right)$$

$$\Delta G_d° = -528.45 + 0 - (-587.44) = 58.99 \text{ MJ}$$

$$K_d = 4.57(10^{-11}) = \frac{m_4 m_6}{m_5} \left(\frac{\gamma_4^\square \gamma_6^\square}{\gamma_5^\square} \right)$$

$$\Delta G_e° = -157.40 + 0 - (-237.4) = 80 \text{ MJ}$$

$$K_e = 1(10^{-14}) = \frac{m_4 m_7}{x_2} \left(\frac{\gamma_4^\square \gamma_7^\square}{\gamma_2} \right)$$

Extremely dilute solutions are expected and it is quite reasonable to take all activity coefficients to be unity. Note that reaction (b) with $K = 1$ simply serves to define $CO_2(ao)$ in terms of the species H_2CO_3; the result $m_1 = m_3$ simplifies the problem. Further simplification results because the magnitudes of K_d and K_e allow us to ignore the presence of $CO_3^=$ and OH^- ions. We have ignored the presence of all ions except H^+ and HCO_3^-; electrical neutrality requires equal molalities, which simplifies the problem to

$$m_3 = m_1 = 0.0380(3.2)(10^{-4}) = 1.21(10^{-5})$$

and

$$4.23(10^{-7}) = \frac{m_4^2}{m_3}$$

We find

$$m_{H^+} = m_{HCO_3^-} = 2.27(10^{-6})$$

and[45]

$$pH = -\log m_{H^+} = 5.64$$

Note: The products $\gamma_4^\square \gamma_5^\square$ and $\gamma_4^\square \gamma_6^\square$ were not written as $\gamma_\pm^2$ because the parent of these ion pairs is only slightly dissociated (a weak electrolyte), and therefore its activity may be referred to its actual molality (the ao standard state).

EXAMPLE 12-19

Crystals, no matter how perfect they may seem, will always contain some vacant lattice sites. Although the number of such sites is minuscule, certain properties, especially those crucial for semiconductors, strongly depend on the vacancy concentration. Show how the concept of chemical equilibrium can be used to advantage by regarding lattice sites as chemical species. Specifically, consider the ionic crystal MX, which at high temperature can equilibrate with a gas containing the species M and X and determine the manner in which the "solubility of vacancies" depends upon the intensive variables of the system.

Solution 12-19

The bulk crystal is in equilibrium with its constituents according to

$$MX(c) = M(g) + X(g)$$

and we write

$$K = p_M p_X$$

---

[45] The pH is defined in terms of hydrogen ion molarity. Here the difference between molarity and molality is justifiably neglected.

There are two types of vacancies: vacant M sites, designated V_M, and vacant X sites, designated V_X. These vacancies "react" with their missing components as in the "reactions"

$$V_M + M(g) = \text{N.O.} \qquad K_M = \frac{a_{\text{N.O.}}}{a_{V_M} p_M}$$

$$V_X + X(g) = \text{N.O.} \qquad K_X = \frac{a_{\text{N.O.}}}{a_{V_X} p_X}$$

where N.O. refers to normal occupancy of the site. Because all sites are overwhelmingly in the normal occupancy mode, $a_{\text{N.O.}}$ is set equal to unity. Conversely, the number of vacant sites is so small that activity can be taken equal to concentration or number N. The system is now described by the following equations:

$$K = p_M p_X$$

$$K_M = \frac{1}{N_{V_M} p_M}$$

$$K_X = \frac{1}{N_{V_X} p_X}$$

which can be combined to yield

$$N_{V_X} \cdot N_{V_M} = K^{-1} \cdot K_X^{-1} \cdot K_M^{-1}$$

The right-hand side is a function only of temperature and is analogous to the solubility product of ionic aqueous solutions.

In addition to vacancies, there can also exist ionized vacancies, free electrons, and holes. The behavior of crystals containing these "chemical species" has been successfully explained through the chemical equilibrium treatment, although, as would be expected, the resulting equations are much more complex.[46]

---

[46] See, for example, N. B. Hannay, *Semiconductors*, Reinhold Publishing Corp., New York, 1959.

12-12 COUPLED REACTIONS

The coupling of chemical reactions is a vital feature of the chemistry of life. Here the energy required to drive an *endergonic* reaction $(\Delta G° > 0)$ is supplied by coupling with an *exergonic* reaction $(\Delta G° < 0)$ to yield an overall reaction which is exergonic. The classic example of this mechanism involves the synthesis of sucrose from glucose and fructose.[47] The reaction

$$\text{glucose} + \text{fructose} = \text{sucrose} + H_2O \tag{a}$$

is endergonic with $\Delta G_a° = 23{,}000 \text{ kJ/kmol}$ and is coupled with the hydrolysis of adenosine triphosphate (ATP) to adenosine diphosphate (ADP)

$$ATP^{4-} + H_2O = ADP^{3-} + HPO_4^{2-} + H^+ \tag{b}$$

which is exergonic with $\Delta G_b° = -29{,}300 \text{ kJ/kmol}$. The coupling is accomplished through the agency of a common intermediate, glucose 1-phosphate, and proceeds via the following enzyme-catalyzed reactions:

$$\text{glucose} + ATP^{4-} = ADP^{3-} + H^+ + \text{glucose 1-phosphate} \tag{c}$$

$$\text{glucose 1-phosphate} + \text{fructose} = \text{sucrose} + HPO_4^{2-} \tag{d}$$

$$\text{glucose} + \text{fructose} + ATP^{4-} = \text{sucrose} + ADP^{3-} + H^+ + HPO_4^{2-} \tag{e}$$

Reaction (e), the sum of reactions (c) and (d), is identical to the sum of reactions (a) and (b), and it is therefore seen that this coupled reaction is exergonic

$$\Delta G_e° = \Delta G_c° + \Delta G_d° = \Delta G_a° + \Delta G_b°$$

$$\Delta G_e° = 23{,}000 + (-29{,}300) = -6300 \text{ KJ / kmol}$$

Within the cell ATP is regenerated by coupling the reverse of reaction (b) (endergonic) with an exergonic reaction such as the oxidation of glucose. Thus, ATP can be considered a carrier of chemical energy—consumed in driving endergonic synthesis reactions and produced from exergonic reactions.

EXAMPLE 12-20

Biochemical oxidation-reduction reactions are often coupled with the following reaction:

---

[47] For details see A. L. Lehninger, *Bioenergetics*, W. A. Benjamin, Inc., New York, 1965, Chap. 4.

EXAMPLE 12-20 CON'T

$$NAD^+(ao) + H_2(g) = NADH(ao) + H^+(ao) \qquad (1)$$

where NAD$^+$ and NADH are, respectively, the oxidized and reduced forms of nicotinamide adenine dinucleotide. Determine $\Delta G°$ for this reaction.

Solution 12-20

Burton[48] has determined equilibrium compositions for the following reaction:

$$NAD^+(ao) + propan-2-ol(ao) = NADH(ao) + acetone(ao) + H^+(ao) \qquad (m)$$

$$\quad (1) \qquad\qquad (2) \qquad\qquad\quad (3) \qquad\qquad (4) \qquad\quad (5)$$

At 9.3°C he reports

$$m_1 = 118 \ \mu M \qquad m_2 = 94.8 \ mM$$

$$m_3 = 148 \ \mu M \qquad m_4 = 9.9 \ mM$$

$$pH = 7.650$$

From the pH we calculate

$$m_5 = 10^{-7.650} = 2.24(10^{-8})$$

and then evaluate the equilibrium constant

$$K = \frac{m_3 m_4 m_5}{m_1 m_2}$$

$$K = \frac{148(10^{-6}) \times 9.9(10^{-3}) \times 2.24(10^{-8})}{118(10^{-6}) \times 94.8(10^{-3})}$$

$$K = 2.93(10^{-9})$$

Burton determined K over a range of temperature and used Eq. (12-19) to interpolate the value at 25°C. Then using Eq. (12-13) he obtained

$$\Delta G_m° = 46,500 \ kJ/kmol$$

---

[48] K. Burton, *Biochem J.*, *143*, 365 (1974).

Reaction (m) can be written as the sum of reactions (l) and (n):

$$\text{propan-2-ol(ao)} = \text{acetone(ao)} + H_2(g) \qquad (n)$$

and we may state

$$\Delta G_m^{\circ} = \Delta G_l^{\circ} + \Delta G_n^{\circ}$$

The ΔG^f values needed to evaluate ΔG_n° can be calculated from the liquid-state values via the following path

acetone(l) $\xrightarrow{\Delta G_1}$ acetone(l) $\xrightarrow{\Delta G_2}$ acetone(g) $\xrightarrow{\Delta G_3}$ acetone(g)

$P = 1$ atm $\qquad\qquad P = P^{\circ} \qquad\qquad f = P^{\circ} \qquad\qquad f = k'$

$\xrightarrow{\Delta G_4}$ acetone(ao)

$m = 1$

yielding

$$\Delta G^f(\text{ao}) - \Delta G^f(\text{l}) = \Delta G_1 + \Delta G_2 + \Delta G_3 + \Delta G_4$$

$$\Delta G^f(\text{ao}) - \Delta G^f(\text{l}) = \Delta G_3 = RT \ln \frac{k'}{P^{\circ}}$$

where, as per Eqs. (11-27a) and (12-36), $k' = 0.0180k$.

Henry's law constants are determined from vapor-liquid equilibrium data. For acetone,[49]

$$k = 1353 \text{ mmHg}$$

$$P^{\circ} = 229.6 \text{ mmHg}$$

$$RT \ln \frac{k'}{P^{\circ}} = -5.56 \text{ MJ} / \text{kmol}$$

For propan-2-ol[50]

$$k = 339 \text{ mmHg}$$

[49] W. G. Beare, G. A. McVicar, and J. B. Ferguson, *J. Phys. Chem.*, **34**, 1310 (1930).
[50] J. A. V. Butler, C. N. Ramchandani, and D. W. Thomson, *J. Chem. Soc. London*, *1935*, 280 (1935).

$$P^\circ = 44.0 \text{ mmHg}$$

$$RT \ln \frac{k'}{P^\circ} = -4.90 \text{ MJ / kmol}$$

Using data from Table D-1, ΔG_n° can now be determined:

$$\Delta G_n^\circ = [-155.5 + (-5.56)] - [-180.5 + (-4.90)]$$

$$\Delta G_n^\circ = 24.34 \text{ MJ / kmol} = 24{,}340 \text{ kJ / kmol}$$

We now find

$$\Delta G_l^\circ = \Delta G_m^\circ - \Delta G_n^\circ$$

$$= 46{,}500 - 24{,}340 = 22{,}160 \text{ kJ / kmol}$$

As each reaction is endergonic, there would appear to be no advantage in coupling reactions (l) and (n). Yet, in spite of an unfavorable equilibrium constant for reaction (m), products are formed in measurable amounts because the concentration of one of the reaction products, H^+, is kept low by regulation of the pH.

Example 12-8 appears to offer an exception to the advantageous coupling of reactions. The reactions (a) and (b) share a common intermediate, $Fe_2O_3(c)$, and the overall reaction is exergonic; however, the overall reaction does not occur. The reason for this is that the common intermediate is a solid phase, which is present when its activity is unity but absent when its activity is less. Thus, for equilibrium with respect to reaction (a) we write

$$K_a = 1^3 \times \frac{p_{H_2O}^9}{1^6}$$

In our system we could reasonably expect to keep the water partial pressure under 1 atm and therefore a value of $K_a = 1$ would ensure that reaction (a) produced $Fe_2O_3(c)$. Thus, a value of zero for ΔG_a° would be just as effective as the actual value of -815 kJ; in fact, the additional 815 kJ are of no help and should not be counted. If we took ΔG_a° to be zero, the overall reaction would be endergonic—a better description of the situation.

EXAMPLE 12-21

The solubility of solid amino acids in water has been found to depend strongly on pH. Determine this dependence for solid tyrosine.

Solution 12-21

Amino acids can exist in aqueous solution in four different forms: the uncharged molecule (A); the charged, but electrically neutral, zwitterion (A±); a positively charged ion (A+); and a negatively charged ion (A−). The formation of these species can be represented by the following reactions:

$$NH_2CHRCOOH = NH_3^+CHRCOO^- \qquad (a)$$

$$A \qquad\qquad\qquad A\pm$$

$$NH_3^+CHRCOOH = H^+ + NH_3^+CHRCOO^- \qquad (b)$$

$$A+ \qquad\qquad\qquad A\pm$$

$$NH_3^+CHRCOO^- = H^+ + NH_2CHRCOO^- \qquad (c)$$

$$A\pm \qquad\qquad\qquad A-$$

In addition, the aqueous phase is saturated with respect to the amino acid, and for this solid-liquid equilibrium we write

$$f_A^S = \hat{f}_A^L$$

Using Eq. (12-36) this becomes

$$f_A^S = k_A' \gamma_A^\square [A]$$

where the brackets indicate molality. The solubility of tyrosine is quite low and we will assume $\gamma_A^\square$ is unity. Thus, we see that the molality of the uncharged molecule, [A], remains constant.

The solubility of tyrosine, S, will be comprised of four contributions:

$$S = [A] + [A\pm] + [A+] + [A-]$$

Because K_a is on the order of 10^5, we can expect [A] to make a negligible contribution to S. The remaining molalities can be expressed in terms of the equilibrium constants for the reactions on the assumption that activity coefficients for all species are unity.

$$[A\pm] = K_a[A]$$

$$[A+] = \frac{[H^+]}{K_b}[A\pm] = \frac{K_a[A][H^+]}{K_b}$$

$$[A-] = \frac{K_c}{[H^+]}[A\pm] = \frac{K_c K_a[A]}{[H^+]}$$

S can now be written as

$$S = K_a[A]\left[1 + \frac{[H^+]}{K_b} + \frac{K_c}{[H^+]}\right]$$

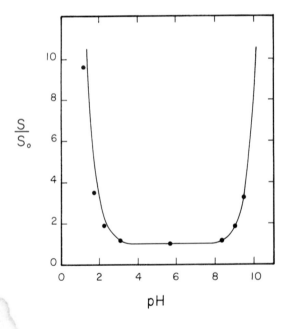

Figure 12-9 Relative solubility of tyrosine vs. pH at 298 K.

At 298 K it is known[51] that $K_b = 4.57(10^{-3})$, $K_c = 7.76(10^{-10})$,and the solubility of tyrosine in pure water is 0.0025 M and results in a pH of 5.6. With these values of K_b and K_c, an examination of the bracketed term shows that $[H^+]/K_b$ becomes important only at low pH and $K_c/[H^+]$

[51] W. B. Guenther, *Chemical Equilibrium*, Plenum Press, New York, 1975, Chap. 11.

becomes important only at high pH. In the pH range 5 to 7, the bracketed term is essentially unity. Therefore, for the saturated solution at a
pH of 5.6 we write

$$S_0 = 0.0025 = K_a[A]$$

which results in

$$S = 0.0025\left[1 + \frac{[H^+]}{4.57(10^{-3})} + \frac{7.76(10^{-10})}{[H^+]}\right]$$

This equation is plotted on Fig. 12-9 as S/S_0 vs. pH along with experimental data. It is seen
that the data are well fitted and that solubility is extremely sensitive to pH.

PROBLEMS

12-1. Hydrogen fluoride associates in the gas phase, and two possibilities are considered:

$$4HF = H_4F_4$$

or

$$6HF = H_6F_6$$

To test these two possibilities, the vapor densities of hydrogen fluoride were measured and found to be
as follows:

$t\,(°C)$	Pressure (atm)	Density (g/liter)
25	1.4	3.800
50	2.8	5.450
50	1.4	1.399
100	5.6	3.978

Do these data agree with either of the foregoing reactions?

12-2. It is desired to carry out the gas-phase hydrogenation of benzene to form cyclohexane. What
conditions of temperature and pressure would result in the maximum conversion? Justify your choice.

12-3.   At equilibrium the following data were taken for the reaction

$$A(g) = B(g) + C(g)$$

$T(°C)$	$P(atm)$	y_A	y_B	y_C
200	3.33	0.2	0.6	0.2
200	6.00	0.333	0.333	0.333
300	2.02	0.005	0.4975	0.4975
300	5.15	0.012	0.6	0.388

Use these data to determine equilibrium mol fractions of A, B, & C that result when A decomposes at 250°C and 10 atm. Assume ideal-gas behavior.

12-4.   For the following gas-phase reaction in equilibrium at 500 K it is found that A is 10% dissociated at 10 atm pressure.

$$A(g) = B(g) + C(g)$$

Assume ideal-gas behavior and calculate the percent dissociation of A at 500 K and 1 atm pressure.

12-5.   The gas-phase reaction

$$A(g) + B(g) = C(g)$$

was studied at 100 and 200°C and at a pressure of 1 atm. The following equilibrium mol fractions were obtained:

	100 °C	200 °C
y_A	0.172	0.642
y_B	0.414	0.179
y_C	0.414	0.179

If equimolar quantities of A and B are reacted at 150°C and at a pressure of 10 atm, what will be the equilibrium composition?

12-6.   A constant-volume bomb contains 2 g mol of gaseous NH_3 at 77°F and 1 atm pressure. The bomb is heated to 500°F. Considering only the species NH_3, N_2, and H_2 to be present, calculate the percentage dissociation of NH_3 when equilibrium is reached. Assume ideal-gas behavior.

12-7. The chemical species A is known to decompose according to

$$A(g) = B(g) + C(g)$$

A rigid container is filled with pure gaseous A at 300 K and 760 mmHg and then heated. The pressure was observed to be 1114 mmHg at 400 K and 1584 mmHg at 500 K. Estimate the pressure for a temperature of 600 K. Assume ideal-gas behavior and chemical equilibrium.

12-8. There is a maximum in the plot of c_p vs. temperature for hydrogen gas at atmospheric pressure. It is known that hydrogen is a mixture of ortho and para hydrogen. These two allotropes differ in whether the electron spins are parallel or opposed and can be considered separate chemical species. Can the maximum in c_p be explained in terms of chemical equilibrium between these two species?

12-9. A dimerization reaction

$$2A = A_2$$

can occur in either the gas or liquid phase. Is it possible for the equilibrium constants K^G and K^L to have opposite temperature dependencies? Explain. The pure component standard state is used for the liquid-phase equilibrium constant.

12-10. The possibility of producing ethanol from ethylene and water in a gas-phase reaction is to be investigated, and you have been asked to comment on the thermodynamics of the reaction

$$C_2H_4(g) + H_2O(g) = C_2H_5OH(g)$$

It is believed that in order to obtain reasonable reaction rates the reaction will have to be carried out at a minimum temperature of 500°F; however, we wish to limit the upper temperature to 1000°F. Someone has suggested that carrying out the reaction at high pressure might be desirable: however, it is felt that 10 atm should be the upper limit. It will be easy to separate unreacted ethylene from ethanol for recycle, and it is felt that a conversion of 10% would be sufficient to justify a plant study. Under these constraints, would the maximum obtainable conversion exceed 10%?

12-11. We are interested in the methanol synthesis reaction

$$CO(g) + 2H_2 = CH_3OH(g)$$

and wish to know the ranges of temperature and pressure for which the equilibrium conversion is at least 10% when stoichiometric ratios of CO and H_2 are used. Delineate this favorable region on a PT plot.

12-12.   Equilibrium with respect to the reaction

$$A(g) + B(g) = C(g)$$

will be studied by measuring the volume change accompanying the reaction. The temperature and pressure are held constant and the initial volume and the final volume of the reacting system are recorded. Three tests were made and are summarized in the table. Has equilibrium been established? If so what is the value of K?

	Initial composition			Volume (cm³)	
P (mmHg)	y_A	y_B	y_C	Initial	Final
500	0.5	0.5	0	200	150
600	0.333	0.667	0	300	233
600	0	0	1	200	293

12-13.   We are interested in producing isopropyl alcohol by the gas-phase hydrogenation of acetone:

$$C_3H_6O + H_2 = C_3H_8O$$

We would like to be able to obtain a 50% conversion of an equimolar feed of acetone and hydrogen. A minimum temperature of 500°F is necessary for a reasonable reaction rate and a maximum temperature of 1000°F is dictated by catalyst stability. Also, an upper limit of 15 atm has been set on the operating pressure.

Is it possible within the temperature range 500° to 1000°F and the pressure range 1 to 15 atm to obtain an equilibrium conversion of at least 50% for an equimolar feed of acetone and hydrogen?

12-14.   Calculate the equilibrium gas-phase composition for the xylene isomers at 140°C and a pressure of 500 mmHg. Use the following thermochemical data for the gaseous state at 298 K.

@ 298K

Species	$\Delta H'$ (kCal/g mol)	$\Delta G'$ (kCal/g mol)
o-Xylene	4.540	29.177
m-Xylene	4.120	28.405
p-Xylene	4.290	28.952

12-15.   We wish to produce formaldehyde, CH₂O, by the gas-phase pyrolysis of methanol, CH₄O, according to

$$CH_4O = CH_2O + H_2$$

The range of conditions available to us is 600°F to 1000°F and 1 to 10 atm. What is the maximum conversion of methanol we can expect within this range of conditions?

12-16. Consider the following reactions between A and B to form D and n isomers of C:

$$v_A A + v_B B = v_C C_1 + v_D D$$
$$\vdots \qquad \vdots \qquad \vdots \qquad \vdots$$
$$v_A A + v_B B = v_C C_i + v_D D$$
$$\vdots \qquad \vdots \qquad \vdots \qquad \vdots$$
$$v_A A + v_B B = v_C C_n + v_D D$$

The free energies of formation are available for components A, B, and D as well as for all n isomers of C. Show how the equilibrium constant for the following reaction can be obtained:

$$v_A A + v_B B = v_C \overline{C} + v_D D$$

where $\overline{C}$ represents the composite of all isomers of C. Simplify for the case where all n isomers have the same free energy of formation.

12-17. A cylinder fitted with a movable piston contains 1 g atom of solid iodine at 10°C. With the temperature kept constant, gaseous hydrogen at 10°C is slowly admitted at a steady rate while the piston is moved back so as to maintain the total pressure at 20 mmHg. How many mols of H_2 are added before the solid phase disappears? At 10°C, $\Delta G° = 0$ for the following reaction:

$$\frac{1}{2}H_2(g) + I(c) = HI(g)$$

The vapor pressure of iodine at 10°C is 0.0808 mmHg. Iodine exists in the vapor phase as I_2.

12-18. A cylinder fitted with a piston initially contains HI gas at a very low pressure. The cylinder and contents are maintained at 10°C, and the pressure in the cylinder is gradually increased until solid iodine is observed to form. The pressure in the cylinder at the instant solid iodine appeared was 8.00 mmHg. Iodine exists in the vapor phase as I_2, and the vapor pressure of pure iodine at this temperature is 0.0808 mmHg. From this information alone, calculate $\Delta G°$ at 10°C for the following reactions:

(a) $HI(g) = \frac{1}{2}H_2(g) + I(c)$

(b) $HI(g) = \frac{1}{2}H_2(g) + \frac{1}{2}I_2(g)$

12-19. Estimate the temperature to which gypsum must be heated at atmospheric pressure in direct contact with a combustion gas containing 10% water vapor in order to produce $CaSO_4$.

12-20. A 2-liter constant-volume bomb is evacuated and then filled with 0.10 gmol of methane, after which the temperature of the bomb and its contents is raised to 1273 K. At this temperature the equilibrium pressure is measured to be 7.02 atm. Assuming that methane dissociates according to the reaction

$$CH_4(g) = C(c) + 2H_2(g)$$

and that ideal gas behavior prevails, calculate K for this reaction from this information.

12-21. It is well known that items made of silver tarnish on exposure to the atmosphere at ambient temperature. Use thermodynamics to show whether this tarnish could be due to the formation of silver oxide.

12-22. For the reaction $2Ca(l) + ThO_2(c) = 2CaO(c) + Th(c)$,

$\Delta G° = -3.4$ kcal at $1000°C$

$\Delta G° = -2.0$ kcal at $1200°C$

Estimate the maximum temperature at which Ca(l) will reduce ThO_2. Assume that all condensed phases are pure.

12-23. Pure ammonium carbamate exerts a pressure of 26.3 atm absolute at $127°C$. The decomposition reaction is

$$NH_4O \cdot CO \cdot NH_2 = 2NH_3 + CO_2$$

The vapor pressure of ammonium carbamate as such is negligible, and the solid has a volume of 56 cm^3/g mol and may be assumed incompressible.

   (a) If ammonia gas is passed over a bed of solid ammonium carbamate at 100 atm and $127°C$, what will be the equilibrium composition of the exit gas?
   (b) What is the equilibrium composition if carbon dioxide is passed over the bed at 100 atm and $127°C$?

12-24. A gas stream contains 2% hydrogen sulfide (H_2S) and 98% inert gas. It is proposed to remove the H_2S by reaction with CaO according to the following reaction

$$CaO(c) + H_2S(g) = CaS(c) + H_2O(g)$$

Assume equilibrium and determine the temperature necessary to insure that the H_2S level in the exit gas is less than one part per million when the reactor is operated at atmospheric pressure. You may use the following data.

			$\log_{10} K^f$		
$T(K)$	CaO(c)	CaS(c)	H_2S	H_2O	SO_2
700	41.925	34.296	3.270	15.583	22.342
900	31.423	26.241	2.657	11.498	17.197
1100	24.737	20.562	1.709	8.883	13.376

12-25. The CaS from Prob. 12-24 is to be regenerated to CaO at atmospheric pressure with air as per the following reaction

$$CaS(c) + \frac{3}{2}O_2(g) = CaO(c) + SO_2(g)$$

Within the range $700 \text{ K} \le T \le 1100 \text{ K}$ is this regeneration possible? Would it be possible to produce a gas containing at least 10 % SO_2?

12-26. As a member of a consumer protection task force you are asked to investigate the stability of household baking soda, $NaHCO_3$, under ambient conditions. This study was prompted when a prominent member of the task force recalled that with a little buttermilk, baking soda is known to raise biscuits by releasing CO_2 and therefore this person reasoned that because buttermilk is innocuous, baking soda might be unstable and constitute a hazard to unsuspecting cooks. There will undoubtedly be many facets to your study, but begin by considering the following reaction

$$2NaHCO_3(c) = Na_2CO_3(c) + H_2O(g) + CO_2(g)$$

and determining whether $NaHCO_3$ could dissociate at room temperature when exposed to air containing 350 parts per million CO_2 and having a relative humidity of 50%. The following data may be useful

	$Na_2CO_3(c)$	$NaHCO_3(c)$
ΔH^f(kcal/g mol)	−269.46	−226.0
ΔG^f(kcal/g mol)	−249.55	−202.66

Vapor pressure of water @ 298 K is 23.76 mmHg.

12-27. In investigating the stability of household baking soda ($NaHCO_3$), we would like to know the temperature at which a can of baking soda would develop a pressure of 10 atm. Ignore the presence of any air that might be in the can and use the following reaction

$$2NaHCO_3(c) = Na_2CO_3(c) + H_2O(g) + CO_2(g)$$

12-28. At 1000 K the equilibrium constant, K, is 1.53 for the following reaction

$$C(c) + H_2O(g) = H_2(g) + CO(g)$$

Calculate the equilibrium gas-phase mol fractions resulting from passing steam at atmospheric pressure through a bed of carbon particles.

12-29. Consideration is being given to the design of shipping containers for calcium carbide, CaC_2. This compound reacts with water to form acetylene as per the following reaction

$$CaC_2(c) + H_2O = CaO(c) + C_2H_2(g)$$

Determine the maximum pressure which could be expected in an enclosed system initially containing calcium carbide and water.

12-30. In a system consisting of $CaCO_3(c)$, $CaO(c)$, and $CO_2(g)$ at 500° C, the equilibrium pressure is 0.11 mmHg. What is the lowest temperature at which $CaCO_3$ will decompose completely in an atmosphere containing 0.03% by volume of CO_2 when the barometer reads 759 mmHg? The mean value of $\Delta H°$ for the reaction

$$CaCO_3 = CaO + CO_2$$

over the temperature range concerned is +41,340 cal.

12-31. It is proposed to clean tungsten wire (remove all traces of the oxide WO_2) by heating the wire to 1000 K in a hydrogen atmosphere (1 atm total pressure) in order to take advantage of the following reaction:

$$WO_2(c) + 2H_2(g) = W(c) + 2H_2O(g)$$

Will this work? If so, what is the maximum water impurity level that can be tolerated in the hydrogen? Free energies of formation are as follows:

$$WO_2(c): \Delta G^f = -131{,}600 + 36.6T$$

$$H_2O(g): \Delta G^f = -58{,}900 + 13.1T$$

where ΔG^f is in cal/g mol and T is in K.

12-32. In a proposed process a flow-type catalytic reactor is to use a cupric oxide catalyst. The reactor is to operate at atmospheric pressure, and the gaseous reaction mixture will have an average oxygen concentration of 5 mol %. It is desired to operate the reactor at as high a temperature as possible in order to increase the reaction rate. Someone has raised a question concerning the stability of the catalyst, and you are asked to find the maximum temperature at which the catalyst can be employed. What is your answer? List all assumptions made and show all calculations.

12-33. Fugacities of ethylene in nitrogen-ethylene mixtures have been measured by a technique employing solid cuprous chloride, CuCl, to form a solid addition compound CuCl $\cdot$ C$_2$H$_4$ with ethylene:

$$CuCl + C_2H_4 = CuCl \cdot C_2H_4$$

In the absence of gas other than C$_2$H$_4$, the dissociation pressure, hence fugacity, of ethylene is a function of temperature only (the phase rule indicates a univariant system). In the presence of an inert gas such as nitrogen, the fugacity of ethylene in the same system is slightly different from the preceding and is also slightly pressure dependent. Explain how one may calculate the fugacity of ethylene in a mixture of C$_2$H$_4$, N$_2$, and CuCl if the following data are available:

1. *PVT* data for pure ethylene,
2. dissociation pressures for ethylene over CuCl as a function of temperature, and
3. molal volume of the two solid compounds (assumed to be independent of pressure).

12-34. We wish to investigate the possibility of producing SO$_2$ and subsequently sulfuric acid by reducing gypsum, CaSO$_4$, with carbon monoxide as per the following reaction:

$$CaSO_4 (c) + CO(g) = CaO(c) + SO_2 (g) + CO_2 (g) \tag{a}$$

However, the following undesirable reaction may occur:

$$CaSO_4 (c) + 4CO(g) + CaS(c) + 4CO_2 (g) \tag{b}$$

And there is always the possibility that calcium carbonate will form via the reaction

$$CaO(c) + CO_2 (g) = CaCO_3 (c) \tag{c}$$

At 1200 K the equilibrium constants for these three reactions are

$$K_a = 0.31; \quad K_b = 7.92; \quad K_c = 1.4$$

For the process carried out at 1200 K and 1 atm pressure, determine

(a) the minimum ratio of CO$_2$ to CO which will suppress reaction (b).
(b) the maximum mol fraction of SO$_2$ which will occur when reaction (b) is suppressed.
(c) whether CaCO$_3$(c) will be present when reaction (b) is suppressed.

12-35. In the vapor phase there are three allotropes of sulfur: S$_2$, S$_6$, and S$_8$. Using the following data,[a] calculate the composition of sulfur vapor in equilibrium with rhombic sulfur at 150 and 200°F:

Product	Reactant	ΔH_0	Δa	$\Delta bx \ 10^3$	I	ΔH°_{298}	ΔG°_{298}
$S_2(g)$	2S(R)	31,360	0.591	−11.6	−38.62	31,020	19,360
$S_6(g)$	6S(R)	29,990	−2.22	−34.90	−71.60	27,780	13,970
$S_8(g)$	8S(R)	30,240	−3.63	−46.40	−86.20	27,090	12,770

ªW. N. Tuller, *The Sulphur Data Book*, McGraw-Hill, New York, 1954.

where

$$\Delta H^{\circ} = \Delta H_0 + \Delta aT + \frac{1}{2}\Delta bT^2$$

and

$$\Delta G^{\circ} = \Delta H_0 - 2.303\Delta aT \log T - \frac{1}{2}\Delta bT^2 + IT$$

T is in K, and ΔH° and ΔG° are in cal.

VAPOR PRESSURE OF RHOMBIC SULFUR				
t (°F)	148.8	158.0	166.8	184.4	192.0
P° (mmHg)	$1.94(10^{-4})$	$3.31(10^{-4})$	$5.43(10^{-4})$	$14.39(10^{-4})$	$20.89(10^{-4})$

12-36. From the data of Problem 12-35, calculate the latent heat of sublimation of rhombic sulfur in the range 150 to 200°F

(a) using the Clausius-Clapeyron equation assuming the vapor phase is S(g) (i.e., we are ignorant of the true nature of the vapor phase).
(b) using the Clausius-Clapeyron equation and assuming the vapor phase is entirely S_8.
(c) using the best possible method.

12-37. The decomposition reaction

$$A = B + C$$

is performed in the liquid state at 100°C. The equilibrium constant for the reaction at 100°C is $K = 2.0$, where all standard states are pure liquids at 100°C. The vapor pressures at 100°C are

$$P_A^{\circ} = 5 \text{ atm}; \quad P_B^{\circ} = 20 \text{ atm}; \quad P_C^{\circ} = 2 \text{ atm}$$

Calculate the equilibrium pressure for this reaction at 100°C and the composition of liquid and vapor phases for the following assumptions: All vapors and their mixtures are ideal, liquid B is completely immiscible with mixtures of liquids A and C, and the A–C liquid mixture is ideal.

12-38. For the reaction

$$A(l) = B(l) + C(l)$$

$K = 2$, and $P_A^\circ = 5$ atm, $P_B^\circ = 20$ atm, and $P_C^\circ = 2$ atm. A and C form ideal-liquid solutions, and liquid B is immiscible with either pure A and C or their mixtures. Find the pressure below which only a gas phase exists when chemical equilibrium prevails and the system is originally comprised of pure A.

12-39. The reaction

$$A + B = C + D$$

is being studied. The temperature at which we desire to carry out the reaction is above the critical temperature of A, B, and C; however, D has a vapor pressure of 5 atm at this temperature. It is known that gaseous A, B, and C are essentially insoluble in liquid D. Would it be possible to increase the conversion by increasing the pressure so that pure liquid D would be present at equilibrium? Explain.

12-40. A dissociates as follows:

$$A = B + C$$

All three species are similar in structure and can be assumed to form ideal-liquid solutions. At a temperature of 400 K and at a pressure high enough to suppress the vapor phase, a liquid phase is found to contain 50 mol % A, 25 mol % B, and 25 mol % C when reaction equilibrium has been established. The vapor pressures are 1 atm, 4 atm, and 6 atm for A, B, and C, respectively. Determine the dew point pressure at 400 K when the system is initially comprised of pure A vapor.

12-41. At 400 K equilibrium with respect to the isomerization, reaction

$$A = B$$

is rapidly established. At this temperature the vapor pressures are:

$$P_A^\circ = 2.0 \text{ atm}; \quad P_B^\circ = 2.5 \text{ atm}$$

We place gaseous A and B in a cylinder fitted with a piston and maintained at 400 K. The cylinder is filled at 1 atm and the volume of the reacting system is slowly decreased by movement of the piston. At 2.2 atm a dew point is observed. Ideal-liquid solutions are expected.

If possible, calculate

    (a)   the equilibrium constant for the reaction $A(g) = B(g)$.
    (b)   the equilibrium constant for the reaction $A(l) = B(l)$.

12-42. A liquid hydrocarbon containing a small amount of water needs to be dried and it is proposed to use calcium sulfate as a drying agent utilizing the following reaction

$$CaSO_4(c) + 2H_2O(l) = CaSO_4 \bullet 2H_2O(c)$$

At 25°C the solubility of water in the hydrocarbon is $2(10^{-3})$ mol %. Estimate the lowest mol % of water to be expected when the hydrocarbon liquid is contacted with an ample quantity of calcium sulfate. Neither calcium sulfate nor its hydrate is soluble in the liquid hydrocarbon.

12-43. When hexaphenylethane is dissolved in benzene, the freezing point depression of a 2 wt % solution of hexaphenylethane is 0.219°C; the boiling point elevation of this solution is 0.135°C. Calculate the heat of dissociation of hexaphenylethane into triphenylmethyl radicals.

12-44. A solute C is distributed between two immiscible solvents A and B. It is believed that the solute is completely dissociated into two fragments in the A phase and that it is completely associated into dimers in the B phase. Using this hypothesis and the data point $x_C^B = 0.01$, $x_C^A = 0.1$, estimate x_C^B when $x_C^A = 0.2$.

12-45. The NBS Circular 500 gives the free energy of formation of $NH_3(g)$ at 298 K as –3.976 kcal, while in the hypothetical $m = 1$ standard state it is -6.37 kcal. Show how you would calculate the latter quantity from the former. What data do you need? Locate the necessary data, and verify these values.

12-46. It is proposed to manufacture hydrogen peroxide by reacting oxygen gas at high pressure with liquid water at 25°C. Do you recommend undertaking a research program to study this proposal? Explain and give details. The following data are available:

$$H_2 + O_2 = H_2O_2(ao) \qquad \Delta G_{298}^\circ = -31{,}470 \text{ cal / g mol}$$

$$H_2 + \frac{1}{2}O_2 = H_2O(l) \qquad \Delta G_{298}^\circ = -56{,}560 \text{ cal / g mol}$$

12-47. The solubility of SO_2 in water can be correlated (J. M. Prausnitz, R. N. Lichtenthaler, E. Gomes de Azevedo, *Molecular Thermodynamics of Fluid-Phase Equilibria*, 2nd ed., Prentice Hall, Englewood Cliffs, NJ, 1986, Chap. 8) by plotting $m/\sqrt{p}$ vs. $\sqrt{p}$. A good linear relationship is obtained for SO_2 partial pressures up to 1 bar. Provide thermodynamic justification for this correlation and estimate the values of slope and intercept when p is expressed in bars. This system should be similar to the CO_2–water system examined in Ex. 12-18.

12-48. For most people the odor threshold for H_2S in air is one part per million. What is the maximum concentration of an aqueous NaHS solution that would be odorless? Complete dissociation is assumed for NaHS and the HS^- ion can undergo the following reactions.

$$HS^- + H^+ \leftrightarrow H_2S(ao)$$

$$HS^- \leftrightarrow S^= + H^+$$

12-49. Scientific investigation [G. W. Kling et al., *Science, 236*, 169 (1987)] attributes a sudden release of CO_2 gas from Lake Nyos as the cause for the deaths of at least 1700 people in Cameroon, West Africa, on August 21, 1986. Lake Nyos has a depth of 200 m and is situated in the core of an inactive volcano. It is believed that some unknown disturbance caused the previously stratified water to turn over and bring bottom water laden with dissolved CO_2 to the surface. The rapid exsolution of large amounts of CO_2 caused the formation of a gas cloud that flowed down the mountainside asphyxiating people in the neighboring villages. The CO_2 is believed to have entered the lake from magmatic sources.

Test this scenario by calculating the maximum amount of CO_2 that could be dissolved at a depth of 200 m expressed as cubic meters of CO_2 at STP per m^3 of water. Assume that pure CO_2 gas entered the bottom of the lake and that the water temperature was 25°C.

12-50. For a solute A show that the relationship between γ_A, based on Eq. (9-35), and $\gamma_A^\square$ based on Eq. (12-36), is

$$\frac{\gamma_A}{\gamma_A^\square} = \frac{\gamma_A^\infty}{1 - x_A}$$

where x_A is the mol fraction of A and γ_A^∞ is the value of γ_A as x_A approaches zero.

12-51. Derive a form of the Gibbs-Duhem equation applicable to an aqueous solution of an electrolyte at constant T and P involving the terms

$$\frac{d \ln \gamma_\pm}{dm} \quad \text{and} \quad \frac{d \ln \gamma_W}{dx_W}$$

With this equation determine

    (a) whether the Debye-Hückel value of $\gamma_\pm$ precludes ideal behavior for water.
    (b) whether a minimum in $\gamma_\pm$ as shown in Fig. 12-7 places any restrictions on the behavior of γ_W.

12-52. The compounds $AB(c)$ and $CD(c)$ are both sparingly soluble electrolytes that form singly charged ions in aqueous solution. At 25°C their solubilities in water are

$$AB\text{: } 0.00400 \text{ molality}$$

$$CD\text{: } 0.00600 \text{ molality}$$

Calculate the solubilities if both compounds are present in the same solution. There are no common ions.

12-53. In Section 12.11 the extent of dissociation of acetic acid was calculated based on the assumption that activities were equal to molalities. Make the best possible estimate of the degree of dissociation, α, when the solution is prepared as 1 mol of acetic acid per kilogram of water. You may wish to consult Fig. 12-8.

12-54. Using only data available in this text, estimate the solubility of $Ca(OH)_2$ in water at 25°C. Comment on the reliability of your estimate.

12-55. Use the information available in Ex. 12-15 to estimate the solubility of PbI_2 in a 0.01 molal solution of $Pb(NO_3)_2$.

12-56. Should the aqueous standard state refer to a hypothetical solution at $m = 1$ or at $m_\pm = 1$? Explain.

12-57. The solubility of $PbCl_2$ in water at 25°C is 1.073 g per 100 g of saturated solution. Use this to determine $\gamma_\pm$ at saturation. Compare with $\gamma_\pm$ from the Debye-Hückel theory.

12-58. Proteins in their native state exist as highly ordered three-dimensional structures. On heating, these structures are converted into random chains resembling ordinary polymers, a process known as denaturation. Despite the many possible intermediates, the process can be understood in terms of a two-state reaction model $N \leftrightarrow D$ where N represents the native state and D the denatured state.

From spectrophotometric measurements in dilute aqueous solution, Brants [*J. Am. Chem. Soc.*, 86, 4291 (1964)] has studied the denaturation of the protein chymotrypsinogen over the temperature range 0–80°C and reports the following standard free energy change for the $N \leftrightarrow D$ reaction.

$$\Delta G^\circ = 121{,}700 - 2248T + 11.57T^2 - 0.01783T^3$$

where ΔG° is in cal/g mol and T is in K.

Calculate

    (a)   the temperature at which the protein exists as 99%N.
    (b)   the temperature at which the protein exists as 99%D.
    (c) ΔH° corresponding to part (b).

12-59. The reaction catalyzed by the enzyme *alcohol dehydrogenase*

$$NAD^+(ao) + ethanol(ao) = NADH(ao) + acetaldehyde(ao) + H^+(ao)$$

has been studied[50] in dilute solution. At 25° C catalytic amounts of enzyme were added to a solution containing NAD^+ and ethanol, each at 0.001M, and a buffer to fix the pH at 8.00. At equilibrium the concentration of NADH was found to be $2.561(10^{-5})M$. If the solution were unbuffered, what equilibrium concentrations and what pH would be expected?

12-60. A study has been made[51] of the following reaction occurring in dilute solution.

$$NAD^+(ao) + ethanol(ao) = NADH(ao) + acetaldehyde(ao) + H^+(ao)$$

Catalyzed by the enzyme *alcohol dehydrogenase*, the reaction yields the following equilibrium molalities.

		EQUILIBRIUM MOLALITIES		
$t(°C)$	pH	NADH = acetaldehyde	NAD⁺	ethanol
10.6	7.680	96.5 μM	225.1 μM	324.2 mM
17.35	7.658	91.6	229.9	194.9
34.6	7.571	162.2	154.3	389.0
42.8	7.568	133.4	188.1	128.9

Use these data to determine $\Delta G°$ and $\Delta H°$ at 298 K for the given reaction.

12-61. Use the result of Prob. 12-60 to calculate $\Delta G°$ at 298 K for reaction (l) of Ex. 12-20. The following data may be useful.

ΔG^f for ethanol(ao):[52] –43,120 cal/g mol
For acetaldehyde in aqueous solution:[53] $k = 4.95$ atm

12-62. Proteins can sometimes be purified from other contaminating proteins by denaturation. If the contaminating protein denatures at a lower temperature than the desired protein, it can be selectively denatured and precipitated from solution. Suppose that chymotrypsinogen is contaminated with a protein in which $\Delta G°$ for the denaturing reaction (see Problem 12-58) is exactly 2000 cal/g mol less than the $\Delta G°$ for chymotrypsinogen at all temperatures, and we wish to remove 99% of the contaminating protein by denaturation. What temperature should be used and what percentage of the chymotrypsinogen will be denatured? The proteins are in dilute aqueous solution.

12-63. In the chemistry of semiconductors, electrons are treated as chemical species. For example, hydrogen gas dissolves in a metal oxide, dissociates, and then ionizes according to

---

[50] K. Burton, *op cit.*
[51] *Ibid.*
[52] From reference 1-E in Table 12-1.
[53] J. L. Kurz, *J. Amer. Chem. Soc., 89,* 3524 (1967).

$$H_2(g) \leftrightarrow 2H^+ + 2e^-$$

Thus an equilibrium is established where the concentration of charge carriers depends on the partial pressure of hydrogen. Where concentrations are low enough so that activities may be replaced with concentrations, determine the manner in which the total concentration of charge carriers (H^+ and e^-) depends on the hydrogen partial pressure.

12-64. The doping of crystals of the compound GaAs with relatively volatile elements can occur at temperatures near $700°C$ by contacting the solid phase with a gas phase containing Ga, As$_4$, and the dopant. There are two types of dopants: type III, which occupies a Ga crystal site and is surrounded by As atoms, and type V, which occupies an As site and is surrounded by Ga atoms. Equilibrium between the dopant in the solid phase, considered an association compound, and the gas phase can be represented as a chemical reaction with equilibrium constants

$$K_{III} = \frac{a_{D \cdot As}}{p_D \, p_{As_4}^{1/4}} \quad \text{and} \quad K_V = \frac{a_{D \cdot Ga}}{p_D \, p_{Ga}}$$

Distribution coefficients for the dopants are defined as

$$k_{III} = \frac{x_{D \cdot As}}{p_D} \quad \text{and} \quad k_V = \frac{x_{D \cdot Ga}}{p_D}$$

where x is the solid-phase concentration of dopant. Equilibrium also exists between the GaAs crystal and the vapor phase according to

$$K = \frac{a_{GaAs}}{p_{Ga} \, p_{As_4}^{1/4}}$$

It has been found that isothermal plots of log k_{III} vs. log p_{As4} are linear with a positive slope, whereas plots of log k_V vs. log p_{As4} are linear with a negative slope. Provide a thermodynamic explanation for this. At constant pressure and gas-phase composition, how would you expect k_{III} and k_V to depend on temperature?

12-65. Most of the newer thermodynamics texts use S.I. units almost exclusively; however, tables of enthalpies and free energies of formation are still based on a standard state defined as a pressure or fugacity of 1 atm. It is argued that to be totally consistent with the S.I. system the pressure unit defining the standard state should be changed from the atmosphere to the bar (1 bar = 10^5 N/m^2, and 1 bar = 0.9869 atm). What problems might be encountered in making this change? Specifically, calculate ΔH_{298}^f and ΔG_{298}^f based on this S.I.-consistent standard state for the following:

(a) $H_2O(g)$

(b) $H_2O(l)$

(c) $CO_2(g)$

12-66. For crystalline solids that obey the third law the Gibbs energy function may be readily calculated by evaluation by either of the following double integrals:

$$\frac{G_T^\circ - H_0^\circ}{T} = -\frac{1}{T} \int_0^T \left(\int_0^T \frac{c_P^\circ}{T} \, dT \right) dT$$

$$= -\int_0^T \left(\frac{\int_0^T c_P^\circ \, dT}{T^2} \right) dT$$

or by

$$\frac{G_T^\circ - H_0^\circ}{T} = \frac{1}{T} \int_0^T c_P^\circ \, dT - \int_0^T \frac{c_P^\circ}{T} \, dT$$

Derive these equations, and show how $(G_T^\circ - H_0^\circ)/T$ could be calculated when a table of c_P° vs. T data is available. Also derive the appropriate equations for calculating $(G_T^\circ - H_0^\circ)/T$ for a substance which is a gas at the temperature T. Determine what data are necessary, and show how the calculation should be made.

12-67. The oxidation of SO_2 to SO_3 with oxygen is an exothermic reaction and therefore the equilibrium conversion decreases with increasing temperature. When the reaction occurs adiabatically, the temperature of the reacting mixture rises with increasing conversion, but at higher temperatures a limiting conversion is reached which is dictated by chemical equilibrium. Using POLYMATH calculate the adiabatic equilibrium temperature and conversion for various reactor feed conditions and prepare a plot showing equilibrium conversion versus reactor feed temperature using the ratio of O_2 to SO_2 in the feed as a parameter. For operation at atmospheric pressure, use reactor feed temperatures ranging from 300 to 500 K with O_2 to SO_2 ratios of 0.4, 0.5, and 0.6.

CHAPTER **1 3**

Complex Chemical Equilibrium

*I*n Chap. 12 the focus was mainly on equilibrium, where only one or two reactions were sufficient to describe the system. While such systems are common, the chemical engineer may also expect to encounter reacting systems of greater complexity. In this chapter we will build upon the previously derived principles of chemical equilibrium to develop a thermodynamic treatment of complex reacting systems. The phase rule is covered first because it provides a framework for viewing and analyzing complex problems. Next, a general approach to the analysis of complex chemical equilibrium problems is outlined. This is followed by the solution of these problems by free energy minimization utilizing POLYMATH. Because of its commercial importance, the Carbon–Hydrogen–Oxygen (CHO) system is studied in some detail with emphasis on conditions of carbon deposition. Finally, by analogy with the CHO system, the Silicon–Hydrogen–Chlorine system is considered.

13-1 THE PHASE RULE FOR REACTING SYSTEMS

In Sec. 9-3 we saw that the phase rule is an accounting procedure for keeping track of variables and equations thereby insuring that the system is described properly in a mathematical sense. We now consider a system of π phases and comprising N identifiable chemical species. In such a system the numbers of intensive variables are as follows:

Composition: $\pi(N-1)$
T and P in each phase: 2π
Total variables: $\pi(N+1)$

At equilibrium there must be equality of temperature, pressure, and chemical potentials between each phase. This supplies the following numbers of equations:

Equality of T: $\pi - 1$
Equality of P: $\pi - 1$
Equality of $\hat{\mu}_i$: $N(\pi - 1)$

In addition to these equations there are equations of chemical equilibrium of the form $\sum v_i \hat{\mu}_i = 0$. Their number R is equal to the number of independent chemical reactions.

Occasionally, there will be additional equations, expressible in terms of intensive variables, which apply to the system. These equations will be called additional constraints and their number will be designated S.

The total number of equations involving intensive variables is therefore

$$(N+2)(\pi-1)+R+S$$

The degree of freedom possessed by the equilibrium system is then

$$\mathscr{F} = \pi(N+1)-[(N+2)(\pi-1)+R+S]$$

$$= N - R - S + 2 - \pi \tag{13-1}$$

When comparing this to the phase rule for nonreacting systems,

$$\mathscr{F} = C + 2 - \pi \tag{1-8}$$

one could write

$$C = N - R - S \tag{13-2}$$

We have been accustomed to characterizing systems in terms of the number of components. For nonreactive systems there is no problem in identifying C, but in complex reactive systems determining C can be a major task. Note that Eq. (13-1) states the phase rule for a reacting system in terms of the quantities N, R, and S and that C is obtained only by analogy with Eq. (1-8). While C does not arise naturally in the analysis leading to Eq. (13-1), it is still convenient to characterize complex reacting systems in terms of the number of phase-rule components—a practice often used in the literature. Although we will continue to use C

to characterize systems, it is not recommended that C be determined from a knowledge of how the system was originally formed except for very simple systems. Instead, the problem should be analyzed in terms of N, R, and S. The key to implementing the phase rule in complex chemically reacting systems is the determination of the number of independent chemical reactions R. This is a problem in stoichiometry.

Stoichiometry.[1] In a reacting system containing N species constituted from m elements, the following generalized notation can be used to describe the ith species:

$$A_i = X^{(1)}_{\beta_{i1}} X^{(2)}_{\beta_{i2}} \cdots X^{(l)}_{\beta_{il}} \cdots X^{(m)}_{\beta_{im}}$$

where $X^{(l)}$ is the lth element ($l = 1, \ldots, m$) and the β's are formula coefficients, which are either zero or positive integers. This representation can be reconciled with the customary expression of a chemical formula when the $X^{(l)}$'s are replaced by their elemental notations and the β's are replaced by integers. For example, in a system comprised of the elements carbon, hydrogen, and oxygen the ith species would be written

$$A_i = C_{\beta_{i1}} H_{\beta_{i2}} O_{\beta_{i3}}$$

and species such as CO_2, H_2O, CH_4, and methanol(CH_4O) would be represented as follows:

$$CO_2: C_1 H_0 O_2$$

$$H_2O: C_0 H_2 O_1$$

$$CH_4: C_1 H_4 O_0$$

$$CH_4O: C_1 H_4 O_1$$

Note that the conventional representation of chemical compounds omits elements with zero β's and $\beta = 1$ is implied but not written.

The governing equations that relate the permissible values of the species mol numbers, n_i's, are conservation equations. The conserved quantity can be an element, electrical charge, or an extensive thermodynamic property. Elemental balances take the form

$$\sum_{i=1}^{N} \beta_{il} n_i = b_l \qquad (l = 1 \cdots m) \tag{13-3a}$$

where n_i is the total number of mols of species i in the system summed over all phases and

---

[1] For a comprehensive treatment of this subject see W. R. Smith and R. W. Missen, *Chemical Reaction Equilibrium Analysis: Theory and Algorithms*, Wiley, New York, 1982, Chap. 2.

b_l is the total atomic weights[2] of element l in the system. The conservation of electric charge can be expressed as

$$\sum n_i z_i = 0 \tag{13-3b}$$

where z_i is the charge on the ith species and is a positive or negative integer. The summation is not indexed because it is applied only to a phase (e.g., an aqueous solution or a plasma) and it is possible that the phase does not contain all N species. Constancy of an extensive thermodynamic property, M (e.g., V, H, S), as per Eq. (11-10), can be written as

$$\sum_{i=1}^{N} n_i \Delta \overline{M}_i = 0 \tag{13-3c}$$

For a set of linear algebraic equations such as this, it is known that the number of independent equations is equal to the rank of a matrix formed with conserved quantities as columns and species as rows. Because systems where electric charge or a thermodynamic property is conserved are not often encountered, this matrix usually contains only formula coefficients, β_{il}'s, and is formed with elements ($l = 1, \ldots, m$) as columns and species ($i = 1, \ldots, N$) as rows. This $N \times m$ matrix is referred to as the formula coefficient matrix and is denoted as β.

$$\beta = \begin{bmatrix} \beta_{11} & \cdots & \beta_{1l} & \cdots & \beta_{1m} \\ \vdots & & \vdots & & \vdots \\ \beta_{i1} & \cdots & \beta_{il} & \cdots & \beta_{im} \\ \vdots & & \vdots & & \vdots \\ \beta_{N1} & \cdots & \beta_{Nl} & \cdots & \beta_{Nm} \end{bmatrix}$$

The rank of a matrix is the order of the largest determinant having a nonzero value which may be formed from the matrix by eliminating any number of complete rows or columns. The order of a determinant, or square matrix, is the number of rows or columns.

In the formula coefficient matrix, β, rows represent species and can be considered as vectors in m-dimensional "formula space." This leads to the interpretation that the rank of β specifies the number of independent vectors in the set of N. If we designate ρ as the rank of β, then there are ρ independent vectors (chemical species) and the remaining $(N - \rho)$ vectors can be expressed as linear combinations of the ρ independent vectors. These linear combinations of the ρ independent chemical species are chemical reactions which form the $(N - \rho)$ dependent species from the ρ independent species. In the notation of Sec. 12-1, the reaction producing each dependent species, A_j, from the independent species, A_i's, will be identified as the jth reaction and written as

---

[2] The number of mols is the number of molecular weights of a compound, however, there is no corresponding abbreviation for the number of atomic weights of an element. Although the use of mols of element l here would be correctly understood, it is technically incorrect.

$$A_j + \sum_{i=1}^{\rho} v_{ij} A_i = 0 \qquad (j = \rho+1, \ldots, N)$$

where v_{ij} is the stoichiometric coefficient of component i in the jth reaction.

The mol numbers can now be expressed in terms of the extents of reaction, α_j's, for the $(N - \rho)$ reactions. We write for the dependent species j which is produced from reaction j

$$n_j = n_{0j} + \alpha_j \qquad (j = \rho+1, \ldots, N) \tag{13-4a}$$

and for the independent species i which may be involved in several of the $N - \rho$ reactions

$$n_i = n_{0i} + \sum_{j=\rho+1}^{N} v_{ij} \alpha_j \qquad (i = 1, \ldots, \rho) \tag{13-4b}$$

Thus, for a given set of n_0's, we have in Eqs. (13-4a) and (13-4b) a set of N algebraic equations relating the N species mol numbers to the extents of conversion, α_j's, in the $(N - \rho)$ independent reactions—a description of the system in terms of extensive variables. This description needs to be reconciled to the phase-rule description which is expressed in terms of intensive variables so that we are confident that results obtained from this stoichiometric analysis are valid for a phase-rule analysis.

The n_i's in Eq. (13-4b) are the total mols of a species summed over all phases and can be used to determine mol fractions only if no species exists in more than one phase. Under these circumstances a knowledge of these n_i's permits the calculation of all mol fractions, and because the α_j's depend on temperature and pressure,[3] their use implies that these two intensive variables have been fixed. Thus, with all intensive variables fixed, the system would be properly described in a phase-rule sense. The restriction that no species exist in more than one phase does not change the application of the phase rule because the elimination of the $N(\pi - 1)$ equations of phase equilibrium is balanced by the elimination of an equal number of composition variables. In a practical sense this restriction should not prove to be a serious limitation to the application of the phase rule as most complex chemical equilibrium problems will involve a fluid phase and one or more condensed phases, usually pure. When a more complex system is encountered, a more detailed phase-rule analysis is required.[4]

Designating the number of independent reactions $(N - \rho)$ as R, we now consider the reaction matrix, $\mathbf{R}$, which is formed from the stoichiometric coefficients, v_{ij}'s with species as rows and reactions as columns:

---

[3] See Ex. 12-4.
[4] See V. J. Lee, *J. Chem. Ed.*, *44*, 164 (1967).

$$\mathbf{R} = \begin{bmatrix} v_{11} & \cdots & v_{1j} & \cdots & v_{1R} \\ \vdots & & \vdots & & \vdots \\ v_{i1} & \cdots & v_{ij} & \cdots & v_{iR} \\ \vdots & & \vdots & & \vdots \\ v_{N1} & \cdots & v_{Nj} & \cdots & v_{NR} \end{bmatrix}$$

The rank of this matrix is the number of independent chemical reactions R.

We see that there are two complementary, and consistent, ways of representing the stoichiometry of chemically reacting systems: as ρ independent species or as R independent reactions. Thus, there have been two approaches to implementing the phase rule: one focusing on β and the other on $\mathbf{R}$.

The β matrix.   Taking the result that $R = N - \rho$, we obtain, on substitution into Eq. (13-2),

$$\rho = C + S$$

Thus, if there are no additional constraints ($S = 0$), the number of phase-rule components C is equal to the rank of the formula coefficient matrix β. The rank of β can be no larger than the number of constituent elements, although it can be smaller ($\rho \le m$).

When conditions are such that complete molecular rearrangement occurs in the reacting system, the rank of the formula coefficient matrix will equal the number of elements m. However, when certain elemental groupings remain intact during reaction, the rank can be less than m. For example, the formula coefficient matrix for a system comprised of HCl, NH_3, and NH_4Cl is

	N	H	Cl
NH_3	1	3	0
HCl	0	1	1
NH_4Cl	1	4	1

This matrix is a third-order determinant with a value of zero, and hence we must test second-order determinants. We can easily find a nonzero second-order determinant and therefore, as expected, determine the number of phase-rule components to be 2.

The method that utilizes β has the advantage that the phase rule can be applied to a complex system without the need of specifying reactions; all that is necessary is to determine the rank of β. This can often be done from only a partial listing of species.

The R Matrix.   When the system is simple or the reactions are known, this appears to be the obvious way to proceed. Starting with a set of possible reactions, Balzhiser, Samuels, and Eliassen.[5] demonstrate the use of the Gauss elimination algorithm for determining both the rank of **R** and a legitimized set of specific reactions.

In another approach, the number of independent reactions, R, is found by writing chemical equations for the formation of each nonelemental species from its constituent elements. Those elements not present in elemental form in the set of N species are then eliminated from this set by combination of the chemical equations. The number of equations in the resulting set is the number of independent reactions, R. It is expected that this set of reactions will produce a reaction matrix of rank R.

EXAMPLE 13-1

In a reacting system at equilibrium the following species are considered to be present in significant amounts: CO_2, CO, C, CH_4, H_2, H_2O, and N_2. Determine the number of phase-rule components and the number of independent reactions by (a) the **R** method and (b) the β method.

Solution 13-1

Seven chemical species, which include three elements, are present. These seven species are constituted from the elements C, H, O, and N.

(a) Formation reactions are

$$C + O_2 = CO_2 \qquad\qquad (a)$$

$$C + \frac{1}{2}O_2 = CO \qquad\qquad (b)$$

$$C + 2H_2 = CH_4 \qquad\qquad (c)$$

$$H_2 + \frac{1}{2}O_2 = H_2O \qquad\qquad (d)$$

Elemental oxygen is not among the N species and is therefore eliminated. Combining Eqs. (a) and (b) and Eqs. (b) and (d) results in three equations:

$$C + CO_2 = 2CO$$

$$C + H_2O = CO + H_2$$

$$C + 2H_2 = CH_4$$

The reaction matrix is

---

[5] R. E. Balzhiser, M. R. Samuels, and J. D. Eliassen, *Chemical Engineering Thermodynamics*, Prentice-Hall, Englewood Cliffs, NJ, 1972, Chap. 11.

	$j = 1$	$j = 2$	$j = 3$
CO_2	−1	0	0
CO	2	1	0
C	−1	−1	−1
CH_4	0	0	1
H_2	0	1	−2
H_2O	0	−1	0
N_2	0	0	0

The first three rows form a nonzero determinant;[6] hence the rank of **R** is 3. Thus, we find $R = 3$, and since $N = 7$,

$$C = N - R = 4$$

(b) For this system the formula coefficient matrix is

	C	O	H	N
CO_2	1	2	0	0
CO	1	1	0	0
C	1	0	0	0
CH_4	1	0	4	0
H_2	0	0	2	0
H_2O	0	1	2	0
N_2	0	0	0	2

From this matrix a fourth-order, nonzero determinant can be formed (e.g., the last four rows), and hence the rank of the formula coefficient matrix ρ is 4. If all possible fourth-order determinants had been zero, then a new set of third-order determinants could be formed by omitting any one column and any one row. If at least one of these determinants had a nonzero value, the rank ρ would have been 3. If no third-order nonzero determinants were found, the procedure would be repeated for second-order determinants.

We have found $\rho = 4$, and since no additional constraints apply, $C = 4$. With $N = 7$, we find $R = N - \rho = 3$.

These methods allow the determination of R or C, the *number* of independent reactions or *number* of phase-rule components; they do not specify *which* reactions or species should be chosen. The choice is usually based on computational considerations which we will not consider here.

---

[6] For an abbreviated set of matrix operations see App. F.

Additional Constraints. There are two types of additional constraints: intrinsic and conditional. Intrinsic constraints are part of the general description of the system and are included in the set of governing conservation equations [Eqs. (13-3a)–(13-3c)]. Conditional constraints arise when the system is prepared in a special way.

Intrinsic constraints. Although each type of conservation equation, Eqs. (13-3a)–(13-3c), must be considered in forming the β matrix, only those that can be expressed in intensive variables qualify as phase-rule equations. As it applies only to species in a particular phase, division of Eq. (13-3b) by the total number of mols of that phase converts the extensive variables, n_i's, into intensive variables and thereby renders Eq. (13-3b) a phase-rule equation which can be included among the number S.

$$\sum x_i z_i = 0$$

If the system contains only one phase, Eq. (13-3c) can, in the same manner, be converted to intensive variables. The resulting statement

$$\sum_{i=1}^{N} x_i \Delta \overline{M}_i = 0$$

specifies that the molar property of the mixture remain constant and therefore qualifies as a legitimate phase-rule equation to be numbered among the S additional constraints.

While division of Eq. (13-3a) by the total number of mols would transform it to an equation involving intensive variables, the right-hand side, b_l taken over the total number of mols, would still have to be specified. This would add an equation and a variable and thus produce no net advantage in a phase-rule description of the system. In the next section it is shown that a ratio of two elemental balances can be used as a phase-rule equation when species containing those two elements are found only in a single phase. Such an equation can be considered either a conditional constraint and numbered among S or as the specification of a variable which results in a reduction in the degrees of freedom, but it does not qualify as an intrinsic constraint.

Conditional constraints. Conditional constraints can be understood in terms of the set of elemental balance equations where certain values of the b_i's can lead to special relationships among the n_i's which under certain conditions result in relationships among the intensive composition variables. An example of a conditional constraint is the condition that the ratio of two species in a phase may be fixed because they are formed in only one reaction and exist in only one phase. Conditional constraints usually apply in relatively simple systems where few chemical reactions occur or distribution of species among two or more phases is not possible.

EXAMPLE 13-2

Apply the phase rule to the following systems:

(a) a mixture of $H_2(g)$ and $I_2(g)$ under conditions such that no reaction occurs.
(b) a mixture originally comprised of $H_2(g)$ and $I_2(g)$ in which the following reaction occurs:

$$H_2(g) + I_2(g) = 2HI(g)$$

(c) a mixture comprised originally of $H_2(g)$ and solid iodine, $I(c)$. Two phases exist, but no reaction occurs.
(d) same original mixture as in part (c) but with the reaction occurring to produce $HI(g)$.
(e) a system originally comprised of $HI(g)$ in which dissociation to $H_2(g)$ and $I_2(g)$ occurs. Only a gas phase exists.
(f) a system originally comprised of $HI(g)$ in which dissociation occurs. Solid iodine, $I(c)$, and a gas phase exist.

Solution 13-2

(a) With no chemical reaction the number of components C is equal to the number of species. We have $C = 2$ and $\pi = 1$, resulting in

$$\mathscr{F} = C + 2 - \pi = 2 + 2 - 1 = 3$$

With three degrees of freedom the system is specified with three variables. They could be temperature, pressure, and hydrogen mol fraction.

(b) In the reacting system the species are H_2, I_2, and HI ($N = 3$). Since both elements are present at equilibrium, only one formation reaction can be written; therefore, there is one independent reaction ($R = 1$). There are no additional constraints ($S = 0$), and therefore

$$C = N - R - S = 3 - 1 - 0 = 2$$

Alternatively, from the formula coefficient matrix,

$$
\begin{array}{c|cc}
 & H & I \\
\hline
H_2 & 2 & 0 \\
I_2 & 0 & 2 \\
HI & 1 & 1 \\
\end{array}
$$

a second-order nonzero determinant can be found ($\rho = 2$), and

$$\rho = 2 = C + S = C$$

With $C = 2$ and $\pi = 1$, we obtain

$$\mathscr{F} = C + 2 - \pi = 2 + 2 - 1 = 3$$

as with the nonreacting system. We see that the same number of variables are required to define the state of the system regardless of whether the reaction occurs. While the two systems are different, each will exhibit a definite and reproducible set of properties in a state of equilibrium when three intensive variables are specified.

(c) As in part (a), $C = 2$, and with $\pi = 2$ there are two degrees of freedom:

$$\mathscr{F} = C + 2 - \pi = 2 + 2 - 2 = 2$$

The system would be defined when, say, the temperature and pressure were specified.

(d) As in part (b), $C = 2$, and with $\pi = 2$ the result is the same as for part (c): $\mathscr{F} = 2$. As with systems (a) and (b), these two-phase systems have the same behavior from the standpoint of the phase rule.

(e) Because only a gas phase exists, the partial pressures (or mol fractions) of H_2 and I_2 must be equal because each species comes only from the dissociation of HI. This constitutes a conditional constraint ($S = 1$).

For this system the two elemental balance equations are

$$b_H = n_{HI} + 2n_{H_2}$$

$$b_I = n_{HI} + 2n_{I_2}$$

When the system is formed from HI we have

$$b_I = b_H = n_{0HI}$$

which results in

$$n_{H_2} = n_{I_2}$$

Because there is only one phase, we are allowed to equate mol fractions or partial pressures.

From the $\boldsymbol{\beta}$ matrix we found $\rho = 2$ and now write

$$\rho = 2 = C + S = C + 1$$

$$C = 1$$

and also from

$$C = N - R - S = 3 - 1 - 1 = 1$$

the degrees of freedom are

$$\mathscr{F} = C + 2 - \pi = 1 + 2 - 1 = 2$$

(f) While the hydrogen and iodine come only from the dissociation of HI and equal numbers of mols are produced, the iodine is distributed between two phases, and thus a statement of equality of intensive variables is not possible. There are no conditional constraints $(S = 0)$, and $C = 2$, resulting in

$$\mathscr{F} = C + 2 - \pi = 2 + 2 - 2 = 2$$

Thus, this system is identical to system (d).

EXAMPLE 13-3

In a reacting system containing the species FeO(c), Fe(c), C(c), $CaCO_3$(c), CaO(c), CO(g), CO_2(g), and O_2(g), it is desired to know the maximum number of solid phases which can coexist with the gas phase.

Solution 13-3

There are eight species comprised of four elements with three of the elements listed among the eight species. The five formation equations are

$$Fe + \frac{1}{2}O_2 = FeO \tag{a}$$

$$Ca + C + \frac{3}{2}O_2 = CaCO_3 \tag{b}$$

$$Ca + \frac{1}{2}O_2 = CaO \tag{c}$$

$$C + O_2 = CO_2 \tag{d}$$

$$C + \frac{1}{2}O_2 = CO \tag{e}$$

Elemental calcium is not present at equilibrium and can be eliminated from this set of equations by subtracting (c) from (b) to give

$$C + O_2 + CaO = CaCO_3 \qquad \text{(f)}$$

The four remaining reactions [(a), (d), (e), and (f)] will now be tested for independence by forming $\mathbf{R}$.

	(a)	(d)	(e)	(f)
FeO	1	0	0	0
Fe	−1	0	0	0
C	0	−1	−1	−1
$CaCO_3$	0	0	0	1
CaO	0	0	0	−1
CO	0	0	1	0
CO_2	0	1	0	0
O_2	−1/2	−1	−1/2	−1

The last four rows form a nonzero determinant; therefore, R = rank $(\mathbf{R})$ = 4.

Alternatively, the number of phase rule components C may be obtained from the formula coefficient matrix:

	Fe	O	C	Ca
FeO	1	1	0	0
Fe	1	0	0	0
C	0	0	1	0
$CaCO_3$	0	3	1	1
CaO	0	1	0	1
CO	0	1	1	0
CO_2	0	2	1	0
O_2	0	2	0	0

A fourth-order nonzero determinant may be formed from the first four rows of this matrix, and we have

$$\rho = 4 = C + S$$

Because the components of the gas phase could be formed from any number of reactions involving the solid species, a conditional constraint does not exist $(S = 0)$, and the number of phase rule components is equal to the number of constituent elements:

$$C = 4$$

This same result is obtained by subtracting R and S from N:

$$C = N - R - S = 8 - 4 - 0 = 4$$

With five solid phases and a gas phase we have

$$\mathscr{F} = C + 2 - \pi = 4 + 2 - 6 = 0$$

The system is invariant. It is possible for all five solid phases and a gas phase to coexist, but we may not specify any intensive variables.

In this example there are many phases, but none which contains all N species. Although the phase rule was derived for the case where each of the π phases contained all N species, the value of $\mathscr{F}$ is unaffected by the absence of any number of species from any phase because the number of variables and the number of equations is reduced equally.

EXAMPLE 13-4

The production of formaldehyde, HCHO, from methanol, CH_3OH, can proceed via either pyrolysis or oxidation. The primary reactions are as follows:

$$\text{Pyrolysis: } CH_3OH(g) = HCHO(g) + H_2(g)$$

$$\text{Oxidation: } CH_3OH(g) + \frac{1}{2}O_2(g) = HCHO(g) + H_2O(g)$$

The pyrolysis is endothermic, and the oxidation is exothermic, and it should be possible to carry out both reactions in the same reactor and thus eliminate the need to supply or remove heat.[7] As a preliminary step to a thorough analysis, assume equilibrium and apply the phase rule to the system. Consider two cases depending on whether oxygen or air is the oxidant.

Because of its large equilibrium constant, the oxidation reaction goes essentially to completion, and in order for the pyrolysis reaction to proceed, the reactor feed should be deficient in oxygen. At equilibrium there will be essentially no oxygen present.

Solution 13-4

When oxygen is the oxidant, the species present at equilibrium will be CH_3OH, HCHO, H_2, and H_2O ($N = 4$). The formula coefficient matrix is

---

[7] This possibility has been thoroughly explored by E. Jones and G. G. Fowlie, *J. Appl. Chem., 3*, 206 (1953).

	C	H	O
CH_3OH	1	4	1
$HCHO$	1	2	1
H_2	0	2	0
H_2O	0	2	1

The rank of this matrix is three ($\rho = 3$); however, the system was actually formed from two components, oxygen and methanol. This leads to a conditional constraint ($S = 1$) because the ratio of hydrogen to carbon will always equal 4. The elemental balance equations for hydrogen and carbon are

$$b_H = 4n_{0CH_3OH} = 4n_{CH_3OH} + 2n_{HCHO} + 2n_{H_2} + 2n_{H_2O}$$

$$b_C = n_{0CH_3OH} = n_{CH_3OH} + n_{HCHO}$$

On simplification they combine to yield

$$n_{HCHO} = n_{H_2} + n_{H_2O}$$

Because there is only one phase, this can be expressed in mol fraction and provides another equation relating the intensive variables of the system. Thus, we have $C = 2$, $S = 1$, and $N = 4$, and there is one independent reaction:

$$C = N - R - S$$

$$2 = 4 - R - 1$$

$$R = 1$$

This result also could have been obtained from the method of determining R. The three formation reactions are

$$C + 2H_2 + \frac{1}{2}O_2 = CH_3OH \qquad\qquad (a)$$

$$C + H_2 + \frac{1}{2}O_2 = HCHO \qquad\qquad (b)$$

$$H_2 + \frac{1}{2}O_2 = H_2O \qquad\qquad (c)$$

Oxygen can be eliminated by combining reactions (a) and (c) and reactions (b) and (c). When combined, these two resulting equations yield a single equation which does not contain carbon ($R = 1$):

$$CH_3OH = HCHO + H_2$$

Only a gas phase exists ($\pi = 1$), and

$$\mathscr{F} = C + 2 - \pi = 2 + 2 - 1 = 3$$

These three degrees of freedom may be specified in terms of temperature, pressure, and a composition variable. For adiabatic operation the requirement is $\Delta H = 0$ or $H = H_0$, and it will be necessary to include this intrinsic constraint in writing the β matrix (see Prob. 13-18).

Because of its inertness, the addition of nitrogen to the system changes little. The number of species increases by one ($N = 5$), and the number of independent reactions remains unchanged ($R = 1$). However, there is another conditional constraint which applies, as can be seen from the following elemental balances:

Oxygen: $n_{0CH_3OH} + 2(0.21)A = n_{CH_3OH} + n_{HCHO} + n_{H_2O}$

Carbon: $n_{0CH_3OH} = n_{CH_3OH} + n_{HCHO}$

Nitrogen: $0.79A = n_{N_2}$

Combining the carbon and oxygen balances gives

$$2(0.21)A = n_{H_2O}$$

Dividing the nitrogen balance by this gives

$$\frac{n_{N_2}}{n_{H_2O}} = \frac{y_{N_2}}{y_{H_2O}} = \frac{0.79}{2(0.21)} = 1.88$$

With two conditional constraints we again find that

$$C = N - R - S = 5 - 1 - 2 = 2$$

This result also could have been obtained from the rank of the formula coefficient matrix which is now $4 (\rho = 4)$:

$$C = \rho - S = 4 - 2 = 2$$

Thus, the system comprised from methanol and air also possesses three degrees of freedom.

If it were desired to consider the presence of undesirable but ubiquitous species such as CO_2 and CO, our phase rule analysis would remain unchanged. N would increase by two; C and S would remain unchanged, which would require R to increase by two; however, $\mathcal{F}$ would be unchanged.

EXAMPLE 13-5

Apply the phase rule to the following systems:
(a) an aqueous solution containing the ions Na^+, K^+, Cl^-, and I^-.
(b) an aqueous solution prepared from NaCl and KI.

Solution 13-5

The species are: H_2O, Na^+, K^+, Cl^-, and I^-. Six elemental balance equations and the conservation of charge constitute seven governing equations. The number of independent governing equations is determined from the rank of the β matrix.

	Na	Cl	K	I	O	H	Charge
Na^+	1	0	0	0	0	0	1
Cl^-	0	1	0	0	0	0	-1
K^+	0	0	1	0	0	0	1
I^-	0	0	0	1	0	0	-1
H_2O	0	0	0	0	1	2	0

The rank of β is 5 ($\rho = 5$). With $N = 5$ and $\rho = 5$ the number of independent chemical reactions is zero ($R = 0$).

(a) The conservation of charge is an intrinsic constraint ($S = 1$) and the number of components is

$$C = \rho - S = 5 - 1 = 4$$

With $C = 4$ and $\pi = 1$ we have

$$\mathcal{F} = C + 2 - \pi = 4 + 2 - 1 = 5$$

The five variables to be specified could be: T, P, and three ion molalities. Any number of components could be used to prepare the system, but four is the minimum number required to prepare all permissible compositions. This can be seen by the selection of H_2O, NaCl, KCl, and KI as components. The elemental balances and the charge balance are

$$n_{0NaCl} = n_{Na^+}$$

$$n_{0\text{NaCl}} + n_{0\text{KCl}} = n_{\text{Cl}^-}$$

$$n_{0\text{KCl}} + n_{0\text{KI}} = n_{\text{K}^+}$$

$$n_{0\text{H}_2\text{O}} = n_{\text{H}_2\text{O}}$$

$$n_{\text{Na}^+} + n_{\text{K}^+} = n_{\text{Cl}^-} + n_{\text{I}^-}$$

These are a set of five governing equations that contain four n_0's which, when fixed, will determine the system.

(b) The only difference between this and the previous system is that we have introduced a conditional constraint which can be seen by re-writing the governing equations

$$n_{0\text{NaCl}} = n_{\text{Na}^+}$$

$$n_{0\text{NaCl}} = n_{\text{Cl}^-}$$

$$n_{0\text{KI}} = n_{\text{K}^+}$$

$$n_{0\text{H}_2\text{O}} = n_{\text{H}_2\text{O}}$$

$$n_{\text{Na}^+} + n_{\text{K}^+} = n_{\text{Cl}^-} + n_{\text{I}^-}$$

Here we see that $n_{\text{Na}^+} = n_{\text{Cl}^-}$ is the conditional constraint. Together with the charge balance equation, this brings the number of additional constraints to two $(S = 2)$. The number of components is

$$C = \rho - S = 5 - 2 = 3$$

and is consistent with our specification of H_2O, NaCl, and KI. With $C = 3$ and $\pi = 1$ we have

$$\mathscr{F} = C + 2 - \pi = 3 + 2 - 1 = 4$$

Four variables which could be specified are: T, P, and two ion molalities.

EXAMPLE 13-6

Perform a phase-rule analysis on the water–carbon dioxide system studied in Ex. 12-18.

Solution 13-6

Seven species were considered present at equilibrium:

$$CO_2 \quad H_2O \quad H_2CO_3 \quad H^+ \quad HCO_3^- \quad CO_3^= \quad \text{and} \quad OH^-$$

Because the statement of electrical neutrality is a governing equation, it must be included when forming the formula coefficient matrix β.

	C	H	O	Charge
CO_2	1	0	2	0
H_2O	0	2	1	0
H_2CO_3	1	2	3	0
H^+	0	1	0	1
HCO_3^-	1	1	3	-1
$CO_3^=$	1	0	3	-2
OH^-	0	1	1	-1

Using the Gauss elimination algorithm[8] the rank of β is found to be 3:

$$\rho = 3 = C + S$$

The condition of electrical neutrality is an intrinsic constraint ($S = 1$) and the number of components is two (e.g., H_2O and CO_2).

The number of independent chemical reactions is $N - \rho = 7 - 3 = 4$. If we choose reactions (b), (c), (d), and (e) of Ex. 12-18, the reaction matrix R is

	(b)	(c)	(d)	(e)
CO_2	-1	0	0	0
H_2O	-1	0	0	-1
H_2CO_3	1	-1	0	0
H^+	0	1	1	1
HCO_3^-	0	1	-1	0
$CO_3^=$	0	0	1	0
OH^-	0	0	0	1

---

[8] Balzhiser et al., *op. cit.*

The first four rows form a nonzero determinant; hence, the matrix has a rank of 4 and all four reactions are independent.

Next, we find $\mathscr{F}$:

$$\mathscr{F} = C + 2 - \pi = 2 + 2 - 2 = 2$$

With two degrees of freedom we could fix the temperature and the partial pressure of CO_2 as was done in Ex. 12-18.

13-2 ANALYZING COMPLEX CHEMICAL EQUILIBRIUM PROBLEMS

The thermodynamic principles previously derived apply to all reacting systems, but the details of treatment of complex chemically reacting systems are specific to the particular system under study. A general approach to the treatment of these systems will be developed, but its implementation will depend on how the system is viewed and the nature of the results desired. It is hoped that the delineation of this general approach and its application to several examples will adequately prepare the reader to deal with this class of problem.

The general approach may be viewed as a four-step process:

1. Determine the species present in significant amounts at equilibrium.
2. Apply the phase rule.
3. Formulate the problem mathematically.
4. Obtain a mathematical solution.

Steps 3 and 4 go together and will be dealt with in the next section. Steps 1 and 2 represent the analysis of the problem. Here we will focus on step 1, which is essentially the delineation of a model. As with all models, a model of complex chemical equilibrium will be judged valuable if its mechanisms are realistic and its predictions are in agreement with available information.

The determination of the species present at equilibrium is the step which characterizes the problem and is therefore the most crucial. Accordingly, it is the step least amenable to generalization. Often a knowledge of reaction kinetics is essential to the intelligent selection of species. Some species may be strongly favored on thermodynamic grounds but may be formed through reactions which are known to proceed quite slowly. In certain situations their omission may be justified on kinetic grounds. However, kinetic considerations for species selection, because of their ad hoc and nonthermodynamic nature, will not be further discussed.

We begin the species selection process by listing ΔG^f or log K^f for every known species which could be formed from the elements constituting the reactants. Those species with the more negative ΔG^f, or larger log K^f, will predominate at equilibrium and are chosen as the N species. Actually, it is not necessary to list every known species because within families of compounds there exist definite trends of ΔG^f or log K^f with molecular weight which may facilitate the selection process. The selection of the species to be included in the calculations will, of course, depend on the intended application; however, it will usually be found that only a few species will predominate at equilibrium. If desired, the concentration of trace species may be calculated after equilibrium concentrations of the originally selected set of N species has been obtained.

EXAMPLE 13-7

For a reacting system comprised of the elements carbon, hydrogen, and oxygen (henceforth referred to as the CHO system), determine the species present in significant amounts at ultimate equilibrium and apply the phase rule. We will be interested in temperatures ranging from 500 to 1000 K at a pressure of 1 atm.

Solution 13-7

Listed here are several chemical species composed of the elements C, H, and O along with their ΔG^f values at 500 and 1000 K. Because of the extreme negative ΔG^f for CO_2 and H_2O, there is no need to consider any substance having a positive value of ΔG^f except as a trace component.

	ΔG^f (kcal / g mol) [a]	
Species	500 K	1000 K
CO	−37.19	−47.95
CO_2	−94.39	−94.61
H_2O	−52.36	−46.04
CH_4	−7.85	4.58
C_2H_6	1.16	26.13
C_4H_{10}	14.55	64.50
CH_3OH	−32.11	−13.46

[a] Source: reference 3 in Table 12-1.

The behavior of the CHO system can be visualized in terms of four composition regions which are identified on the triangular plot shown in Fig. 13-1. The scale is in atom percent, and the overall atomic composition of any system comprised of these elements can be represented by a point on the diagram. In this figure are shown points representing the dominant chemical species, and the areas labeled I, II, and III represent the regions where oxygen, hydrogen, and carbon, respectively, are in excess. When sufficient oxygen is present for the formation of H_2O and CO_2, these two compounds, because of their extreme negative

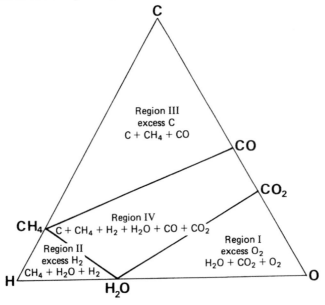

Figure 13-1 Regions of the CHO diagram.

ΔG^f values, will be present almost exclusively. With an excess of hydrogen one may expect the oxygen to exist as H_2O and the carbon as CH_4. Similarly, with an excess of carbon the species CO and CH_4 will predominate. In these three regions the system is rather easily described; however, a point located in region IV would represent a system in which $C(c)$, CO, CO_2, H_2O, CH_4, H_2, and O_2 are the major species which could exist at equilibrium. This is the region with the most interesting behavior and to which we will direct our attention.

Because of the extremely large equilibrium constants for the formation of CO_2 and H_2O, the partial pressure of oxygen in region IV will be driven to extremely low levels, and O_2 need not be considered as one of the N species. This extremely low oxygen partial pressure will force equilibrium concentrations of oxygen-containing organic compounds such as methanol to also be quite low even though ΔG^f values appear to favor their formation. Consider the formation of methanol:

$$C(c) + 2H_2(g) + \frac{1}{2}O_2(g) = CH_3OH(g)$$

$$K = \frac{p_{CH_3OH}}{p_{H_2}^2 \, p_{O_2}^{1/2} \, \hat{a}_C}$$

$$p_{CH_3OH} = K p_{H_2}^2 \, p_{O_2}^{1/2} \, \hat{a}_C$$

At 500 K the equilibrium constant is

$$K = \exp\left[\frac{32{,}110}{1.987(500)}\right] = 1.1(10^{14})$$

Now also consider the formation reaction for CO_2:

$$C(c) + O_2(g) = CO_2(g)$$

$$K = \frac{p_{CO_2}}{\hat{a}_C p_{O_2}}$$

At 500 K the equilibrium constant is

$$K = \exp\left[\frac{94{,}390}{1.987(500)}\right] = 1.8(10^{41})$$

and we may write

$$\hat{a}_C p_{O_2} = \frac{p_{CO_2}}{K}$$

The partial pressure of CO_2 may not exceed 1 atm, and the maximum value of the product $a_C p_{O_2}$ is $5.5(10^{-42})$. To estimate the partial pressure of methanol, the product $a_C \, p_{O_2}^{1/2}$ is needed. This product cannot exceed $2.3(10^{-21})$, and the maximum partial pressure of methanol would be

$$p_{CH_3OH} \leq 1.1(10^{14})(1^2)(2.3)(10^{-21}) \leq 2.5(10^{-7}) \text{ atm}$$

A similar approach allows us to rule out the presence of other oxygen-containing organic compounds. The species considered present in significant amounts at equilibrium are $C(c)$, CO, CO_2, H_2, H_2O, and CH_4 ($N = 6$).

Insight into the behavior of the system may be obtained by applying the phase rule. To determine the number of phase rule components C, we construct the formula coefficient matrix:

	C	H	O
CO	1	0	1
CO_2	1	0	2
H_2O	0	2	1
CH_4	1	4	0
C	1	0	0
H_2	0	2	0

Several combinations of rows yield nonzero third-order determinants $(\rho = 3)$, and because there are no additional constraints $(S = 0)$, the system is composed of three components $(C = 3)$. This will be true no matter how many additional species we decide to include among the set of N, and from the standpoint of the phase rule such an addition will not affect the behavior of the system. The computational aspect of the problem is, of course, dependent on the number of species included in N.

For this system there are two cases to consider depending on the presence or absence of solid carbon. When only a gas phase exists,

$$\mathscr{F} = C + 2 - \pi = 3 + 2 - 1 = 4$$

and when solid carbon is also present,

$$\mathscr{F} = C + 2 - \pi = 3 + 2 - 2 = 3$$

The system will possess the maximum degrees of freedom, four, when only a gas phase exists and could be defined when temperature, pressure, and two elemental ratios are specified. The specification of an elemental ratio provides a relationship among species mol fractions and therefore serves to reduce the degrees of freedom. The oxygen-hydrogen elemental ratio may be expressed as

$$\frac{O}{H} = \frac{y_{CO} + 2y_{CO_2} + y_{H_2O}}{2y_{H_2} + 2y_{H_2O} + 4y_{CH_4}}$$

and the carbon-hydrogen ratio is

$$\frac{C}{H} = \frac{y_{CO} + y_{CO_2} + y_{CH_4}}{2y_{H_2} + 2y_{H_2O} + 4y_{CH_4}}$$

When solid carbon is present, only one elemental ratio, O/H, can be utilized; however, because there is one less degree of freedom, only one such ratio is needed. Thus, we conclude that elemental ratios may be used to specify the state of an equilibrium system. For given values of these ratios the state of the system, and hence the value of all intensive variables, remains the same regardless of the nature of the reactants.

EXAMPLE 13-8

When coal is liquefied, the resulting fuel will contain small amounts of compounds of nitrogen and sulfur which on combustion could produce, respectively, the pollutants NO_x and SO_x. Sulfur compounds obviously must be removed from the fuel; however, because the combustion process occurs with air, considerable nitrogen will be present in the combustion zone, and therefore the need for removal of nitrogen compounds from the fuel is questioned. Based on the assumption of equilibrium within the combustion zone, assess the efficacy of removing the nitrogen compounds from the fuel.

Solution 13-8

Coal, and therefore the fuel, contains the five elements carbon, hydrogen, oxygen, nitrogen, and sulfur, and of course, the elements nitrogen and oxygen are present in the combustion air. We have seen that the number of phase rule components may equal but not exceed the number of elements comprising the system; therefore,

$$C \leq 5$$

Since only a gas phase will exist, application of the phase rule requires

$$\mathscr{F} = C + 2 - \pi = C + 1$$

$$\leq 6$$

When the temperature and pressure are specified, a maximum of four degrees of freedom remain. They may be specified in terms of four elemental ratios. Because the composition of fuel and air are fixed, these elemental ratios are in turn fixed when the air to fuel ratio is specified.

Two cases should be considered depending on whether nitrogen compounds are absent or present in the fuel. For each case at a fixed temperature and pressure the equilibrium mol fractions will depend on the elemental ratios, say [C/O], [H/O], [N/O], and [S/O], and the question becomes the following: Is there a significant difference in these elemental ratios between the two cases? Because the overwhelming share of nitrogen comes from air, one would expect little ratio difference and hence little difference in equilibrium NO_x concentrations when nitrogen compounds are removed from the fuel.

This analysis is based on the assumption of equilibrium in the combustion zone. While this assumption is routinely made for combustion problems, there is always the possibility that in actuality kinetic factors could influence the composition of the combustion gas. Only research can answer this question; however, thermodynamic analysis provides a convenient framework for visualizing and analyzing the behavior of such a system.

EXAMPLE 13-9

Hydrogen sulfide, a noxious pollutant, is removed from gaseous waste streams through the Claus process which is based on partial oxidation of sulfides to SO_2 followed by a lower-temperature catalytic reaction between SO_2 and the remaining H_2S to form sulfur. The reactions may be written as

$$H_2S + \frac{3}{2}O_2 = H_2O + SO_2$$

$$2H_2S + SO_2 = 3S + 2H_2O$$

Because H_2S and SO_2 react in a 2:1 ratio, it is desirable to oxidize only one-third of the incoming H_2S. This is often carried out in a waste heat boiler.

A refinery has several sulfide-containing gas streams and wishes to consider employment of the Claus process. The combined gas stream also contains some light hydrocarbons, mercaptans, carbon dioxide, and ammonia which will complicate the determination of the combustion air requirement and which may also cause the presence of toxic or otherwise undesirable species. Preparatory to undertaking a complete study of this problem, it is desired to carry out a preliminary analysis for the oxidation step. The species likely to be present in significant amounts at equilibrium will be identified, and the phase rule will be applied for insight.

Solution 13-9

With five constituting elements (carbon, hydrogen, oxygen, nitrogen, and sulfur) many species are possible. The temperature range of interest is 1600 to 2200°F, and log K^f values at these temperatures are tabulated for many of the possible compounds comprised of these elements:

	log $K^{f\,a}$	
	1600°F	2200°F
CO_2	18.0768	14.0072
CO	9.7373	8.5651
SO_2	12.711	8.984
SO_3	12.349	7.680

	log K^{f} [a]	
	1600° F	2200° F
COS	9.99	7.85
H_2O	8.4261	5.8574
HCHO	3.765	2.460
H_2S	1.551	0.622
CS_2	0.971	0.836
N_2	0	0
H_2	0	0
S_2	0	0
C(c)	0	0
CH_4	−1.6008	−2.5438
NH_3	−3.608	−4.182
NO	−3.741	−2.539
NO_2	−4.783	−4.438
N_2O	−7.576	−6.690
HCN	−4.195	−2.878
CH_3SH [b]	−3.6	−4.5

[a] Based on $S_2(g)$ as the standard state for sulfur. Source: reference 7 in Table 12-1.
[b] Extrapolated.

Because this is a partial oxidation process, the formation of species such as CO_2, H_2O, and SO_2 with extremely large K^f's will drive the oxygen partial pressure down to extremely low values. Under similar conditions as shown in Ex. 13-7, the partial pressure of oxygen-containing organic species such as HCHO will be negligibly small even though their K^f's are reasonably large.

The K^f for SO_3 is almost as large as that of SO_2; however, this species may be neglected by consideration of the following reaction:

$$SO_2 + \frac{1}{2}O_2 = SO_3$$

$$K = \frac{p_{SO_3}}{p_{SO_2}\, p_{O_2}^{1/2}}$$

$$\frac{p_{SO_3}}{p_{SO_2}} = K p_{O2}^{1/2}$$

For this reaction $K = 0.435$ at 1600°F, and $K = 0.050$ at 2200° F. Because $p_{O_2}^{1/2}$ is quite small, p_{SO_3} / p_{SO_2} will be also.

The species COS also contains oxygen but because of its large K^f will be included.

Sulfur vapor may exist in several allotropic forms, but at high temperatures S_2 is the predominant form.

With the exception of SO_3, oxygen-containing organic species, and solid carbon, all species with nonnegative log K^f values will be considered present at equilibrium. Solid carbon was omitted because the formation of species such as CO_2, CO, and COS with extremely large K^f values should prevent its formation even though the oxygen partial pressure is extremely low. Omission of solid carbon also facilitates the calculations. A test for the presence of this and any other omitted species can be made after the calculations have been made.

We have chosen CO_2, CO, SO_2, COS, H_2O, H_2S, CS_2, N_2, H_2, and S_2 as the 10 species present in significant amounts at equilibrium. The formula coefficient matrix is

Index l		1	2	3	4	5
	Element	N	O	S	C	H
Index i	Species					
1	N_2	2	0	0	0	0
2	H_2	0	0	0	0	2
3	S_2	0	0	2	0	0
4	CO_2	0	2	0	1	0
5	SO_2	0	2	1	0	0
6	CO	0	1	0	1	0
7	COS	0	1	1	1	0
8	H_2O	0	1	0	0	2
9	H_2S	0	0	1	0	2
10	CS_2	0	0	2	1	0

and has a rank of 5 $(\rho = 5)$. Because there are no additional constraints $(S = 0)$, there are five phase rule components $(C = 5)$, and for a single-phase system,

$$\mathscr{F} = C + 2 - \pi = 5 + 2 - 1 = 6$$

These six degrees of freedom could be satisfied by specification of temperature, pressure, and four elemental ratios. Because the feed gas and air are of fixed composition, these four elemental ratios are fixed when the air to feed gas ratio is specified.

We desire that the ratio H_2S/SO_2 be close to 2 and therefore might inquire as to whether it is legitimate to specify this ratio as one of the system variables. Since all compositions have been shown to depend only on the air to feed gas ratio, which in turn is equivalent to fixing four elemental ratios, specification of the H_2S/SO_2 ratio is tantamount to fixing four phase rule variables. This would leave two others, such as temperature and pressure, to be fixed. While the phase rule tells us that specification of the H_2S/SO_2 ratio is legitimate, we have no way of knowing a priori which air to feed gas ratio corresponds to the desired H_2S/SO_2 ratio, and therefore this is not a practical way to specify the system from a computational standpoint. However, we know that there is a good possibility of finding the desired H_2S/SO_2 ratio at some selected temperature and pressure.[9]

13-3 FORMULATING COMPLEX CHEMICAL EQUILIBRIUM PROBLEMS

Basically, there are two approaches to the formulation of complex chemical equilibrium problems: use of equilibrium constants and free energy minimization. It has been shown that the criterion of equilibrium is that the Gibbs free energy has its minimum value at constant temperature and pressure. Because this provides the basis for defining the equilibrium constant, it is obvious that both methods are but variations of the same thermodynamic principle.

Equilibrium Constant Methods. It has already been shown that for equilibrium in systems described by multiple chemical reactions all reactions are in equilibrium and their equilibrium constants must be satisfied. There will be R independent equilibrium constant expressions relating the N species-concentration variables. Additional equations are provided by m elemental balances and S additional constraints. Essentially the computational problem is one of solving the set of N equations in N unknowns. This calls for numerical procedures which are best executed by means of a digital computer. The problem is not trivial or routine, as is evidenced by the quantity of work published. While these problems are challenging and an understanding of thermodynamic principles is essential to the intelligent implementation of the computational techniques, such matters are beyond the scope of this work.[10]

Free Energy Minimization Methods. The Gibbs free energy of a system consisting of N species can be written

$$G = \sum_{i=1}^{N} n_i \hat{\mu}_i \tag{13-5}$$

---

[9] Because the phase rule merely states what is possible, there is no guarantee that a specific ratio can be attained. As always, some knowledge and understanding of the system are necessary for the intelligent application of the phase rule.

[10] For a detailed and comprehensive treatment, the reader is referred to the review by F. S. Zeleznik and S. Gordon [*Ind. Eng. Chem.*, 60(6), 27 (1968)] or the monograph by F. van Zeggeren and S. H. Storey, *The Computation of Chemical Equilibria*, Cambridge University Press, New York, 1970.

and dividing by RT produces the more convenient dimensionless expression

$$g = \frac{G}{RT} = \sum_{i=1}^{N} n_i \frac{\hat{\mu}_i}{RT} \tag{13-5a}$$

We wish to find the set of mol numbers n_i's which produces a minimum in G or g, however, all the mol numbers are not independent because elemental balance constraints must also be satisfied. We have written these constraints as

$$\sum_{i=1}^{N} \beta_{il} n_i - b_l = 0 \qquad (l = 1, \ldots, m) \tag{13-3a}$$

where the β_{il}'s are the formula coefficients and b_l is the total number of atomic weights of element l present in the system. This problem can be alleviated by introducing m additional variables, the Lagrangian multipliers λ_l's so that the N mol numbers can be considered independently variable. This is done by adding the term

$$\sum_{l=1}^{m} \lambda_l \left(\sum_{i=1}^{N} \beta_{il} n_i - b_l \right)$$

to g to obtain a new function g':

$$g' = \sum_{i=1}^{N} n_i \frac{\hat{\mu}_i}{RT} + \sum_{l=1}^{m} \lambda_l \left(\sum_{i=1}^{N} \beta_{il} n_i - b_l \right)$$

Because the added term is equal to zero, g and g' are identical and will exhibit the same minimum even though their partial derivatives with respect to the n_i will be different.[11] The conditions for a minimum are determined by obtaining the partial derivatives of g' with respect to n_i and setting them equal to zero. This results in

$$\frac{\hat{\mu}_i}{RT} + \sum_{l=1}^{m} \lambda_l \beta_{il} = 0 \qquad (i = 1, \ldots, N) \tag{13-6}$$

Because $\hat{\mu}_i$ will be a known function of composition, or mol numbers, this set of N equations when combined with the m elemental balances [Eq. (13-3a)] is sufficient to determine the N mol numbers and the m Lagrangian multipliers λ_l's. As with the equilibrium constant method, much effort has been devoted to computational techniques for solving these equations.[12]

---

[11] At the minimum all partial derivatives must equal zero, and hence they are equal for g and g' at this point.

[12] See Zeleznik and Gordon, *op. cit.*, and van Zeggeren and Storey, *op. cit.*

While the λ_l's were introduced as a mathematical device, they can be given physical meaning. Substitution of Eq. (13-6) into Eq. (13-5a) yields

$$g = \sum_{i=1}^{N} n_i \left(-\sum_{l=1}^{m} \lambda_l \beta_{il} \right)$$

which can be written as

$$g = -\sum_{l=1}^{m} \lambda_l \sum_{i=1}^{N} n_i \beta_{il}$$

From Eq. (13-3a) this is seen to be

$$g = -\sum_{l=1}^{m} \lambda_l b_l \tag{13-7}$$

λ_l is dimensionless and represents the contribution of the lth element to the dimensionless free energy of the system. These Lagrangian multipliers can be used through Eq. (13-6) to calculate the quantity of trace components not included among the N species in the original problem formulation. Values of the λ_l's determined from the original problem formulation are used for this calculation. While the λ_l's are expected to depend on composition, this is permissible because inclusion or omission of trace components will have a negligible effect on the system composition.

Before Eqs. (13-3a) and (13-6) can be solved, values of $\hat{\mu}_i$ must be obtained. For component i in a mixture we write

$$\hat{\mu}_i = \mu_i^\circ + RT \ln \hat{a}_i \tag{12-10}$$

where μ_i° is the chemical potential of component i in its standard state and $\hat{a}_i$ is its activity in the mixture. While an absolute value of μ_i° can never be known, it can be shown that setting μ_i° equal to ΔG_i^f will produce the same minimum in the system free energy as would μ_i°. For a gas phase activities are numerically equal to fugacities and at pressures low enough for fugacities to be replaced with partial pressures we could write

$$\frac{\hat{\mu}_i}{RT} = \frac{\Delta G_i^f}{RT} + \ln p_i = -\ln K_i^f + \ln P + \ln y_i \tag{12-10a}$$

Thus, for components in a perfect gas mixture Eq. (13-6) becomes

$$\ln P + \ln \frac{n_i}{n} - \ln K_i^f + \sum_{l=1}^{m} \lambda_l \beta_{il} = 0 \tag{13-6a}$$

where n_i is the mols of component i in the mixture and n is the total number of mols of the mixture. This equation is written for each component with the β's in the summation being evaluated from the component's row in the β matrix.

For a liquid mixture activities would be replaced with the product of mol fraction and activity coefficient to yield

$$\ln \hat{a}_i - \ln K_i^f + \sum_{l=1}^{m} \lambda_l \beta_{il} = 0 \tag{13-6b}$$

where the activity is a product of a composition variable and an activity coefficient as per Eq.(12-33) or (12-37).

As we have seen in Sec. 12-8, when pure condensed phases are present, their μ_i values can be replaced by μ_i°, or ΔG_i^f, unless the pressure is large. This results in

$$-\ln K_i^f + \sum_{l=1}^{m} \lambda_l \beta_{il} = 0 \tag{13-6c}$$

for pure solids.

EXAMPLE 13-10

Use free energy minimization to determine compositions at ultimate equilibrium for the carbon–hydrogen–oxygen(CHO) system consisting of a gas phase at 1000 K and 1 atm. The system was comprised originally from one mol of methane and 3 mols of water.

Solution 13-10

The CHO system has been analyzed in Ex. 13-7 and was found to possess four degrees of freedom. In addition to temperature and pressure, specifying the ratio of water to methane fixes the elemental ratios and thus determines the system. We have

$$b_C = n_{0CH_4} = 1$$

$$b_H = 4n_{0CH_4} + 2n_{0H_2O} = 10$$

$$b_O = n_{0H_2O} = 3$$

At this low pressure it is reasonable to replace fugacities by partial pressures and use Eq. (13-6a). This equation is written for each component with the β's in the summation being evaluated from the component's row in the β matrix. The ln K' data obtained from the spreadsheet LOGKF(T).WQ1(or .XLS) are:

Species	ln K'
CO_2	47.8774
CO	24.4572
H_2O	23.1579
CH_4	-1.9774

The non-linear-equation feature of the software package POLY-MATH on the accompanying CD-ROM was used to obtain the solution. This example is stored in the POLYMATH library as files EX13-10 and EX13-10G. The following is a facsimile of the POLYMATH worksheet

POLYMATH PROBLEM EX13-10

The equations: Initial Guess:

f(nCO) = nCH4 + nCO + nCO2 – 1 $nCO_0 = 0.5$

f(nCH4) = 4*nCH4 + 2*nH2O + 2*nH2 – HtoC $nCH4_0 = 0.25$

f(nH2O) = nH2O + nCO + 2*nCO2 – OtoC $nH2O_0 = 2.0$

f(lamC) = 1.9774 + ln(nCH4/n) + lamC + 4*lamH $lamC_0 = 1$

f(lamH) = –23.1579 + ln(nH2O/n) + 2*lamH + lamO $lamH_0 = 1$

f(lamO) = –24.4572 + ln(nCO/n) + lamC + lamO $lamO_0 = 1$

f(nCO2) = –47.8774 + ln(nCO2/n) + lamC + 2*lamO $nCO2_0 = 0.25$

f(nH2) = ln(nH2/n) + 2*lamH $nH2_0 = 2.5$

n = nCH4 + nH2O + nCO + nCO2 + nH2

HtoC = 10

OtoC = 3

Note that the functions are written so that the expression on the right-hand side of the equation should be zero for a valid solution and that each variable is identified with a function even though the right-hand expression may not contain that variable. The first three equations are elemental balances, the next five are statements of Eq.(13-6a); these are the variable statements. The next three lines are auxiliary equations: the ninth defines n, and the last two give the elemental ratios and add flexibility to the program. The initial guesses for the mol numbers satisfy elemental balances and were obtained from the following set of linear equations.

$$(1)n_{CO} + (1)n_{CO_2} + (1)n_{CH_4} + (0)n_{H_2O} + (0)n_{H_2} = b_C$$

$$(0)n_{CO} + (0)n_{CO_2} + (4)n_{CH_4} + (2)n_{H_2O} + (2)n_{H_2} = b_H$$

$$(1)n_{CO} + (2)n_{CO_2} + (0)n_{CH_4} + (1)n_{H_2O} + (0)n_{H_2} = b_O$$

$$(0)n_{CO} + (1)n_{CO_2} + (0)n_{CH_4} + (0)n_{H_2O} + (0)n_{H_2} = xb_C$$

$$(0)n_{CO} + (0)n_{CO_2} + (0)n_{CH_4} + (0)n_{H_2O} + (2)n_{H_2} = yb_H$$

The first three equations are elemental balances for carbon, hydrogen, and oxygen respectively. The last two equations simply specify the mols of CO_2 and H_2 where x is the fraction of elemental carbon as CO_2 and y is the fraction of elemental hydrogen as H_2. For $x = 0.25$ and $y = 0.5$ this set of equations was solved with POLYMATH's linear equation solver. The problem, EX13-10G, is in the POLYMATH library and is shown below as it appears on the worksheet.

POLYMATH PROBLEM EX13-10G

matrix of coefficients   vector of

column number   constants

		1	2	3	4	5	6
	1	1	1	1	0	0	1
row	2	0	0	4	2	2	10
number	3	1	2	0	1	0	3
	4	0	1	0	0	0	0.25
	5	0	0	0	0	2	5

The solution to problem EX13-10G is shown as the initial guess listed on the worksheet for problem EX13-10. The mol numbers found from the solution to problem EX13-10 are listed below and used to calculate gas-phase mol fractions.

RESULTS OF FREE ENERGY MINIMIZATION WITH POLYMATH
GAS-PHASE COMPOSITIONS IN THE C–H–O SYSTEM

Component	number of mols	mol fraction
CO	0.6026	0.1009
CH_4	0.0145	0.00243
H_2O	1.6317	0.2733
CO_2	0.3828	0.0641
H_2	3.3392	0.5592

The Lagrangian multipliers are dimensionless and were found to be

$$\lambda_C = 2.8766; \quad \lambda_H = 0.2906; \quad \lambda_O = 23.874$$

When solving a problem by numerical methods, it is always a good idea to check the solution by using different initial guesses. This was done with satisfactory results. For this particular problem there is another test that can be applied: the equilibrium constants for all possible reactions involving the species must be satisfied as shown in Sec. 12.2. We will consider the following reactions

Reaction 1: $H_2O + CO = H_2 + CO_2$

Reaction 2: $3H_2 + CO = CH_4 + H_2O$

Reaction 3: $CH_4 + CO_2 = 2CO + 2H_2$

with equilibrium constants calculated from $\ln K^f$ values:

$$K_1 = 1.30; \quad K_2 = 0.0377; \quad K_3 = 20.4$$

For $P = 1$ atm, the equilibrium constants determined from the calculated mol fractions are

$$K_1 = \frac{y_{H_2} y_{CO_2}}{y_{H_2O} y_{CO}} = \frac{0.5592(0.0641)}{0.2733(0.1009)} = 1.30$$

$$K_2 = \frac{y_{CH_4} y_{H_2O}}{y_{CO} y_{H_2}^3} = \frac{0.00243(0.2733)}{0.1009(0.5592)^3} = 0.0377$$

$$K_3 = \frac{y_{CO}^2 y_{H_2}^2}{y_{CH_4} y_{CO_2}} = \frac{(0.1009)^2 (0.5592)^2}{0.00243(0.0641)} = 20.4$$

and are seen to be in excellent agreement.

We have solved the problem for a gas-phase system and before the solution can be considered valid we must check for the absence of solid carbon. This is done by considering any reaction involving our species and solid carbon, e.g.,

$$CO_2(g) + C(c) = 2CO(g)$$

where the equilibrium constant is

$$\ln K = 2 \ln K^f_{CO} - \ln K^f_{CO_2} = 2(24.4572) - 47.8774$$

$$K = 2.821 = \frac{y_{CO}^2 P}{y_{CO_2} a_C}$$

Rearranging for $P = 1$ atm we have

$$a_C = \frac{y_{CO}^2}{K y_{CO_2}}$$

$$a_C = \frac{(0.1009)^2}{2.821(0.0641)} = 0.0563$$

and note that because a_C is less than one, there is no solid carbon present.

In Ex. 13-7 where this problem was analyzed it was determined that methanol would not be present in significant amount. We now test that assumption using Eq. (13-6a) and the Lagrangian multipliers listed above.

$$-\ln K_{CH_4O}^f + \ln y_{CH_4O} + \lambda_C + 4\lambda_H + \lambda_O = 0$$

Using the ΔG^f value of -13.46 kcal/g mol from Ex. 13-7 and rearranging we write

$$y_{CH_4O} = \exp\left(\frac{-(-13,460)}{1.987(1000)} - 2.8766 - 4(0.2906) - 23.874\right) = 6.6(10^{-10})$$

and note that our assumption was justified.

EXAMPLE 13-11

Use free energy minimization to determine compositions at ultimate equilibrium for the CHON system where solid carbon and nitrogen are present at 1000 K and 1 atm. The overall system composition can be expressed in terms of the following elemental ratios

$$\frac{C}{O} = 10; \quad \frac{H}{O} = 1; \quad \frac{N}{O} = 3.76$$

Nitrogen can be considered inert and present as N_2.

Solution 13-11

This system was analyzed in Ex. 13-1 where it was determined that $C = 4$. With two phases there are four degrees of freedom and after fixing temperature and pressure two remain to be fixed. These can be the elemental ratios H/O and N/O which when fixed determine all the intensive properties of the system. Because we have formulated free energy minimization problems in terms of mol numbers, it is necessary to also fix the C/O ratio.

Choosing $b_C = 10$ as our basis, we have

$$b_C = 10; \quad b_H = 1; \quad b_O = 1; \quad b_N = 3.76$$

Because nitrogen is present only as N_2, we need not write an elemental balance for nitrogen and need only consider its presence when determining the total number of gas-phase mols, n. If the pressure is not too high, for solid carbon Eq. (13-6c) becomes

$$\lambda_C = \frac{\Delta G_C^f}{RT} = 0$$

Thus, according to Eq. (13-7) solid carbon makes no contribution to the free energy of the system and as the phase rule states, the quantity of solid carbon will not affect the intensive properties of the system (i.e., gas-phase mol fractions). The needed $\ln K^f$'s were obtained from the ΔG^f's of Ex. 13-7. The problem was solved using POLYMATH and is stored in the POLYMATH library as EX13-11 and EX13-11G. Problem EX13-11G solves a set of linear equations to obtain the initial values for the Problem EX13-11 which determines the equilibrium values. Facsimiles of these worksheets are shown below as well as the results of the calculation.

POLYMATH PROBLEM EX13-11G

		matrix of coefficients column number						vector of constants
		1	2	3	4	5	6	7
	1	1	1	1	1	0	0	10
Row	2	0	0	0	4	2	2	1
Number	3	0	1	2	0	1	0	1
	4	1	0	0	0	0	0	9
	5	0	0	0	0	0	2	0.3
	6	0	0	1	0	0	0	0.15

In POLYMATH problem EX13-11G the matrix columns 1 through 6 correspond to the species C, CO, CO_2, CH_4, H_2O, and H_2 respectively and rows 1, 2, and 3 are elemental balances for carbon, hydrogen, and oxygen respectively. Rows 4, 5, and 6 set the amounts of solid carbon, H_2, and CO_2 respectively. The solution to this problem provided the initial guesses for the free energy minimization executed in POLYMATH problem EX13-11.

POLYMATH PROBLEM EX13-11

Equations: Initial guess

$f(nCO) = -24.4572 + \ln(nCO/n) + lamO$ 0.6833

$f(nCO2) = -47.8774 + \ln(nCO2/n) + 2*lamO$ 0.15

$f(nCH4) = 1.9774 + \ln(nCH4/n) + 4*lamH$ 0.1667

$f(lamO) = -23.1579 + \ln(nH2O/n) + 2*lamH + lamO$ 1

$f(lamH) = \ln(nH2/n) + 2*lamH$ 1

$f(nH2O) = nCO + 2*nCO2 + nH2O - 10*OtoC$ 0.0167

$f(nH2) = 4*nCH4 + 2*nH2O + 2*nH2 - 10*HtoC$ 0.15

$lamN = -0.5*\ln(nN2/n)$

$nN2 = 1.88$

$n = nCO + nCO2 + nCH4 + nH2O + nH2 + nN2$

$nC = 10 - nCO - nCO2 - nCH4$

$HtoC = 0.1$

$OtoC = 0.1$

The solution of POLYMATH problem EX13-11 is summarized below.

RESULTS OF FREE ENERGY MINIMIZATION WITH POLY-
MATH

COMPOSITIONS IN THE CHON SYSTEM

Component	number of mols	Gas-phase mol fraction
CO	0.8223	0.2517
CH_4	0.00865	0.00265
H_2O	0.0310	0.00949
CO_2	0.0734	0.02247
H_2	0.4517	0.1383
N_2	1.88	0.5754
C(c)	9.0957	—

The Lagrangian multipliers are:

$$\lambda_O = 25.837; \quad \lambda_H = 0.9893; \quad \lambda_N = 0.2763$$

Note that even though nitrogen is inert, λ_N is not zero and Eq. (13-7) shows that it makes a contribution to the free energy of the system. This is attributable to a free energy change due to mixing.[13] The calculations may be checked by considering the following reactions

---

[13] See Sec. 14-3.

Reaction 1: $CO_2 + C = 2CO$

Reaction 2: $C + 2H_2 = CH_4$

Reaction 3: $C + H_2O = CO + H_2$

with K's determined from the $\ln K^f$ data

$$K_1 = 2.821; \quad K_2 = 0.1384; \quad K_3 = 3.667$$

The corresponding K's evaluated from the computed mol fractions are

$$K_1 = \frac{y_{CO}^2}{y_{CO_2} a_C} = \frac{(0.2517)^2}{0.02247(1)} = 2.820$$

$$K_2 = \frac{y_{CH_4}}{y_{H_2}^2 a_C} = \frac{0.00265}{(0.1383)^2 (1)} = 0.1385$$

$$K_3 = \frac{y_{CO} y_{H_2}}{y_{H_2O} a_C} = \frac{0.2517(0.1383)}{0.00949(1)} = 3.667$$

and are seen to be in excellent agreement.

If we wished, we could use the Lagrangian multipliers to compute mol fractions of trace components. For example, we will consider NH_3 and find[14] $\ln K^f = -3.2233$. On writing (13-6a) we have

$$\ln y_{NH_3} = \ln K_{NH_3}^f - \lambda_N - 3\lambda_H$$

$$\ln y_{NH_3} = -3.2233 - 0.2763 - 3(0.9893)$$

$$y_{NH_3} = 0.00155$$

The same result can be obtained by using $K^f = 0.03982$ and the mol fractions of H_2 and N_2 found from the POLYMATH solution.

$$y_{NH_3} = K^f y_{N_2}^{1/2} y_{H_2}^{3/2} P$$

$$y_{NH_3} = 0.03982(0.5754)^{1/2} (0.1383)^{3/2} (1) = 0.00155$$

---

[14] From the spreadsheet LOGKF(T).WQ1 (or .XLS).

EXAMPLE 13-12

For the Claus process oxidation step studied in Ex. 13-9, determine equilibrium mol fractions by the free energy minimization method. The pressure is 1 atm, and the temperature is 2200°F.

Solution 13-12

Because it would be undesirable to deposit solid carbon, the calculation will be made on the assumption that only a gas phase exists. A test for the existence of solid carbon will be made when the species mol fractions have been obtained. The pressure is 1 atm, and perfect gas mixture behavior can be safely assumed. Thus, it is reasonable to replace fugacities by partial pressures and use Eq. (13-6a) with the β's in the summation being evaluated from the species' row in the β matrix.

The composition of the feed gas is as follows:

Component	Flow rate (lb mol h^{-1})
H_2S	31.60
CO_2	8.29
CH_4	0.50
C_2H_6	0.33
C_3H_8	0.38
C_3H_6	3.58
C_4H_{10}	0.13
C_5H_{12}	0.70
CH_3SH	5.17
C_2H_5SH	0.69
H_2O	12.97
NH_3	8.68

For each component the terms needed in Eq. (13-6a) are as follows:[15]

	Species	$\ln K^f$	$\Sigma \lambda_j \beta_{ij}$
1	N_2	0	$2\lambda_1$
2	H_2	0	$2\lambda_5$
3	S_2	0	$2\lambda_3$
4	CO_2	32.2529	$2\lambda_2 + \lambda_4$

---

[15] Values of $\ln K^f$ were determined from log K^f values in Ex. 13-9.

i	Species	$\ln K^d$	$\Sigma \lambda_l \beta_{il}$
5	SO_2	20.6902	$2\lambda_2 + \lambda_3$
6	CO	19.7222	$\lambda_2 + \lambda_4$
7	COS	18.0761	$\lambda_2 + \lambda_3 + \lambda_4$
8	H_2O	13.4880	$\lambda_2 + 2\lambda_5$
9	H_2S	1.4322	$\lambda_3 + 2\lambda_5$
10	CS_2	1.9242	$2\lambda_3 + \lambda_4$

The elemental balances, Eq. (13-3a), can be written using the tabulated values of b_l and $\Sigma \beta_{il} n_i$. The latter term is evaluated from a column of the formula coefficient matrix:

Element	l	$\Sigma \beta_{il} n_i$	b_l
Nitrogen	1	$2n_1$	$8.68 + 2(0.79)A$
Oxygen	2	$2n_4 + 2n_5 + n_6 + n_7 + n_8$	$29.55 + 2(0.21)A$
Sulfur	3	$2n_3 + n_5 + n_7 + n_9 + 2n_{10}$	37.46
Carbon	4	$n_4 + n_6 + n_7 + n_{10}$	31.90
Hydrogen	5	$2n_2 + 2n_8 + 2n_9$	178.40

When the mols of oxidation air A were set at 275, the set of 15 equations and 15 variables was solved using POLYMATH's non-linear algebraic equation solver. This problem is stored in the POLYMATH library as EX13-12 and EX13-12G. The initial values of mol numbers satisfy elemental balances and were obtained from elemental balances and the following additional equations.

$$n_{SO_2} = xb_3$$

$$n_{COS} = 1$$

$$n_{H_2O} = \frac{yb_5}{2}$$

$$n_{H_2S} = zb_3$$

$$n_{CS_2} = 1$$

These five equations when combined with the five elemental balances yield a set of 10 linear equations with three adjustable parameters (x, y, z) that can be solved with POLYMATH's linear algebraic equation solver to obtain a set of 10 initial values satisfying the elemental balances. In this set of equations the number of mols of COS and CS_2 have been set at low values and the variables x, y, and z denote respectively the fraction of sulfur present as SO_2, the fraction of hydrogen present as water, and the fraction of sulfur present as H_2S. The solution of this problem (EX13-12G) provides the initial guess for the solution of the set of nonlinear equations (EX13-12).

Results of the POLYMATH solution are:

Combustion Gas Composition[a]

Species	mols	mol %
N_2	221.59	61.2
H_2	9.043	2.5
S_2	11.317	3.1
CO_2	23.96	6.6
SO_2	7.867	2.2
CO	7.681	2.1
COS	0.262	0.1
H_2O	73.46	20.3
H_2S	6.697	1.9
CS_2	0.00040	<0.1
H_2S/SO_2		0.852

[a] Air rate = 275 lb mol h^{-1}, t = 2200°F, P = 1 atm.

LAGRANGIAN MULTIPLIERS

Element	l	λ_l
Nitrogen	1	0.2452
Oxygen	2	11.3932
Sulfur	3	1.7325
Carbon	4	12.1816
Hydrogen	5	1.8447

The low H_2S/SO_2 ratio indicates that the air rate is too high. To test for the presence of solid carbon, the following reaction is considered:

$$CO_2(g) + 2H_2(g) = C(c) + 2H_2O(g)$$

The equilibrium constant for this reaction at 2200°F is

$$\ln K = 2(13.4880) - 32.2529$$

$$K = 0.00510 = \frac{a_C p_{H_2O}^2}{p_{CO_2} p_{H_2}^2}$$

Rearranging, for $P = 1$ atm we have

$$a_C = 0.00510 y_{CO_2} \frac{y_{H_2}^2}{y_{H_2O}^2}$$

$$a_C = 0.00510(0.066) \frac{0.025^2}{0.203^2} = 5.1(10^{-6})$$

The activity of carbon in equilibrium with this combustion gas is seen to be considerably removed from unity, and we conclude that no solid carbon exists.

Because HCN is extremely toxic, it is prudent to calculate the mol fraction of this trace component. From Eq. (13-6a), at $P = 1$ atm we have

$$\ln y_{HCN} - \ln K_{HCN}^f + \lambda_1 + \lambda_4 + \lambda_5 = 0$$

$$y_{HCN} = \exp\left[\ln K_{HCN}^f - \lambda_1 - \lambda_4 - \lambda_5\right]$$

Using the previously calculated and tabulated λ_i's and we obtain

$$y_{HCN} = 8.4(10^{-10})$$

Thus, the presence of HCN can be ignored.

The heat released to the waste heat boiler for these conditions has been calculated in Ex. 6-11.

13-4 THE CHO SYSTEM AND CARBON DEPOSITION BOUNDARIES

A major part of the chemical industry is concerned with manufacturing chemicals comprised of the elements carbon, hydrogen, and oxygen; therefore, it is reasonable to award this system special consideration. Accordingly, much effort has been devoted to calculating equilibrium compositions and determining whether solid carbon is present for ultimate equilibrium in this system.[16] Except for high-temperature operations such as combustion or pyrolysis, ultimate equilibrium is not approached and, in fact, is usually not desired. However, it is

---

[16] For example, see E. J. Cairns and A. D. Tevebaugh, *J. Chem. Eng. Data, 9*, 453 (1964); R. E. Baron, J. H. Porter, and O. G. Hammond, *Chemical Equilibria in Carbon–Hydrogen–Oxygen Systems*, MIT Press, Cambridge, MA, 1976; and S. Mohnot and B. G. Kyle, *Ind. Eng. Chem. Process Des. Dev., 17*, 270 (1978).

desirable to know the conditions under which solid carbon could deposit because the prudent engineer would not design for such conditions.

We have seen that for the CHO system there are a maximum of four degrees of freedom. Therefore, at a specified temperature and pressure two elemental ratios will define the system. When solid carbon is in equilibrium with the gas phase, specification of the O/H elemental ratio is sufficient. This situation can be exploited by means of a carbon deposition boundary plotted on a triangular diagram, such as shown in Fig. 13-2. The carbon deposition boundary is the solid curve passing through the points X, V, and Y and represents gas phase compositions, expressed on an elemental basis, which are in equilibrium with solid carbon. Any CHO system may be represented by a point on this triangular diagram, and any

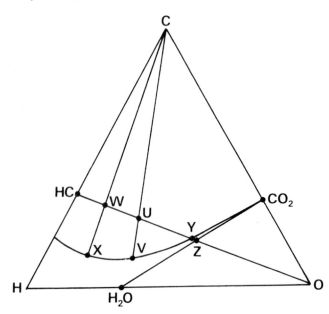

Figure 13-2 Carbon deposition boundary for the CHO system.

point lying above the carbon deposition boundary represents a system comprised of a gas phase and solid carbon. This provides a very convenient way of ascertaining whether there is a tendency for the formation of solid carbon. The phase rule suggests that such a representation is possible; however, the boundary must be established from calculated equilibrium compositions.

To illustrate the utility of this approach, consider a system originally comprised of any combination of reactants in which overall elemental composition is represented by point U in Fig. 13-2. Because U lies above the carbon deposition boundary, the system at ultimate equilibrium consists of solid carbon and a gas phase in which elemental composition is given by point V. The line CUV may be regarded as a tie line, and the relative proportions of each phase on an elemental basis may be obtained by applying the lever-arm principle. The effect of varying the hydrogen or oxygen content of any system containing carbon as in hydrogenation or oxidation can also be visualized. For example, if a hydrocarbon whose composition is given by point HC is oxidized, the overall composition of the reaction mixture moves from HC toward O. At W the system at equilibrium is composed of solid carbon and a gas phase of elemental composition represented by X. Point Y represents the minimum amount of oxygen necessary to prevent carbon formation, and Z corresponds to a state of complete oxidation when only CO_2 and H_2O are present.

Calculations of ultimate gas-phase equilibrium compositions in the presence of solid carbon have been executed over the temperature range 500 to 1500 K and the pressure range of 1 to 25 atm. At 1 atm these compositions are presented in Fig. 13-3. At 1300 K and above all species except H_2 and CO exist at mol fractions less than 10^{-2}, and for most purposes the system can be assumed to consist only of $C(c)$, $CO(g)$, and $H_2(g)$.

This information has been transformed into carbon deposition boundaries. Figure 13-4 shows these boundaries with temperature as a parameter for a pressure of 1 atm, and Fig. 13-5 shows them at 25 atm. The effect of pressure is not great but is most pronounced in the range of 1000 K.

EXAMPLE 13-13

Methanol is produced from CO and H_2 according to the reaction

$$CO(g) + 2H_2(g) = CH_3OH(g)$$

The reaction is exothermic and is equilibrium-limited at high temperature. The reaction is carried out catalytically at as low a temperature as possible, 250°C, and at pressures between 50 and 100 atm. As carbon deposition on the catalyst would be undesirable, it is desired to estimate the range of reactant ratio which would be safe.

Solution 13-13

A carbon deposition boundary is not available at 50 atm; however, a comparison of results at 1 and 25 atm should provide an estimate. A graphical procedure illustrated in Fig. 13-6 is employed. Mixtures of CO and H_2 will lie on a straight line connecting them, and where this line intersects the carbon deposition boundary determines the minimum ratio of H_2/CO. Ratios less than this minimum represent system points above the carbon deposition boundary where the possibility for carbon formation exists. For the 500 K isotherm at both pressures the intersection point lies at 73% hydrogen, 13.5% carbon, and 13.5% oxygen, and it appears safe to ignore the pressure effect. The minimum H_2/CO ratio is

$$\frac{H}{C} = \frac{0.73}{0.135} = \frac{2y_{H_2}}{y_{CO}}$$

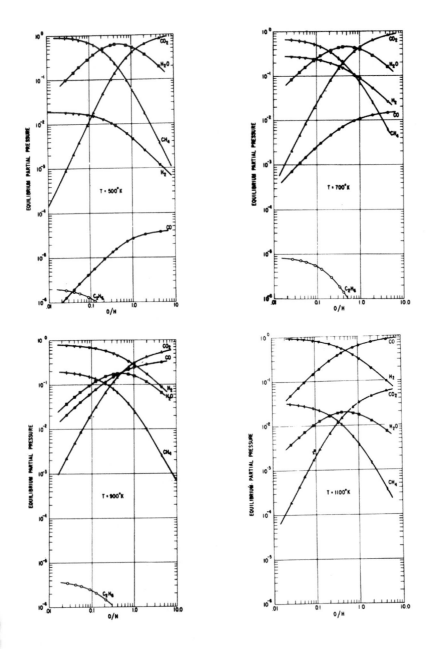

Figure 13-3 Partial pressures of gaseous species in equilibrium with carbon in the CHO system at 1 atm. [Reprinted with permission from E. J. Cairns and A. D. Tevebaugh, *J. Chem. Eng. Data, 9*, 453 (1964), copyright 1964 by the American Chemical Society.]

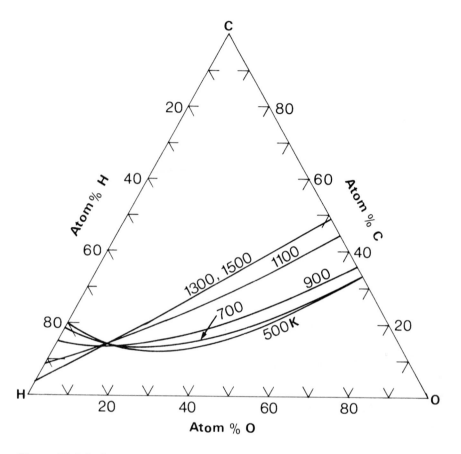

Figure 13-4 Carbon deposition boundaries for the CHO system at 1 atm, T as a parameter.

$$\frac{H_2}{CO} = \frac{y_{H_2}}{y_{CO}} = \frac{0.73}{2(0.135)} = 2.70$$

If the stoichiometric ratio of 2.0 were specified, there would be a danger of carbon deposition on the catalyst.

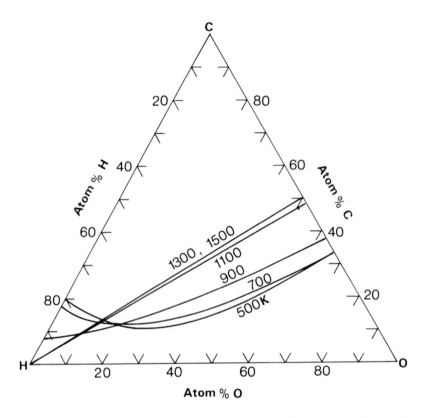

Figure 13-5 Carbon deposition boundaries for the CHO system at 25 atm, *T* as a parameter.

EXAMPLE 13-14

The initial step in coal gasification involves the reaction between coal, here assumed to be carbon, and steam:

$$C(c) + H_2O(g) = CO(g) + H_2(g)$$

Because this reaction is endothermic, the heat will be supplied by adding oxygen so that exothermic reactions such as

$$C(c) + \frac{1}{2}O_2(g) = CO(g)$$

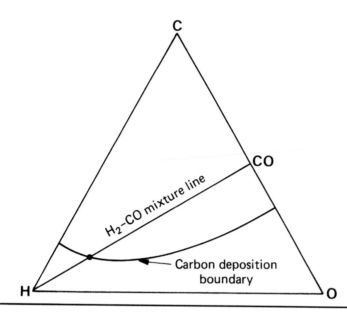

Figure 13-6

EXAMPLE 13-14 CON'T

may occur. The gasifier will operate adiabatically at 1800°F and 1 atm. The product gas is next subjected to a shift reaction at 600°F and in cooling to this temperature will be used to preheat the gasifier feed. Assuming equilibrium is attained within the gasifier, determine the quantities of steam and oxygen required based on 1-lb atom of carbon. Carbon and oxygen are available at 77°F, and steam is at 250°F. The following data are available:[17]

$$H° - H_0° - \Delta H_0^f$$

Species	t (°F)	(Btu/lb mol)
CO	600	−41,534
H_2	600	7,289
H_2O	250	−97,129
O_2	77	3,726
C	77	453

---

[17] Source: reference 7 in Table 12-1.

Solution 13-14

At 1800°F (1256 K) Fig. 13-3 shows that for practical purposes all gaseous species except CO and H_2 may be neglected, and a material balance determines the equilibrium compositions. The process is visualized as shown in Fig. 13-7. Elemental balances are as follows:

Carbon: $1 = Ny_{CO}$

Hydrogen: $W = Ny_{H_2}$

Oxygen: $W + 2X = Ny_{CO}$

where N is the total mol of product gas. These equations together with the relationship

$$y_{CO} + y_{H_2} = 1$$

yield

$$N = \frac{1}{y_{CO}}$$

$$W = N - 1$$

$$X = 1 - \frac{N}{2}$$

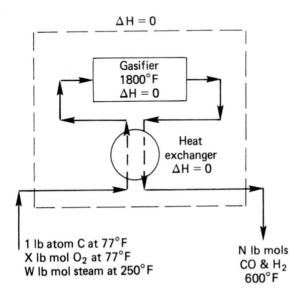

Figure 13-7 Schematic diagram of a coal gasifier.

Once a value of y_{CO} has been chosen, all quantities in the material balance may be calculated. The correct value of y_{CO} satisfies the energy balance

$$\Delta H = 0 = -41{,}534 + 7289W - 453 - 3726X + 97{,}129W$$

This is found to be $y_{CO} = 0.709$:

> N = 1.410 lb mol product gas
> W = 0.410 lb mol steam
> X = 0.295 lb mol oxygen

For simplicity, coal has been assumed to be carbon. However, had the ultimate analysis specifying the percentages of carbon, hydrogen, and oxygen been supplied, the problem could be solved. Only the material balances would change because only CO and H_2 would be present at equilibrium.

13-5 THE Si–Cl–H SYSTEM AND SILICON DEPOSITION BOUNDARIES

At high temperature epitaxial silicon films can be grown on single crystals of silicon by deposition from a gas phase containing $SiCl_4$ and H_2. As it is important to know the range of conditions under which silicon will deposit, a direct visual representation, such as the deposition boundary, should prove useful. Here we extend the deposition boundary concept to the Si–Cl–H system.

Through free energy minimization, gas-phase partial pressures have been computed for 14 gaseous species in equilibrium with solid silicon.[18] At atmospheric pressure and over the range 300 to 1700 K, the calculations showed the major species to be H_2, HCl, $SiCl_4$, $SiCl_2$, $SiHCl_3$, and SiH_2Cl_2. From these equilibrium compositions the silicon deposition boundary at 1500 K has been plotted in Fig. 13-8. The region near the hydrogen apex has been expanded and is shown in Fig. 13-9. The heavy line is the deposition boundary and the lighter line connecting $SiCl_4$ and the hydrogen apex represents the composition of a feed gas comprised of these two species. In Fig. 13-8 these two lines intersect at point a, which corresponds to a silicon-free composition marked by b. To the left of point a the feed gas line lies above the deposition boundary and silicon can deposit. A feed gas composition to the right of point a will etch the substrate by removing silicon.

---

[18] L. P. Hunt and E. Sirtl, *J. Electrochem. Soc.*, *119*, 1741 (1972).

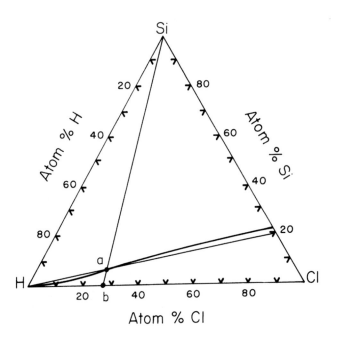

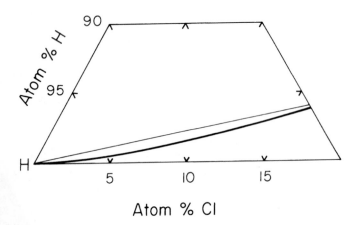

Figure 13-8 Silicon deposition boundary for the Si–Cl–H system at 1500 K and 1 atm.

Figure 13-9 Magnification of Fig. 13-8.

PROBLEMS

13-1. In a system containing $CaSO_4(c)$, $CaS(c)$, $CaO(c)$, $CaCO_3(c)$, $CO(g)$, $CO_2(g)$, and $SO_2(g)$, is it possible for all four pure solid phases and a gas phase to coexist at equilibrium? If so, how many variables may we fix? If we include $COS(g)$ and $CS_2(g)$ among the species, how will your answer be affected?

13-2. A furnace operating at 1800°F is charged with FeO, C, and O_2 and produces liquid iron, Fe. Assuming that ultimate equilibrium prevails and that the components present in significant amounts are $FeO(c)$, $Fe(l)$, $C(c)$, $CO(g)$, and $CO_2(g)$, is it possible to fix any other intensive variable in addition to the temperature?

13-3. In a system comprised of calcium, sulfur, and oxygen (Ca–S–O) the following species are expected to be present at equilibrium

$$CaSO_4(c), \ CaS(c), \ CaO(c), \ SO_2(g), \ O_2(g)$$

(a) Is it possible for all of these species to be present at equilibrium? Explain.
(b) If all species were present, would we have any control over the system? Explain.
(c) If $SO_3(g)$ were included among the species, how would this change your response to parts a and b? Explain.

13-4. For ultimate equilibrium in a system comprised of the elements C, Si, H, and Cl the following species may be present: $H_2(g)$, $HCl(g)$, $CH_4(g)$, $SiCl_4(g)$, $C(c)$, $Si(c)$, and $SiC(c)$. Would it be possible for all three solid phases and a gas phase to be present at equilibrium? If so, could any intensive variables be fixed?

13-5. A process for removing oxides of nitrogen from stack gas has been developed in which ammonia reacts catalytically with the nitrogen oxides to produce nitrogen and water. Five chemical reactions have been identified as occurring in the process. How many independent reactions are possible for this system?

13-6. In the literature one finds a plot of calculated equilibrium compositions for dissociating water vapor vs. temperature at a pressure of 1 atm. The temperature range is 1000–4000 K, and the species considered are H_2O, H_2, O_2, O, H, and OH. Apply the phase rule to this system, and determine whether it is properly defined.

13-7. One step in the manufacture of zinc involves roasting zinc sulfide, ZnS, in oxygen. Substances which could be present at equilibrium are $ZnS(c)$, $ZnO(c)$, $ZnSO_4(c)$, $O_2(g)$, $SO_2(g)$, and $SO_3(g)$. Zinc sulfate is an undesirable species, and an old company report is uncovered which shows the conditions where $ZnSO_4(c)$ is absent [a system consisting of $ZnS(c)$, $ZnO(c)$, $O_2(g)$, $SO_2(g)$, and $SO_3(g)$] as an area plotted on pressure-temperature coordinates. Is such a representation consistent with the phase rule? Explain.

13-8. In a partially miscible ternary system a heterogeneous azeotrope is defined as the condition that the vapor composition lies on a tie line connecting the two liquid phases. How many degrees of freedom would such a system possess? Would the same conclusion apply to a partially miscible binary system?

13-9. Solid ammonium bicarbonate, $NH_4HCO_3(c)$, on dissociation produces equal amounts of $NH_3(g)$, $H_2O(g)$, and $CO_2(g)$. In this two-phase system the pressure is found to depend only on temperature. However, when water is added to the system, the pressure in the three-phase system consisting of vapor, aqueous solution, and $NH_4HCO_3(c)$ is found to depend on the relative amounts of the fluid phases. Is this behavior anomalous or can it be explained by the phase rule?

13-10. In a homogeneous gas-phase reaction phenol, C_6H_6O, reacts with bromine to produce ortho, meta, and para bromophenols as per the general reaction

$$C_6H_6O(g) + Br_2(g) = C_6H_5OBr(g) + HBr(g)$$

(a) Apply the phase rule to this system when it is comprised originally of phenol and bromine.
(b) Using either the free-energy-minimization method or the equilibrium-constant method, write the equations which describe the system at equilibrium.
(c) At a specified T and P what additional variables would you specify in order to solve for all the equilibrium mol fractions?
(d) Show how the relative amounts of the isomers can be determined.

13-11. Ultimate chemical equilibrium in the Carbon–Silicon–Hydrogen–Chlorine system was determined by free energy minimization and reported in a recently published article. In this system there are three distinct solid phases—Si, C, and SiC—which can coexist with the gas phase. In the gas phase the major species are H_2, HCl, CH_4, and $SiCl_4$. At a specified temperature and pressure and Cl/Si ratio, the H/Si ratio is plotted on the ordinate and the C/Si ratio is plotted on the abscissa. On this plot four areas are shown where the following solid phases are present: Si, SiC, Si + SiC, and SiC + C. Is such a representation consistent with the phase rule? Explain.

13-12. For the Claus oxidation system discussed in Exs. 13-9, 13-12, and 6-11, answer the following:

(a) If we had desired to consider more species present at equilibrium, how would this have affected the computations?
(b) Would it be possible to find the air rate that will yield a H_2S/SO_2 ratio of 2 in an adiabatic process? Explain.

13-13. When the Claus oxidation system discussed in Exs. 13-9, 13-12, and 6-11 is carried out adiabatically

(a) Write the governing equations and formulate β.
(b) Determine C.
(c) Write the equations for free energy minimization equivalent to Eq. (13-6).
(d) Write the entire set of equations to be solved in finding the free energy minimum.

13-14. When carbon dioxide and ammonia dissolve in water, the species are: H_2O, CO_2, NH_3, H^+, OH^-, NH_4^+, HCO_3^-, $CO_3^=$, and $NH_2CO_3^-$. Apply the phase rule to this two-phase system. Specifically, determine:

    (a) ρ, the rank of the β matrix.
    (b)   a set of R reactions.
    (c)   the rank of the **R** matrix.
    (d)   the number of components, C.
    (e)   the degrees of freedom, .

13-15. Consider the addition of a gaseous inert substance (e.g., N_2) to the CHO system. Where ideal-gas behavior is allowable, show that the effects of pressure and quantity of inert gas can be combined into a single term, $P/(N_0 + N)$, where N_0 is the mols of inert gas and N is the total mols of gas phase components in the CHO system. Is this consistent with the phase rule? Show that for the same temperature and b_i values [Eq. (13-3a)] the inert-free mol fractions in the CHO inert system will be equal to the mol fractions in the CHO system when

$$\frac{P'}{(N_0 + N)} = \frac{P}{N}$$

Where P' is the pressure of the CHO inert system and P is the pressure of the CHO system.

13-16. Suppose that we are interested in ultimate equilibrium in the CHO system under conditions where liquid water could be present and we desire to delineate water-condensation boundaries. Assuming that the water phase is pure, write the equations that describe this two-phase system. For a given temperature, sketch condensation boundaries on a triangular diagram showing the effect of pressure. Is it possible for the water-condensation and carbon-deposition boundaries to intersect? Write the equations that describe the three-phase system of $C(c)$, $H_2O(l)$, and gas. Sketch these boundaries on a triangular diagram labeling all regions, lines, and points for conditions where three phases can coexist.

13-17. When fuel is burned with air, oxides of nitrogen, mainly NO, are formed. It has been proposed to reduce the NO concentration of flue gas by the addition of NH_3. The desired reaction is

$$NO + NH_3 + \frac{1}{4}O_2 = N_2 + \frac{3}{2}H_2O$$

Many other reactions may be possible; however, of most concern is the oxidation of ammonia.

    (a)   Assuming that only the desired reaction occurs, determine the equilibrium concentration of NO in parts per million (ppm) when a flue gas containing 68% N_2, 15% CO_2, 15% H_2O, 2% O_2, and 200 ppm of NO is reacted at 1200 K and 1 atm with ammonia. Two mol of ammonia are added per mol of NO in the incoming flue gas.
    (b)   If ultimate chemical equilibrium can be assumed, what species would be present in significant amounts? (For oxides of nitrogen, 10 ppm will be considered significant.)

(c)   If the flue gas also contains 500 ppm of SO_2, what might be expected for ultimate chemical equilibrium?

13-18.   For the formaldehyde production scheme described in Ex. 13-4, perform a phase-rule analysis and calculate equilibrium compositions at 1000°F when the oxygen to methanol ratio in the reactor feed is adjusted so that there is no heat requirement for the reaction ($\Delta H = 0$).

13-19.   Show that the minimum in G is unaffected by the choice of reference state used for μ_i°. Test specifically (a) the ΔG^f of elements taken to be zero at the temperature of the system or (b) the ΔG^f of elements taken to be zero only at a single reference temperature (e.g., 0 or 298 K).

13-20.   In Sec. 13-3 the free energy minimization method was delineated for a gas phase system. Write the corresponding equations which describe the CHO system with solid carbon present. Would it be computationally advantageous to use a particular type of free energy data for μ_i° ?

13-21.   In Ex. 13-12 calculate the equilibrium mol fraction of O_2.

13-22.   Write the required equations for the application of the free energy minimization method   to a system comprised of aqueous ions where it is necessary to include the constraint of electrical neutrality. In your formulation let z_i stand for the charge on ionic species i, where z_i can be either positive or negative. Make sure that there is an equal number of equations and unknowns.

13-23.   Hydrogen is to be produced by the steam reforming of methane as per the reaction

$$CH_4 + H_2O = CO + 3H_2$$

The reaction will take place at 1000 K and 1 atm. Deposition of carbon inside the reactor tubes is undesirable, and we wish to know the minimum steam to methane ratio which will ensure no carbon deposition.

13-24.   Carbon black, finely divided solid carbon, is to be produced by passing ethane (C_2H_6) into a reactor operating at atmospheric pressure and a high temperature. Assume ultimate chemical equilibrium and answer the following:

(a)   What temperature should produce the maximum yield of carbon black? Practical considerations limit the maximum temperature to 900 K.

(b)   What is the maximum yield?

(c)   If a plant were designed and operated with ethane as a feedstock, could the reactor performance be duplicated with a feedstock of methane (CH_4) and propane (C_3H_8)? If so, what ratio of methane to propane should be used? Explain.

13-25.   A good catalyst has been found for producing benzene, C_6H_6, from cyclohexane, C_6H_{12}, via the reaction

$$C_6H_{12}(g) = C_6H_6(g) + 3H_2(g)$$

The best operating conditions seem to be 600°F and atmospheric pressure.

Our enthusiasm for the project has been at least temporarily dampened by the concern of one member of our group that carbon deposition on the catalyst is possible. However, another group member has responded that carbon deposition may be circumvented by the mixing of a recycle hydrogen stream with the cyclohexane feed. You are asked to explore these possibilities and provide answers to the following questions.

(a) Is carbon deposition possible?
(b) If mixing hydrogen with the feed would prevent carbon deposition, how much hydrogen, H_2, is needed per mol of cyclohexane fed?
(c) If this amount of hydrogen were added to the feed, how would the conversion in the reaction be affected?

13-26. Formaldehyde is to be produced by the pyrolysis of methanol (see Ex. 13-4) at 700 K and atmospheric pressure and we are concerned that solid carbon might deposit in the reactor.

(a) Is this a possibility and if so, is there a range of operating temperatures where we are assured that carbon will not deposit?
(b) Noting that formaldehyde can also be formed by the oxidation of methanol, it has been suggested that if carbon deposition were a problem, it might be eliminated by feeding the reactor with methanol and oxygen. If you think this is a viable suggestion, determine how much oxygen should be added to prevent carbon deposition when operating at 700 K.

13-27. Isopropanol (C_3H_8O) is to be produced by reacting acetone (C_3H_6O) and H_2 in a gas-phase catalytic reactor operating at 700 K and one atmosphere. If the stoichiometric ratio is employed, is there any danger of carbon deposition? If so, what ratio can be used to avoid this possibility?

13-28. For the catalytic reaction

$$C_6H_{12}(g) = C_6H_6(g) + 3H_2(g)$$

occurring at 700 K there is the possibility of solid carbon deposition and it has been suggested that adding steam, an inert substance, to the reactor feed could prevent this possibility. If you agree with this, determine the ratio of steam to C_6H_{12} that would prevent carbon deposition.

13-29. As shown in Ex. 13-4, the production of formaldehyde from methanol can proceed by either an endothermic pyrolysis or an exothermic oxidation reaction. At a temperature of 1000°F and at atmospheric pressure, a calculation shows that a feed containing 0.158 mol O_2 per mol of methanol would result in a net heat of reaction of zero. Under these conditions is there be any possibility of carbon formation?

13-30. Use the phase rule to justify use of a silicon deposition boundary.

13-31. The composition at point b of Fig. 13-8 is $x_H = 0.738$, $x_{Cl} = 0.262$. Determine the minimum ratio of H_2 to $SiCl_4$ required to deposit silicon at 1500 K.

13-32.   When the ratio of H_2 to $SiCl_4$ in the feed is 31.33 to 1, calculate the percentage of silicon in the feed that could be deposited at 1500 K.

13-33.   By revising either of the POLYMATH programs, EX13-10 or EX13-11, obtain a program to determine ultimate chemical equilibrium in a gas-phase system containing the elements carbon, hydrogen, oxygen, and nitrogen (the CHON system) at a temperature of 1000 K and at a pressure P. Use your program to determine ultimate chemical equilibrium when the system is at a pressure of 2 atm and is originally comprised of 1 mol of CH_4, 1 mol of NH_3, and 2 mol of O_2. Assume nitrogen to be inert, but check this assumption with your calculation results by considering the presence of NH_3 and NO. Also, test for the presence of solid carbon.

13-34.   By revising either of the POLYMATH programs, EX13-10 or EX13-11, obtain a program to determine ultimate chemical equilibrium in the two-phase CHO system at a temperature of 1000 K and at a pressure P.

   (a)   For O to H elemental ratios of 0.5, 1, and 2 determine equilibrium compositions at a pressure of 1 atm and use this information to obtain points on the carbon deposition boundary.
   (b)   Repeat part *a* for a pressure of 10 atm (assume ideal-gas behavior).

13-35.   Simplify the Claus partial oxidation example (Ex. 13-12) to a stream of pure H_2S and air and write a POLYMATH program to determine ultimate equilibrium compositions at 2200°F and 1 atm. Use this program to determine the ratio of air to H_2S, A, that yields a ratio of H_2S to SO_2 of 2.

13-36.   Write a POLYMATH program to determine ultimate chemical equilibrium at 1500 K in   a two-phase system comprised of the elements Silicon, Hydrogen and Chlorine (the Si–H–Cl system) and containing the species for which the log K' values are listed below.

Species	log K'
$SiCl_2$	7.068
SiH_2Cl_2	6.052
$SiHCl_3$	10.708
$SiCl_4$	16.082
HCl	3.615
H_2	0
Si(c)	0

Use a Cl to H elemental ratio of 3 to determine equilibrium compositions and use these to calculate a point on the silicon deposition boundary.

13-37.   Using the free energy minimization method with POLYMATH, calculate ultimate chemical equilibrium when the reactants are water and methane. For your assigned temperature and pressure, prepare a plot of species mol fractions versus water to methane ratio (W). This plot should resemble Fig. 13-3 of your text with a log scale for the ordinate but a linear scale for the abscissa. Cover a wide range of W including the range where solid carbon is present and label the one-phase and two-phase regions. Also, calculate the mol fraction of the trace components formaldehyde, CH_2O, and oxygen, O_2, and tabulate these values for the range of water to methane ratios. Choose one temperature and one pressure from the following set. Assume ideal-gas behavior.

T (in K): 700, 800, 900, 1000, 1100, 1200, 1300, 1400, 1500
P (atm): 1, 2, 3, 4, 5

The necessary log K^f data can be obtained from the spreadsheet LOGKF(T).WQ1(or .XLS). Include in your submission:

(a) the plot of x_i's vs. H_2O to CH_4 ratio (W).
(b) a table containing carbon activity, and m.f. of O_2 and CH_2O for values of W used to construct item 1.
(c) a sample calculation for item b.
(d) copies of your POLYMATH worksheet for one value of W in each of the one-phase and two-phase regions.

CHAPTER 14

Thermodynamic Analysis of Processes

$\mathcal{T}$he utility of the first and second laws for analyzing processes involving exchanges of heat and work with the surroundings has already been demonstrated. In this chapter these principles will be applied to a wide variety of problems, and formal means of comparing an actual process against the ideal thermodynamic process, the reversible process, will be developed. Special attention is given to the work of separation of mixtures and to the evaluation of processes involving chemical transformation.

14-1 WORK AND FREE ENERGY FUNCTIONS

For a process occurring in a closed system surrounded by a heat reservoir at the temperature T, the Clausius inequality was shown to apply:

$$\Delta U - T\Delta S - W \leq 0 \tag{5-5}$$

where ΔU and ΔS are changes in the system properties, T is the temperature of the surroundings, and W is the total work. If the initial and final temperatures of the process are also T, the first two terms in Eq. (5-5) are equal to the Helmholtz free energy change,[1]

$$\Delta U - T\Delta S = \Delta A_T$$

and the inequality becomes

$$\Delta A_T \leq W \qquad\qquad (14\text{-}1a)$$

or

$$-\Delta A_T \geq -W \qquad\qquad (14\text{-}1b)$$

The equality sign applies to a reversible process. For a process requiring work $(W > 0)$, Eq. (14-1a) shows that ΔA_T is the minimum value, while for a work-producing process $(W < 0)$ Eq. (14-1b) shows ΔA_T to be the maximum value. Thus, for an isothermal process occurring in a closed system, the total reversible work is equal to the Helmholtz free energy change.

If the process occurring in the closed system is carried out such that $T_1 = T_2 = T$ and $P_1 = P_2 = P$, where P is the pressure of the surroundings, it will be convenient to use Eq. (5-6), which involves the useful work:

$$\Delta U - T\Delta S + P\Delta V - W' \leq 0 \qquad\qquad (5\text{-}6)$$

The useful work W' is defined as the total work minus any work exchanged with the constant-pressure surroundings due to a system volume change:

$$W' = W - (-P\Delta V) = W + P\Delta V$$

For the prescribed change, the first three terms in Eq. (5-6) represent the Gibbs free energy change for the system,

$$\Delta U - T\Delta S + P\Delta V = \Delta G_{TP}$$

and we have

$$\Delta G_{TP} \leq W' \qquad\qquad (14\text{-}2)$$

This inequality is interpreted in the same manner as Eqs. (14-1). It is therefore seen that the Gibbs free energy change is the maximum useful work obtainable, or the minimum required, for such a process.

For a steady-state flow process in which kinetic and potential energy effects are negligible, the first law is

$$\Delta H = Q + W_s$$

---

[1] Because A is a state function and ΔA_T depends only on initial and final states, the following relationship is not restricted to processes which proceed isothermally. All that is required is that initial and final temperatures be equal.

If the system exchanges heat only with a reservoir at T, the second law can be stated as

$$\Delta S \geq \frac{Q}{T}$$

These statements can be combined to obtain

$$\Delta H - T\Delta S \leq W_s \qquad (14\text{-}3)$$

When the process is carried out so that inlet and exit temperatures equal T, the left-hand side equals the isothermal Gibbs free energy change, and we see that

$$\Delta G_T \leq W_s \qquad (14\text{-}4)$$

Thus, for an open system ΔG is the maximum work obtainable, or the minimum work required, for a steady-state isothermal process.

Because it should always be remembered that work may also be associated with heat exchanges, the net work realizable from a reversible process should include this as well as that calculated from the free energy change. Therefore, in addition to the restriction to isothermal processes the free energy changes do not give us the total work potential for a process. A thermodynamic function not subject to these restrictions would be of considerable utility. The availability is such a function.

14-2 THE AVAILABILITY

Consider any process occurring in a steady-state flow system such as shown in Fig. 14-1a. There may be heat exchanged with the surrounding Q_i's at several temperatures T_i's, including exchange with the medium[2] Q_0 at the temperature T_0. The net shaft work exchanged with the surroundings is W_s. Within the process a definite change of state occurs, and it will be of interest to know the maximum amount of work which could be obtained from the process (or the minimum work required to drive the process) when heat is exchanged only with the medium. The reason for considering heat to be exchanged only with the medium is twofold: First, the medium is an unlimited source or sink for heat, and second, the thermodynamic value of heat at T_0 is zero. This allows the evaluation of a process to be standardized by expressing its performance in terms of a single energy form—work. The first law with its demand for conservation makes no distinction between the energy forms of heat and work. The second law, however, provides a quantitative measure of the value of energy. Work is the highest form, and the value of heat is determined by the amount of work it could produce in a reversible heat engine that rejects heat to the medium. Because we wish to judge processes

---

[2] The medium is any naturally occurring and readily available heat reservoir of infinite capacity (e.g., air or groundwater).

by the extent to which useful energy is conserved, work should be the currency of our energy accounting.

A system in which heat is exchanged only with the medium is shown in Fig. 14-1b. The process itself still receives or rejects heat Q_i's at the various temperatures T_i's, but heat

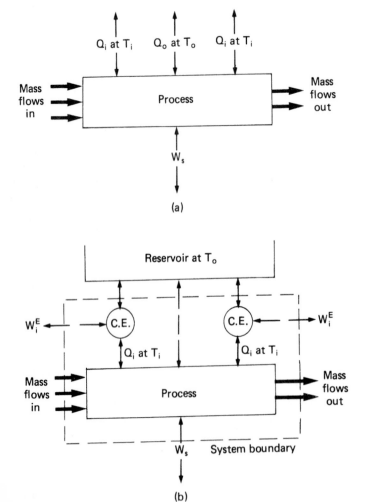

(a)

(b)

Figure 14-1 (a) Steady-state flow process exchanging heat and work with surroundings. (b) Steady-state flow process exchanging heat with the medium at T_0 and work with surroundings.

engines or heat pumps are used to remove or deliver the heat by operating between T_0 and the various T_i's.[3] These devices are included within the bounds of the system shown by the dashed envelope, and because they are at steady state, they undergo no property changes, and therefore the property changes for the system are simply those of the process. The process heat exchanges Q_i's are now expressed as the equivalent amount of work W_i^E crossing the boundaries of the system. The Carnot factor relates Q_i and W_i^E:

$$W_i^E = Q_i \left(\frac{T_i - T_0}{T_i} \right)$$

(14-5)

Applying the steady-state form of the first law for an open system in which kinetic and potential energy effects are ignored yields

$$\Delta H = Q_0 + \left(W_s + \sum W_i^E \right)$$

(14-6)

where Q_0 is the total heat exchange with the medium and the parenthesized term is the total work exchanged with surroundings, denoted by **W**, which includes work done on or by the process and the equivalent work of the heat exchanges. Equation (14-6) can be simplified to

$$\Delta H = Q_0 + \mathbf{W}$$

(14-6a)

Application of the second law yields[4]

$$\Delta S + \left(\frac{-Q_0}{T_0} \right) \geq 0$$

(14-7)

which can be expressed as an equality by introducing a term σ for the created entropy:

$$\Delta S + \frac{-Q_0}{T_0} = \sigma$$

(14-8)

σ is always greater than or equal to zero; it is zero for a reversible process.
Eliminating Q_0 between Eqs. (14-6a) and (14-8) gives

$$\Delta H - T_0 \, \Delta S + T_0 \, \sigma = \mathbf{W}$$

(14-9)

For a reversible process this becomes

$$\Delta H - T_0 \, \Delta S = \mathbf{W}_{\text{rev}}$$

(14-10)

---

[3] If the concept of utilizing heat engines for exchange with the medium seems abstract and remote, one could consider the process exchanging heat reversibly with heat reservoirs at various temperatures and visualize the W_i^E's as the reversible work done or required to maintain the reservoirs.

[4] In Eqs. (14-6) and (14-6a) the term Q_0 refers to the system; therefore the heat exchange for the medium is $-Q_0$. and the entropy change of the medium is $-Q_0/T_0$.

and we see that reversible work is equal to the change in a state function whose value depends on the specification of T_0. This state function is called the availability, or sometimes the exergy, and is designated by the symbol B:

$$\Delta B = \Delta H - T_0 \, \Delta S \tag{14-11}$$

$$= \mathbf{W}_{rev} \tag{14-12}$$

Equation (14-9) may now be written as

$$\Delta B + T_0 \, \sigma = \mathbf{W}_{rev} + T_0 \, \sigma = \mathbf{W} \tag{14-13}$$

and it can be seen that $T_0 \sigma$ is the dissipated work due to irreversibilities in the system. When $\Delta B < 0$, the system is capable of doing work, and $\mathbf{W}$ is always less than the maximum possible value $\mathbf{W}_{rev}$. Conversely, when $\Delta B > 0$, work must be done on the system, and this work is always greater than $\mathbf{W}_{rev}$. For the special case where the process occurs with inlet and exit temperatures of T_0, the change in availability ΔB is equal to the Gibbs free energy change ΔG_T. There are, of course, no restrictions on the use of the availability other than that heat exchange be only between the system and the medium at T_0.

When applied to closed systems,[5] the first and second laws are

$$\Delta U = Q_0 + \mathbf{w} \tag{14-14}$$

$$\Delta S - \frac{Q_0}{T_0} = \sigma \tag{14-8}$$

Again, eliminating Q_0 gives

$$\Delta U - T_0 \, \Delta S + T_0 \, \sigma = \mathbf{w} \tag{14-15}$$

$\mathbf{w}$ is the total work and includes the equivalent work of heat exchanges and work exchanged with the surroundings including work by or against the medium at the constant pressure P_0. The total work may be written as

$$\mathbf{w} = \mathbf{w}' + (-P_0 \, \Delta V) \tag{14-16}$$

where $\mathbf{w}'$ is the useful work. Equation (14-15) now becomes

$$\Delta U + P_0 \, \Delta V - T_0 \, \Delta S + T_0 \sigma = \mathbf{w}' \tag{14-17}$$

For a reversible process,

$$\Delta U + P_0 \, \Delta V - T_0 \, \Delta S = \mathbf{w}'_{rev} \tag{14-18}$$

---

[5] Again, the system is considered to be the primary system plus the heat engines or heat pumps necessary to affect the reversible heat exchanges with the medium.

The three terms on the left-hand side represent the change in a state function. Unfortunately, this function is also called the availability or exergy, and to avoid confusion[6] here, it will be designated by B' or $\Delta B'$:

$$\Delta B' = \Delta U + P_0 \, \Delta V - T_0 \, \Delta S \tag{14-19}$$

$$\Delta B' = \mathbf{w}'_{rev} \tag{14-20}$$

Equation (14-17) may now be rewritten as

$$\mathbf{w}' = \mathbf{w}'_{rev} + T_0 \, \sigma$$

As with Eq. (14-13), $T_0\sigma$ is dissipated work and is a measure of how close the actual process approaches the thermodynamic ideal process. For the special case where $T_1 = T_2 = T_0$ and $P_1 = P_2 = P_0$ the left-hand side of Eq. (14-18) is seen to be the Gibbs free energy change.

Because it determines the maximum work available from a process or the minimum work to drive the process, the availability is obviously a very important thermodynamic function. It serves as a basis for measuring the performances of actual processes and therefore should be an essential tool of the engineer whose responsibility is to bring the actual process as close to the thermodynamic ideal as is permitted by economic and other constraints.

EXAMPLE 14-1

In many plants steam used for process heat is raised by combustion of a fuel (say methane) in a boiler. If the steam is saturated at 25 psia, estimate the efficiency of the operation assuming the stoichiometric amount of air for combustion and no heat losses from the boiler.

Solution 14-1

The maximum work obtainable from the oxidation of methane[7] is given by the availability change, but if reactants enter and combustion products leave at 298 K, the Gibbs free energy change is equivalent. This change is calculated via the path diagram shown in Fig. 14-2 and is $-814,079$ kJ/kmol CH_4 oxidized. This quantity of work is capable of delivering heat at the condensing steam temperature, 388.9 K, by means of a reversible heat pump taking heat from the medium. By Eq. (14-5) the maximum heat which can be delivered is

---

[6] In practice there should be no confusion because we will surely know whether we are dealing with an open or closed system.

[7] This normally irreversible combustion can be visualized as carried out in a reversible fuel cell.

$$Q = W^E \left(\frac{T}{T - T_0} \right) = 814,079 \left(\frac{388.9}{388.9 - 298} \right)$$

$$= 3,482,900 \text{ kJ}$$

When fuel is burned in the boiler, the most heat which can be delivered at 388.9 K is the heat of combustion. This process, idealized from a first law perspective, is sketched in Fig. 14-3. The heat of combustion of methane is $-890,000$ kJ/kmol, and the efficiency of the process can be expressed as

$$\eta = \frac{890,900}{3,482,900} = 0.256$$

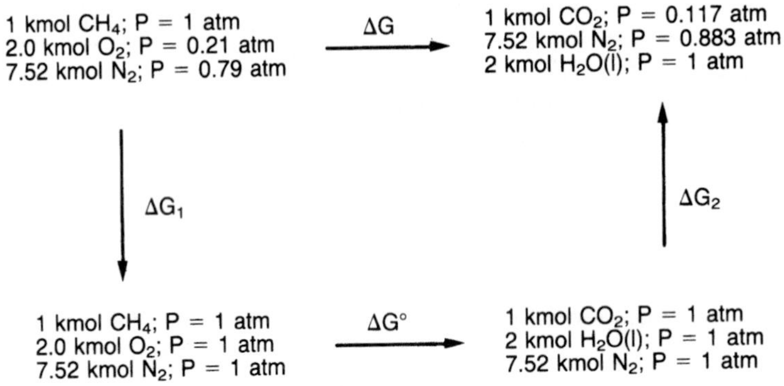

1 kmol CH_4; P = 1 atm
2.0 kmol O_2; P = 0.21 atm
7.52 kmol N_2; P = 0.79 atm

$\xrightarrow{\Delta G}$

1 kmol CO_2; P = 0.117 atm
7.52 kmol N_2; P = 0.883 atm
2 kmol $H_2O(l)$; P = 1 atm

ΔG_1 ↓

ΔG_2 ↑

1 kmol CH_4; P = 1 atm
2.0 kmol O_2; P = 1 atm
7.52 kmol N_2; P = 1 atm

$\xrightarrow{\Delta G^\circ}$

1 kmol CO_2; P = 1 atm
2 kmol $H_2O(l)$; P = 1 atm
7.52 kmol N_2; P = 1 atm

$\Delta G = \Delta G_1 + \Delta G^\circ + \Delta G_2$

$$\Delta G_1 = (8.3143)(298)\left(2 \ln \frac{1}{0.21} + 7.52 \ln \frac{1}{0.79} \right) = 12,125 \text{ kJ}$$

$\Delta G^\circ = 2(-237,400) + (-394,600) - (-50,830) = -818,570 \text{ kJ}$

$\Delta G_2 = (8.3143)(298)(\ln 0.117 + 7.52 \ln 0.883) = -7,634 \text{ kJ}$

$\Delta G = 12,125 - 818,570 - 7634 = -814,079 \text{ kJ}$

Figure 14-2 Computational path and evaluation of ΔG for the oxidation of methane. [Data are from references 1 and 2 in Table 12-1.]

Any heat losses or inefficient heat exchanges will, of course, reduce this efficiency; however, it can be increased by raising the temperature (saturation pressure) of the steam. Many well-managed petroleum refineries and chemical manufacturing complexes make efficient use of energy by using fuel combustion only to supply high-temperature process heat. Heat exchange is employed to recover the heat of exothermic reactions and high-temperature sensible heat of process streams. This recovered heat is used to raise steam or supply process heat needed at lower temperatures. Cogeneration of electricity and steam is another operation which results in effective fuel utilization. This will be discussed in Chap. 15.

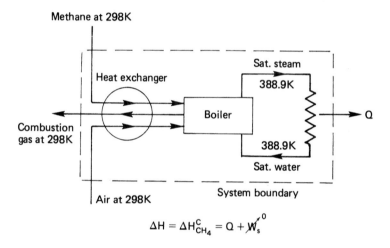

$$\Delta H = \Delta H^C_{CH_4} = Q + \cancel{W_s}^{0}$$

Figure 14-3 Boiler system idealized from first law perspective. (The heat capacity of combustion gas is assumed equal to that of air plus methane.)

The inefficiency of supplying low-level heat by combustion of fuel has long been recognized by serious students of thermodynamics; however, only recently has this become widely appreciated. Apropos of U.S. energy usage, it has been estimated that the efficiency of conventional residential and commercial space heating by fuel combustion averages about 6%.[8]

[8] W. D. Metz, *Science, 188,* 820 (1975).

EXAMPLE 14-2

A means of reducing the energy requirement for distillation is to compress the overhead vapor to a sufficiently high pressure so that it can condense and deliver heat to the column reboiler at a higher temperature. Consider a distillation column operating at atmospheric pressure with essentially pure water as the overhead vapor. The vapor should be compressed to a pressure such that it will condense in the reboiler at a temperature of 150°C.
(a) Assuming reversible adiabatic compression, find the ratio of heat delivered in the reboiler to the work required.
(b) If the plant generates its own electricity for the compressor at an efficiency of 33%, what is the ratio of heat delivered to the reboiler to fuel used in the plant boiler?
(c) Repeat parts (a) and (b) for a compressor efficiency of 70%.

Solution 14-2

(a) At 150°C the saturation pressure of water is 0.476 MPa, and the vapor will be compressed adiabatically and reversibly from 1 bar to this pressure. The vapor leaving the compressor will be superheated with its final temperature being determined by the condition $\Delta s = 0$. From saturated vapor at 1 bar a constant-entropy process to a pressure of 0.476 MPa results in a vapor with an enthalpy of 2993.6 kJ/kg. For saturated vapor at 1 bar the enthalpy is 2676.0 kJ/kg, and for a steady-state adiabatic operation the reversible work of compression is

$$W_s = \Delta h = 2993.6 - 2676.0 = 317.6 \text{ kJ / kg}$$

On condensing to saturated liquid at 0.476 MPa (where its enthalpy is 632.1 kJ/kg) the compressed vapor delivers

$$2993.6 - 632.1 = 2361.5 \text{ kJ / kg}$$

Thus, the units of heat delivered per unit of work expended are

$$\frac{2361.5}{317.6} = 7.43$$

(b) With a generating efficiency of 33%[9] it is found that

$$\frac{317.6}{0.33} = 953.7 \text{ kJ}$$

---

[9] The generating efficiency is usually taken as the electricity generated divided by the heat of combustion of the fuel consumed.

of fuel combustion heat is required to generate the electricity to power the compressor. Thus, units of heat delivered per unit of fuel combustion heat expended are

$$\frac{2361.5}{953.7} = 2.48$$

(c) For a 70% compressor efficiency the work required is

$$W_s = \frac{317.6}{0.70} = 454 \text{ kJ / kg} = \Delta h$$

and the fuel consumption is

$$\frac{454}{0.33} = 1376 \text{ kJ}$$

The final enthalpy of the adiabatically compressed vapor at 0.476 MPa is

$$h = 2676.0 + 454 = 3130 \text{ kJ / kg}$$

The heat delivered by the vapor in the reboiler is now

$$3130 - 632.1 = 2498 \text{ kJ / kg}$$

and the ratios are

$$\frac{2498}{454} = 5.50 \text{ kJ of heat delivered / kJ of work supplied}$$

$$\frac{2498}{1376} = 1.82 \text{ kJ of heat delivered / kJ of fuel combustion heat supplied}$$

If the steam which would have ordinarily supplied the reboiler heat were raised in a fuel-fired boiler, the fuel savings would be

$$\frac{2361.5 - 953.7}{2361.5} \times 100 = 60\%$$

for reversible compression and

$$\frac{2498 - 1376}{2498} \times 100 = 45\%$$

for 70% compressor efficiency.

14-3 MIXING AND SEPARATION PROCESSES

We have just seen that the availability change specifies the work in any reversible process. While this includes mixing and separation processes, it is worthwhile to single out these processes for special study because of their importance to chemical engineers. That work is involved in these processes can be seen from Fig. 14-4, which illustrates how a gas mixture (e.g., air) may be separated in an idealized steady-state flow process. The process uses semi-permeable membranes[10] to separate the components followed by compression to the original pressure.

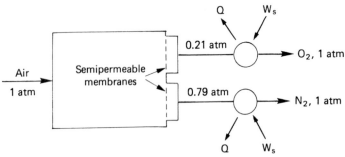

Figure 14-4 Isothermal, reversible, steady-state process for separating air.

The separation step occurs with equality of partial pressures across the membranes and is easily reversed by an infinitesimal change of pressure. No work or heat is exchanged with the surroundings; therefore,

$$\Delta H = 0$$

$$\Delta S = 0$$

Hence,

$$\Delta G = 0$$

The last statement also follows from the condition of equilibrium across the membrane,

$$\hat{f}_i = f_i$$

or

$$\hat{\mu}_i = \mu_i$$

---

[10] This is not as far-fetched as first appears because of progress in designing hollow polymer fibers and the recent availability of commercial devices for concentrating hydrogen by membrane permeation.

and Eq. (10-8).

The compressions are isothermal and reversible, and for each component, assumed to be an ideal gas, the work is

$$W_s = \int v \, dP = RT \ln \frac{P_2}{P_1}$$

For each component, P_1 is its partial pressure in the original mixture p_i, and P_2 is the original pressure of the mixture. Thus,

$$\frac{P_2}{P_1} = \frac{P}{p_i} = \frac{P}{y_i P} = \frac{1}{y_i}$$

The total work of separation can now be expressed as

$$W_s = RT \sum n_i \ln \frac{1}{y_i} = -RT \sum n_i \ln y_i \tag{14-21}$$

From Table 11-1 the quantity $RT \sum n_i \ln y_i$ is seen to be the Gibbs free energy change on forming an ideal gaseous solution; with the negative sign it represents the Gibbs free energy change for the separation step.

Because ideal gas behavior has been assumed, $\Delta H = 0$ for the isothermal compression steps. This results in no enthalpy change for the entire separation process, and therefore the total entropy change is[11]

$$\Delta S = -R \sum n_i \ln \frac{1}{y_i} = R \sum n_i \ln y_i \tag{14-22}$$

It is interesting to note that the entropy change is associated only with the compression step with no contribution from the membrane separation step. Because of the tendency to regard the entropy as a measure of randomness or disorder, zero entropy change for this step may seem paradoxical; however, this may be visualized as offsetting contributions due to un-mixing (negative) and pressure reduction (positive).

The idealized separation of a liquid mixture in an isothermal, isobaric, steady-state flow process is illustrated in Fig. 14-5. The liquid is passed into a chamber where vapor is formed by the addition of heat. The vapor is in equilibrium with the liquid and is separated by passing through semipermeable membranes. Each component vapor is compressed iso-thermally from its partial pressure to its vapor pressure and is subsequently condensed to a pure liquid by removal of heat. In the vaporization and condensation chambers the pressure is kept constant at P by the presence of an inert, noncondensable, and insoluble gas. While the vapor composition is different from the liquid, the liquid composition in the vaporization

---

[11] This entropy change for the separation process is seen from inspection of Table 11-1 to be the negative of that for the forming of an ideal solution.

chamber is maintained at the feed composition by the proper adjustment of the vapor flow rates.

As before, the work requirement for the entire process is due only to isothermal compression and can be expressed as

$$W_s = RT \sum n_i \ln \frac{P_i^\circ}{p_i} = -RT \sum n_i \ln \hat{a}_i \qquad (14\text{-}23)$$

where Eq. (12-9) shows the pressure ratio p_i / P_i° to be expressible as the activity $\hat{a}_i$. By comparison with Eq. (11-12) W_s is seen to equal the Gibbs free energy change of unmixing. For an ideal solution $\hat{a}_i = x_i$, and

$$W_s = RT \sum n_i \ln \frac{1}{x_i} = -RT \sum n_i \ln x_i \qquad (14\text{-}24)$$

And it is observed that the minimum work required to separate ideal gaseous solutions and ideal liquid solutions is identical, as is the entropy change.

The preceding illustrations were based on ideal solutions because of the ease of calculation. For gaseous solutions we have seen that this assumption is not too restrictive; however, for liquid solutions it rarely holds, and therefore Eq. (14-23) will have more utility than Eq. (14-24). The calculation of ΔS for the mixing or separation of a nonideal liquid solution requires ΔG and ΔH or ΔG at several temperatures (see Sec. 11-2).

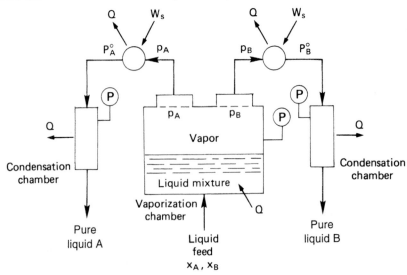

Figure 14-5 Isothermal, reversible, steady-state process for separating a liquid mixture.

As long as the separation process occurs at temperatures close to that of the medium T_0, the Gibbs free energy change adequately reflects the minimum energy requirement. For an isothermal process ΔG always gives the minimum work of separation, but when there is a heat effect and the process is carried out at a temperature removed from T_0, there is an equivalent work effect associated with that heat effect which must also be considered. This is included in the total work when the availability change is employed, and the difference between ΔG and ΔB often can be sizable. Thus, the use of ΔG to estimate the separational energy requirement can sometimes give misleading results. Consider the separation of air into pure oxygen and pure nitrogen, where

$$\Delta H = 0$$

$$\Delta S = R \sum n_i \ln y_i$$

$$= 1.987(0.21 \ln 0.21 + 0.79 \ln 0.79)$$

$$= -1.021 \text{ cal / g mol air} \cdot \text{K}$$

These two property changes are independent of temperature, as is the availability change:

$$\Delta B = \Delta H - T_0 \Delta S = 1.021 T_0$$

With $T_0 = 298$ K,

$$\Delta B = 304 \text{ cal / g mol air}$$

The Gibbs free energy change, however, is dependent on the temperature at which the process is carried out:

$$\Delta G = \Delta H - T \Delta S = 1.021 T$$

If air were separated at T_0, there would be, of course, no difference between ΔG and ΔB; however, if the separation were carried out at 77.8 K, the atmospheric boiling point of air, we would find that

$$\Delta G = 1.021(77.8) = 79.4 \text{ cal}$$

It would therefore appear that less work is required at the lower temperature; however, the 79.4 cal of heat requiring removal in the isothermal compression steps has to be pumped to the medium. The work necessary for this task may be calculated for a Carnot engine operating between 77.8 and 298 K using Eq. (14-5):

$$W^E = -79.4 \left(\frac{77.8 - 298}{77.8} \right) = 224.6 \text{ cal}$$

The total work required is

$$\mathbf{w} = 224.6 + 79.4 = 304 \text{ cal}$$

as is specified by the availability change.

All separation processes require the expenditure of work $(\Delta B > 0)$, and conversely, all mixing processes are capable of performing work $(\Delta B < 0)$. Compared to chemical changes, the minimum work associated with separations is rather small; however, the efficiency of actual separational processes (e.g., distillation) is very low, and consequently a sizable amount of energy is expended for this purpose in the chemical and petroleum industries.[12] Because of the small amount of energy available and the expected low efficiencies, until recently there has been little interest in harnessing the energy available in mixing processes.[13]

EXAMPLE 14-3

An aqueous solution containing 3 mol % ethanol is fractionated in a distillation column operating at atmospheric pressure into a distillate containing 61 mol % ethanol and a bottoms containing 0.01 mol % ethanol. This separation uses 3.5 kg of saturated 0.2 MPa steam per kmol of feed. Calculate a thermodynamic efficiency for this process.

Solution 14-3

A schematic diagram of the distillation column with the attendant material balances is shown in Fig. 14-6, and the thermodynamic property changes for this process are calculated via the computational path shown in Fig. 14-7. The feed enters, and the distillate and bottoms each leave the system at essentially the temperature of the medium. Thus, the process can be regarded as isothermal, and ΔG is the minimum work required for the indicated separation. Based on 1 kmol of feed and by utilizing Eq. (14-23), the ΔG's of Fig. 14-7 are

$$\Delta G_I = RT\left[0.03 \ln \frac{1}{0.03\gamma_E \ (x_E = 0.03)} + 0.97 \ln \frac{1}{0.97}\right]$$

$$\Delta G_{II} = RT\left[0.03 \ln 0.61\gamma_E \ (x_E = 0.61) + 0.019 \ln 0.39\gamma_W \ (x_E = 0.61)\right]$$

$$\Delta G_{III} = RT(0.000095 \ln 0.0001 \ \gamma_E^\infty + 0.951 \ln 0.9999)$$

---

[12] For a detailed thermodynamic analysis of specific separational processes, see M. Benedict, *Trans. AIChE, 43,*41 (1947).

[13] Recently, there have been some proposals for utilizing the mixing of fresh river water with saline ocean water to generate power. See R. S. Norman, *Science, 186,* 350 (1974); B. H. Clampitt, and F. E. Kiviat, *Science, 194,* 719 (1976); M. Olsson, G. L. Wick, J. D. Isaacs, *Science, 206,* 452 (1979); and G. L. Wick and J. D. Isaacs, *Science, 199,* 1436 (1978).

The necessary activity coefficients are obtained from vapor-liquid equilibrium data for the ethanol-water system at 25°C[14] and are

$$\gamma_E \, (x_E = 0.03) = 3.67$$

$$\gamma_E \, (x_E = 0.61) = 1.08$$

$$\gamma_E^\infty = 4.05$$

$$\gamma_W \, (x_E = 0.61) = 1.82$$

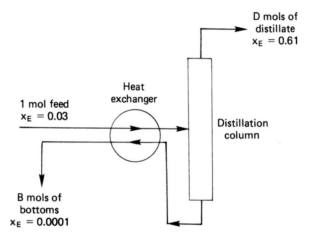

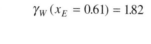

Total material balance: 1 = D + B
Ethanol balance: 0.03 = 0.61 D + 0.0001 B
     D = 0.049
     B = 0.951

Figure 14-6

[14] Source: reference B2 in Table 10-8.

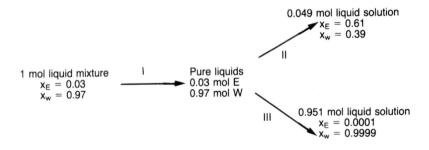

0.049 mol liquid solution
$x_E = 0.61$
$x_w = 0.39$

II

1 mol liquid mixture   I   Pure liquids
$x_E = 0.03$ 0.03 mol E
$x_w = 0.97$ 0.97 mol W

III   0.951 mol liquid solution
$x_E = 0.0001$
$x_w = 0.9999$

$$\Delta G = \Delta G_I + \Delta G_{II} + \Delta G_{III}$$
$$\Delta H = \Delta H_I + \Delta H_{II} + \Delta H_{III}$$

Figure 14-7

The minimum work of separation is

$$W_s = \Delta G = \Delta G_I + \Delta G_{II} + \Delta G_{III}$$

$$= 8.3143(298)(0.0957 - 0.0190 - 0.0008)$$

$$= 188 \text{ kJ / kmol feed}$$

We now evaluate the availability, or equivalent work, of the 3.5 kg of steam actually used in the distillation. Because the practice is to utilize only the latent heat of steam, the availability change is computed for saturated liquid at 0.2 MPa as the final state.[15] Per kg of steam the availability change is

$$\Delta b = h^L - h^G - 298\left(s^L - s^G\right)$$

$$= 504.7 - 2706.7 - 298(1.5301 - 7.1271)$$

$$= -534.1 \text{ kJ / kg}$$

---

[15] It is, of course, possible to extract the maximum possible work from the steam by allowing the condensate to reach the medium temperature. This would give

$$\Delta b = 105.0 - 2706.7 - 298(0.3678 - 7.1271)$$

$$= -587.4 \text{ kJ / kg}$$

and would result in a somewhat lower separation efficiency.

The total availability change is

$$\Delta B = 3.5(-534.1) = -1869 \text{ kJ}$$

and the separation efficiency may be stated as

$$\eta = \frac{188}{1869} = 0.101$$

14-4 HEAT EXCHANGE

Because effective heat exchange is usually significant and often vital to the efficiency or profitability of a process, the subject merits detailed thermodynamic analysis. Figure 14-8 shows a sketch of a countercurrent heat exchanger—the most common type of heat exchange device.

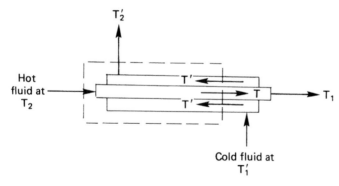

Figure 14-8
Countercurrent heat exchanger.

The assumption of adiabatic operation is usually quite reasonable and leads to the following first law statement for the heat exchanger:

$$\Delta H = 0 \qquad (14\text{-}25)$$

The enthalpy change is the sum of hot and cold stream contributions:

$$\Delta H = 0 = \Delta H_H + \Delta H_C \qquad (14\text{-}26)$$

For fluids of constant heat capacity that undergo no phase change or chemical reaction, we may write

$$\Delta H_H = \dot{m}_H c_{PH} (T_1 - T_2) \qquad (14\text{-}27a)$$

$$\Delta H_C = \dot{m}_C c_{PC} (T_2' - T_1') \qquad (14\text{-}28a)$$

where T and T' are the temperatures and $\dot{m}_H$ and $\dot{m}_C$ are the mass flow rates of the hot and cold streams. It is understood that Eqs. (14-27a) and (14-28a) apply over a unit time interval. These equations result from an energy balance around the entire heat exchanger. Similar equations that result from a balance around the dashed envelope of Fig. 14-8 are

$$\Delta H_H = \dot{m}_H c_{PH}(T - T_2) \tag{14-27b}$$

$$\Delta H_C = \dot{m}_C c_{PC}(T_2' - T') \tag{14-28b}$$

Use of these equations in Eq. (14-26) yields

$$\dot{m}_H c_{PH}(T - T_2) = \Delta H_H = \dot{m}_C c_{PC}(T' - T_2') \tag{14-29}$$

and it is seen that plots of T or T' vs. ΔH_H are linear with slopes of $1/\dot{m}_H c_{PH}$ and $1/\dot{m}_C c_{PC}$, respectively. These plots provide a convenient means of representing the performance of a heat exchanger.

Figure 14-9 shows such a plot for a countercurrent heat exchanger for the special case where $\dot{m}_H c_{PH} = \dot{m}_C c_{PC}$. This condition results in a constant temperature difference $T - T'$ throughout the heat exchanger. The condition for reversible heat transfer is that $T - T'$ should approach zero. Because the lines in Fig. 14-9 are parallel, it would be possible to realize this condition throughout the heat exchanger, and as the two lines become juxtaposed, reversible operation is approached. The infinitesimal driving force required for reversibility unfortunately leads to an infinitely large heat exchanger to accomplish the task.

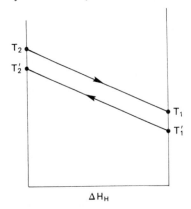

Figure 14-9 Performance of counter-current heat exchanger with

$$\dot{m}_H c_{PH} = \dot{m}_C c_{PC}.$$

It is only for the special case of equal $\dot{m} c_p$ that a heat exchanger can approach reversibility. Unequal $\dot{m} c_p$ leads to lines of different slope which permits $T - T'$ to approach zero only at one point in the heat exchanger. When $\dot{m}_H c_{PH} > \dot{m}_C c_{PC}$, the lines can touch only at the hot end T_2, as shown in Fig. 14-10. Conversely, $\dot{m}_H c_{PH} < \dot{m}_C c_{PC}$ leads to touching at the cold end T_1, as shown in Figure 14-11. While the touching of T and T' lines illustrated in these figures requires infinite heat exchange area, the exchangers do not operate reversibly because $T - T'$ is not zero throughout.

Some measure of the thermodynamic efficiency of a heat exchanger is desirable, and we can expect that the availability is the appropriate function for this assessment. Each stream entering and leaving the heat exchanger may possess the ability to do work by virtue of its temperature. This is the availability change for the process in which the stream at the temperature T is allowed to come to thermal equilibrium with the medium at T_0. For the ith stream,

$$\Delta B_{0i} = \dot{m}_i \, \Delta b_{0i} \qquad (14\text{-}30)$$

$$\Delta b_{0i} = h_{0i} - h_i - T_0 \left(s_{0i} - s_i \right) \qquad (14\text{-}31)$$

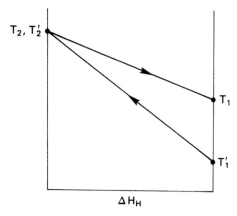

Figure 14-10 Limiting performance of a countercurrent heat exchanger with

$\dot{m}_H c_{P_H} > \dot{m}_C c_{P_C}.$

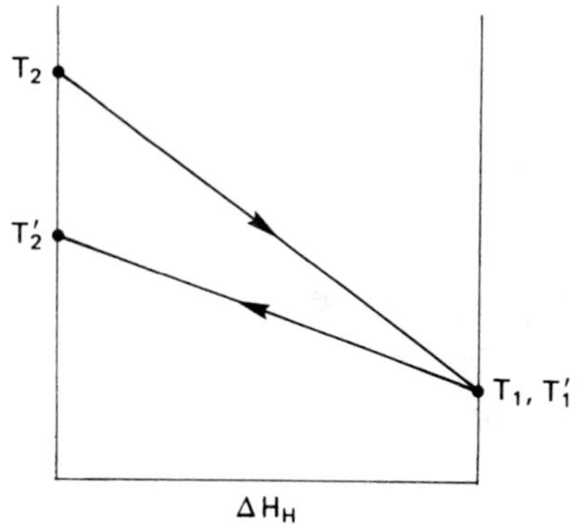

Figure 14-11 Limiting performance of a countercurrent heat exchanger

$\dot{m}_H c_{P_H} < \dot{m}_C c_{P_C}.$

Because we are concerned only with thermal energy, h_{0i} and s_{0i} are evaluated at T_0 and the pressure of the stream. For a fluid of constant heat capacity Eq. (14-31) is written as

$$\Delta b_{0i} = c_{P_i} \left(T_0 - T_i - T_0 \ln \frac{T_0}{T_i} \right) \qquad (14\text{-}32)$$

Because availability is always conserved in a reversible process, a reasonable definition of efficiency is

$$\eta = \frac{\underset{\substack{\text{exit} \\ \text{streams}}}{\sum} \Delta B_{0i}}{\underset{\substack{\text{inlet} \\ \text{streams}}}{\sum} \Delta B_{0i}} \tag{14-33}$$

This efficiency compares the performance of a heat exchanger against the best possible performance—that of an unspecified reversible device. Only for the case of equal $\dot{m}c_P$ can the efficiency of a heat exchanger become unity. Other definitions of efficiency are possible. For example, Dodge[16] uses the ratio of the actual temperature rise of a stream to the maximum possible temperature rise—an efficiency which can approach unity for all types of heat exchange. It compares the performance of a heat exchanger to the best possible performance of a heat exchanger and is therefore a relative, rather than an absolute, efficiency.

In actual practice the close approach of stream temperatures in a heat exchanger is undesirable. So far we have considered only streams of constant heat capacity where T and T' lines are straight and touching is discernible from the terminal conditions. In cases where the heat capacity is not constant or phase changes occur, it is possible for T and T' curves to touch or intersect even though $T > T'$ at both ends of the heat exchanger. Such a situation obviously corresponds to an unworkable design and is a possibility the design engineer should be alert to. The stream temperature vs. enthalpy change plot is a useful tool for avoiding such mistakes.

EXAMPLE 14-4

Examine the efficiency of a countercurrent heat exchanger with equal $\dot{m}c_P$ in which a cold stream enters at 298 K. Make a parametric study for several levels of entering hot stream temperature T_2 and approach $T - T'$.

Solution 14-4

A sample calculation for $T_2 = 600$ K and $T - T' = 50$ K is shown. The first law statement reduces to

$$T_2 - T_1 = T_2' - T_1'$$

With $T_2 = 600$ K, $T_1' = 298$ K, and $T - T' = 50$ K, we obtain

$$T_1 = 348 \text{ K and } T_2' = 550 \text{ K}$$

---

[16] B. F. Dodge, *Chemical Engineering Thermodynamics*, McGraw-Hill, New York, 1944, Chap. 9.

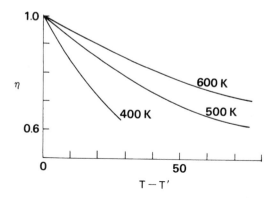

Figure 14-12 Efficiency of a countercurrent heat exchanger with $\dot{m}_H c_{PH} = \dot{m}_C c_{PC}$ for three values of T_2.

With equal $\dot{m} c_P$ the efficiency from Eqs. (14-30)–(14-33) is

$$\eta = \frac{T_0 - T_1 - T_0 \ln T_0/T_1 + T_0 - T_2' - T_0 \ln T_0/T_2'}{T_0 - T_2 - T_0 \ln T_0/T_2}$$

with the preceding temperatures we obtain $\eta = 0.783$. For other values of T_2 and $T - T'$ the efficiencies have been calculated and are plotted in Fig. 14-12. As expected, the efficiency is seen to decrease with increasing $T - T'$, and the effect of a given $T - T'$ is seen to be smaller as T_2 increases.

EXAMPLE 14-5

In a process for liquefying nitrogen a high-pressure nitrogen stream is to be cooled by countercurrent heat exchange with a colder, low-pressure stream of nitrogen. Nitrogen at 800 psia will enter the hot end of the heat exchanger at 300°R and will be cooled to 160°R by nitrogen at 150°R and 14.7 psia entering the cold end.
(a) Find the minimum ratio $\dot{m}_C/\dot{m}_H$ necessary to accomplish this task.
(b) Using 1.2 times the minimum $\dot{m}_C/\dot{m}_H$, determine the final temperature of the low-pressure gas.
(c) Determine the thermodynamic efficiency for the operating conditions of part (b).
(d) For a reversible adiabatic process (unspecified), find the value of $\dot{m}_C/\dot{m}_H$ and the final low-pressure gas temperature.

Solution 14-5

Thermodynamic property data for nitrogen are as follows:

THERMODYNAMIC PROPERTIES of NITROGEN[a]

| T(°R) | P = 800 psia | | P = 14.7 psia | |
	h (Btu/lb)	s (Btu/lb °R)	h (Btu/lb)	s (Btu/lb °R)
160	−41.71	0.7359	38.50	1.3338
180	−30.73	0.7962	43.67	1.3597
200	−20.68	0.8463	48.78	1.3838
220	−9.07	0.8999	53.85	1.4063
240	8.35	0.9746	58.90	1.4274
260	32.86	1.0728	63.93	1.4474
280	46.25	1.1228	68.95	1.4661
300	55.64	1.1557	73.95	1.4838
537	128.01	1.3429	132.95	1.6345

[a] From K. E. Starling, *Fluid Thermodynamic Properties for Light Petroleum Systems*, Gulf Publ. Co., Houston, 1973.

Enthalpy changes per pound of high-pressure nitrogen calculated from these data are used to establish the $T - \Delta H_H$ curve shown in Fig. 14-13. This curve, labeled AB, is definitely not linear. For low-pressure nitrogen c_p is very nearly constant at 0.254 Btu lb^{-1} °R^{-1}, and a linear $T' - \Delta H_H$ relationship results. This linearity allows a graphical solution to parts (a) and (b).

(a) Point C in Fig. 14-13 represents the low-temperature condition for the entering cold fluid, and its $T' - \Delta H_H$ curve will be a straight line terminating at this point. The slope of this line will be $1/\dot{m}_C c_{PC}$ or $3.937/\dot{m}_C$, where $\dot{m}_C$ is understood to be the pounds of cold fluid per pound of hot fluid. The minimum value of $\dot{m}_C$ is determined by the touching of the two $T - \Delta H_H$ curves at point D. From the slope of the line $\overline{CD}$ the minimum value of $\dot{m}_C$ is found to be 2.706. In this case it is observed that while an infinitely large heat transfer area is required there is a positive temperature difference at each end of the exchanger. Had we considered only the terminal temperature differences, we would have been unaware of this condition. Fortunately, this type of situation is not commonly encountered.

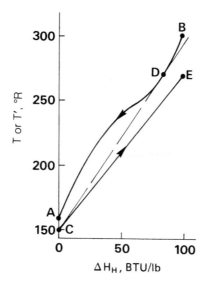

Figure 14-13

(b) When 1.2 times the minimum rate is used, we have

$$\dot{m}_C = 1.2(2.706) = 3.25$$

A line drawn through C with slope $3.937/3.25$ terminates at point E where $T_2' = 268°R$. This answer could also have been obtained by application of the first law:

$$\Delta H = 0 = h_1 - h_2 + \dot{m}_C c_{PC}(T_2' - T_1')$$

$$0 = -41.71 - 55.64 + 3.25(0.254)(T_2' - 150)$$

$$T_2' = 267.9°R$$

(c) The availabilities required for evaluation of the efficiency are determined from the given enthalpy and entropy data. For the high-pressure stream h_0 and s_0 refer to 537°R but at the pressure of 800 psia. The availabilities are as follows:

Low-pressure inlet: $\Delta b_0 = 132.95 - 35.92 - 537(1.6345 - 1.3209)$:

$$\Delta b_0 = -71.37 \quad \Delta B_0 = 3.25(-71.37) = -231.75 \text{ Btu}$$

Low-pressure exit: $\Delta b_0 = 132.95 - 65.92 - 537(1.6345 - 1.4548)$:

$$\Delta b_0 = -29.47 \quad \Delta B_0 = 3.25(-29.47) = -95.77 \text{ Btu}$$

High-pressure inlet: $\Delta b_0 = 128.01 - 55.64 - 537(1.3429 - 1.1557)$:
$$\Delta b_0 = \Delta B_0 = -28.16 \text{ Btu}$$

High-pressure exit: $\Delta b_0 = 128.01 - (-41.71) - 537(1.3429 - 0.7359)$:
$$\Delta b_0 = \Delta B_0 = -156.24 \text{ Btu}$$

The efficiency is

$$\eta = \frac{-95.77 + (-156.24)}{-231.75 + (-28.16)} = 0.970$$

(d) Using the preceding Δb_0's for a reversible adiabatic process ($\eta = 1$) we write

$$71.37 \dot{m}_C + 28.16 = 156.24 - \dot{m}_C \Delta b_0 \qquad \text{(a)}$$

The first law statement is

$$-41.71 - 55.64 + \dot{m}_C (h - 35.92) = 0 \qquad \text{(b)}$$

In Eqs. (a) and (b), h and Δb_0 refer to the exit low-pressure stream at T_2'. By using the supplied thermodynamic data, these two equations are solved to obtain

$$\dot{m}_C = 2.024 \text{ lb}$$

$$T_2' = 339.4°\text{R}$$

From this example we see that the proposed heat exchanger is quite efficient. If the task could be performed reversibly, considerably less of the low-pressure nitrogen would be required, but its exit temperature would be higher, in this case higher than that of the inlet high-pressure stream.

14-5 SYSTEMS INVOLVING CHEMICAL TRANSFORMATIONS

Chemical processes are often analyzed by means of efficiencies based on the first law. They measure only the extent of realizable conservation of energy and ignore energy quality or availability. Although the first law is a useful tool for process analysis, they are not true thermodynamic efficiencies and will not be considered here. There are two approaches based on the second law which may be used for a rigorous thermodynamic analysis of systems that involve chemical transformations: (1) the use of the availability change and (2) the assignment of a work equivalent to each chemical species. While both approaches are valid, each views the process from a different perspective, and efficiencies based on these methods are usually different.

Availability-Change Method. Because there were no restrictions on the derivation of Eqs. (14-12) and (14-20), the availability change gives the maximum work obtainable or the minimum work required in any process, including those in which chemical transformations occur. Once the conditions of the medium are set (T_0 and P_0), the availability change depends only on the change in state of the system and can be calculated when this is specified. As before, all actual heat exchanges between the system and the various heat sources and sinks Q_i at temperatures T_i are reduced to equivalent work W_i^E using the Carnot factor. In the case of heat released by the system at T_i the W_i^E term is easily visualized as the maximum work which could be realized by a heat engine operating reversibly between T_i and the medium at T_0. When heat is delivered to the system at T_i, it can be visualized as the minimum work required to pump the heat to the system from the medium. All exchanges between the system and the surroundings are now expressible in terms of work, and when compared to the change in availability, the net work provides the measure of efficiency. The application of this method will be illustrated with the analysis of the ammonia oxidation process based on the work of Denbigh.[17]

There are two major steps in producing nitric acid from ammonia: (1) the oxidation of ammonia with air over a platinum catalyst at 850°C and 1 atm and (2) the dissolution of the resulting nitric oxide in water to form aqueous nitric acid. The latter step occurs at essentially 25°C and 1 atm. The reactions can be written as

$$NH_3 + 9.6 \text{ air} = NO + \frac{3}{2}H_2O + \frac{3}{4}O_2 + 7.6N_2 \qquad (a)$$

and

$$1.34H_2O(l) + NO + \frac{3}{2}H_2O(g) + \frac{3}{4}O_2 + 7.6N_2 = HNO_3 \text{ (60 wt \%)} + 7.6N_2 \qquad (b)$$

When added, these reactions yield the overall reaction for the process:

$$1.34H_2O(l) + NH_3 + 9.6 \text{ air} = HNO_3 \text{ (60 wt \%)} + 7.6N_2 \qquad (c)$$

The maximum work obtainable from this process is determined by evaluating the availability change for this reaction. All participants in the reaction are assumed to enter or leave the process at 25°C, and ammonia is assumed to enter as a saturated liquid. First, ΔH and ΔG are calculated via the path illustrated in Fig. 14-14.

Step I. Because ideal-gas behavior can be assumed, there is no enthalpy change accompanying the separation of air and the isothermal compression of O_2 and N_2 to 1 atm. For

---

[17] K. G. Denbigh, *Chem. Eng. Sci.*, **6**, 1 (1956).

ammonia the enthalpy change is the heat of vaporization. The free energy change for the separation of air is given by Eq. (14-21). For ammonia the calculation is similar to that of Ex. 12-3:

$$\Delta H_1 = \Delta H_{NH_3}^{vap} = 4738 \text{ cal / g mol}$$

$$\Delta G_I = RT\left(7.6 \ln \frac{1}{0.79} + 2 \ln \frac{1}{0.21} + \ln \frac{1}{9.89} \right)$$

$$= 1552 \text{ cal / g mol}$$

Step II.   This step constitutes the following reaction:[18]

$$NH_3(g) + 2O_2(g) = HNO_3(l) + H_2O(l)$$

$$\Delta H_{II} = \Delta H_{298}^{\circ} = -286.0 + (-173.4) - (-46.22) = -413.2 \text{ MJ / kmol}$$

$$\Delta H_{II} = -98,690 \text{ cal / g mol}$$

$$\Delta G_{II} = \Delta G_{298}^{\circ} = -237.4 + (-79.97) - (-16.6) = -300.8 \text{ MJ / kmol}$$

$$\Delta G_{II} = -71,840 \text{ cal / g mol}$$

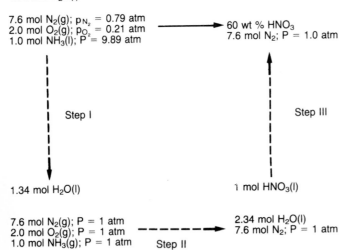

Figure 14-14
Computational path for ΔH and ΔG for ammonia oxidation.

---

[18] Data from Table D-2 in the Appendix.

Step III. In this step 1 mol HNO_3 and 2.34 mol H_2O are mixed to form a 60 wt % HNO_3 solution. From Fig. 6-5, Δh_s is found to be -5250 cal/g mol HNO_3:

$$\Delta H_{III} = \Delta h_s = -5250 \text{ cal}$$

From Eq. (11-12) the free energy change for this mixing step is

$$\Delta G_{III} = RT\left(\ln x_{HNO_3}\, \gamma_{HNO_3} + 2.34 \ln x_{H_2O}\, \gamma_{H_2O}\right)$$

$$= RT\left(\ln \hat{a}_{HNO_3} + 2.34 \ln \hat{a}_{H_2O}\right)$$

The activities may be evaluated from vapor-liquid equilibrium data.[19] The partial pressures of 60 wt % HNO_3 are

$$p_{HNO_3} = 1.21 \text{ mmHg} \qquad p_{H_2O} = 7.7 \text{ mmHg}$$

The vapor pressures are

$$P^{\circ}_{HNO_3} = 57 \text{ mmHg} \qquad P^{\circ}_{H_2O} = 23.76 \text{ mmHg}$$

The activities are

$$\hat{a}_{HNO_3} = \frac{p_{HNO_3}}{P^{\circ}_{HNO_3}} = \frac{1.21}{57} = 0.0212$$

$$\hat{a}_{H_2O} = \frac{p_{H_2O}}{P^{\circ}_{H_2O}} = \frac{7.7}{23.76} = 0.324$$

and

$$\Delta G_{III} = 1.987(298)(\ln 0.0212 + 2.34 \ln 0.324)$$

$$= -3843 \text{ cal}$$

For the three-step path we have

$$\Delta H = 4738 - 98{,}690 - 5250 = -99{,}202 \text{ cal}$$

$$\Delta G = 1552 - 71{,}840 - 3843 = -74{,}131 \text{ cal}$$

---

[19] J. H. Perry, *Chemical Engineers' Handbook*, 3rd ed., McGraw-Hill, New York, 1950, p. 169.

With the medium taken to be 298 K and with all streams entering and leaving at 298 K, the availability change is equal to the Gibbs free energy change. The maximum work obtainable from the process is therefore 74.1 kcal/g mol of HNO_3 produced. Denbigh reports for the operation of a typical ammonia oxidation plant that each gram mol of HNO_3 produced requires 4.1 kcal of electricity and supplies 31.9 kcal of heat for the generation of 50-psia saturated steam $(T_i = 411.7$ K). The equivalent work of the steam is

$$W^E = -31.9\left(\frac{411.7 - 298}{411.7}\right) = -8.8 \text{ kcal}$$

The net work obtained from the process is

$$-8.8 + 4.1 = -4.7 \text{ kcal}$$

When compared to the ideal, or reversible, process, the efficiency of the ammonia oxidation process is

$$\frac{-4.7}{-74.1} = 0.063$$

One of the reasons for this low efficiency can be attributed to the irreversible manner in which chemical reactions occur under industrial conditions. Reversibility implies balanced potentials, and for a chemical reaction this condition can be approached only in an electrochemical cell—a device encountered more frequently in theory than in practice.

Work-Equivalent Method.[20]   Instead of evaluating an availability change for a process, a work equivalent **E** may be assigned to each stream such that the difference of the total output work equivalents and the total input work equivalents represents the availability change for the process. The work equivalent measures the ability of a stream to do work by virtue of its temperature, pressure, or composition being different from the medium. In addition to physical work, a stream also possesses a work equivalent due to the ability of its chemical species to do work in undergoing chemical transformations.

In a reversible process there is no degradation of energy or change in the ability to do work; therefore, availability is conserved. This suggests the following definition of efficiency:

$$\eta = \frac{\mathbf{E}_{out} + \mathbf{W}_{out}}{\mathbf{E}_{in} + \mathbf{W}_{in}} \tag{14-34}$$

where $\mathbf{W}_{in}$ and $\mathbf{W}_{out}$ represent total work (including the availability of heat flows) into and out of the system and $\mathbf{E}_{in}$ and $\mathbf{E}_{out}$ represent the work equivalents of matter flows into and out of the system.[21] This efficiency is unity for a reversible process and less for an irreversible one.

---

[20] Based on the work of L. Riekert, *Chem. Eng. Sci.*, 29, 1613 (1974).

[21] In Eq. (14-34) all work terms are taken positive. The **E**'s are always equal to or greater than zero and are extensive properties. The intensive work equivalent is designated **e**.

The work equivalent of a stream $\mathbf{E}$ is defined as the maximum work received by the surroundings by allowing the stream to come to equilibrium with the medium. Thus,

$$\mathbf{E} = -\Delta B_0 = H - H_0 - T_0(S - S_0) \tag{14-35}$$

where ΔB_0 is the availability change between the state of the stream and a *dead state* denoted by zero subscripts. The dead state is a state which is in equilibrium with the medium and therefore is one in which a system is incapable of performing work. Consider now an unspecified process occurring in a steady-state flow system such as that shown in Fig. 14-1. There may be several inflowing streams with flow rates $\dot{n}_l$ having associated work equivalents $\mathbf{e}_l$ and several outflowing streams with flow rates $\dot{n}_k$ and work equivalents $\mathbf{e}_k$. During a unit time interval the change in work equivalents for the process is

$$\sum \dot{n}_k \mathbf{e}_k - \sum \dot{n}_l \mathbf{e}_l$$

From Eq. (14-35) this is seen to be

$$\sum \dot{n}_k \mathbf{e}_k - \sum \dot{n}_l \mathbf{e}_l = \sum \dot{n}_k h_k - \sum \dot{n}_l h_l - T_0\left(\sum \dot{n}_k s_k - \sum \dot{n}_l s_l\right)$$

$$-\left[\sum \dot{n}_k h_{0k} - \sum \dot{n}_l h_{0l} - T_0\left(\sum \dot{n}_k s_{0k} - \sum \dot{n}_l s_{0l}\right)\right] \tag{14-36}$$

which simplifies to

$$\sum \dot{n}_k \mathbf{e}_k - \sum \dot{n}_l \mathbf{e}_l = \Delta H - T_0\,\Delta S = \Delta B \tag{14-37}$$

In Eq. (14-36) the term in brackets represents the availability change between dead states and is thus zero.[22] Equation (14-37) shows that the difference in stream work equivalents is equal to the availability change. This, as we have seen, is the total reversible work associated with the process.

The stream work equivalent as defined in Eq. (14-35) is an extensive thermodynamic property. The contribution of each component in the stream to this property is, from Eq. (11-4), expressible as

$$\mathbf{E} = \sum n_i \bar{\mathbf{e}}_i \tag{14-38}$$

where the species work equivalent is

$$\bar{\mathbf{e}}_i = -\Delta \bar{B}_{0i} = \bar{H}_i - \bar{H}_{0i} - T_0(\bar{S}_i - \bar{S}_{0i}) \tag{14-39}$$

---

[22] The dead states must be at T_0 and P_0 but could correspond to different compositions.

The species work equivalent $\bar{e}_i$ is taken to be the sum of a physical component e_i^P and a chemical component e_i°:

$$\bar{e}_i = e_i^\circ + e_i^P \qquad (14\text{-}40)$$

The physical part, e_i^P includes work the species is capable of performing by virtue of its temperature, pressure, or composition being different from the medium. The chemical part, e_i° is the work obtainable from pure species i at T_0 and P_0 by virtue of chemical transformation.

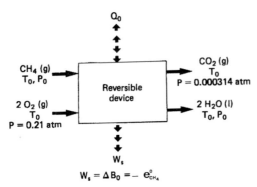

Figure 14-15

To conceptualize e_i° we must extend our perception of the medium. In addition to an infinite heat source or sink at T_0, it will also be regarded as an infinite source or sink for certain materials, such as air and liquid water, which will be referred to as *datum level materials*. The chemical work equivalents for datum level materials are taken to be zero, and it follows that the work equivalent of any species in equilibrium with the medium is also zero. We now define the chemical work equivalent of any species i to be the work which could be done in a reversible process involving only pure species i and datum level materials that exchanges heat only with the medium and ends in a dead state. This can be illustrated by considering a specific substance, say methane. Figure 14-15 shows a reversible process where methane is oxidized with oxygen taken from the medium at a pressure of 0.21 atm and entering the device through a semipermeable membrane. The oxidation products are CO_2 and $H_2O(l)$. The latter is a datum level material and is simply released to the medium. The partial pressure of CO_2 in the atmosphere is in the neighborhood of 0.000314 atm, and therefore the CO_2 leaving the device is released to the medium at this pressure through another semipermeable membrane. With the exception of methane, every substance or stream entering or leaving the process is in equilibrium with the medium and thus possesses zero work equivalent. Application of Eq. (14-37) shows that the negative of the availability change for this process, the work received by the surroundings, is equal to the chemical work equivalent of methane.[23]

The chemical work equivalent of any substance can be determined in a systematic manner. As pictured in Fig. 14-16, consider the formation of compound $C_\alpha H_\beta O_\gamma N_\delta$ from α mol of carbon, $\beta/2$ mol of hydrogen, $\gamma/2$ mol of oxygen, and $\delta/2$ mol of nitrogen. The process occurs reversibly at T_0 and P_0 with heat exchange only with the medium. For this isothermal process at T_0 the availability change equals the Gibbs free energy change. Because P_0 is the pressure which defines the standard state, this Gibbs free energy change is the free energy of formation of the compound. We now write and rearrange Eq. (14-37) to obtain

$$\mathbf{e}^\circ_{C_\alpha H_\beta O_\gamma N_\delta} = \alpha\, \mathbf{e}^\circ_C + \frac{\beta}{2}\, \mathbf{e}^\circ_{H_2} + \frac{\gamma}{2}\, \mathbf{e}^\circ_{O_2} + \frac{\delta}{2}\, \mathbf{e}^\circ_{N_2} + \Delta G^f_{C_\alpha H_\beta O_\gamma N_\delta} \qquad (14\text{-}41)$$

and note that when the chemical work equivalents for the constituent elements are known, the chemical work equivalent for any compound can be determined from its free energy of formation. With air and liquid water chosen as datum level materials,[24] we now proceed to calculate the chemical work equivalents of the elements carbon, hydrogen, oxygen, and nitrogen. Pure oxygen and nitrogen are obtained from air, and $\mathbf{e}^\circ$ for these elements is

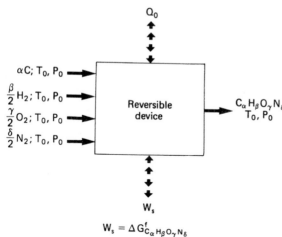

Figure 14-16

[23] It should be obvious that the numerical value of $\mathbf{e}^\circ$ will depend on the choice of datum level materials; however, this choice does not affect the value of ΔB for the process, as is shown by Eq. (14-37).

[24] These substances are obvious choices; however, any choice which conforms to the reality of the problem under study could be made. The choice is subject to the restriction that no work be obtainable from a process involving only datum level materials. For example, liquid water and solid sodium chloride would represent an improper choice because work could be obtained by forming a solution.

merely the work of separation

$$e^{\circ}_{O_2} = -RT \ln 0.21 = 924 \text{ cal}$$

$$e^{\circ}_{N_2} = -RT \ln 0.79 = 140 \text{ cal}$$

Air also contains 314 ppm of CO_2, and

$$e^{\circ}_{CO_2} = -RT \ln 0.000314 = 4776 \text{ cal}$$

Utilizing Eq. (14-41), we write

$$e^{\circ}_{CO_2} = e^{\circ}_{C} + e^{\circ}_{O_2} + \Delta G^{f}_{CO_2}$$

and hence may obtain e°_{C}. The work equivalent of hydrogen may also be obtained through application of Eq. (14-41):

$$e^{\circ}_{H_2O(l)} = 0 = e^{\circ}_{H_2} + \frac{1}{2} e^{\circ}_{O_2} + \Delta G^{f}_{H_2O(l)}$$

Because liquid water is a datum level component, its work equivalent is zero.

The work equivalents of these four elements are tabulated below. From these values and the free energy of formation of the species, one may calculate e°_{i} for the species via Eq. (14-41). Many fuels and feedstocks are not simple compounds for which ΔG^{f} values are available; however, the heat of combustion is usually available, and it has been shown[25] that this closely approximates e°_{i} for high-molecular-weight compounds comprised of carbon, hydrogen, and oxygen.

Element	e°_{i} (cal / g mol)
O_2	924
N_2	140
H_2	56,230
C	98,100

The work equivalent of a species e_i may be visualized as shown in Fig. 14-17 where the physical component e^{P}_{i} is shown to be a sum of separate terms accounting for the temperature, pressure, and composition effects e^{PT}_{i}, e^{PP}_{i}, and e^{Px}_{i}, respectively. Based on the defining equation [Eq. (14-39)], these terms are

---

[25] T. L. Unruh and B. G. Kyle, Paper 117c, San Francisco meeting, American Institute of Chemical Engineers, Nov. 1979.

$$\mathbf{e}_i^{PT} = \int_{T_0}^{T} c_{Pi}\left(1 - \frac{T_0}{T}\right) dT \tag{14-42}$$

$$\mathbf{e}_i^{PP} = \int_{P_0}^{P} \left[v - (T - T_0)\left(\frac{\partial v}{\partial T}\right)_P \right]_i dP \tag{14-43}$$

$$\mathbf{e}_i^{Px} = \Delta \overline{G}_i = RT \ln \hat{a}_i \tag{14-44}$$

For an ideal gas Eq. (14-43) simplifies to

$$\mathbf{e}_i^{PP} = RT_0 \ln \frac{P}{P_0} \tag{14-43a}$$

For ideal solutions Eq. (14-44) simplifies to

$$\mathbf{e}_i^{Px} = RT \ln x_i \tag{14-44a}$$

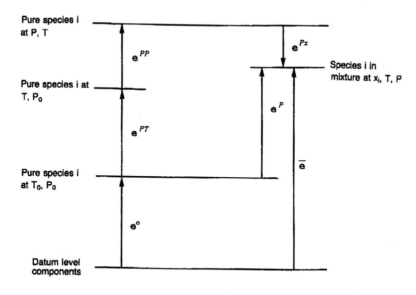

Figure 14-17   Illustration of work equivalent of a substance.

Note that the composition effect e_i^{Px} is always negative, as is indicated in Fig. 14-17. For a stream this effect, designated E^{Px}, is given by

$$E^{Px} = \sum n_i e_i^{Px} = RT \sum n_i \ln \hat{a}_i = \Delta G \qquad (14\text{-}45)$$

and is seen to be the free energy of mixing. This is the only contribution to the stream work equivalent which is negative. The reason for this can be seen by the contrasting of a stream of air at T_0 and P_0 with separate streams of nitrogen and oxygen each at T_0 and P_0 and having flow rates in the ratio 79/21. The air stream is in equilibrium with the medium and can do no work, while the separate streams can, in principle, be expanded isothermally to partial pressures of 0.79 and 0.21 atm, respectively, and exhausted to the medium through semipermeable membranes. Mixtures, therefore, have less thermodynamic value than their pure constituents.

The application of this approach also will be demonstrated with the ammonia oxidation process. With the previously stated utility requirements, we have

$$W_{in} = 4.1 \text{ kcal}$$

$$W_{out} = 8.8 \text{ kcal}$$

The chemical work equivalent of ammonia which enters as a saturated liquid at 298 K is

$$e_{NH_3(l)}^{\circ} = \frac{1}{2}e_{N_2}^{\circ} + \frac{3}{2}e_{H_2}^{\circ} + \Delta G_{NH_3(l)}^f$$

Using the method illustrated in Ex. 12-3 with $\Delta G_{NH3(g)}^f = -3976$ cal/g mol $P_{NH3}^{\circ} = 9.89$ atm. and we obtain[26]

$$\Delta G_{NH_3(l)}^f = -2619 \text{ cal / g mol}$$

Thus,

$$e_{NH_3(l)}^{\circ} = 0.5(140) + 1.5(56,230) - 2619$$

$$= 81,796 \text{ cal / g mol}$$

If the small pressure contribution is neglected, this is the total work equivalent of liquid ammonia.

The chemical work equivalent of pure nitric acid is (reference 1, Table 12-1)

$$e_{HNO_3(l)}^{\circ} = \frac{1}{2}e_{H_2}^{\circ} + \frac{1}{2}e_{N_2}^{\circ} + \frac{3}{2}e_{O_2}^{\circ} + \Delta G_{HNO_3(l)}^f$$

---

[26] ΔG^f from reference 1 in Table 12-1.

$$= 0.5(56,230) + 0.5(140) + 1.5(924) - 19,100$$

$$= 10,471 \text{ cal} / \text{g mol}$$

The mixing contribution is simply the free energy of mixing as previously calculated. The work equivalent of 60 wt % HNO_3 is

$$\mathbf{e}_{HNO_3(60\text{ wt\%})} = \mathbf{e}^{\circ}_{HNO_3(l)} + 2.333\mathbf{e}^{\circ}_{H_2O(l)} + \Delta G_{mixing}$$

and with $\mathbf{e}^{\circ}_{H_2O(l)} = 0$ we obtain

$$\mathbf{e}_{HNO_3(60\text{ wt\%})} = 10,471 - 3843$$

$$= 6628 \text{ cal} / \text{g mol } HNO_3$$

The efficiency of the process is[27]

$$\eta = \frac{\mathbf{e}_{HNO_3(60\text{wt\%})} + \mathbf{W}_{out}}{\mathbf{e}_{NH_3(l)} + \mathbf{W}_{in}}$$

$$= \frac{6628 + 8800}{81,796 + 4100}$$

$$= 0.18$$

Comparison of Availability-Change and Work-Equivalent Methods. It is observed that for the nitric acid process the work-equivalent efficiency is considerably larger than the availability-change efficiency (0.18 vs. 0.063). The reason for this is that with the work-equivalent method the value of the product is considered. It is recognized that the 60 wt % HNO_3 possesses some ability to do work or that part of the chemical energy stored in the NH_3 now resides in the product. The availability-change efficiency judged the process only in terms of how well its potential to do work was realized. To emphasize this point, consider what happens if a nitric acid plant is so poorly managed that the equivalent work of the generated steam is less than the input electrical energy of 4.1 kcal. In this case the net equivalent work would be positive, and for this process, marked by a negative availability change, a meaningful efficiency could not be obtained by the availability-change method. This problem does not arise with the work-equivalent efficiency as it is seen from Eq. (14-34) that this will always lie between zero and unity.

---

[27] As indicated by reaction (c) and Fig. 14-4, nitrogen in a pure or concentrated state is a product of this process. While this product possesses a small work equivalent by virtue of its composition, this work equivalent is not counted since this stream is not recovered or utilized for a practical purpose.

Because a value can be assigned to the chemical energy of influent and effluent streams, the work-equivalent method should be superior when one desires to compare alternative processes for producing the same product. Each process may use different feedstocks and produce different by-product streams. The availability-change method would tell us how effectively the process was carried out in terms of work required or work available but would have nothing to say about the chemical energy value of raw materials or by-products.

While Eq. (14-34) will always give an efficiency between zero and unity, it is not useful for evaluating the efficiency of a physical process such as compression or separation because often the chemical work equivalents will overshadow the physical values. Consider the efficiency of a compressor as expressed by Eq. (14-34):

$$\eta = \frac{\mathbf{e}_{out}}{\mathbf{e}_{in} + \mathbf{W}_{in}} = \frac{\mathbf{e}^{\circ}_{out} + \mathbf{e}^{P}_{out}}{\mathbf{e}^{\circ}_{in} + \mathbf{e}^{P}_{in} + \mathbf{W}_{in}}$$

With no change in composition $\mathbf{e}^{\circ}_{out} = \mathbf{e}^{\circ}_{in}$. If air is being compressed $\mathbf{e}^{\circ}_{out} = \mathbf{e}^{\circ}_{in} = 0$; however, if methane is being compressed, $\mathbf{e}^{\circ}$ is much larger than the $\mathbf{e}^{P}$ terms, and the efficiency will be close to unity. Thus, with the work-equivalent approach the efficiency of a physical step is strongly dependent on the nature of the material processed. Therefore, we may conclude that the change in availability provides the best basis for a realistic *device* efficiency and that the work-equivalent approach should be used only where chemical transformations occur.

Both efficiencies have merit; it is merely the perspective that is different. The availability-change efficiency measures how well a specific task has been accomplished, while the work-equivalent efficiency measures the conservation of availability. Some may lament the lack of a single "true" thermodynamic prescription for efficiency and the absolute certainty which would accompany it; however, whatever certitude appears lost because of this imagined deficiency is more than compensated for by the flexibility of the thermodynamic approach.

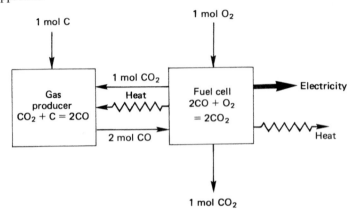

Figure 14-18

While thermodynamics can be a valuable tool for process analysis and synthesis, in practice the actual thermodynamic constraints on a process often are not inherent but arise from the limitations imposed by current technology. This condition has been thoroughly explored recently in application to synthetic natural gas processes.[28]

EXAMPLE 14-6

A proposed process for utilizing coal (assumed for the purposes of calculation to be pure carbon) to produce electricity in a fuel cell is shown in Fig. 14-18. The fuel cell produces electricity by oxidizing CO to CO_2 with oxygen. One-half of the CO_2 produced is discarded and the other half sent to a gas producer where it reacts with carbon to form CO. The reactions are as follows:

fuel cell: $\qquad 2CO + O_2 = 2CO_2 \qquad\qquad$ (a)

gas producer: $\qquad \underline{CO_2 + C = 2CO} \qquad\qquad$ (b)

overall: $\qquad\qquad C + O_2 = CO_2$

The sum of these two reactions is simply the oxidation of carbon to carbon dioxide.

The reaction in the gas producer is endothermic, while that of the fuel cell is exothermic, and it is hoped that the process can be operated so that heat needed in the gas producer can be supplied by the fuel cell and that the process would require no outside inputs of energy. Determine if this is possible and if so the best conditions for operation. Also determine (a) the maximum work obtainable and (b) the efficiency when the fuel cell efficiency is 50%.

Solution 14-6

The gas producer will be operated at a temperature high enough that the equilibrium gas composition is at least 98% CO. This will allow feed directly to the fuel cell without the necessity of a separation step. At 1700°F we find that[29]

$$\log K^f_{CO} = 9.499 \qquad \log K^f_{CO_2} = 17.2425$$

which for reaction (b) yields

$$\log K_b = 2(9.499) - 17.2425 = 1.7555$$

---

[28] R. Shinnar, G. Fortuna, and D. Shapira, *Ind. Eng. Chem. Process Des. Dev.*, *21*, 728 (1982).
[29] Source: from reference 7 in Table 12-1.

$$K_b = 57 = \frac{\hat{f}_{CO}^2}{\hat{f}_{CO_2} a_C}$$

In the presence of carbon and at a pressure of 1 atm we write

$$K_b = 57 = \frac{y_{CO}^2}{1 - y_{CO}}$$

and obtain $y_{CO} = 0.983$. Thus, a temperature of 1700°F is adequate. At this temperature the following enthalpy data are available,[30]

$$\left(H° - H_0° + \Delta H_0^f \right)_{CO} = -33,004 \text{ Btu / lb mol}$$

$$\left(H° - H_0° + \Delta H_0^f \right)_{CO_2} = -145,975$$

$$\left(H° - H_0° \right)_C = 7432$$

$$\left(H° - H_0° \right)_{O_2} = 16,528$$

and we obtain

$$\Delta H_b° = 2(-33,004) - (-145,975) - 7432$$

$$= 72,535 \text{ Btu / lb mol } CO_2$$

or

$$\Delta H_b° = 40,297 \text{ cal / g mol } CO_2$$

This is the heat required to sustain the reaction in the gas producer, and in order that this heat be supplied by the fuel cell it is necessary that it be operated at a temperature of at least 1700°F.[31]

Turning our attention to the fuel cell reaction, we find at 1700°F that

$$\log K_a = 2(17.2425) - 2(9.499) = 15.487$$

---

[30]Source: reference 7 in Table 12-1.
[31] Reversible operation would prescribe a temperature of $1700 + dt$ for the fuel cell.

$$\Delta G_a^\circ = -RT \ln K_a$$

$$= -1.987(2160)(2.303)(15.487)$$

$$= -153,078 \text{ Btu} / 2 \text{ lb mol CO}$$

or

$$\Delta G_a^\circ = -85,043 \text{ cal} / 2 \text{ g mol CO}$$

$$\Delta H_a^\circ = 2(-145,975) - 2(-33,004) - 16,528$$

$$= -242,470 \text{ Btu} / 2 \text{ lb mol CO}$$

$$= -134,706 \text{ cal} / 2 \text{ g mol CO}$$

$$T\Delta S_a^\circ = \Delta H_a^\circ - \Delta G_a^\circ = -134,706 - (-85,043)$$

$$= -49,663 \text{ cal} / 2 \text{ g mol CO}$$

For the fuel cell the standard thermodynamic property changes are equal to the actual property changes because pure O_2 and CO enter at 1700°F and 1 atm, and pure CO_2 leaves at 1700°F and 1 atm. For reversible operation of the fuel cell and reversible heat exchange with a reservoir at the same temperature, we write

$$\Delta S_a^\circ + \left(\frac{-Q_a}{T}\right) = 0$$

$$Q_a = T\Delta S_a^\circ = -49,663 \text{ cal}$$

Thus, when the fuel cell reversibly consumes 2 g mol CO at 1700°F, there will be 49,663 cal of heat liberated which may be transferred to the gas producer. The electrical work obtainable from the fuel cell is

$$W_s = \Delta G_a^\circ = -85,043 \text{ cal}$$

Since the gas producer requires only 40,297 cal, there remain 9366 cal of heat available at 1700°F. The equivalent work of this heat is

$$W^E = -9366\left(\frac{2160 - 537}{2160}\right) = -7038 \text{ cal}$$

Heat is required to raise the temperature of carbon and oxygen from 77 to 1700°F and is available from the cooling of carbon dioxide from 1700 to 77°F. The equivalent work of this heat is expressible in terms of the availability change. For streams requiring heat and undergoing no phase changes,

$$\Delta b = \Delta h - T_0 \, \Delta s = \int_{T_0}^{T} c_P \left(1 - \frac{T_0}{T}\right) dT$$

Similarly, for streams supplying heat,

$$\Delta b = \Delta h - T_0 \Delta s = \int_{T}^{T_0} c_P \left(1 - \frac{T_0}{T}\right) dT$$

Using c_p data from reference 7 in Table 12-1 and graphical evaluation of these integrals, we obtain

C: $W^E = \Delta b = 2259$ cal

O_2: $W^E = \Delta b = 3944$ cal

CO_2: $W^E = \Delta b = -5960$ cal

The maximum net work obtainable from the process is

$$W_{max} = -85,043 - 7038 - 5960 + 2259 + 3944$$

$$= -91,838 \text{ cal}$$

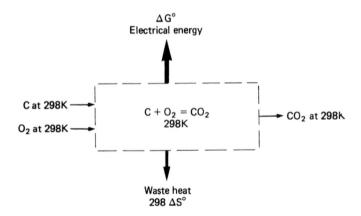

Figure 14-19

Because provision for heating reactants and cooling the CO_2 product has been included in the calculations, the overall process may be visualized as shown in Fig. 14-19 where entering and exit streams are at 298 K. For such a process the maximum available work is

$$\Delta G_{298}^{\circ} = \left(\Delta G_{CO_2}^{f} \right)_{298} = -94{,}258 \text{ cal}$$

The maximum work obtainable from the process is somewhat less than the 94,258 cal expected for a completely reversible process. The difference, 2420 cal, is significant and represents irreversibilities inherent in the operation of the gas producer.

If the fuel cell were operated so that the efficiency were 50%, we would obtain electrical energy in the amount

$$W_s = 85{,}043 \times 0.5 = 42{,}522 \text{ cal}$$

Application of the first law for steady-state operation of the fuel cell yields

$$\Delta H = Q + W_s$$

$$Q = \Delta H - W_s = -134{,}706 - (-42{,}522) = -92{,}184 \text{ cal}$$

and we observe that the lost work appears as heat.

Of this heat 40,297 cal are supplied to the gas producer, and some will be used to heat reactants to 1700°F. There is sufficient heat available in the exit CO_2 stream to heat the incoming O_2 stream to 1700°F. Heating the C from 77 to 1700°F would require

$$\Delta h = \left(H^{\circ} - H_0^{\circ} \right)_{1700^{\circ}} - \left(H^{\circ} - H_0^{\circ} \right)_{77^{\circ}} = 7432 - 453 = 6979 \text{ Btu / lb atom}$$

$$\Delta h = 3877 \text{ cal / g atom}$$

The remaining fuel cell heat is

$$92{,}184 - 40{,}297 - 3877 = 48{,}010 \text{ cal}$$

While this heat is available at 1700°F and is therefore capable of considerable work, modern power plants operate at much lower temperatures with efficiencies on the order of 35 to 40%. A power generating efficiency of 37.5% would set the value of this excess heat at

$$48{,}010(0.375) = 18{,}004 \text{ cal}$$

and the total electrical energy (work) obtained from the process would be

$$W_s = 42,522 + 18,004 = 60,526 \text{ cal}$$

The efficiency of the process would be

$$\frac{60,526}{94,258} = 0.641$$

or using Eq. (14-34), we would obtain

$$\eta = \frac{\mathbf{W}_{out}}{\mathbf{e}_C^\circ + \mathbf{e}_{O_2}^\circ} = \frac{60,526}{98,100 + 924} = 0.611$$

Thus, if a fuel cell could be operated at this high temperature at the reasonable efficiency of 50% with excess heat converted to electricity in a conventional power cycle, the overall efficiency would be much better than that expected from a coal-fueled conventional power plant. The work-equivalent efficiency judges the process more severely than the availability-change efficiency mainly because the effluent CO_2 stream represents a potential but unexploited source of work.

It is possible to find an operating temperature in the range of 1500°F where the heat required by the gasifier ΔH_b° is just balanced by the heat liberated by the reversible fuel cell $T\Delta S_a^\circ$. However, in this temperature range the equilibrium concentration of CO in the feed to the fuel cell will drop, and work of separation would be necessary to remove unreacted CO_2.[32]

PROBLEMS

14-1.   The following scheme has been proposed for the solar air conditioning of a house. The air in the house will be contacted with a glycol-water solution resulting in a drying of the air and an increase in the water content of the glycol solution. The dry air is then contacted adiabatically with water and is cooled and humidified, but the temperature is low enough so that the humidification is not undesirable. Solar energy is used to evaporate the water from the glycol solution (the glycol can be assumed nonvolatile), and it is concentrated from 60 mol % water to 40 mol % water. Estimate the minimum circulation rate of glycol if the outside air is 95°F and 12,000 Btu/h must be removed to maintain the house temperature at 75°F. Assume that glycol-water solutions are ideal.

14-2.   Derive an expression for the heat input required to separate an ideal, k-component, equimolal mixture into pure components in a reversible process. A heat source at T_B and a sink at T_C are avail-

---

[32] We could, of course, calculate the maximum electrical work expected from a fuel cell fed with a mixture of CO and CO_2.

able. Assume that the enthalpy of the products relative to the feed is zero and that the entropy change of products relative to the feed is due only to the unmixing process. Also comment on the reasonableness of these assumptions.

14-3. Following are equilibrium data for the adsorption of water vapor by wheat at 60°F given as weight percent moisture in the wheat vs. relative humidity. Use these data to calculate the minimum energy requirement, based on a pound of water removed, for drying wheat at 60°F from a moisture content of 15 wt % to a moisture content of 10 wt % when the ambient air has a relative humidity of 70%. Contrast this energy with that required when heat is used for drying.

wt % moisture	10.0	12.5	15.0
Relative humidity (%)	37	50	64

14-4. A farmer interested in making ethanol for on-farm use has been told by a traveling salesperson that his firm's apparatus can recover pure ethanol from fermentation beer (about 3 mol % ethanol) with an energy expenditure of 1000 Btu of 10-psig steam per gallon of pure ethanol. The farmer seeks your advice concerning this claim. Determine the minimum amount of 10-psig steam required for this separation assuming that a waste stream of essentially pure water is also produced. The following vapor-liquid equilibrium data for the ethanol-water system are available. The density of liquid ethanol is 0.79 g/cc.

mol % ethanol			
Liquid	Vapor	$t(°C)$	Pressure (mmHg)
12.2	47.4	25	41.8
16.3	53.1	25	45.2
22.6	56.2	25	47.9
32.0	58.2	25	50.7
33.7	58.9	25	51.0
43.7	62.0	25	52.7
44.0	61.9	25	52.8
57.9	68.5	25	54.8
83.0	84.9	25	58.4

14-5.   A slush of 25% ice and 75% water is to be formed in a countercurrent heat exchanger. Water enters at 70°F and leaves as the slush at 32°F and a glycol-water solution (c_p = 0.87 Btu/lb °F) entering at 20°F is to be the coolant.

    (a)   Determine the minimum ratio of coolant flow to water flow.
    (b)   Use 1.2 times this minimum ratio and determine the exit coolant temperature.

    c_p of H_2O(c) = 0.50 Btu/lb °F
    c_p of H_2O(l) = 1.0 Btu/lb °F
    Heat of fusion of water = 143 Btu/lb

14-6.   A solid is being continuously melted in a vertical kiln. The solid enters the top at 100°F and melts at 2100°F; the molten liquid leaves the bottom of the kiln at 2100°F. The hot burner gases enter the bottom of the kiln at 2500°F, in direct contact with the stock, and leave at the top. The average c_p of the stock and gas, considered constant, are 0.20 and 0.28 Btu/lb °F, respectively. The stock has a heat of fusion of 200 Btu/lb.

    Calculate the minimum pounds of burner gas necessary per ton of stock fed and the corresponding outlet temperature of the gas.

14-7.   Two streams of water pass through a device which exchanges no work or heat with the surroundings. A hot stream is cooled, and a cold stream is warmed. Is it possible for the stream being warmed to leave the device at a temperature higher than the entering temperature of the stream being cooled? Explain.

14-8.   For the process shown in Fig. 14-18, show that if the gas producer could be operated reversibly, the maximum net work available from the process would be $(\Delta G^f_{CO_2})_{298}$.

14-9.   The electrolysis of water to produce H_2 and O_2 is being studied as a possible means of storing electrical energy. It is proposed that the hydrogen produced would be liquefied and stored as a liquid. Because the liquefaction step involves first pressurizing the gas, it has been suggested that the compression step could be avoided if the electrolysis cell were allowed to operate at the desired higher pressure. To answer this question, compare the work requirements for the following two processes operating at room temperature:

    Process a.   Liquid water is electrolyzed at 1 atm to produce H_2 and O_2 gas. The H_2 is compressed isothermally to a pressure of 500 atm.
    Process b.   Liquid water is pumped into an electrolysis cell operating at 500 atm, and H_2 and O_2 gas leave at 500 atm. Work is recovered by expanding O_2 gas isothermally to 1 atm.

Assume that the gases behave ideally. Assume that the efficiency of electrolysis is 60% and that because of reduced ohmic losses due to smaller bubble size it increases to 66% at the higher pressure. The efficiency of compressors and expanders will be assumed to be 70%.

14-10.   It has recently been demonstrated [C. G. Vayenas and R. D. Farr, *Science, 208*, 593 (1980)] that NH_3 can be oxidized to NO in a fuel cell operating at temperatures in the range of 1400°F. The half-cell reactions are believed to be as follows:

Anode: $$2NH_3 + 5O^{2-} = 2NO + 3H_2O + 10e^-$$

Cathode: $$\frac{5}{2}O_2 + 10e^- = 5O^{2-}$$

$$\overline{2NH_3 + \frac{5}{2}O_2 = 2NO + 3H_2O}$$

While thermochemical data indicate that NO is not stable at this temperature, no appreciable dissociation occurs until much higher temperatures. Therefore, leaving the anode side will be a mixture of NO and H_2O in stoichiometric proportions. For such a fuel cell operating at 1400°F and 1 atm and essentially 100% conversion, calculate the following:

(a) the maximum electrical work which could be obtained per mol of NH_3 oxidized.
(b) the heat which should be added or removed in part (a).
(c) When only 60% of the maximum electrical work can be obtained due to irreversibilities, calculate the work and the heat effect.
(d) For part (c), convert any available heat into electricity at a thermal efficiency of 35% and compute the efficiency of the total process. Assume efficient heat exchange is possible between products and reactants so that products leave the system at the temperature of entering reactants, 298 K.

14-11. It has recently been shown [R. W. Coughlin and M. Farooque, *Ind. Eng. Chem. Process Des. Dev., 19*, 211 (1980)] that the following reaction between coal (here assumed to be carbon) and water can be carried out in an electrochemical cell:

$$C(c) + 2H_2O(l) \rightarrow 2H_2(g) + CO_2(g)$$

The CO_2 leaves at the anode and H_2 at the cathode.

(a) For reversible operation of the cell, calculate the electrical energy requirement and the heat requirement when 1 g atom of carbon is oxidized at the temperatures of 25 and 200°C.
(b) For a cell efficiency of 67%, repeat the calculations of part (a) assuming that all cell irreversibilities appear as heat generated within the cell.
(c) For the irreversible operation ($\eta = 0.67$) at 200°C, calculate a second-law-based efficiency for the process if any heat required is supplied by combustion of coal (assumed to be carbon).

14-12. Calculate the chemical work equivalent $e°$ of elemental sulfur based on the following choices of datum level materials:

(a) flue gas containing 500 ppm SO_2.
(b) gypsum, $CaSO_4 \cdot 2H_2O(c)$, and limestone, $CaCO_3(c)$.

Calculate the chemical work equivalent of $SO_3(g)$ for each set of datum level materials, and comment on the results.

14-13.   In the desert water is scarce, and we desire to use as a datum water vapor corresponding to a relative humidity of 5%. Calculate $e°$ for CH_4 using this dead state.

14-14.   Use what you consider to be an appropriate second law efficiency and determine which of the three processes of Fig. 15-24 is most efficient.

14-15.   A newly constructed ammonia synthesis plant uses 0.712 mols of methane to produce one mol of liquid ammonia. The methane is used as a feedstock and also as a fuel to produce all the electricity and steam required by the plant. Raw materials for the process include methane, water, and air and the only product is liquid ammonia. Calculate a second-law efficiency for this plant. See Sec.14-5 for ammonia data.

14-16.   In the production of ammonia, hydrogen is produced via the reaction

$$CH_4(g) + 2H_2O(g) = CO_2(g) + 4H_2(g)$$

and subsequently reacts with nitrogen to form ammonia

$$N_2(g) + 3H_2(g) = 2NH_3(g)$$

Thus, air, water, and methane are the raw materials.

Determine the efficiency of an ammonia plant that requires the following inputs to produce a ton (2000 lbs) of liquid ammonia

   47 lb mols CH_4

   $7(10^6)$ Btu of electricity

   $14(10^6)$ Btu of heat delivered by saturated steam at 10 MPa

14-17.   Considering both chemical and physical energy, what is the maximum power that could be obtained from a gaseous, equimolar mixture of C_2H_6, H_2, and CO at 700 K and 2.0 atm flowing at the rate of 1 kmol/sec?

14-18.   A newly constructed ammonia synthesis plant is stated to have an energy requirement of $30(10^6)$ Btu/ton of liquid ammonia. This is based on natural gas with a heating value of 1000 Btu/SCF. This is essentially the total energy input since the plant raises its own steam to drive the compressors. Raw materials for the process include natural gas, water, and air. Estimate a second law efficiency for this process.

14-19.   A news item in the May 5, 1980, issue of *Chemical and Engineering News* states that the Westinghouse Electric Corp. has submitted a proposal to the U.S. Department of Energy for the construction of a prototype plant which would produce methanol from bituminous coal. It is stated that this plant could ultimately produce 100,000 bbl of methanol per day from 14,000 tons of coal a day. No mention of other products or of utility consumption is made. Estimate the thermodynamic efficiency of this process on the assumption that no utilities are required or produced and that coal, air, and water are inputs and methanol is the only output. The heat of combustion of bituminous coal is estimated to be 12,500 Btu/lb.

14-20. Westinghouse Electric Corp. has developed a coal gasification process and has coupled this with electricity generation using the gasifier product gas in a combustion turbine-steam turbine system. In addition to power generation, by-products of sulfur and ammonia are recovered. Use the following reported performance data [C. W. Schwartz, L. K. Rath, and M. D. Freier, *Chem. Eng. Prog., 78,* 55 (1982)] to determine the thermodynamic efficiency of this process:

	10^3 kg/h
Inputs	
Illinois #6 coal	218
Air	5080
Water	803
	6101
Outputs	
Ash	24
Sulfur	29
Ammonia	4
Solid Waste	4
Exhaust gas	5312
Water by Evaporation	728
	6101

Electricity produced: 671 MW.

Heating value of Illinois #6 coal: 28,420 kJ/kg.

For sulfur, $e° = 139,600$ cal/g atom.

14-21. Recently published information on a process to convert coal to methanol [*Chem. Eng. Prog.* April 1982] is given here. Determine a second law efficiency for this process.

Inputs		Outputs	
Coal	1880 T/day	Methanol (CH_3OH)	1487 T/day
Pure oxygen	864 T/day	Ammonia	3 T/day
Electricity	$9.12(10^8)$ Btu/day	Sulfur	28 T/day

The heat of combustion of the coal is 13,750 Btu/lb. For sulfur $e°$ is 139,600 cal/g atom. 1 T = 2000 lb.

14-22. A new process for fixing nitrogen has been reported (*Chemical and Engineering News,* Dec. 11, 1989) where high temperatures produced by an electric arc cause the formation of NO from air. It is reported that 107 g of NO are produced per kWhr of electricity. Calculate a second law efficiency for the process

     (a)   assuming that the product is pure NO gas.

     (b)   assuming that the product is a mixture of 5% NO and 95% air.

14-23. The oxyhydrochlorination process for converting methane into hydrocarbon liquids has been recently evaluated [J. M. Fox, T. P. Chen, and B. D. Degen, *Chem. Eng. Prog., 86*(4), 42 (1990)]. The process is designed to be self-sufficient in power and produce 2304 m^3/day of liquid hydrocarbon from methane. The reported annual cost for methane was $23.59 (10^6) yr^{-1} based on $0.50/$10^6$ Btu. Assume the liquid hydrocarbon to be *n*-octane and estimate a second law efficiency for the process.

14-24. The June 9, 1986, issue of *Chemical and Engineering News* reports that Sandia National Laboratories conducted a successful test of a thermochemical energy transport system utilizing the following chemical reaction

$$CO_2(g) + CH_4(g) = 2CO(g) + 2H_2(g)$$

Heat is absorbed by the endothermic reaction of CO_2 and CH_4 carried out catalytically at a high temperature, T_2. The product gas, rich in CO and H_2, can be transported to another location or stored until needed. Heat is liberated by reacting CO and H_2 catalytically at the temperature T_1.

     (a)   At 1 atm and with reactants in stoichiometric proportions, estimate the temperatures T_1 and T_2 if it is assumed that the reactions come to equilibrium and the gas contains 2.6% methane at high temperature and 40.9% methane at low temperature.

     (b)   Compute the quantities of heat absorbed at the higher temperature and released at the lower temperature and assess the thermodynamic efficiency of the process.

     (c)   Is there any possibility of carbon deposition at either temperature?

Physicomechanical Processes

*I*n this chapter we consider several widely used operations such as compression and expansion of gases as well as several somewhat standardized processes such as gas liquefaction, refrigeration, and power generation. The emphasis here is on the practical application of the laws of thermodynamics in order to determine the efficiency and identify sources of inefficiency in the processes. Only sufficient details required for the description and understanding of the operations are supplied. Mechanical details and design complexities are mainly the concern of mechanical engineers and are considered beyond the scope of this text.

15-1 COMPRESSION AND EXPANSION OF GASES

In Chap. 3 equations were derived for the reversible compression and expansion of ideal gases. Two types of processes were considered: isothermal and adiabatic. The reversible shaft work for an isothermal compression or expansion of an ideal gas was found to be

$$W_s = RT \ln \frac{P_2}{P_1} \tag{3-16}$$

while for an adiabatic process it is

$$W_s = \Delta h = c_P (T_2 - T_1) \tag{3-23}$$

or

$$W_s = \frac{\gamma RT_1}{\gamma - 1} \left[\left(\frac{P_2}{P_1} \right)^{(\gamma-1)/\gamma} - 1 \right] \tag{3-25}$$

Compression.   For compression the isothermal path requires less work than the adiabatic path. Shown in Fig. 15-1 are a reversible adiabatic path and a reversible isothermal path for the compression of an ideal gas from an initial pressure P_1 to a final pressure P_2. Because the shaft work is $\int v\, dP$, the work of isothermal compression is represented by the area $1BP_2P_1$, while the work of adiabatic compression is represented by the area $1AP_2P_1$. Clearly, less work is required for isothermal compression.

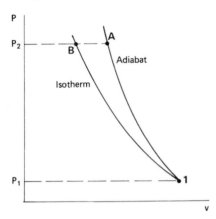

Figure 15-1 Adiabatic and isothermal paths for a reversible compression.

The path of an actual compression lies between these two limiting paths but is more nearly adiabatic. However, by compressing in stages with constant-pressure cooling between stages, it is possible, at least in principle, to approach the isothermal path. This is illustrated

in Fig. 15-2, which shows three adiabats intersecting a single isotherm. A three-stage operation is shown with reversible adiabatic compression steps $1a$, bc, and de and two constant-
pressure interstage cooling steps ab and cd which return the gas to its original temperature.
Obviously, by increasing the number of stages with interstage cooling, an isothermal compression path can be approximated.

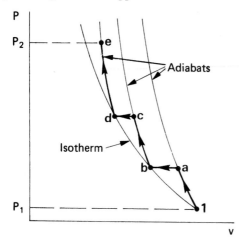

Figure 15-2 Path of a three-stage
compression.

For reciprocating compressors mechanical considerations dictate multistage compression when the ratio of discharge to intake pressures is large. Multistage reciprocating compressors are usually designed with a pressure ratio per stage of between 4 and 6. Centrifugal
compressors are only capable of a pressure ratio of about 1.6 per stage; however, it is common practice to join several stages in series by mounting the impellers on a common shaft.
As with reciprocating machines it is desirable to provide fluid cooling when the pressure
ratio reaches 4 to 6. This can be accomplished by water-jacketing the compressor or use of
interstage cooling. In recent years design improvements have made the centrifugal compressor competitive with and even preferred over the reciprocating compressor for all except
extremely high-pressure applications.

For determination of the actual work of compression several problems arise: the actual
path is unknown, the process is not reversible, and the gas may not behave ideally. While it
is possible to calculate the reversible isothermal or adiabatic work for a nonideal gas from
thermodynamic diagrams or property tables, this generally is not done. Instead, one usually
finds efficiencies based on either the reversible isothermal or reversible adiabatic compression of an ideal gas. The isothermal efficiency is defined as

$$\eta_T = \frac{RT_1 \ln P_2/P_1}{\text{actual work of compression}} \tag{15-1}$$

and the adiabatic efficiency is defined as

$$\eta_S = \frac{\left[\gamma R T_1/(\gamma - 1)\right]\left[(P_2/P_1)^{(\gamma-1)/\gamma} - 1\right]}{\text{actual work of compression}} \qquad (15\text{-}2)$$

In both expressions T_1 is the intake temperature. Because the reversible work of isothermal compression is smaller, the isothermal efficiency is always less than the adiabatic efficiency. Usually, the former ranges between 50 and 70%, while the latter lies between 70 and 90%. These figures are suitable for estimates, but the most reliable efficiencies are those supplied by the manufacturer.[1]

While it is possible for reversible and irreversible isothermal compressions (or expansions) to occur between the same initial and final states, the same cannot be said of adiabatic compressions (or expansions). For a given initial state T_1, P_1 and final pressure P_2, the final temperature in an irreversible adiabatic compression will be higher than for the corresponding reversible compression. However, because η_S is defined in terms of P_1, T_1, and P_2, the use of this efficiency should introduce no ambiguity.

Expansion.   There are two major reasons for expanding gases: to obtain work and to produce a lower temperature in processes such as liquefaction and refrigeration. In all cases the gas is capable of performing work, but for the latter application the emphasis is on attaining a low temperature. An extreme example of this is the adiabatic throttling process $(\Delta h = 0)$, which lowers the gas temperature but produces no work. At the other extreme is the steam turbine found in modern power plants. This is a highly engineered device capable of extracting close to 85% of the maximum work available from a stream of high-pressure, high-temperature steam.

To obtain the most work from a gas expansion, an isothermal process would be preferred over an adiabatic one. This can be seen from Fig. 15-3, which shows that the area

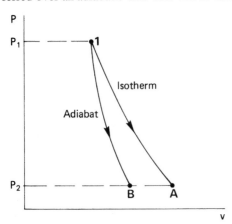

Figure 15-3 Adiabatic and isothermal paths for a reversible expansion.

---

[1] For details of compressor selection and sizing, consult L. F. Scheel, *Gas Machinery*, Gulf Publishing Co., Houston, 1972.

under the isothermal curve $\int v \, dP$ exceeds that under the adiabatic curve. In a multistage expansion with interstage heating it would be possible to approach the isothermal process when each stage operated adiabatically. This would be analogous to the path illustrated in Fig. 15-2 and would be desirable when the objective is to maximize the obtainable work. However, if attainment of the lowest possible temperature is the objective, the adiabatic process terminating in state B is the obvious choice.

The efficiencies of expanders and turbines are defined in a manner analogous to those for compressors:

$$\eta_T = \frac{\text{actual work delivered}}{RT \ln P_2/P_1} \tag{15-3}$$

$$\eta_S = \frac{\text{actual work delivered}}{\left[\gamma \, RT_1/(\gamma - 1)\right]\left[(P_2/P_1)^{(\gamma-1)/\gamma} - 1\right]} \tag{15-4}$$

As with compressor efficiencies, the most reliable values for expander or turbine efficiencies are those supplied by the manufacturer. However, for estimation purposes one can expect adiabatic efficiencies in the range of 50 to 80%.

15-2 THE JOULE-THOMSON EXPANSION

The adiabatic throttling of a gas is often called a Joule-Thomson expansion in honor of the investigators who were the first to extensively study this phenomenon. The process is characterized by constant enthalpy and may be visualized with the aid of Fig. 15-4, which shows a line of constant enthalpy, an isenthalp, for methane. The path of a Joule-Thomson expansion is traced by such a curve.

Although Joule-Thomson expansions are often used for cooling and liquefying gases, there are conditions where a temperature rise occurs. For example, from Fig. 15-4 it is seen that methane gas initially in state A will experience a temperature rise as the pressure is reduced. This occurs until state B is reached where it is seen that further pressure reduction results in a temperature decrease. The slope of the isenthalp $(\partial T / \partial P)_H$ characterizes the Joule-Thomson expansion and is called the Joule-Thomson coefficient. From Fig. 15-4 it is seen to be negative between A and B, zero at B, and positive for pressures less than B. Also, in the ideal-gas region it will be zero. A portion of the Joule-Thomson inversion curve for methane is also shown in Fig. 15-4 as the dashed line. This curve is the locus of isenthalp maxima. Along this curve the Joule-Thomson coefficient is zero, to the right of the curve it is negative, and to the left it is positive. A complete inversion curve for nitrogen is shown in Fig. 15-5. The region bounded by the curve and the temperature axis represents conditions under which nitrogen will cool on being subjected to a Joule-Thomson expansion.

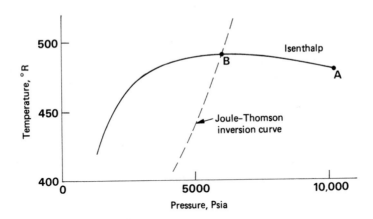

Figure 15-4 An isenthalp for methane. [Data from K. E. Starling, *Fluid Thermodynamic Properties for Light Petroleum Systems*, Gulf Publishing Co., Houston, 1973.]

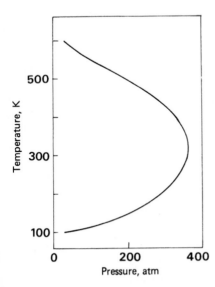

Figure 15-5 Joule-Thomson inversion curve for nitrogen. [Data of J. R. Roebuck and H. Osterberg, *Phys. Rev., 48*, 450 (1935).]

The Joule-Thomson coefficient can be directly measured or calculated from other measurable quantities. From Eq. (5-31c) it is seen that a knowledge of the heat capacity and the *PVT* behavior of the gas allows the determination of the Joule-Thomson coefficient:

$$\left(\frac{\partial T}{\partial P}\right)_H = \frac{1}{c_P}\left(T\left(\frac{\partial v}{\partial T}\right)_P - v\right) \tag{5-31c}$$

With the use of Eqs. (3-26) and (5-41) the Joule-Thomson coefficient can be expressed in terms of the compressibility factor. This results in

$$\left(\frac{\partial T}{\partial P}\right)_H = \frac{RT^2}{c_P P}\left(\frac{\partial Z}{\partial T}\right)_P \tag{15-5}$$

where it is seen that the sign of $(\partial Z/\partial T)_P$ determines the sign of the Joule-Thomson coefficient. Examination of the compressibility factor chart, Fig. 3-6, shows that $(\partial Z/\partial T)_P$ is positive for P_r less than about 9 and negative at higher reduced pressures. The Joule-Thomson coefficient should therefore be negative at pressures greater than about $9P_c$.

15-3 LIQUEFACTION OF GASES

A simple gas-liquefaction process is sketched in Fig. 15-6. In this process fresh gas (1) and recycle gas (8) are compressed (2) and then cooled by ambient air or cooling water (3). This high-pressure gas is cooled further (4) by heat exchange with unliquefied low-pressure gas (7) and then passed through a throttling valve (5) where a fraction f liquefies.
Idealized operation of the process can be traced on a Ts diagram such as Fig. 15-7. A three-stage reversible adiabatic (isentropic) compression with constant-pressure interstage cooling is shown. All heat exchanges are shown as constant-pressure steps, and the throttling step is isenthalpic. Point 5 represents a mixture of liquid 6 and vapor 7, and the lever-arm rule can be used to determine the relative amounts of these phases.

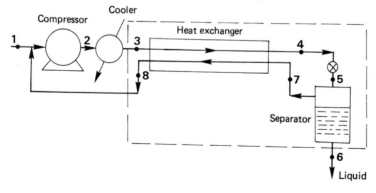

Figure 15-6 Simple throttling liquefaction process.

For steady-state operation where the fraction f of the gas passing through the throttle valve is liquefied, the application of the first law to the dashed envelope of Fig. 15-6 results in

$$f\, h_6 + (1 - f)\, h_8 = h_3 \tag{15-6}$$

or

$$f = \frac{h_8 - h_3}{h_8 - h_6} \tag{15-7}$$

To transfer heat in the heat exchanger, it is necessary that $T_3 > T_8$; however, according to Eq. (15-7), it is necessary for $h_8 > h_3$ in order to produce liquid. Thus, the gas at position 3 must be compressed to pressures where there is a sizable effect of pressure on the enthalpy.

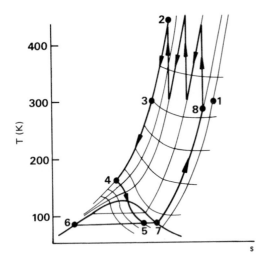

Figure 15-7 Ts diagram for a simple throttling liquefaction process.

EXAMPLE 15-1

Nitrogen is to be liquefied using the process shown in Figs. 15-6 and 15-7. The high-pressure gas will be at 200 atm, and the low-pressure gas will be at 2 atm. On leaving the compressor, the high-pressure gas will be cooled to 300 K, and the approach in the hot end of the heat exchanger, $T_3 - T_8$, will be 10 K. What fraction of gas entering the separator will be liquefied? What will be the gas temperature on the high-pressure side of the throttle valve?

Solution 15-1

When the enthalpies at points 3, 6, and 8 are known, Eq. (15-7) can be used to calculate the fraction liquefied. These enthalpies are determined from Fig. 15-8, a Ts diagram for nitrogen:

Condition 3, $P = 200$ atm and $T = 300$ K: $h_3 = 431$ J / g

Condition 6, Saturated liquid at 2 atm: $h_6 = 45$ J / g

Condition 8, $P = 2$ atm and $T = 290$ K: $h_8 = 452$ J / g

$$f = \frac{452 - 431}{452 - 45} = 0.052$$

To determine T_4, the first law is applied to the heat exchanger:

$$h_4 - h_3 + (1 - f)(h_8 - h_7) = 0$$

Condition 7 refers to saturated vapor at 2 atm and from Fig. 15-8, h_7 is found to be 234 J/g. Using the previously determined values of h_3, h_8, and f, we find that

$$h_4 = 431 - (1 - 0.052)(452 - 234) = 224 \text{ J / g}$$

A pressure of 200 atm and this enthalpy determine the state of the fluid at condition 4: From Fig. 15-8 one finds $T_4 = 172$ K. Alternatively, h_4 can be found from the condition

$$h_4 = h_5$$

where

$$h_5 = f\, h_6 + (1 - f)\, h_7$$

$$= 0.052(45) + (1 - 0.052)234 = 224 \text{ J / g}$$

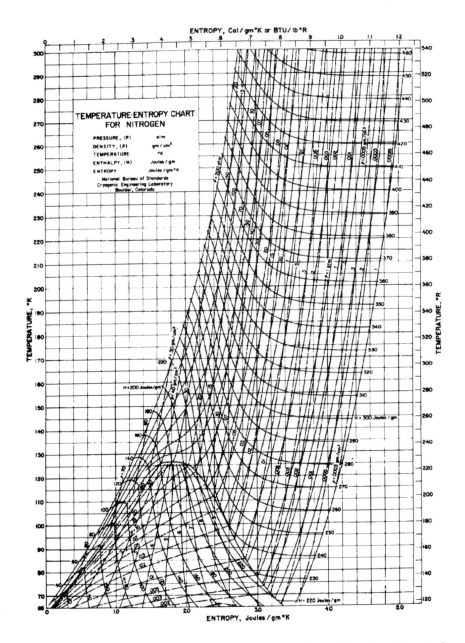

Figure 15-8 [Reprinted from *ASHRAE Handbook of Fundamentals*, 1981, with permission of the American Society of Heating, Refrigerating, and Air-Conditioning Engineers, Inc., Atlanta, GA.]

EXAMPLE 15-2

It is desired to assess the thermodynamic efficiency of the liquefaction process described in Ex. 15-1, and to estimate this, it is assumed that the compressor operates reversibly and adiabatically. Fresh nitrogen enters the process at 298 K and 1 atm, and ample cooling water is available at 298 K.

Solution 15-2

The process is visualized as shown in Fig. 15-9 where for each gram of nitrogen compressed, 0.052 g of gaseous nitrogen enter and 0.052 g of liquid nitrogen leave.

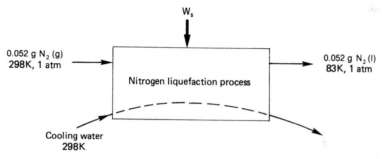

Figure 15-9

The work W_s is the work of compressing a gram of nitrogen from 2 to 200 atm in a three-stage isentropic process with interstage cooling that returns the gas to 300 K. The liquid nitrogen produced has the ability to do work because of its low temperature (e.g., a heat engine could receive heat at 298 K and reject heat to the low-temperature nitrogen). A reasonable efficiency would therefore be

$$\eta = \frac{\Delta B}{W_s}$$

where ΔB is the availability change of the 0.052 g of nitrogen. For each gram of nitrogen we have

$$\Delta b = h_6 - h_1 - 298(s_6 - s_1)$$

Using Fig. 15-8 to evaluate the thermodynamic properties gives

$$\Delta b = 45 - 460 - 298(0.65 - 4.42) = 708 \text{ J} / \text{g}$$

and

$$\Delta B = 0.052\, \Delta b = 0.052(708) = 36.8 \text{ J / g}$$

The work of compression could also be determined from Fig. 15-8; however, Eq. (3-25) should yield an acceptable estimate. With three stages the pressure ratio per stage is 4.65, and the work per stage is

$$(W_s)_s = \frac{1.40(8.3143)(300)}{1.40 - 1}\left(4.65^{(1.40-1)/1.40} - 1\right) = 4823 \text{ J / g mol}$$

The total work is

$$W_s = 3(4823) = 14{,}470 \text{ J / g mol} = 517 \text{ J / g}$$

and the efficiency is

$$\eta = \frac{36.8}{517} = 0.071$$

There are two major reasons for this low efficiency: the irreversible throttling process that wastes the ability of the gas to perform work and the failure to recover heat from the cooling water exiting the compressor interstage coolers and the after cooler.

The efficiency of the liquefaction process could be improved if the gas could be expanded isentropically (say in a turbine) instead of isenthalpically through a throttling valve. In Fig. 15-7 an isentropic process would be represented by a vertical line with point 5 lying directly below point 4 on the horizontal line connecting points 6 and 7. While this corresponds to a higher fraction liquefied, it presents a serious practical problem: the operation of a turbine with a two-phase fluid. With some sacrifice in potential process efficiency this problem may be circumvented by expanding a portion of the high-pressure stream through a turbine to a low temperature and using this gas to cool the remaining high-pressure gas prior to throttling.

A process utilizing this concept is shown in Fig. 15-10 and diagrammed in Fig. 15-11. A fraction x of the high-pressure stream at point 3 is expanded isentropically until the saturation curve is reached at point 7. This gas along with the uncondensed gas from the separator is used to cool the remaining fraction of high-pressure gas $1 - x$ by countercurrent heat exchange. The high-pressure gas is thereby cooled to point 4 and is subsequently throttled. A large fraction of the high-pressure stream is liquefied.

In analyzing this process, we write the first law for the dashed envelope of Fig. 15-10,

$$W_s = f\, h_6 + (1 - f)\, h_8 - h_3 \qquad (15\text{-}8)$$

and for the expansion turbine,

$$W_s = x(h_7 - h_3) \tag{15-9}$$

These equations are combined to give

$$f = \frac{h_8 - h_3}{h_8 - h_6} + x\left(\frac{h_3 - h_7}{h_8 - h_6}\right) \tag{15-10}$$

From comparison with Eq. (15-7) the first term is seen to be the fraction liquefied in a simple throttling process. Although Eq. (15-10) predicts increasing f with increasing x, there is a material balance constraint which requires that

$$f \leq 1 - x$$

In Fig. 15-11 point 4 has been located to ensure adequate temperature differences within the heat exchanger.[2] In comparing Figs. 15-11 and 15-7, it should be observed that a lower pressure is required at point 3 when the isentropic expansion is employed.

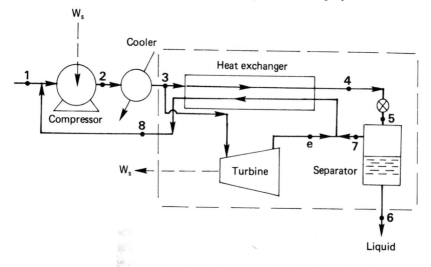

Figure 15-10 Liquefaction process with expansion turbine.

---

[2] See Ex. 14-5.

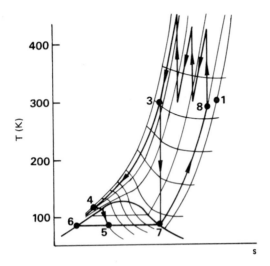

Figure 15-11 *Ts* diagram for liquefaction process with expansion turbine.

EXAMPLE 15-3

Nitrogen is to be liquefied using the process shown in Fig. 15-10 and diagrammed in Fig. 15-11. The high-pressure gas will be at 150 atm, and the low-pressure gas will be at 2 atm. On leaving the compressor the high-pressure gas will be cooled to 300 K, and the approach in the hot end of the heat exchanger T_3-T_8 will be 10 K. The unexpanded portion of the high-pressure gas will be cooled to 118 K in the heat exchanger, and the expanded gas leaves the turbine saturated at 2 atm. Based on a gram of nitrogen compressed, how much liquid is obtained? What fraction of the high-pressure gas is expanded?

Solution 15-3

Because Eq. (15-10) contains both f and x as unknowns, another equation relating these variables is required and is obtained by applying the first law to the countercurrent heat exchanger:

$$(1-x)(h_4-h_3)+(1-f)(h_8-h_7)=0 \qquad (15\text{-}11)$$

The enthalpies required for the solution of these equations are obtained from Fig. 15-8:

$$h_3=436 \quad h_4=120 \quad h_6=45 \quad h_7=234 \quad h_8=452$$

With these values, from the solution of Eqs. (15-10) and (15-11) we obtain

$$\ell = 0.294 \quad \text{and} \quad x = 0.513$$

EXAMPLE 15-4

Assess the thermodynamic efficiency of the liquefaction process described in Ex. 15-3. Assume reversible adiabatic compression and expansion.

Solution 15-4

The efficiency applied to Ex. 15-2 will be used, and again we have $\Delta b = 708$ J / g. Thus,

$$\Delta B = \ell \, \Delta b = 0.294(708) = 208 \text{ J}$$

The compressor operates between 2 and 150 atm, and with three stages the compression ratio is 4.22. The work of compression is

$$W_s = \frac{3(1.40)(8.3143)(300)}{1.40 - 1}\left(4.22^{(1.40-1)/1.40} - 1\right)$$

$$= 13,352 \text{ J / g mol} = 477 \text{ J / g}$$

The work of expansion is determined from Eq. (15-9):

$$W_s = 0.513(234 - 436) = -104 \text{ J}$$

The net work is

$$477 - 104 = 373 \text{ J}$$

and the efficiency of the liquefaction process is

$$\eta = \frac{208}{373} = 0.558$$

Many processes have been developed and are used to liquefy gases. These processes are comprised of the operations of compression, heat exchange, throttling, and expansion—they differ only in the selection and arrangement of these basic operations. Refrigeration processes are similarly constituted; they differ from liquefaction processes mainly in that they are cyclic and the working fluid generally remains at a temperature below its critical temperature.

15-4 REFRIGERATION

A refrigerator can be considered a heat engine operating in reverse, and we have already seen that the performance of reversible heat engines can be calculated independently of the actual cyclic process or working fluid. Thus, the ultimate performance of a refrigerator can be easily estimated. Here we wish to consider the nature of a few particular cycles and perform more detailed thermodynamic analyses in order to gain understanding of their operation and estimate their actual performances. Two broad types of refrigeration processes will be considered depending on whether the motive force is mechanical or thermal.

Mechanical Refrigeration.   The most common refrigeration cycle is the vapor-compression cycle. This process is sketched in Fig.15-12 and also represented on the Ts diagram of Fig.15-13. In step $4 \rightarrow 1$, heat is removed at the temperature T_1 from the system being refrigerated by the evaporation of a liquid under the pressure P_1. The saturated vapor at P_1 is then compressed isentropically to P_2, where it becomes superheated vapor, point 2. Removal of heat from this vapor leads to cooling followed by condensation at T_2, step $2 \rightarrow 3$. The cycle is closed by throttling the liquid to the lower pressure, point 4.

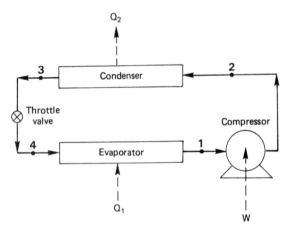

Figure 15-12
Vapor-compression refrigeration cycle.

The Ts diagram is convenient for visualizing the performance of heat engines because for reversible heat transfer $Q = \int T \, ds$, and a heat effect is the area under the curve representing the path. In Fig. 15-13 the heat rejected by the refrigerator is the area 2-3-a-c, which is negative. The heat removed from the system being refrigerated is the area 4-1-c-b and is positive. For the cyclic process the first law requires that

$$Q_1 + Q_2 = -W$$

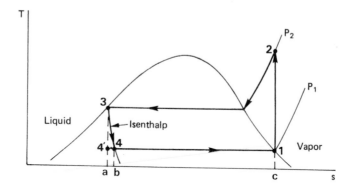

Figure 15-13 *Ts* diagram for a vapor-compression refrigeration cycle.

and the required work is seen to be the areas 1 - 2 - 3 - 4 and 3 - 4 - *b* - *a*. The traditional measure of efficiency, the coefficient of performance, was defined in Chap. 4 as

$$COP = \frac{|Q_1|}{W} \tag{4-25}$$

and can be seen to be the ratio of the area 4-1-*c*-*b* to the sum of the areas 1 - 2 - 3 - 4 and 3 - 4 - *b* - *a*.

Some improvement of the cycle efficiency could be gained by replacing the isenthalpic throttling step with an isentropic expansion. The cycle would then appear in Fig. 15-13 as 1 - 2 - 3 - 4′, and the coefficient of performance, represented now by the ratio of the area 4′ - 1 - *c* - *a* to the area 1 - 2 - 3 - 4′, would increase somewhat. While such an improvement is theoretically possible and is therefore appealing, usually practical considerations argue convincingly that the low cost, simplicity, reliability, and freedom from maintenance of the throttle valve justify its use. Also, compared with the work of compressing the vapor, the work lost in throttling the liquid is small. Thus, in all but large and specialized refrigeration processes the throttle valve is a ubiquitous element.

So far we have considered a cycle where all steps, save throttling, were reversible—a semireversible cycle. Let us now consider the effects of irreversible heat exchange and compression. These effects are illustrated in Fig. 15-14 where both the semireversible cycle (solid lines) and an irreversible cycle (dashed lines) are sketched on a *Ts* diagram. Heat is to be removed from the refrigerated system at T_1, but because a finite temperature difference is necessary for heat exchange, the actual working fluid temperature must be lower by the amount ΔT. Similarly, to reject heat to a sink at T_2, the working fluid temperature must be $T_2 + \Delta T$. The compression is still expected to be adiabatic, and therefore irreversible operation increases the entropy of the vapor and is manifested as additional superheat. As expected, Fig. 15-14 shows that the irreversible cycle requires more work to achieve a smaller cooling effect.

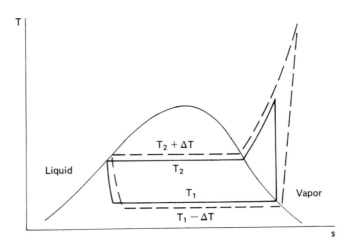

Figure 15-14 Effect of irreversibilities on the vapor-compression cycle.

EXAMPLE 15-5

A vapor-compression refrigeration process using ammonia as the working fluid is to operate between 20 and 80°F. Determine the coefficient of performance for (a) semi-reversible operation and (b) the case where a ΔT of 10°F is required and the adiabatic compression efficiency is 75%. Compare the results with the COP of a Carnot refrigerator.

Solution 15-5

Data for the solution of the problems will be taken from the Ts diagram for ammonia, Fig. 8-5, and all calculations are based on a pound of ammonia. From Eq. (4-25) the COP of a Carnot refrigerator is

$$\text{COP} = \frac{T_1}{T_2 - T_1} = \frac{460 + 20}{60} = 8.00$$

From Fig. 8-5, for the semireversible cycle we find that

$$h_4 = 136 \text{ Btu}$$

$$h_1 = 620 \text{ Btu}$$

$$h_2 = 684 \text{ Btu}$$

and calculate

$$Q_1 = -(h_1 - h_4) = 136 - 620 = -484 \text{ Btu}$$

$$W = h_2 - h_1 = 684 - 620 = 64 \text{ Btu}$$

$$\text{COP} = \frac{|Q_1|}{W} = \frac{484}{64} = 7.56$$

For the irreversible cycle where $T_1 = 10°\text{F}$ and $T_2 = 90°\text{F}$, from Fig. 8-5 we find that

$$h_4 = 150 \text{ Btu}$$

$$h_1 = 616 \text{ Btu}$$

which yields $Q_1 = -(h_1 - h_4) = -466 \text{ Btu}$. An isentropic compression between the saturation pressure at 10°F (40 psia) and the saturation pressure at 90°F (180 psia) requires that

$$W = \Delta h = 716 - 616 = 100 \text{ Btu}$$

and with a compression efficiency of 75% the actual work is

$$\frac{100}{0.75} = 133 \text{ Btu}$$

The COP for the irreversible process is then

$$\text{COP} = \frac{|Q_1|}{W} = \frac{466}{133} = 3.50$$

The closeness of the COP for the reversible and semireversible processes suggests that little work is wasted in throttling the liquid; irreversible heat exchange and vapor compression are seen to be much more detrimental to the efficiency.

Absorption Refrigeration. The work required to drive a refrigeration process can be generated within the process itself by a "built-in" engine which receives and rejects heat. Such a refrigeration process, the ammonia absorption process, is shown in Fig. 15-15 where it is seen that to the left of the dashed line the process is identical to the vapor-compression process. An absorber-generator arrangement replaces the compressor. Low-pressure ammonia vapor from the evaporator is absorbed in the absorber to form concentrated aqueous ammonia solution, and the heat of absorption Q_1' is removed. This solution is sent to the generator where its temperature is raised and ammonia vapor is stripped out of solution at a higher pressure. Due to the pressure difference, a small amount of mechanical energy is required to pump the solution from the absorber to the generator. Some of the heat needed for stripping comes from heat exchange with stripped solution flowing back to the absorber, but most is supplied externally, Q_2'.

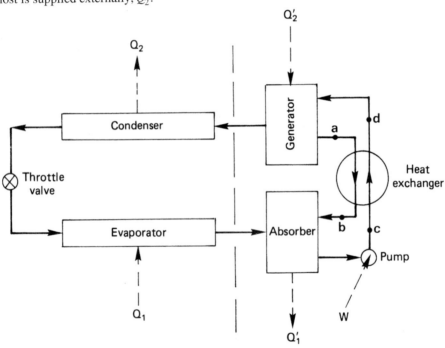

Figure 15-15 Ammonia absorption refrigeration process.

Ammonia is so much more volatile than water that the vapors from the generator are practically pure ammonia. The pressure of the ammonia vapor entering the condenser is the partial pressure of ammonia over the aqueous solution. The aqueous ammonia concentrations in the absorber and generator are determined by their pressures, which are set by the temperatures in the evaporator and condenser, respectively, and by their temperatures, which are set by available cooling and heating resources. While a material balance around the generator dictates that the aqueous ammonia concentration must be higher in the absorber, the partial pressure of ammonia (and hence the total pressure) in the generator is higher due to its higher temperature. As an illustration we will set conditions in the absorber and generator to be 40 psia, 80°F and 180 psia, 190°F, respectively. In Fig. 15-16 the partial pressure isotherms for 80 and 190°F are shown with point *a* representing conditions in the generator and point *c* the absorber. These concentrations are 40 and 44%, respectively. For a given flow of ammonia through the evaporator these concentrations determine the circulation rate of solution in the absorber-generator unit. When these concentrations are close, a large circulation rate is needed.

The requirement that the aqueous ammonia concentration in the absorber exceeds that in the generator imposes a limit to the operation of the process which corresponds to a minimum generator temperature or a minimum temperature difference between generator and absorber. As an illustration, consider the pressures to be 40 and 180 psia, and assume that the absorber can be maintained at 80°F. Figure 15-16 shows the 173°F isotherm where at a pressure of 180 psia (point *a'*) the concentration in the generator is 44%—the same as that in the absorber. Thus, 173°F is the minimum required temperature for the generator

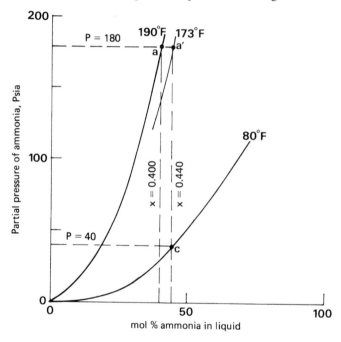

Figure 15-16 Partial pressures of ammonia over aqueous solutions. [Data from J. H. Perry and C. H. Chilton, *Chemical Engineers' Handbook*, 5th ed., McGraw-Hill, New York, 1973.]

for these conditions. It should be noted that this limitation is not of a general thermodynamic origin but is specific to this particular device selected to perform the task. A similar situation was observed in Sec. 14-4 where the limitations of a particular device, a heat exchanger, were found to be more stringent than those imposed by the laws of thermodynamics.

By means of an ingenious design the need for the pump between the absorber and generator can be eliminated. The Servel-Electrolux ammonia absorption refrigeration system shown in Fig. 15-17 has no moving parts. Ammonia vapor, stripped from solution in the generator, entrains slugs of liquid by vapor-lift action to a higher elevation where it flows by gravity through the absorber and back to the generator. The total pressure in the system is maintained constant by the presence of an inert gas (hydrogen). However, the partial pressure of ammonia, which determines the temperature for the phase change, is low in the evaporator and high in the air-cooled condenser. Circulation of gas through the evaporator-absorber loop as shown is by natural convection. This is caused by the gas leaving the absorber being leaner in ammonia and hence less dense than the gas entering the absorber.

Another system with no moving parts uses water as the working fluid and an aqueous salt solution, usually lithium bromide, as an absorbent. In the generator heat is provided to

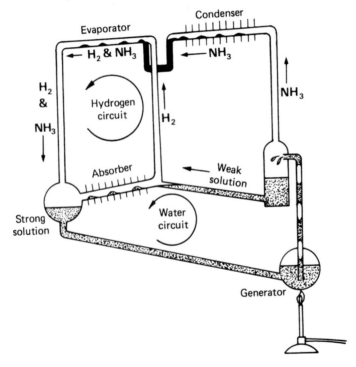

Figure 15-17 Servel-Electrolux refrigerator.

vaporize water from the solution at a pressure high enough to allow subsequent condensation at ambient temperature. This liquid water is then vaporized in the evaporator at a lower pressure which is maintained by the temperature and concentration of solution in the absorber. This system operates at pressures considerably below atmospheric, and flows are induced by vapor-lift action, pressure difference, and hydrostatic head.

15-5 HEAT PUMPS

As shown in Chap. 4, a heat pump is a heat engine operated in reverse manner. It differs from a refrigerator (see Fig. 15-12) only in that the high-temperature reject heat is desired for heating rather than the low-temperature input heat being desired for cooling. Thus, any of the devices or cycles discussed and analyzed in the previous section can also function as heat pumps. For residential winter heating the pump would take its low-temperature heat for the evaporator from the outside air or well water and would supply the house with heat from the condenser. Some units can function as either a heat pump or an air conditioner. Figure 15-18 shows how such a unit might operate.

Industrial applications of the heat pump principle are likely to involve open rather than closed cyclic systems. Vapor compression, illustrated in Ex. 14-2, is sometimes used to reduce energy requirements for evaporation and distillation. This same principle applied to upgrading low-pressure steam is illustrated in Fig. 15-19. In this process the compressor prime mover is a diesel engine, and the waste heat in the engine exhaust and coolant is effectively utilized.

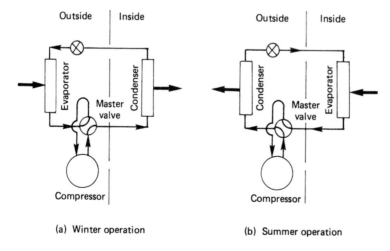

(a) Winter operation   (b) Summer operation

Figure 15-18 Operation of a combination heat pump–air conditioner.

15-6 POWER GENERATION

The generation of power can be viewed from two perspectives: as a cyclic process that converts heat into work or as a steady-state flow process with inflows of fuel and air and outflows of flue gas, power, and waste heat. We have already seen that the Carnot cycle is the paragon by which other cycles are judged; however, this perspective suffers a limitation arising from an indifference to the origin of the input heat source. Thus, while a cyclic process might be developed that would convert the heat of combustion of a fuel into work at an efficiency approaching the Carnot limit, it is still appropriate to take the broader perspective and examine the advisability of the irreversible oxidation of the fuel.

If the fuel could be oxidized in a reversible manner, the amount of work obtained would equal $-\Delta B$ for the oxidation reaction. If, however, combustion occurs irreversibly, only the heat of combustion is available for use and in a Carnot cycle would provide work equal to $-\Delta H°(T-T_0)/T$. The difference between these two quantities is the lost work, $T_0\sigma$, due to combustion of the fuel and is

$$T_0\sigma = -\frac{T_0\Delta G}{T}$$

In arriving at this expression, it was assumed that while the reaction proceeds irreversibly, the heat of reaction is utilized in a reversible manner. The maximum work realizable from the fuel is its chemical work equivalent $\mathbf{e}°$, and the fraction of this work that is lost due to irreversible execution of the reaction is

$$\frac{T_0|\Delta G|}{T\mathbf{e}°}$$

This fraction has been calculated[3] for the oxidation of methane, and its complement, the fraction of available work still realizable from the irreversible execution of the reaction, is plotted vs. reaction temperature in Fig. 15-20. In this figure it is seen that this fraction increases rapidly with increasing temperature up to about 1000 K and then rises more gradually at higher temperatures. At 1000 K approximately 72% of the maximum possible work could still be realized from the irreversible combustion of methane.

---

[3] Included in $\mathbf{e}°$ is work that could be realized from concentration differences. In computing the lost work $T_0\sigma$, this was ignored, as was the work available from cooling the reaction products to T_0. Because these terms are small in comparison to the heat of reaction, little error arises from their omission, and the quantity $T_0|\Delta G|/T\mathbf{e}°$, while not exact, is nevertheless quite useful.

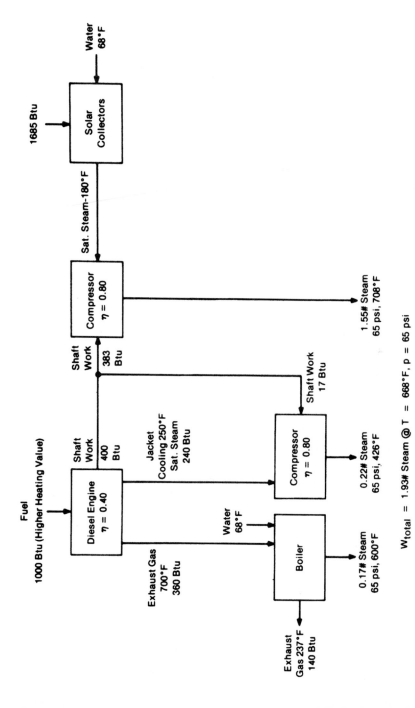

Figure 15-19 Diesel-solar steam generation process. (From E. P. Gyftopoulous, L. J. Lazaridis, and T. F. Widmer, *Potential Fuel Effectiveness in Industry*, Ballinger Publishing Co., 1974, with permission of the Ford Foundation.)

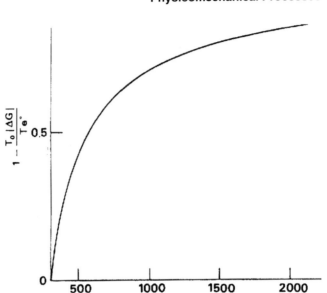

Figure 15-20 Fraction of fuel work equivalent realizable as work after irreversible combustion.

Clearly, there is great potential for power generation from fuel combustion if the cycle can accept heat at very high temperatures. In modern power plants the working fluid accepts heat over a temperature range rather than at a constant temperature. Materials of construction and economic considerations usually impose a practical upper temperature limit of about 900 K, and consequently the average temperature of the working fluid is considerably lower than 900 K. Thus, when operated under optimum conditions, a modern power plant does well to convert 40% of the available work of the fuel (i.e., $e°$) into work (electrical energy).

The Rankine Cycle. Most large modern power plants generate power from fuel combustion heat by means of a Rankine cycle with water as the working fluid.[4] There are many variations on this cycle, but basically the equipment consists of the four elements shown in Fig. 15-21, and the cycle can be delineated in four steps as shown on the Ts diagram of Fig. 15-22.

---

[4] The tutorial RANKINE that runs in the software EQUATIONS OF STATE on accompanying CD-ROM describes a modern-day power cycle.

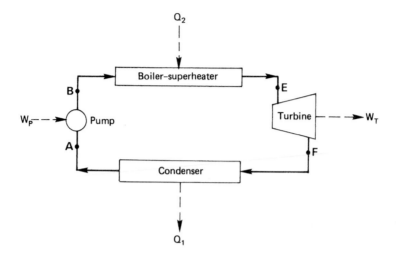

Figure 15-21 Rankine cycle.

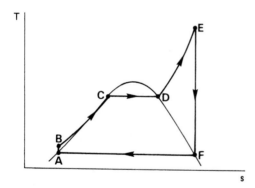

Figure 15-22 *Ts* diagram for the Rankine cycle.

Input heat is used to produce high-pressure superheated steam from liquid water. This heating follows approximately a constant-pressure path and is shown in Fig. 15-22 as $B \to C$, heating the liquid to saturation; $C \to D$, vaporizing the liquid; and $D \to E$, superheating the vapor. The steam then passes through a turbine, $E \to F$, and produces work. This step is shown as isentropic and ending at saturation, although it is not uncommon for the exit steam to be slightly wet. Heat necessary to condense the steam is removed in the condenser, $F \to A$. The turbine exhaust pressure is determined by the pressure in the condenser, which is the saturation pressure of water at the condenser temperature; this in turn is determined by the temperature of the available coolant. A pump is required to deliver liquid to the boiler against the pressure difference, $A \rightarrow B$. The work required for pumping is quite small compared to the work obtained from the turbine.[5]

On the Ts diagram the inflowing heat Q_2 is represented by the area under the path $BCDE$ and is positive and the outflowing heat Q_1 by the area under AF which is negative. For a complete cycle,

$$\Delta U = 0$$

which leads to

$$Q_1 + Q_2 = -(W_T + W_P) = -W_{net}$$

and it is seen that the net work is represented by the area enclosed by the cycle $ABCDEF$. Note that while the working fluid attains a high temperature at E, Fig. 15-22 shows that a major portion of the input heat goes to produce saturated steam at a much lower temperature. This is the major reason for the rather low efficiency of this cycle—about 40% as opposed to 67% for a Carnot cycle operating between the maximum temperature of 900 K and a sink at 300 K or 68% as determined at 900 K from Fig. 15-20. Modifications such as turbine staging with interstage reheating produce only slight increases in efficiency.

Uncertainty always surrounds the use of the term *efficiency* because there is no unique thermodynamic prescription for it, nor is there a single widely accepted convention. Yet it would seem that if the efficiency serves to compare performance against the thermodynamically permissible limit, then it should be based on the availability. For power plant efficiency this would suggest that

$$\eta = \frac{\text{power output}}{\text{rate of fuel comsumption} \times \mathbf{e}^{\circ}_{fuel}}$$

where $\mathbf{e}^{\circ}_{fuel}$ is the work equivalent of the fuel or the availability change between the fuel in its original state and the oxidation products in a dead state. Because $\mathbf{e}^{\circ}_{fuel}$ is approximately equal to the heat of combustion (cf. Sec. 14-5), the denominator will approximate the rate of heat input to the working fluid, and, as we have just seen, the Carnot efficiency based on the maximum temperature and the availability-based efficiency will be essentially equal. Despite this concurrence, it should be remembered that applying the Carnot efficiency as a measure of power plant performance implies the degradation of fuel energy to heat and thus imposes a rather restricted perspective on energy utilization.

The Gas Turbine.   For smaller power plants and for meeting peak power demand in larger power plants, the gas turbine is often used. This device functions as an open system, as shown in Fig. 15-23, with inflows of air and fuel and outflows of combustion gas and power. A high-temperature, pressurized gas is produced by supplying compressed air and fuel to the combustion chamber. This gas does work in expanding through the turbine. Some

---

[5] For illustrative purposes the $A \rightarrow B$ step has been greatly exaggerated in Fig. 15-22. In actuality, the path ABD would be practically indistinguishable from the liquid saturation curve.

of the work is used to drive the compressor, which is usually mounted on a common shaft with the turbine. Greater efficiency is obtained by adding a countercurrent heat exchanger, known as a regenerator, to heat the compressed air with exhaust gases, as shown in Fig. 15-23(b).

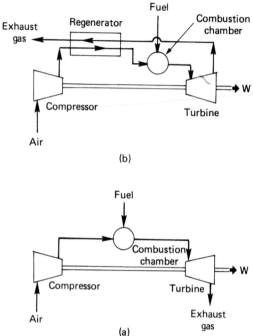

(b)

(a)

Figure 15-23 Gas turbine (a) with regenerator (b).

15-7 COGENERATION OF STEAM AND POWER

With the awareness of relentlessly rising energy cost, the chemical, petroleum, and related process industries are beginning to consider and exploit opportunities for more effective energy utilization. While each industry has its own unique energy requirements and conservation opportunities, here we will consider a few conservation measures of general applicability under the rather broad heading of cogeneration.[6]

---

[6] This section is based largely on the work of E. P. Gyftopoulos, L. J. Lazaridis, and T. F. Widmer, *Potential Fuel Effectiveness in Industry*, Ballinger Publishing Co., Cambridge, MA, 1974. They also present detailed energy use analyses for several specific industries.

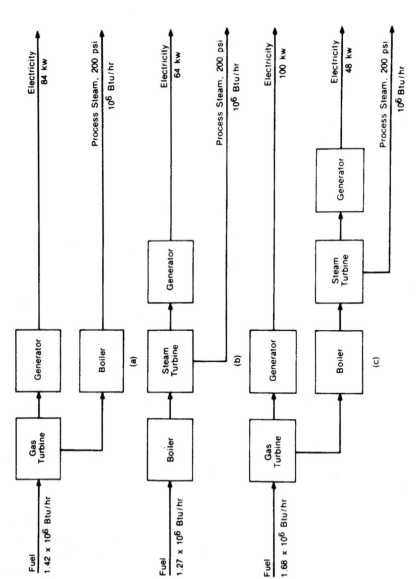

Figure 15-24 Three topping cycles. (From E. P. Gyftopoulos, L. J. Lazaridis, and T. F. Widmer, *Potential Fuel Effectiveness in Industry*, Ballinger Publishing Co., Cambridge, Mass., 1974, with permission of the Ford Foundation.)

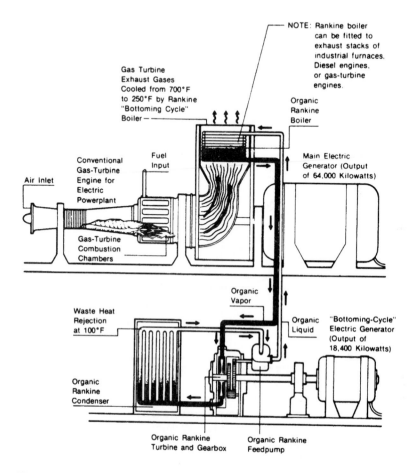

Figure 15-25 Gas turbine with organic Rankine bottoming cycle. [From E. P. Gyfto-
polos, L. J. Lazaridis, and T. F. Widmer, *Potential Fuel Effectiveness in
Industry*, Ballinger Publishing Co., Cambridge, MA, 1974, with permission
of the Ford Foundation.]

Topping Cycles. The practice of raising steam by the combustion of fuel in a boiler
is widespread, and often the efficiency and profitability of a plant or process can be improved by generating power as well as steam. Many processes do not require high-pressure
steam, and it is possible to *top* the high-quality energy from the fuel combustion by generating power. Three possible topping schemes are shown in Fig. 15-24. In schemes *b* and *c*,
steam is extracted from a turbine at the point where its pressure is that required for process
steam. Such turbines are often called extraction turbines. In this figure the data are based on
raising 10^6 Btu/h of 200-psi steam for process requirements. Of course, more electricity
could be generated if the process could accept lower-pressure steam.

Bottoming Cycles.   A bottoming cycle uses relatively low-level heat, often reject process heat, to generate power, usually by means of a Rankine cycle with an organic working fluid. While the organic Rankine engine has only recently been developed and therefore may not yet have gained wide acceptance, there are many opportunities where it could be profitably employed.[7] One such application, an organic Rankine engine operating on the exhaust of a gas turbine, is shown in Fig. 15-25. This combination is reported[8] to have an efficiency of 47%.

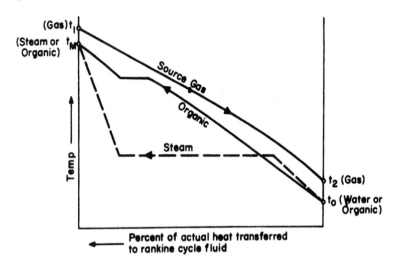

Figure 15-26 Source gas and Rankine cycle temperature profiles. [From E. P. Gyftopoulos, L. J. Lazaridis, and T. F. Widmer, *Potential Fuel Effectiveness in Industry*, Ballinger Publishing Co., Cambridge, MA, 1974, with permission of the Ford Foundation.]

One reason for the efficacy of the organic Rankine engine is discernible from Fig. 15-26, which shows temperature profiles in the boiler-superheater for the organic working fluid and, for comparison, water. Because of a relatively small latent heat, it is observed that the temperature profile of the organic fluid can be closely matched with that of the source gas. Thus, the heat transfer is efficient.

---

[7]By the year 2000 Israel plans to generate 2000 MW of electricity via organic Rankine engines with heat supplied from salt gradient solar ponds [T. M. Maugh, *Science, 216*, 1213 (1982)].
[8] Gyftopolous et al., *ibid.*

PROBLEMS

15-1. For a three-stage compression of oxygen from 1 atm and 80°F to a pressure of 1500 psia with equal pressure ratios per stage, calculate the reversible work, and compare it to that of an isothermal compression to the same pressure. Assume the compression steps for each stage are reversible and adiabatic and that interstage cooling returns the gas to its original temperature. Make the calculations for

    (a)   ideal-gas behavior with a constant c_p of 7 Btu/lb mol °F.
    (b)   the real gas, using Fig. 8-6.

15-2. Using the corresponding states, generalized property correlations, what generalizations can be made regarding the work of compression of a real gas compared to that of an ideal gas in each case?

    (a)   for a reversible isothermal compression.
    (b)   for a reversible adiabatic compression.

15-3. In liquefying nitrogen by the process shown in Fig. 15-10, determine the value of x which produces the maximum fraction liquefied. Is this also the most efficient operating condition?

15-4. Determine the efficiency of the liquefaction process described in Ex. 15-4 when the adiabatic efficiencies of the compressor and turbine are 70%. *Note*: An irreversible adiabatic expansion to a pressure of 2 atm will not terminate on the saturation curve.

15-5. Show that the sign of the slope of an isenthalp on a Ts diagram depends on the sign of the Joule-Thomson coefficient and that a maximum in the isenthalp occurs at a point on the Joule-Thomson inversion curve.

15-6. Use Fig. 8-5 to answer the following:

    (a)   Ammonia gas at 1500 psia and 265°F is throttled to a pressure of 50 psia. What is the final state?
    (b)   Ammonia gas at 100 psia and 360°F is throttled to a pressure of 5 psia. What is the final state?
    (c)   Calculate the minimum work required to compress ammonia vapor adiabatically from $P_1 = 20$ psia, $t_1 = 100$°F to a pressure of 100 psia.

15-7. It has been suggested that ammonia can be liquefied in a steady-state flow process by compressing the gas followed by a constant-pressure cooling and condensation step. With the available cooling water it is estimated that the ammonia could reach a temperature of 100°F. Refer to Fig. 8-5 and answer the following.

    (a)   To what pressure must we compress the ammonia in order to obtain liquid in the cooling-condensation step?
    (b)   If we start with gaseous ammonia at 50 psia and 100°F, how much work would be required in a reversible, single-stage, adiabatic compression to the pressure of part (a)?
    (c)   How much heat must be removed in the cooling-condensation step?

15-8. In a liquefaction process nitrogen gas at $-100°F$ and 200 atm is expanded reversibly through an adiabatic turbine to saturation and is then allowed to pass through a throttle valve to atmospheric pressure.

    (a)   What fraction of the nitrogen is liquefied?
    (b)   A heat exchanger will be placed between the turbine and throttle valve so that the uncondensed low-pressure stream from the separator can cool the high-pressure stream. The heat exchanger will operate with a hot-end approach temperature of $10°F$. What fraction of the nitrogen can now be liquefied?

15-9. One of the processes in our plant uses ammonia as a raw material. It is stored as a liquid at ambient temperature and is vaporized in a steam-heated heat exchanger before being fed into the process at essentially atmospheric pressure. Someone has suggested that this is an untapped source of work and that the ammonia could be vaporized and heated at its saturation pressure and then expanded through a turbine to atmospheric pressure. You are asked to explore this possibility.

Consult the Ts diagram for ammonia, Fig. 8-5, and design a system consisting of a steam-heated heat exchanger followed by a turbine. The exhaust from the turbine should be saturated vapor at 15 psia and the turbine is assumed to be both reversible and adiabatic. Determine the following:

    (a)   the temperature and pressure of the ammonia vapor prior to entering the turbine.
    (b)   the amount of work available from the turbine in Btu per lb of ammonia.
    (c)   the heat added in the heat exchanger in Btu per lb of ammonia.

15-10. Refine your design from Prob. 15-9 by using a turbine efficiency, η_s, of 70% and a condensing steam temperature $20°F$ higher than the vapor stream entering the turbine. Redetermine parts (a), (b), and (c) of Prob. 15-9 and also determine a thermodynamic efficiency for your refined design.

15-11. Your officemate, noticing your interest in utilizing the work potential of the liquid ammonia mentioned in Prob. 15-9, reminds you that the company plans a new office complex adjacent to the plant that will require a source of chilled water for air conditioning. Explore this possible use for the liquid ammonia by sketching out a process for producing chilled water at $50°F$ from water taken from the cooling tower at $80°F$. How many gallons of chilled water could be obtained per pound of liquid ammonia?

15-12. Ammonia is to be liquified in a two-step, steady-state process. In the first step ammonia gas at $75°F$ and 30 psia is compressed adiabatically and reversibly and in the second step it is cooled and condensed at constant pressure in a heat exchanger. With the available cooling water it is estimated that the condensation temperature will be $80°F$. Refer to Fig. 8-5 and answer the following.

    (a)   To what pressure must the ammonia be compressed in order that it can be liquified at a temperature of $80°F$?
    (b)   How much work is required in the compression step per lb of ammonia?
    (c)   What will be the temperature of the gas leaving the compressor?
    (d)   How much heat (Btu/lb ammonia) must be removed in order to condense the gaseous ammonia leaving the compressor?

15-13. The following process has been suggested for air conditioning a house. Air from the house at $75°F$ is compressed adiabatically, cooled to $100°F$ by heat exchange with outside air, and then expanded adiabatically through a turbine. The work from the turbine is applied to the compression step. The air leaves the turbine at 55°F and enters the house's air-handling ductwork. Calculate a COP for this system for the following cases:

(a) the compressor and turbine operate reversibly and air is an ideal gas.
(b) the compressor and turbine each have an efficiency η_s of 70% and air is an ideal gas.

15-14. Recalculate the efficiency of the heat pump in Ex. 4-7 for an ammonia vapor-compression cycle with $\Delta T = 15°F$ and a compressor efficiency of 75%.

15-15. We have a source of flue gas at 420°F and a plentiful supply of cooling water at 80°F and want to consider the possibility of operating a Rankine cycle using ammonia as the working fluid. In terms of Fig. 15-21 we plan to have $P = 1500$ psia, $t = 400°F$ at point E, and a temperature of 100°F at point A. The turbine should operate adiabatically at an efficiency of 75%, and in the boiler (see Fig. 15-26) we require that the Δt be no less than 15°F at any location. Estimate the available horsepower obtainable from 640 ft³/s of flue gas. For the flue gas assume c_p is constant at 7.00 Btu/lb mol °F. Properties of ammonia can be obtained from Fig. 8-5.

15-16. A Brayton cycle consists of two constant-pressure heat exchanges and two reversible adiabatic changes of pressure with an ideal gas as the working fluid (see Fig. 15-27). After being compressed adiabatically and reversibly from P_1 to P_2, the gas receives heat at the constant pressure P_2 as its temperature increases from T_2 to T_3. Next, the gas is expanded reversibly and adiabatically to P_1, where its temperature becomes T_4. Finally, the gas is cooled from T_4 to T_1 at the constant pressure P_1.

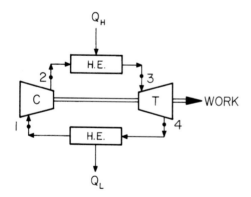

Figure 15-27 Vapor-compression refrigeration cycle.

(a) Sketch this cycle on Pv and Ts coordinates and compare with a Carnot cycle.
(b) If possible, determine the efficiency of this cycle in terms of characteristic temperatures.

15-17. A gas turbine of the type shown in Fig. 15-23(a) uses n-C_8H_{18}(l) as a fuel with gases leaving the combustion chamber at 1200°F. Air is compressed from 80°F and 14.7 psia to 90 psia prior to entering the combustion chamber and the combustion gas is expanded from 90 to 14.7 psia. Assume reversible adiabatic operation of compressor and turbine and neglect the work required to pump the liquid fuel into the combustion chamber. Assume air and combustion gas have a constant c_p of 7 Btu/lb mol °F and calculate an efficiency

(a) based on the chemical work equivalent, $e°$, of the fuel.

(b) based on the maximum amount of work available after combustion (Fig. 15-20).

15-18. Recalculate the efficiency of the gas turbine of Prob. 15-17 if the compression and expansion are adiabatic but the efficiency of each is 80%.

15-19. Determine a second-law efficiency for the diesel-solar steam generation process shown in Fig. 15-19. Count only steam as output and consider solar energy to be free.

15-20. For the combustion chamber of a gas turbine operating at a specified pressure and fuel rate, the combustion gas temperature decreases as its flow rate (and the air flow rate) increases. Accompanying this increased flow rate, however, is a decrease in the work done by a mol of combustion gas. The net work based on a unit of fuel depends on both of these opposing factors and it seems likely that there may be a combustion gas temperature that maximizes the efficiency of the gas turbine. Determine if this is the case for the system described in Prob. 15-17.

15-21. A gas turbine with regenerator (see Fig. 15-23b) uses liquid methanol, $CH_4O(l)$, as a fuel with a combustion chamber operating at 90 psia and 1600°F. The regenerator operates with a hot-end Δt of 20°F. Air enters the compressor at 80°F and 14.7 psia where it is compressed to 90 psia.

(a) Calculate the work delivered by the gas turbine in Btu's per lb mol of fuel.

(b) Calculate the maximum work available from one lb mol of the fuel and use this figure to evaluate the efficiency of the gas turbine.

Assume no pressure drops in the combustion chamber or regenerator, adiabatic operation of compressor and turbine each with an efficiency of 75%, ideal-gas behavior, and c_p constant at 7 Btu/lb mol °R for all gas streams and c_p constant at 0.5 Btu/lb °R for the liquid fuel. The fuel enters the combustion chamber at 80°F and the work of pumping it into the chamber can be ignored.

15-22. A simple gas turbine, such as shown in Fig. 15-23 (a), uses liquid $n\text{-}C_8H_{18}$ as fuel. For a fixed combustion gas temperature of 1400°F, determine the effect of combustion chamber pressure on the work output per mol of fuel. Assume reversible adiabatic operation of compressor and turbine, ideal-gas behavior, and c_p constant at 7 Btu/mol °F for both air and combustion gas.

15-23. A gas turbine with regenerator [see Fig. 15-23(b)] uses liquid $n\text{-}C_8H_{18}$ as a fuel with a combustion chamber operating at 90 psia and 1600°F. The regenerator operates with a hot-end temperature difference of 20°F. Air enters the compressor at 80°F and 14.7 psia. Calculate the work produced per lb mol of fuel. Assume no pressure drops in combustion chamber or regenerator, reversible adiabatic operation of compressor and turbine, ideal-gas behavior, and c_p constant at 7 Btu/lb mol °F for both air and combustion gas.

15-24. Recognizing that combustion gas leaves the regenerator at a fairly high temperature, it has been suggested that more heat could be recovered if heat were exchanged between gas leaving the turbine and air before it enters the compressor (at 14.7 psia and 80°F). The combustion chamber will operate at 90 psia and 1600°F using liquid $n\text{-}C_8H_{18}$ as fuel. The regenerator will be operated with a hot-end temperature difference of 20°F. Comment on the efficiency of this arrangement using the same assumptions as in Prob. 15-23.

15-25. A desert pumping station for a natural gas pipeline uses a gas turbine to drive the compressor. The gas, assumed to be pure methane, is compressed from 30 to 90 psia in a compressor that operates adiabatically with an efficiency of 80%. The pipeline gas is the fuel for the gas turbine which operates with a combustion chamber at 1600°F and 90 psia and a hot-end temperature difference of 20°F in the regenerator [see Fig. 15-23(b)]. Ambient air and methane entering the compressor are both at 110°F. Assume no pressure drops in the combustion chamber or regenerator, adiabatic operation of compressor and turbine at 80% efficiency, ideal-gas behavior, and constant heat capacities, and calculate the gas rate leaving the pumping station as a fraction of the entering gas rate. Use the following c_p's:

air and combustion gas: 7.0 Btu/lb mol °F.
methane: 9.0 Btu/lb mol °F .

15-26. Consideration is being given to the use of propane, C_3H_8, in an absorption refrigeration system with a nonvolatile hydrocarbon as solvent. Ideal-liquid solutions are expected and the following operating temperatures apply.

Condenser: 80°F
Evaporator: 20°F
Generator: 200°F
Absorber: 80°F

Based on 1000 Btu's removed in the evaporator, calculate

(a)   the solvent circulation rate.
(b)   the heat supplied to the generator.
(c)   the pumping work.

Physical properties of the solvent are

molecular weight: 170
liquid density: 0.75 g/cm^3
liquid c_p: 0.50 cal/g °C

Thermodynamic properties of propane are available in R. H. Perry, D. W. Green, and J. O. Maloney, eds., *Perry's Chemical Engineers' Handbook*, 6th ed., McGraw-Hill, New York, 1984, Chap. 3.

C H A P T E R **1 6**

Compressible Fluid Flow

It has already been shown in Chaps. 2 and 4 that certain problems involving flowing fluids can be solved by the application of the first and second laws of thermodynamics. These problems are characterized by the use of thermodynamic property changes which depend only on initial and final state and hence by an implied indifference to the exact nature of processes occurring within the system. In contrast, the emphasis of fluid mechanics is on a detailed description of these processes in terms of the laws of physics and the use of these laws to synthesize a solution to the problem under study. The connection between these two subjects arises from the inclusion of the first and second laws of thermodynamics among those laws of physics which find application in fluid mechanics. It is the object of this chapter to demonstrate the role played by the laws of thermodynamics in establishing some of the basic equations of fluid mechanics and in providing insight into various fluid-flow phenomena. Problems involving high-velocity flow of a compressible fluid require more thermodynamic insight than the more familiar problems involving incompressible fluids and therefore will be the focus of this chapter. Fluid mechanics can be a very sophisticated subject, requiring considerable mathematical proficiency; nevertheless, it is possible to demonstrate the role of thermodynamic reasoning with the use of some rather unsophisticated fluid mechanics.

16-1 THE BASIC EQUATIONS OF FLUID MECHANICS

The four basic equations which form the foundation of fluid mechanics are

1. The first law of thermodynamics
2. The second law of thermodynamics
3. The law of conservation of mass
4. Newton's second law of motion

When written in differential form and applied to a small volume of fluid, these laws are referred to as microscopic balances. When the microscopic balances are integrated over the volume and area of an open system, the resulting equations are identified as macroscopic balances. Equation (2-17) applies to an open system and is therefore a macroscopic energy balance.

To complete the mathematical description of a fluid mechanics problem, it is often necessary to employ certain ancillary relationships such as Newton's law of viscosity, Fourier's law of heat conduction, and an equation of state for the fluid.

The problems chosen for study in this chapter deal with steady-state flow and are of such a nature that the assumptions of one-dimensional flow and negligible body forces (e.g., gravity and buoyancy) are appropriate. Rather than tackle the formidable task of deriving the most general form of the macroscopic balances for subsequent simplification, we choose the easier task of deriving the much simpler equations which are specific to these systems.

The Equation of Continuity.   For the systems we have selected for study, the equation of continuity, or macroscopic mass balance, can be written as

$$\dot{m} = \rho \mathbf{v} A = \text{constant} \tag{16-1a}$$

or

$$\rho_1 \mathbf{v}_1 A_1 = \rho_2 \mathbf{v}_2 A_2 \tag{16-1b}$$

where $\dot{m}$ is the mass flow rate, ρ is the fluid density, A is the cross-sectional area for flow, and $\mathbf{v}$ is velocity. The subscripts 1 and 2 identify two arbitrarily located planes normal to flow.

The Momentum Balance. Newton's second law of motion states that the sum of forces on a body is equal to the time rate of change of momentum. Figure 16-1 illustrates the application of this law for the case of steady-state, one-dimensional flow in a conduit. The system is the fluid contained between planes 1 and 2 which experiences surface forces due to pressure and friction but no body forces. The pressure force acts on the cross-sectional area A, while the frictional force $\mathbf{F}_f$ acts on the wetted area of the conduit (πDL for a circular conduit), and the sum of forces is

$$P_1 A_1 - P_2 A_2 - \mathbf{F}_f$$

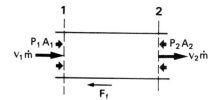

Figure 16-1 Illustration of momentum balance.

The rate of change of momentum is the difference in the outflow and inflow of momentum through the space between planes 1 and 2. And for this simple system we write Newton's second law of motion as

$$P_1 A_1 - P_2 A_2 - \mathbf{F}_f = \dot{m}\mathbf{v}_2 - \dot{m}\mathbf{v}_1 \tag{16-2}$$

This equation is often termed a macroscopic momentum balance.

Mechanical Energy Balance. The first and second laws of thermodynamics may be combined to yield a macroscopic mechanical energy balance. We begin with the steady-state version of the first law,

$$\Delta H + \Delta E_P + \Delta E_K = Q + W_s \tag{2-18}$$

and utilize Eqs. (2-22)–(2-24) to write

$$\Delta h + \mathbf{g}\,\Delta z + \frac{\Delta \mathbf{v}^2}{2} = q + w_s \tag{16-3a}$$

where it is understood that q and w_s are heat and work effects per unit mass of flowing fluid. It is convenient to write this in differential form,

$$dh + \mathbf{g}\,dz + \frac{1}{2}d\mathbf{v}^2 = dq + dw_s \qquad (16\text{-}3b)$$

and make the substitutions

$$dh = T\,ds + v\,dP \qquad (5\text{-}9)$$

$$T\,ds = dq + T\,d\sigma \qquad (16\text{-}4)$$

to obtain

$$v\,dP + \mathbf{g}\,dz + \frac{1}{2}d\mathbf{v}^2 - dw_s + T\,d\sigma = 0 \qquad (16\text{-}5)$$

Equation (16-4) is the differential form of Eq. (14-8), where σ is created entropy due to irreversibilities and is always greater than or equal to zero. Integration of Eq. (16-5) yields

$$\int v\,dP + \mathbf{g}\,\Delta z + \frac{\Delta \mathbf{v}^2}{2} - w_s + \int T\,d\sigma = 0 \qquad (16\text{-}6a)$$

The term $\int T\,d\sigma$ is nonnegative and represents the lost work or mechanical energy degraded to heat due to irreversibilities in the system; it will be replaced by the symbol $l.w.$

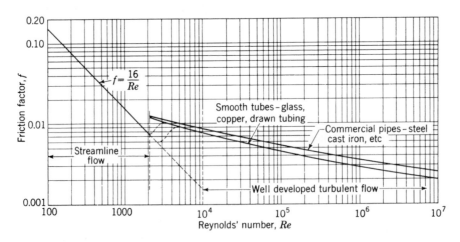

Figure 16-2 Friction factor for flow in circular pipes. [From W. H. McAdams, *Heat Transmission*, 3rd ed., copyright © 1954 by McGraw-Hill. Used with permission of McGraw-Hill Book Company, New York.]

$$\int v\, dP + \mathbf{g}\, \Delta z + \frac{\Delta \mathbf{v}^2}{2} - w_s + l.w. = 0 \qquad (16\text{-}6b)$$

When frictionless flow cannot be assumed, it is necessary to evaluate the l.w. term. For laminar flow involving simple geometries, it is possible to calculate l.w. directly from integration of the microscopic balances; however, for turbulent flow only dimensional analysis makes the problem tractable.

Application of dimensional analysis to the flow of fluids inside straight tubes yields

$$l.w. = \frac{2\, \mathbf{v}^2\, Lf}{D} \qquad (16\text{-}7)$$

where D and L are the tube diameter and length. The friction factor f is dimensionless and depends on a dimensionless surface roughness factor and Reynolds number. Dimensional analysis does not prescribe the nature of this relationship but does allow the number of variables to be considerably reduced by the appropriate grouping of $D, \mathbf{v}, \rho$, and μ into a single variable, the Reynolds number $(D\rho\, \mathbf{v}/\mu)$. The relationship between friction factor and Reynolds number must be established from experimental measurements. Figure 16-2 shows this relationship,[1] which is based on considerable experimental data.

The application of the macroscopic mechanical energy balance, Eq. (16-6b), to a given problem also requires the evaluation of $\int v\, dP$. For this it is necessary that both a path and an equation of state be specified. Usually liquids are assumed incompressible, and it is permissible to write

$$\int v\, dP = v\, \Delta P = \frac{\Delta P}{\rho} \qquad (16\text{-}8)$$

16-2 SONIC VELOCITY

That which is perceived as sound is in actuality pressure vibrations acting on the ear. These vibrations travel as pressure waves, alternate expansions and contractions, through the surrounding fluid; hence, the velocity of sound is the same as the velocity of a small pressure fluctuation.

Consider a plane, infinitesimal pressure wave traveling through a fluid in a conduit of uniform cross section. The wave front moves steadily with the velocity $\mathbf{c}$. The fluid into which the wave front moves has the properties P and ρ and is motionless. Behind the front the fluid has properties $P + dP$ and $\rho + d\rho$ and moves with a velocity $d\mathbf{v}$ due to the differences in P and ρ across the front. It is convenient to consider a coordinate system which

---

[1] For laminar flow the relationship $f = 16/\text{Re}$ is obtained from a solution of the microscopic balance equations.

moves with the wave front so that there is now a steady flow of fluid through the front, as pictured in Fig. 16-3.

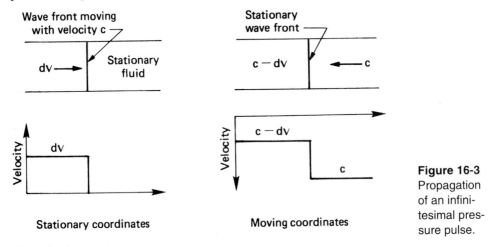

Figure 16-3 Propagation of an infinitesimal pressure pulse.

We now apply the mechanical energy balance, Eq. (16-6b), to the wave front of infinitesimal thickness with the assumption that dissipative processes are absent:

$$v \, dP + \mathbf{v} \, d\mathbf{v} = 0 \tag{16-9}$$

To eliminate $d\mathbf{v}$, we now write the equation of continuity as

$$\rho \, d\mathbf{v} + \mathbf{v} \, d\rho = 0 \tag{16-10}$$

and combine Eqs. (16-9) and (16-10) to obtain

$$dP = \mathbf{v}^2 \, d\rho \tag{16-11}$$

or

$$c^2 = \left(\frac{\partial P}{\partial \rho} \right)_s \tag{16-12}$$

The condition of constant entropy is imposed because of the assumption of reversibility ($l.w.= 0$) and adiabaticity. The latter follows from the rapidity of the compressions and expansions and also because the smallness of the pressure changes would produce correspondingly small temperature changes between system and surroundings.

The right-hand side of Eq. (16-12) is obviously a thermodynamic property and, if our assumptions are valid, so is the sonic velocity. In any thermodynamic property determination from experimental data one must always contend with errors arising from the limitations of the measurements; however, the sonic velocity appears to be a thermodynamic property only under certain restricted conditions, and this calls for additional experimental precautions.

These conditions have been delineated[2] and are realizable. In fact, it has been shown[3] that sonic velocity measurements in helium gas can be used as a basis for a thermometer in the temperature range 2 to 20 K.

For an ideal gas undergoing an isentropic process, we have seen that

$$Pv^\gamma = \text{constant} \qquad (3\text{-}21)$$

or

$$\frac{P}{\rho^\gamma} = \text{constant}$$

from which we obtain

$$\left(\frac{\partial P}{\partial \rho}\right)_S = \frac{\gamma P}{\rho} = \gamma RT \qquad (16\text{-}13)$$

and

$$c^2 = \gamma RT$$

From this it is seen that the temperature and heat capacity ratio determine the sonic velocity in an ideal gas.

For a truly incompressible fluid $d\rho = 0$, and Eq. (16-12) suggests infinite sonic velocity. While no fluid is completely incompressible, sonic velocity in liquids is several times larger than in gases.

16-3 ISENTROPIC FLOW

Reversible, adiabatic flow is often approximated in nozzles and diffusers and in that part of a flowing stream lying outside the boundary layer that exists at solid surfaces.

A nozzle increases fluid velocity, and a diffuser decreases it. For isentropic flow the first law, Eq. (16-3), simplifies to

$$\Delta h + \frac{\Delta \mathbf{v}^2}{2} = 0$$

and it is seen that velocity and enthalpy changes are complementary. Therefore, the temperature of a gas should decrease in a nozzle and increase in a diffuser. In nozzles and diffusers large pressure, velocity, and density changes occur over a short distance, and therefore

---

[2] A. B. Cambel, D. P. Duclos, and T. P. Anderson, *Real Gases*, Academic Press, New York, 1963, Chap. 1.
[3] H. H. Plumb and G. Cataland, *Science*, *150*, 155 (1965).

as a first approximation frictional effects may be neglected in comparison with other energy changes.

For one-dimensional, steady-state frictionless flow we write the mechanical energy balance as

$$dP = -\rho \mathbf{v}\, d\mathbf{v} \tag{16-14}$$

The continuity equation in differential form is

$$\frac{d\rho}{\rho} + \frac{dA}{A} + \frac{d\mathbf{v}}{\mathbf{v}} = 0 \tag{16-15}$$

Eliminating $d\mathbf{v}$ and rearranging, we obtain

$$\frac{dA}{A} = \frac{dP}{\rho}\left(\frac{1}{\mathbf{v}^2} - \frac{d\rho}{dP}\right) \tag{16-16}$$

Because the flow is isentropic,

$$\frac{dP}{d\rho} = \left(\frac{\partial P}{\partial \rho}\right)_S = \mathbf{c}^2 \tag{16-17}$$

and Eq. (16-16) may be written as

$$\frac{dA}{A} = \frac{dP}{\rho\mathbf{v}^2}\left(1 - \frac{\mathbf{v}^2}{\mathbf{c}^2}\right) \tag{16-18}$$

It is convenient to work with the Mach number $\mathbf{M}$, defined as

$$\mathbf{M} = \frac{\mathbf{v}}{\mathbf{c}} \tag{16-19}$$

and Eq. (16-18) may now be written as

$$\frac{dA}{dP} = \frac{\left(1 - \mathbf{M}^2\right)A}{\rho\mathbf{v}^2} \tag{16-20}$$

When combined with Eq. (16-14), the result is

$$\frac{dA}{d\mathbf{v}} = \frac{-\left(1 - \mathbf{M}^2\right)A}{\mathbf{v}} \tag{16-21}$$

Equations (16-14), (16-17), (16-20), and (16-21) are the equations which describe isentropic flow and apply to any fluid.

We now consider the case of a gas being accelerated from subsonic to supersonic velocity in a nozzle and inquire as to how the cross-sectional area A should vary in the direction of flow. We let l represent distance in the direction of flow and note that $d\mathbf{v}/dl > 0$. This allows the qualitative dependence of A on l to be discerned from the sign of the derivative $dA/d\mathbf{v}$ which has the same sign as dA/dl [4] and is evaluated from Eq. (16-21). We find that for $\mathbf{M} < 1$,

$$\frac{dA}{d\mathbf{v}} < 0 \quad \text{and} \quad \frac{dA}{dl} < 0$$

and for $\mathbf{M} = 1$,

$$\frac{dA}{d\mathbf{v}} = 0 \quad \text{and} \quad \frac{dA}{dl} = 0$$

and for $\mathbf{M} > 1$,

$$\frac{dA}{d\mathbf{v}} > 0 \quad \text{and} \quad \frac{dA}{dl} > 0$$

This means that to continuously accelerate the gas beyond Mach 1, the cross-sectional area of the nozzle must first decrease but later increase in the direction of flow. At the nozzle throat where $dA/dl = 0$ and $\mathbf{M} = 1$ the velocity is sonic. A subsonic gas can be accelerated in a converging nozzle but can only attain sonic velocity. A diverging section is required to accelerate the gas to supersonic velocities. From Eq. (16-14) it is seen that the pressure continuously decreases in the direction of flow.

16-4 ISENTROPIC FLOW THROUGH NOZZLES

The purpose of a nozzle is to convert thermal or mechanical energy into kinetic energy. We consider here steady-state, one-dimensional isentropic flow of an ideal gas and begin by integrating the mechanical energy balance [Eq. (16-14)] between the entrance plane where $P = P_0$ and $\mathbf{v} = 0$ and any arbitrarily chosen plane normal to flow within the nozzle:

$$\frac{\mathbf{v}^2}{2} = -\int_{P_0}^{P} \frac{dP}{\rho} \tag{16-22}$$

For an ideal gas undergoing an isentropic process,

$$\frac{P}{\rho^\gamma} = \text{constant} \tag{3-21}$$

---

[4] Note that $dA/dl = (dA/d\mathbf{v})(d\mathbf{v}/dl)$.

and after integration we obtain

$$\frac{\mathbf{v}^2}{2} = \frac{\gamma P_0}{(\gamma - 1)\rho_0}\left[1 - \left(\frac{P}{P_0}\right)^{(\gamma-1)/\gamma}\right] = \frac{\gamma RT_0}{\gamma - 1}\left[1 - \left(\frac{P}{P_0}\right)^{(\gamma-1)/\gamma}\right] \tag{16-23}$$

The right-hand expression, on comparison with Eq. (3-25), is seen to be the mechanical energy (work) available from the isentropic expansion. Thus, the nozzle converts this mechanical energy into kinetic energy.

Equation (16-23) relates velocity to pressure at any location within the nozzle and when combined with the continuity equation

$$\frac{\dot{m}}{A} = \rho \mathbf{v} \tag{16-1a}$$

and the relationship

$$\rho = \rho_0\left(\frac{P}{P_0}\right)^{1/\gamma} \tag{3-21}$$

yields the following relationship between $\dot{m}/A$ and P/P_0:

$$\frac{\dot{m}}{A} = \sqrt{\frac{2P_0\rho_0\gamma}{\gamma - 1}\left(\frac{P}{P_0}\right)^{2/\gamma}\left[1 - \left(\frac{P}{P_0}\right)^{(\gamma-1)/\gamma}\right]} \tag{16-24}$$

Figure 16-4 shows the functional nature of this relationship. The maximum in this curve is found by differentiating Eq. (16-24) with respect to P/P_0, setting the derivative equal to zero, and solving for P/P_0 to obtain

$$\frac{P_c}{P_0} = \left(\frac{2}{\gamma + 1}\right)^{\gamma/(\gamma-1)} \tag{16-25}$$

This so-called critical pressure ratio is seen to depend only on γ, the heat capacity ratio of gas.

Substitution of the critical pressure ratio into Eq. (16-24) yields

$$\left(\frac{\dot{m}}{A}\right)_{max} = \sqrt{\gamma P_0\rho_0\left(\frac{2}{\gamma + 1}\right)^{(\gamma+1)/(\gamma-1)}} \tag{16-26}$$

It is seen that the maximum mass flow per unit area, which occurs at the throat and determines the capacity of the nozzle, depends on entrance conditions and γ but not on downstream conditions.

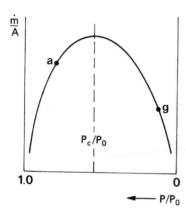

Figure 16-4 Graphical representation of Eq. (16-24).

Substitution of the critical pressure ratio into Eq. (16-23) yields the throat velocity at maximum flow:

$$\mathbf{v}_t = \sqrt{\frac{2\gamma\, RT_0}{\gamma + 1}} \tag{16-27}$$

The temperature at the throat T_t can be shown equal to $2T_0/(\gamma + 1)$ by substituting the critical pressure ratio into the isentropic property relationship, Eq. (3-19):

$$\frac{T_t}{T_0} = \left(\frac{P_c}{P_0}\right)^{(\gamma-1)/\gamma} = \left[\left(\frac{2}{\gamma + 1}\right)^{\gamma/(\gamma-1)}\right]^{(\gamma-1)/\gamma} = \frac{2}{\gamma + 1} \tag{16-28}$$

Using Eq. (16-28), Eq. (16-27) may now be written as

$$\mathbf{v}_t = \sqrt{\gamma\, RT_t} \tag{16-29}$$

and simplified by Eq. (16-13) to

$$\mathbf{v}_t = \mathbf{c}$$

Thus, the throat velocity at maximum flow is sonic velocity at the temperature prevailing in the throat, which in turn is seen to depend only on T_0 and γ.

Converging Nozzles. In Sec. 16-3 it was shown that a diverging section is needed in order to accelerate a supersonic gas, and therefore the maximum velocity attainable in a converging nozzle is sonic. This can also be seen from the application of Eqs. (16-23)–(16-29). Figure 16-5 shows a converging nozzle, its pressure profiles, and its flow behavior. For a given nozzle and external pressure ratio, $\dot{m}$ is fixed and Fig. 16-4 therefore relates P/P_0 to

A. The pressure profile can then be determined from the nozzle design (*A* vs. length). Curve *a* of Fig. 16-5b shows a pressure profile when the external pressure ratio is greater than the critical. It can be visualized in terms of tracking the curve in Fig. 16-4 from $P/P_0 = 1$, the entrance, to point *a*, the tip. Profile *b* would represent traversing the curve in Fig. 16-4 from $P/P_0 = 1$ past point *a* to the maximum where the flow at the tip of the nozzle is sonic. The flow rates corresponding to these two cases are indicated by points *a* and *b* in Fig. 16-5c where flow rate is plotted vs. external pressure ratio.

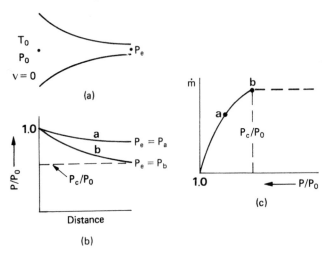

Figure 16-5 Characteristics of a converging nozzle.

While it is possible to impose an external pressure ratio less than the critical, the velocity at the nozzle tip can never exceed sonic velocity, and the pressure ratio there will always be P_c/P_0. Irreversible adjustments in the flow downstream from the nozzle occur in response to external pressure ratios less than P_c/P_0. Thus, in Fig. 16-5c the curve for P/P_0 less than P_c/P_0 is shown dashed because the imposed pressure is not the nozzle exit pressure. This curve suggests that when operated at external pressure ratios less than the critical a converging nozzle can be used to deliver a constant flow of gas independent of downstream pressure. This phenomenon is called *choking*.

Converging-Diverging Nozzles. Figure 16-6 shows a converging-diverging nozzle with its performance curves. Again, the pressure profiles may be visualized with the aid of Fig. 16-4. For low flow rates the throat velocity is subsonic, and the diverging section acts as a diffuser which increases the pressure and decreases the velocity [cf. Eqs. (16-20) and (16-14)]. Profile *a* of Fig. 16-6b represents operation in this flow regime and can be visualized in terms of traversing the curve of Fig. 16-4 from P/P_0 of unity up to point *a*, which would represent conditions in the throat. Conditions downstream from the throat would be represented by moving back down the curve toward a P/P_0 of unity. As the downstream pressure is decreased and flow increases, this flow regime will prevail until sonic velocity is attained in the throat. For this condition there are two possible isentropic solutions shown by profiles *b* and *g*. Again, Fig. 16-4 proves useful in understanding these profiles. Profile *b* corresponds to moving along the curve in the direction of decreasing P/P_0 until the maximum is reached and then reversing direction. Profile g corresponds to a continual decrease in P/P_0 ending at point g.

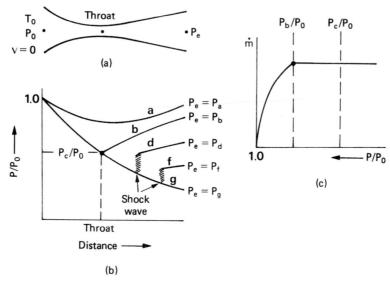

Figure 16-6 Characteristics of a converging-diverging nozzle.

For downstream pressures equal to or greater than P_b and equal to P_g, isentropic flow obtains, and for pressures between P_b and P_g and less than P_g the flow is nonisentropic. Nonisentropic flow is characterized by the presence of standing shock waves and will be discussed in Sec. 16-5. Operation corresponding to g is desired because the flow is both isentropic and supersonic. For all external pressures less than P_b, flow through the throat is sonic, and therefore Fig. 16-6c shows the flow rate constant in this range, and the flow is said to be *choked.*

16-5 NONISENTROPIC FLOW

Here we will consider two situations in which the flow is adiabatic but not reversible: flow with friction and shock waves. For flow in ducts irreversibilities can be attributed to friction and the changes are gradual. On the other hand, shock waves serve to relieve instabilities and are characterized by very abrupt changes.

Flow with Friction.   We now consider the steady-state, adiabatic, one-dimensional flow of a gas in a duct of constant cross section and write the first law, Eq. (16-3):

$$\Delta h + \frac{\Delta \mathbf{v}^2}{2} = 0 \tag{16-30}$$

The equation of continuity in the form

$$\mathbf{G} = \frac{\dot{m}}{A} = \rho \mathbf{v} = \text{constant} \tag{16-31}$$

can be used to eliminate the velocity, and the first law is now stated as

$$h - h_1 + \frac{\mathbf{G}^2}{2}\left(\frac{1}{\rho^2} - \frac{1}{\rho_1^2}\right) = 0 \tag{16-32}$$

It should be noted that the variables h and ρ are thermodynamic properties, and for any given $\mathbf{G}$ and upstream condition specified by h_1 and ρ_1 Eq. (16-32) traces the path followed by the gas as it flows down the duct. Together with this equation, the fixing of either h or ρ will define the state of the gas. The remaining properties may be determined, and therefore the path may be plotted on other coordinates. If an hs diagram is chosen, the path would appear as shown in Fig. 16-7 and is called a *Fanno line*. The line shown corresponds to a particular set of $\mathbf{G}$, h_1, and ρ_1 and is the locus of states through which the gas will pass.

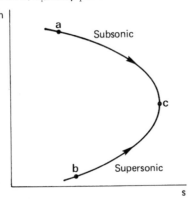

Figure 16-7 Fanno line.

There are two features of the Fanno line that should be noted. First, because the entropy can never decrease in an adiabatic process, it is possible only to move along the Fanno line toward point c. Second, at point c any change occurs under conditions of constant entropy, and it will be shown that the velocity is sonic. To demonstrate this, we begin with the fundamental equation [Eq. (5-9)] and write

$$dh = T\,ds + v\,dP = v\,dP = \frac{dP}{\rho}$$

Substituting this into the differential form of Eq. (16-32) yields

$$dh = \frac{dP}{\rho} = \frac{\mathbf{G}^2}{\rho^2} \frac{d\rho}{\rho}$$

which reduces to

$$\frac{\mathbf{G}^2}{\rho^2} = \mathbf{v}^2 = \left(\frac{\partial P}{\partial \rho}\right) = \mathbf{c}^2$$

and it is seen that point c represents sonic velocity.

The upper portion of the Fanno line between a and c is characterized by decreasing h and hence increasing $\mathbf{v}$ [cf. Eq. (16-30)]. Along ac the velocity increases and approaches sonic velocity. Conversely, h increases along the curve segment bc, and the flow is supersonic. Thus, the effect of friction is to accelerate subsonic flows and decelerate supersonic flows. In both cases sonic velocity is the limit and can occur only at the end of the conduit.

Choking also occurs for flow in conduits. For a given Fanno line, characterized by a specified $\mathbf{G}$ and upstream condition, the entropy maximum (point c of Fig. 16-7) will occur at a definite pressure. At this pressure sonic velocity is attained at the end of the conduit. When lower exit pressures are imposed, the exit velocity remains sonic, and the gas expands irreversibly to this pressure as it leaves the conduit.

Shock Waves. Under certain conditions a flowing fluid is unable to accommodate itself to change in a gradual and continuous manner and undergoes a drastic and abrupt adjustment. Such a discontinuity is called a shock wave. In supersonic flow schlieren photography shows the presence of shock waves with thickness on the order of 10^{-6} m across which pressure and velocity change by several fold.

In applying the macroscopic balances across the shock wave, the extreme thinness of the front allows us to neglect area changes and wall friction and write the following:

first law:
$$h_1 + \frac{\mathbf{v}_1^2}{2} = h_2 + \frac{\mathbf{v}_2^2}{2} \qquad (16\text{-}30)$$

momentum balance:
$$P_1 + \frac{\dot{m}}{A}\mathbf{v}_1 = P_2 + \frac{\dot{m}}{A}\mathbf{v}_2 \qquad (16\text{-}2)$$

continuity equation:
$$\mathbf{G} = \rho_1\mathbf{v}_1 = \rho_2\mathbf{v}_2 \qquad (16\text{-}31)$$

Equations (16-30) and (16-31) were shown to form the basis for the Fanno line and in a similar way Eqs. (16-2) and (16-31) can be combined to yield

$$\mathbf{G}^2\left(\frac{1}{\rho} - \frac{1}{\rho_1}\right) = P_1 - P$$

Like Eq. (16-32), this equation relates thermodynamic properties and can also be represented as a curve on an hs diagram. Such a curve is called a *Rayleigh line* and is sketched in Fig. 16-8 along with a Fanno line. Like the Fanno line, a Rayleigh line represents a particular **G** and initial state.

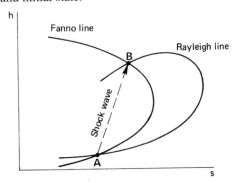

Figure 16-8 Graphical representation of shock wave.

For a given upstream condition and **G**, Eqs. (16-30), (16-2), and (16-31) along with the necessary thermodynamic property relationships (e.g., an hs diagram) define the system and determine conditions downstream from the shock wave. Equivalent to solving these equations simultaneously is the determination of the intersection of the Fanno and Rayleigh lines. There are two points of intersection shown in Fig. 16-8 as A and B—one representing the upstream and the other the downstream condition. The equations are satisfied regardless of whether A or B is the upstream condition; however, for this adiabatic system the second law requires a nonnegative entropy change, and $A \rightarrow B$ is the only acceptable solution. Thus, a shock wave occurs only in supersonic flow and always produces a change to subsonic flow. Equation (16-2) shows that in the direction of flow the pressure increases across a shock wave.

Shock Waves in Nozzles.   The behavior of the converging-diverging nozzle shown in Fig. 16-6 for downstream pressures between P_b and P_g can be explained in terms of shock waves. Isentropic flow with sonic throat velocity exists only for P_b and P_g, with standing shock waves contributing the irreversibilities for nonisentropic flow at other pressures. Consider the pressure profile for an external pressure P_d where supersonic velocities are attained in the diverging section. Within the diverging section a standing shock wave occurs which increases the pressure and decelerates the fluid to subsonic velocities. The remainder of the diverging section then acts as a diffuser for the subsonic flow, and the pressure increases. As the downstream pressure is lowered, the shock wave is located farther away from the throat and eventually moves out of the nozzle.

PROBLEMS

16-1. Show that

$$\left(\frac{\partial P}{\partial \rho}\right)_S = \gamma \left(\frac{\partial P}{\partial \rho}\right)_T$$

and express the velocity of sound for a van der Waals gas in terms of critical properties.

16-2. Use Eq. (16-6) to derive Eqs. (3-16) and (3-25) for the isothermal and adiabatic work of compressing an ideal gas. Show why less work is required to pump a liquid than a gas. Note that the use of Eq. (16-6) requires the usual assumptions to be explicitly stated.

16-3. It was shown for a Fanno line that the maximum in entropy corresponds to sonic velocity. What can be said about the entropy maximum for a Rayleigh line?

16-4. An ideal gas ($c_p = 30$ kJ/kmol · K and MW = 28) flows through a converging-diverging nozzle. The gas enters the nozzle at 10 bar and 500 K at a very low velocity and at the rate of 1 kg/s. For supersonic operation with no standing shock waves, calculate at the throat the

    (a)   area.
    (b)   pressure.
    (c)   temperature.
    (d)   velocity.

16-5. For the nozzle described in Prob. 16-4 calculate the external pressure and the area at the exit end of the nozzle required to produce an exit velocity of 600 m/s.

16-6. The nozzle described in Prob. 16-4 has an exit area five times the throat area. Determine the

    (a)   exit pressure.
    (b)   Mach number of the gas at the nozzle exit.
    (c)   density of the gas at the nozzle exit.

16-7. Show how $\mathbf{v}$ and $\dot{m}/A$ can be related to P for the isentropic flow of a nonideal gas for which a Ph or Ts diagram is available.

16-8. Show how velocity and $\dot{m}/A$ can be related to P/P_0 for the isentropic flow of superheated steam through the use of the steam tables. Provide sufficient detail.

16-9. Compressed air is to be passed through a converging-diverging nozzle that feeds a supersonic wind tunnel operating at atmospheric pressure. The tunnel is 0.15 m in diameter and operates at Mach 1.5. The temperature of the air in the wind tunnel is to be 300 K. What should be the temperature and pressure of the air fed to the nozzle? Take $c_p = 30$ kJ/kmol · K and MW = 28.8 for air and assume ideal-gas behavior.

16-10.   Air taken from the atmosphere at 300 K is to be compressed and passed through a converging-diverging nozzle that feeds a supersonic wind tunnel operating at atmospheric pressure. The tunnel is 0.15 m in diameter and capable of producing Mach 1.5. A multistage compressor with interstage cooling which lowers the gas temperature to 350 K will be used. Because air temperatures between 280 and 310 K are desired in the tunnel, it may be necessary to heat or cool the gas leaving the final stage of the compressor. Specify the number of compressor stages, the power requirement, and the rate at which heat must be added or removed to operate the wind tunnel as desired. Assume ideal-gas behavior, and use an adiabatic efficiency of 80% for the compressor.

16-11.   Superheated steam at 6.0 MPa and 500°C enters a converging-diverging nozzle, where the downstream pressure is 2.0 MPa. Assuming isentropic flow with no standing shock waves, find the following for the exiting stream:

(a)   velocity.
(b)   temperature.

16-12.   For the situation in Prob. 16-11 we wish to know the Mach number of the exit stream and therefore need the sonic velocity. Calculate this

(a)   assuming the steam is an ideal gas.
(b)   using data from the steam table.

16-13.   Equation (16-24), the major equation for isentropic compressible flow, derives from the application of Eq. (3-21) and therefore applies to an ideal gas. Suppose we have reason to believe that a certain gas would not behave ideally under the specified conditions of a particular nozzle. How would we determine the counterpart of Eq. (3-21), an isentropic path expressed in terms of P and ρ , if the gas did not behave ideally? Demonstrate by calculating ρ for a pressure of 42 atm when propane (C_3H_8) enters a nozzle at a temperature of 555 K and a pressure of 168 atm. In this temperature range we may take c_p constant at 15.8 cal/g mol K.

16-14.   It is desired to deliver an ideal gas ($c_p = 30$ kJ/kmol · K and MW = 28) at a constant rate of 1 kg/s to a vessel that is maintained at a pressure of 1 bar. To accomplish this, a converging nozzle is attached to the end of the supply line, and the upstream pressure will be set at a value that ensures choked flow. The upstream temperature is 300 K. (a) Determine the nozzle tip area and upstream pressure. (b) What is the temperature of the gas leaving the nozzle? Neglect friction.

16-15.   In our treatment of flow with friction the pipe length did not appear as a variable. If we desire to relate fluid properties to pipe length, we combine Eqs. (16-6) and (16-7) and write the mechanical energy balance as

$$\mathbf{v}d\mathbf{v} + v\, dP + \frac{2f\, \mathbf{v}^2\, dL}{D} = 0$$

(a)   Show that this equation can be integrated to yield

$$\mathbf{G}^2 \ln \frac{v_2}{v_1} + \int \frac{dP}{v} + \frac{2f\,\mathbf{G}^2 L}{D} = 0$$

(b) When the temperature change is small enough to neglect and the gas behaves ideally, show that the preceding leads to

$$\mathbf{G}^2 \ln \frac{v_2}{v_1} + \frac{MW}{2RT}\left(P_2^2 - P_1^2\right) + \frac{2f\,\mathbf{G}^2 L}{D} = 0$$

where MW is molecular weight.

(c) The path of the irreversible adiabatic flow is specified by the Fanno line [Eq. (16-32)], and there exists a definite relationship between P and v. Show that this relationship is

$$P = \left(\frac{\mathbf{G}^2 v_1^2}{2} - \frac{\mathbf{G}^2 v^2}{2} + \frac{kPv_1}{k-1}\right)\frac{k-1}{kv}$$

where $k = c_p MW/R$.

(d) Further, show that for irreversible adiabatic flow the mechanical energy balance integrates to

$$\frac{(k+1)\mathbf{G}^2}{2k}\ln\frac{v_2}{v_1} + \frac{1}{2}\left[\frac{(k-1)\mathbf{G}^2 v_1^2}{2k} + P_1 v_1\right]\left(\frac{1}{v_2^2} - \frac{1}{v_1^2}\right) + \frac{2f\,\mathbf{G}^2 L}{D} = 0$$

CHAPTER **17**

Thermodynamics and Models

*I*n Sec. 1-1 it is stated that the results of thermodynamics are "independent of theories of matter and are thus respected and confidently accepted." While this statement holds without exception for the laws of thermodynamics, we have seen many instances throughout this text where it was necessary to employ a theory or a model or otherwise simplify a complex problem in order to apply thermodynamics. For example, in applying the first law to an open system we may begin with the exhaustive description provided by Eq. (2-17) and proceed to simplify by dropping non-contributing or negligible terms such as kinetic or potential energy or we might assume that the process is adiabatic ($Q = 0$) or that the enthalpy of a particular stream is constant. In doing this we have modeled the system. These simplifications, approximations, and assumptions provide a physical description of our model and the resulting equation is a mathematical description of the model. Often we do not bother to distinguish between these two descriptions and will refer to the equation or set of equations as the model rather than the model equation or equations. This sometimes occurs in thermodynamics because for some problems it may not be possible to delineate a physical description of the system.

In phase equilibrium we have seen that there are many equations that satisfy the Gibbs-Duhem equation and would therefore be thermodynamically acceptable. When we choose a particular one to represent our system, we are in effect proposing a model for the behavior of our system. There is no guarantee that the chosen equation will represent our system and therefore we feel more secure when we are able to test the fit with experimental data. Additionally, we know that the various equations have been tested on a large number of systems and only those proven to be effective have survived. Still, of the survivors—the Margules, the van Laar, and the Wilson equations—one never knows *a priori* which will do the better job for a particular system, and results will be dependent on the model chosen. Another aspect to modeling systems with a vapor phase is our usual assumption of ideal-gas mixtures. Here it is usually the case at low pressure that gas-phase deviations from ideal behavior are small particularly in comparison to the uncertainties associated with representation of activity coefficients.

Similarly, in computing chemical equilibrium we assume that only a single reaction of the set of possible reactions progresses to any extent and reaches equilibrium. Thus, we have tacitly assumed the presence of an effective catalyst that accelerates only the chosen reaction and in doing so we have fashioned a model of our reacting system. This is much more obvious when dealing with complex reacting systems as in Sec. 13-2 where the focus is on ways in which the problem of representing a complex reacting system can be simplified (i.e., a reasonable model can be constructed).

With the easy accessibility of computing resources, there has been, and surely will continue to be, considerable attention given to treating ever more complex thermodynamic problems and using methods of property estimation of greater mathematical complexity. In short, there will continue to be increased emphasis on modeling. This chapter is intended to be an introduction to the subject of modeling thermodynamic systems. Here models are arbitrarily classified as *standard* or *ad hoc*. A standard model is usually a tool for determining or correlating thermodynamic properties (e.g., the Peng-Robinson equation of state, the Wilson equation, or the UNIFAC method) and is often used "as is" as part of an *ad hoc* model. As modeling is usually an *ad hoc* process, several examples will be presented that are intended to illustrate the nature of the process.

17-1 STANDARD MODELS

So far we have encountered two models that are used extensively for estimating thermodynamic properties: a generalized van der Waals–type equation of state, such as the Peng-Robinson equation, and the UNIFAC method of estimating activity coefficients. While the Peng-Robinson equation of state has been used almost exclusively throughout this text, we saw in Sec. 3-2 that it is only one of several widely used models for fluid *PVT* behavior.

As the mathematical complexities associated with these models are formidable, the use of a computer is practically essential for their application. Accompanying this text are two computer programs for using the Peng-Robinson equation of state: PREOS.EXE which calculates Z, v, ϕ, Δh, and Δs for pure gases or liquids and PRVLE.EXE which calculates fugacity coefficients, and hence vapor-liquid equilibrium, for binary and multicomponent systems. The program UNIFAC.EXE estimates activity coefficients via the UNIFAC method which treats a solution as a mixture of groups rather than a mixture of molecules. This *group contribution* approach has also been used to estimate thermochemical properties of pure substances based on the number and type of groups constituting the molecule.[1] These two models—the generalized van der Waals equation of state and group contribution correlations—are used in much of the commercially available process simulation software for the estimation of thermodynamic and thermochemical properties.

For the calculation of vapor-liquid equilibrium, the equation of state method (EOS method) has proven effective for mixtures where the molecules are relatively small and nonpolar. Conversely, activity coefficients are usually employed for substances with more complex molecular structure—the excess free energy (G^e) method. The EOS method can be applied over a wide range of temperature and pressure whereas the G^e method is usually applied at moderate temperatures and low pressures. Recent work has indicated that these two methods may be combined (EOS + G^e) so that G^e information can be used to determine a more effective set of EOS parameters and thereby extend the applicability of the EOS method to more diverse systems.[2] Several approaches for the EOS + G^e method have been proposed, all involving a modified van der Waals equation of state. Early work focused on equating G^e determined from the equation of state G^e_{EOS} to G^e derived from activity coefficients G^e_γ.

In employing the EOS method in Sec. 9-12, the condition of phase equilibrium for component i was expressed as

$$Py_i\hat{\phi}_i^G = Px_i\hat{\phi}_i^L$$

with $\hat{\phi}_i$ given by Eq. (9-62). This leads to

$$\frac{y_i}{x_i} = \frac{\hat{\phi}_i^L}{\hat{\phi}_i^G}$$

which forms the basis for calculating phase equilibria via an equation of state. When Eq. (9-63), the basic equation of vapor-liquid equilibrium, is written in terms of liquid-phase fugacity coefficient we have

---

[1] R. C. Reid, J. M. Prausnitz, and B. E. Poling, *Properties of Gases and Liquids*, 4th ed., McGraw-Hill, New York, 1987, Chap. 6.

[2] For a comprehensive review see S. I. Sandler, H. Orbey, and B. I. Lee, "Equations of State" in *Models for Thermodynamic and Phase Equilibria Calculations*, S. I. Sandler, ed., Marcel Dekker, New York, 1994 and H. Orbey and S. I. Sandler, *AIChEJ.*, **42**, 2327 (1996).

$$Py_i \hat{\phi}_i^G = P\phi_i^L x_i \gamma_i$$

which simplifies to

$$\gamma_i = \frac{\hat{\phi}_i^G}{\phi_i^L} \times \frac{y_i}{x_i}$$

On substituting for y_i/x_i from above we obtain

$$\gamma_i = \frac{\hat{\phi}_i^G}{\phi_i^L} \times \frac{\hat{\phi}_i^L}{\hat{\phi}_i^G} = \frac{\hat{\phi}_i^L}{\phi_i^L}$$

Using Eq. (10-14) the excess free energy change as calculated from an equation of state becomes[3]

$$G_{EOS}^e = RT \sum x_i \ln \frac{\hat{\phi}_i^L}{\phi_i^L}$$

and is seen to be a function of temperature, pressure, and liquid composition. Because the excess free energy change calculated from activity coefficients G_γ^e is usually evaluated at low pressure and is seldom considered a function of pressure, it is necessary to choose a pressure at which to equate G_γ^e and G_{EOS}^e. Limiting pressures of zero and infinity have been used, but these choices are subject to either theoretical or calculational difficulties.

Wong and Sandler[4] eliminated these difficulties by calculating the excess Helmholtz free energy of mixing at a pressure approaching infinity $A_{EOS}^e(P \to \infty)$ from the equation of state. The Helmholtz free energy change is nearly independent of pressure and is related to the Gibbs free energy change through the excess volume change on mixing[5]

$$A^e = G^e - P\Delta v^e \tag{17-1}$$

When the pressure is low, the second right-hand term is small and to a good approximation A^e and G^e are equal. At a low pressure P^*, we write

$$A^e(P \to \infty) = A^e(P^*) \doteq G^e(P^*)$$

and see that it is reasonable to state

$$A_{EOS}^e(P \to \infty) = G_\gamma^e \tag{17-2}$$

---

[3] For consistency with the current literature, G^e is used instead of Δg^e.
[4] D. S. H. Wong and S. I. Sandler, *AIChEJ.*, *38*, 671 (1992).
[5] Because there is no volume change on forming an ideal solution, this is also the volume change on mixing.

Wong and Sandler used Eq. (17-2) and a mixing rule consistent with that of the virial equation to obtain binary interaction parameters. This approach yields good results but involves considerable computational effort to determine the mixture parameters a and b. The computational effort can be lessened without excessive loss of efficacy by using the usual combining rule for b

$$b = \sum_i x_i b_i \tag{17-3}$$

together with Eq. (17-2) applied to the Peng-Robinson equation

$$\frac{G_\gamma^e}{RT} = -\sum_i x_i \ln\left(\frac{b}{b_i}\right) - 0.62323\left(\frac{a}{bRT} - \sum_i x_i \frac{a_i}{b_i RT}\right) \tag{17-4}$$

Equations (17-3) and (17-4) constitute a simpler model which Orbey and Sandler[6] have identified as the HVOS model. In using this model one may determine the Peng-Robinson mixture parameter a at liquid mol fractions where G_γ^e is known thereby establishing a relation between a and x, or one may fit the interaction parameter k_{12} to the a-x data using the relation

$$a = x_1^2 a_1 + 2x_1 x_2 (1 - k_{12})\sqrt{a_1 a_2} + x_2^2 a_2 \tag{17-5}$$

EXAMPLE 17-1

Use the UNIFAC model combined with the HVOS model and the Peng-Robinson equation of state to estimate VLE data for the acetone–methanol system at temperatures of 100, 150, and 200°C. Compare the results to available experimental VLE data.[7]

Solution 17-1

The program UNIFAC.EXE was used to calculate values of γ_1 and γ_2 for this system at a temperature of 55°C where the system pressure ranges close to 1 atm. The calculation of b via Eq. (17-3), of G_γ^e from the γ's, and of a via Eq. (17-4) were performed for each liquid composition in the spreadsheet EX17-1.WQ1.[8] This results in a set of a-x_1 data which was fitted to Eq. (17-5) with the spreadsheet's optimize function to obtain the best value of k_{12}. A value of $k_{12} = -0.00353$ produced an excellent fit as can be seen from Fig. 17-1. This k_{12} was used in the program PRVLE.EXE to calculate the xyP

---

[6] Ibid.
[7] Source: reference B2 in Table 10-8.
[8] Spreadsheets EX17-1.WQ1 and EX17-1D.WQ1 are in the BACKUP directory of the companion CD-ROM.

data at 100, 150, and 200°C. The results of these calculations can be
viewed on the spreadsheet EX17-1D.WQ1 and are here summarized by
the average absolute difference in y, $\Delta \overline{y}$, and the average percent absolute pressure difference $\Delta \overline{P}$ defined as

$$\Delta \overline{P} = \frac{100 \sum\limits_{i=1}^{N} \left| P_{cal} - P_{exp} \right|_i}{N \overline{P}}$$

where N is the number of data points and $\overline{P}$ is the average experimental
pressure.

Temperature(°C)	$\Delta \overline{y}$	$\Delta \overline{P}(\%)$
100	0.0094	1.9
150	0.023	2.9
200	0.035	3.7

The value of k_{12} was determined at 55°C and as expected, the agreement
between calculated and experimental values deteriorates the more the
temperature differs from this value. Considering the polar nature of
these substances and the wide range of temperature, these results, based
only on pure-component properties, are remarkably good and attest to
the efficacy of EOS + G^e models.

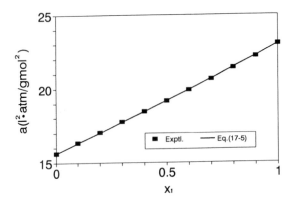

Figure 17-1

17-2 AD HOC MODELS

Included here are several models used to describe specific systems of a fairly complex nature. Example 17-2 involves only phase equilibrium and Ex. 17-3 involves only chemical equilibrium while Ex. 17-4 involves both phase and chemical equilibrium.

EXAMPLE 17-2

Fuel mixtures of gasoline and ethanol, gasohol, can be produced directly by extraction of aqueous ethanol using gasoline as a solvent.[9] The gasohol so produced is saturated with water and will require removal of some of the water before it can be marketed and to this end it has been proposed to dry the gasohol in a stripping column. Before the feasibility of this operation can be evaluated, VLE data suitable for a preliminary design will be needed. Show how the available liquid-liquid equilibrium (LLE) data for gasoline–ethanol–water could be used to estimate the needed VLE data.

Solution 17-2[10]

Because gasoline is a mixture of innumerable components, it is characterized by standardized measurements such as API gravity, octane number, and ASTM distillation. API gravity is simply a special scale for representing liquid density. The ASTM distillation is a batch process carried out in a simple distilling flask using standardized equipment and procedure that produces an ASTM distillation curve—temperature vs. volume percent distilled. Correlations have been established relating other properties of petroleum fractions to these standardized measurements. One such correlation[11] was used to estimate the molecular weight of the gasoline used for the LLE measurements from its API gravity and ASTM distillation curve. This value, 96, was needed to convert the LLE data from weight fraction to mol fraction.

For the operation of extraction it is reasonable to consider gasoline as a single component because while it is a mixture of many components, they are very similar chemically and should exhibit similar solution behavior as characterized by similar activity coefficients. However, for the operation of distillation, vapor pressures as well as activity coefficients play an important role.

---

[9] U.S. Patent #4,297,172.

[10] This example is based on the work of S. T. Chou, MS Thesis, Kansas State University, 1982.

[11] O. A. Hougen, K. M. Watson, R. A. Ragatz, *Chemical Process Principles Charts*, 2nd ed., Wiley, New York, 1960.

A triangular diagram representative of this system is shown in Fig. 17-2. The produced "wet" gasohol is located on the right-hand

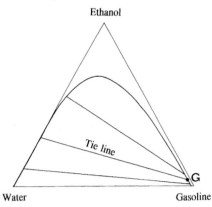

Figure 17-2

saturated liquid curve and is shown on Fig. 17-2 as the point G. The dried product will lie in the one-phase region to the right of point G and the design of the stripping column will require VLE data in this composition region—the gasoline-rich region.

For the liquid-liquid equilibrium data, we write for each tie line

$$x'_{Es}\gamma'_{Es} = x''_{Es}\gamma''_{Es}$$

$$x'_{Ws}\gamma'_{Ws} = x''_{Ws}\gamma''_{Ws}$$

$$x'_{Gs}\gamma'_{Gs} = x''_{Gs}\gamma''_{Gs}$$

where the primes and double primes refer respectively to the water-rich and the gasoline-rich phases and the subscripts E, W, and G refer to ethanol, water, and gasoline respectively. If the small quantity of gasoline in the water-rich phase can be ignored, this phase can be regarded as an ethanol–water binary system for which activity coefficients can be evaluated from the ample quantity of binary VLE data available for this system. Thus, we can write

$$\gamma''_{Es} = \frac{x_E \gamma_E}{x''_{Es}}$$

$$\gamma''_{Ws} = \frac{x_W \gamma_W}{x''_{Ws}}$$

where x_E and x_W are respectively the mol fractions of ethanol and water in the water-rich phase on a gasoline-free basis and γ_E and γ_W are the corresponding activity coefficients of ethanol and water in the ethanol–water binary system.

Activity coefficients in the ethanol–water binary system were obtained by fitting the Wilson equation to combined isothermal data sets at 40, 55, and 70°C[12] and determining the parameters a_{12} and a_{21} using a computer program similar to WEQ-ISOP.EXE. The values determined were 532.5 and 852.1 cal/gmol for a_{12} and a_{21} respectively; these values produced a reasonably good fit with a $\Delta \bar{y} = 0.010$ for the combined data set.

At 50°C we take the following tie line

$$x'_{Es} = 0.2041 \qquad x''_{Es} = 0.0727$$
$$x'_{Ws} = 0.7937 \qquad x''_{Ws} = 0.0064$$
$$x'_{Gs} = 0.0022 \qquad x''_{Gs} = 0.9209$$

and find the gasoline-free mol fraction of ethanol in the aqueous phase to be

$$x_E = \frac{0.2041}{(0.2041 + 0.7937)} = 0.2046$$

From Eqs. (9-58d) and (9-58e) at this temperature G_{12} and G_{21} are

$$G_{12} = \left(\frac{18}{58.4} \right) \exp\left(\frac{-532.5}{1.987 \cdot 323} \right) = 0.1344$$

$$G_{21} = \left(\frac{58.4}{18} \right) \exp\left(\frac{-852.1}{1.987 \cdot 323} \right) = 0.8601$$

and from Eqs.(9-58a) and (9-58b) the activity coefficients at $x_E = 0.2046$ are

$$\gamma_E = 2.237; \qquad \gamma_W = 1.130$$

This tie line yields the following values of the activity coefficients of ethanol and water in the gasoline-rich phase

---

[12] Source: reference B2 in Table 10-8.

$$\gamma''_{Es} = \frac{x_E \gamma_E}{x''_{Es}} = \frac{0.2046(2.237)}{0.0727} = 6.30$$

$$\gamma''_{Ws} = \frac{x_W \gamma_W}{x''_{Ws}} = \frac{1.130(1-0.2046)}{0.0064} = 140$$

at the known liquid composition

$$x_E = 0.0727; \quad x_W = 0.0064; \quad x_G = 0.9209$$

Using this procedure on the remaining tie lines results in a set of activity coefficients at known liquid compositions and at the temperatures of 25 and 50°C. These activity coefficients were used to obtain parameters in the UNIQUAC activity coefficient equation by minimizing the objective function OFSQ with a computerized pattern search technique:

$$\text{OFSQ} = \sum_{j=1}^{n} \sum_{i=1}^{2} (\gamma_{\text{exp}} - \gamma_{\text{cal}})_{i,j}^{2}$$

where j indexes the tie lines and i indexes the components. The UNIQUAC equation has been shown effective in dealing with systems exhibiting limited liquid miscibility and, like the Wilson equation, requires only two parameters per binary pair to describe multicomponent systems. It also resembles the Wilson equation in the specified temperature dependence of its parameters. Because of its algebraic complexity it will not be written here.[13] With the fitted parameters the UNIQUAC equation yielded activity coefficients that agreed reasonably well with those derived from the tie lines.

For the purpose of distillation calculations the gasoline was treated as a mixture of 20 components. A correlation by Edmister[14] allowed the estimation of the true boiling point (TBP) curve from the ASTM curve. The TBP curve is represented as boiling temperature vs. mol % distilled and was partitioned into 20 equispaced segments. Each 5 mol % segment was regarded as a pseudocomponent with a boiling point equal to the average value over that segment. Another correlation by Edmister[15] allowed K values to be estimated for the 20 pseudocomponents.

We now have vapor pressures or K values for this 22-component system but have activity coefficients only for ethanol, water, and gaso-

---

[13] See J. M. Prausnitz, R. N. Lichtenthaler, and E. Gomes de Azevedo, *Molecular Thermodynamics of Fluid-Phase Equilibria*, 2nd ed., Prentice-Hall, Englewwood Cliffs, NJ, 1986, Chap. 6.
[14] W. C. Edmister, *Applied Hydrocarbon Thermodynamics*, Gulf Publishing Co., Houston, 1961, Chap. 12.
[15] *Ibid*, Chap. 17.

line. Thus, we need to assign activity coefficients to each of the pseudo-components in gasoline. It seems reasonable to assume that the activity coefficient of each pseudocomponent γ_{Gi} is the same. Further, it also seems reasonable to assume that the activity of the gasoline is equal to the sum of the activities of the pseudocomponents (as would be the case for ideal-liquid solutions)

$$a_G = x_G \gamma_G = \sum_i x_{Gi} \gamma_{Gi}$$

because the γ_{Gi}'s are equal we can write

$$x_G \gamma_G = \gamma_{Gi} \sum_i x_{Gi}$$

which, because the sum of x_{Gi}'s is equal to x_G, results in

$$\gamma_{Gi} = \gamma_G$$

This convenient result seems reasonable, but we should test it for ther-modynamic consistency. For the ternary ethanol–water–gasoline system the Gibbs-Duhem equation requires

$$x_E \, d \, \ln \gamma_E + x_W \, d \, \ln \gamma_W + x_G \, d \, \ln \gamma_G = 0$$

This condition is met because activity coefficients have been calculated by the UNIQUAC equation, a thermodynamically consistent expression. In the 22-component system the Gibbs-Duhem equation becomes

$$x_E \, d \, \ln \gamma_E + x_W \, d \, \ln \gamma_W + \sum x_{Gi} \, d \, \ln \gamma_{Gi} = 0$$

and thermodynamic consistency is ensured if

$$x_G \, d \, \ln \gamma_G = \sum x_{Gi} \, d \, \ln \gamma_{Gi}$$

This condition is satisfied because

$$\gamma_G = \gamma_{Gi}$$

leads to

$$d \, \ln \gamma_G = d \, \ln \gamma_{Gi}$$

which results in the definition

$$x_G = \sum x_{Gi}$$

Thus, it is legitimate to use the activity coefficient determined for gasoline for each of the pseudocomponents. As the mol fraction of gasoline is close to unity, these activity coefficients will also be close to unity but will still be calculated from the UNIQUAC equation.

Vapor-liquid equilibria suitable for a preliminary design can now be estimated with

$$y_E = \frac{P_E^\circ}{P}\gamma_E\, x_E$$

$$y_W = \frac{P_W^\circ}{P}\gamma_W\, x_W$$

$$y_{Gi} = K_{Gi}\gamma_{Gi}\, x_{Gi}$$

where P is the system pressure, P°'s are vapor pressures, K_{Gi}'s are the K values for the pseudocomponents, and the activity coefficients are evaluated from the ternary UNIQUAC equation. These equations were used in a computerized plate-to-plate calculation of a stripping column fed with wet gasohol. Test runs in a laboratory column containing between six and seven ideal stages were performed and yielded overhead and bottoms water mol fractions reasonably close to those calculated by the plate-to-plate method using the above VLE correlation. Thus, this uncomplicated, thermodynamically consistent model was adequate for the task.

EXAMPLE 17-3

Measurements of equilibrium compositions for the liquid-phase reaction of isobutene and methanol to produce methyl *tert*-butyl ether (MTBE),

$$i\text{-}C_4H_8(l) + CH_3OH(l) = t\text{-}C_4H_9OCH_3(l)$$

have been reported by Colombo *et al.*[16] A commercial cut containing butenes was reacted with methanol in the liquid phase using an acidic resin as a catalyst. Outline the model developed by these authors to determine the equilibrium constant for the reaction.

---

[16] F. Colombo, L. Corl, L. Dalloro, and P. Delogu, *Ind. Eng. Chem. Fundam., 22*, 219 (1983).

Solution 17-3

The model will be detailed using a typical reported liquid-phase equilibrium composition at 75°C.

Component	mol %
isobutane	0.87
n-butane	1.04
trans-2-butene	11.07
isobutene	5.80
1-butene & 2-butene	34.55
dimethyl ether	0.09
isobutene dimer	2.70
MTBE	40.66
tert-butyl alcohol	0.07
methanol	2.60
others	0.55

The equilibrium constant for this reaction can be written

$$K^L = \frac{x_{MTBE}}{x_{C_4H_8} x_{CH_3OH}} \times \frac{\gamma_{MTBE}}{\gamma_{C_4H_8} \gamma_{CH_3OH}}$$

and requires activity coefficients for its evaluation. These can be calculated by the program UNIFAC.EXE after the liquid composition has been simplified by lumping the butenes and the butanes, ignoring trace components, and normalizing to 100%. Activity coefficients will be estimated for a liquid of the following composition

Component	mol %
total butenes	54.51
total butanes	1.92
MTBE	40.95
methanol	2.62

The butanes were assumed to be *n*-butane and the butenes were assumed to be 2-butene. The following table shows the groups, group numbers, and number of each group in each molecular species.

	Number of group *k* in molecule *i*					
	$k = 1$	$k = 2$	$k = 4$	$k = 6$	$k = 16$	$k = 25$
Component	CH_3	CH_2	C	CH=CH	CH_3OH	CH_3O
$i = 1$: 2-C_4H_8	2	0	0	1	0	0
$i = 2$: *n*-C_4H_{10}	2	2	0	0	0	0
$i = 3$: methanol	0	0	0	0	1	0
$i = 4$: MTBE	3	0	1	0	0	1

At 75°C (348 K) estimates from the program UNIFAC.EXE are

Component	γ
2-C_4H_8	1.06
n-C_4H_{10}	1.13
methanol	5.45
MTBE	1.02

With these activity coefficients and the reported equilibrium mol fractions the experimental equilibrium constant is

$$K^L = \frac{0.4066}{0.058(0.026)} \times \frac{1.02}{1.06(5.45)} = 47.6$$

This result was obtained from a model described as follows: activity coefficients via UNIFAC with butanes represented by *n*-butane and butenes represented by 2-butene. As a check on the efficacy of the model, the equilibrium constant can be calculated from thermochemical data via Eqs. (12-15) and (12-16)

$$\frac{d \ln K}{dT} = \frac{\Delta H^\circ}{RT^2} \tag{12-15}$$

$$\Delta H^\circ = \Delta H^\circ_{298} + \int_{298}^{T} \left(\sum v_i c^\circ_{Pi} \right) dT \tag{12-16}$$

The following thermochemical data is available for the components in the gaseous state

Component	ΔH°_{298} cal/g mol	ΔG°_{298} cal/g mol	c°_P cal/g mol · K
isobutene	−4040[a]	13,880[a]	21.30[a]
methanol	−48,050[a]	−38,810[a]	10.49[a]
MTBE	−67,751[b]	−28,040[b]	32.49[c]

[a] Source: reference 2 in Table 12-1.
[b] J. O. Fenwick, D. Harrop, and A. Head, *J. Chem. Thermodyn.*, 7, 943 (1975).
[c] Estimated by group contribution method. See S. W. Benson, *Thermochemical Kinetics*, 2nd ed., Wiley, New York, 1976, Appendix.

At 298 K we have

$$\Delta H^{\circ} = -67,751 - (-48,050) - (-4040) = -15,661$$

$$\Delta G^{\circ} = -28,040 - (-38,810) - 13,880 = -3110$$

$$\sum v_i c^{\circ}_{Pi} = 32.49 - 10.49 - 21.30 = 0.70$$

Heat capacities are assumed constant over the temperature range 25 to 75°C which allows us to obtain on integrating Eq. (12-16)

$$\Delta H^{\circ} = -15,661 + 0.70(T - 298) = 0.70T - 15,870$$

Equation (12-15) can now be written

$$\frac{d \ln K^G}{dT} = \frac{0.70T - 15,870}{RT^2}$$

and integrated to obtain

$$\ln \frac{K^G}{K^G_{298}} = 0.352 \ln \frac{T}{298} + 7987\left(\frac{1}{T} - \frac{1}{298}\right)$$

which results in $K^G = 4.29$ at 75°C.

Vapor pressures are needed in order to calculate the equilibrium constant in the liquid state K^L as illustrated in Sec. 12-5. These P°'s can be determined from the Clausius-Clapeyron equation

$$\ln P^{\circ} = A - \frac{B}{T}$$

using the following parameters for P in atm and T in Kelvin.[17]

Component	A	B
isobutene	10.42892	2776.60
methanol	13.51821	4564.92
MTBE	11.07472	3634.82

From these parameters we obtain at 75°C vapor pressures of 11.59, 1.493, 1.877 atm respectively for isobutene, methanol, and MTBE. We ignore the pressure effect on the liquid phase and write for each component

$$\Delta G_i^f(l) - \Delta G_i^f(g) = RT \ln \frac{P_i^\circ}{1}$$

and for the reaction

$$\Delta G^\circ(l) = \Delta G^\circ(g) + RT \ln \frac{P_{MTBE}^\circ}{P_{C_4H_8}^\circ P_{CH_3OH}^\circ}$$

This leads to

$$-\ln K^L = -\ln K^G + \ln \frac{1.877}{11.59(1.493)}$$

and with the previously determined value of 4.29 for K^G we find $K^L = 39.5$.

   When compared to the model value of 47.6 this result is judged reasonable because of inherent uncertainties. In addition to uncertainties associated with the model itself, the experimental value is subject to any limitations in the experimental determination of liquid mol fractions. Because the reactant mol fractions are small, even small errors in these could cause large errors in the value of K^L. Additionally, the assumption of constant heat capacity and the use of an estimated heat capacity for MTBE contribute to the uncertainty in the value of K^L determined from the thermochemical data. Colombo et al. measured equilibrium mol fractions over a range of reactant ratios and temperatures and found good agreement despite a slight scattering of the data. The conclusion is that this model can be quite useful.

---

[17] F. Colombo et al., *op. cit.*

EXAMPLE 17-4

Vapor-liquid equilibrium data in the water–acetic acid system cannot be correlated in the usual manner because the activity coefficients are not thermodynamically consistent. Show that this system can be successfully modeled by considering association of acetic acid molecules.

Solution 17-4

Experimental studies such as vapor density measurements have shown that acetic acid does associate in the gas phase. For this system, thermo-dynamically consistent activity coefficients have been obtained from experimental VLE data using three different models: (1) based on the formation of acetic acid dimers and tetramers in the gas phase, (2) based on the formation of dimers and trimers in the gas phase, and (3) based on the formation of dimers in both phases.

(1) **The Dimer-Tetramer, Gas-Phase Model.**[18] The model is delineated by assuming the liquid phase to contain only water and acetic acid monomers while the vapor is assumed to be a perfect gas mixture containing water and monomers, dimers, and tetramers of acetic acid. Chemical equilibrium prevails in the gas phase with respect to the reactions

$$2A = A_2 \qquad K_2 = \frac{p_2}{p_1^2}$$

$$4A = A_4 \qquad K_4 = \frac{p_4}{p_1^4}$$

and phase equilibrium prevails between the liquid and vapor phases for water and acetic acid monomer.

The partial pressures of the dimer p_2 and tetramer p_4 are related to the partial pressure of the monomer p_1 through the chemical equilibrium relations, and for pure acetic acid the vapor pressure P_A° is

$$P_A^\circ = p_1 + p_2 + p_4 = p_1 + K_2 p_1^2 + K_4 p_1^4 \qquad (17\text{-}6)$$

The value of p_1 obtained by solving this equation will be designated P_1°, a fictitious monomer vapor pressure. Values of K_2, K_4, and P_A° are known as functions of temperature.

---

[18] Based on the work of E. Sebastiani and L. Lacquaniti, *Chem. Eng. Sci.*, 22, 1155 (1967).

$$\log K_2 = \frac{3164}{T} - 10.4184$$

$$\log K_4 = \frac{5884}{T} - 23.4824$$

$$\log P_A^\circ = 14.39756 - \frac{9399.86}{T + 424.9}$$

In the binary system the apparent mol fraction of acetic acid in the vapor y_A is given in terms of mol numbers by

$$y_A = \frac{n_1 + 2n_2 + 4n_4}{n_1 + 2n_2 + 4n_4 + n_W}$$

or in terms of partial pressures by

$$y_A = \frac{p_1 + 2K_2 p_1^2 + 4K_4 p_1^4}{p_1 + 2K_2 p_1^2 + 4K_4 p_1^4 + p_W}$$

The total pressure is

$$P = p_1 + p_2 + p_4 + p_W$$

and this expression can be used to eliminate p_W from the preceding equation with the result

$$y_A P = p_1 + (2 - y_A)K_2 p_1^2 + (4 - 3y_A)K_4 p_1^4 \qquad (17\text{-}7)$$

The above equations are the mathematical description of our model and are used with *xyt* data at a pressure of 760 mmHg to obtain activity coefficients for acetic acid and water.

For a single VLE data point the known values of P and y_A are used in Eq. (17-7) along with K_2 and K_4 evaluated at the known temperature to obtain p_1. At this temperature P_1° is obtained from the vapor pressure and Eq. (17-6). Because we have assumed that acetic acid exists in the liquid phase only as monomers ($x_A = x_1$), we may determine its activity coefficient from a statement of phase equilibrium

$$p_1 = x_1 P_1^\circ \gamma_1 = x_A P_1^\circ \gamma_A$$

and similarly for water

$$p_W = x_W P_W^\circ \gamma_W$$

where p_w is obtained from the total pressure relation using p_1 and the equilibrium values of p_2 and p_4.

From the VLE data for a liquid composition of 50 mol % water the equilibrium temperature and vapor composition are 104.3°C and 64.0 mol % water respectively. At this temperature we have

$$K_2 = 0.00921 \text{ mmHg}^{-1}; \; K_4 = 1.28(10^{-8}) \text{ mmHg}^{-3}; \; P_A^\circ = 481.0 \text{ mmHg}$$

Using these values along with $y_A = 0.360$ and $P = 760$ mmHg yields

$$p_1 = 104.44; \qquad p_2 = 100.44; \qquad p_4 = 1.52 \text{ mmHg}$$

At this temperature $P_w^\circ = 884.36$ mmHg and from Eq. (17-6) we find $P_1^\circ = 177.66$ mmHg. With these values we obtain

$$\gamma_A = \frac{p_1}{x_A P_1^\circ} = \frac{104.44}{0.5(177.66)} = 1.176$$

$$\gamma_W = \frac{P - p_1 - p_2 - p_4}{x_W P_W^\circ} = \frac{760 - 104.44 - 100.44 - 1.52}{0.5(884.36)}$$

$$\gamma_W = 1.252$$

The VLE data was processed in this manner to obtain a thermodynamically consistent set of γ_A and γ_W. Available heat of mixing data were used to adjust these activity coefficients to a temperature of 25°C[19] so that comparison with other models could be made. The heat of mixing data could be fitted to

$$\Delta h = x_W(1 - x_W)[a + b(2x_W - 1) + c(2x_W - 1)^2 + d(2x_W - 1)^3]$$

with

$$a = 294; \qquad b = -198; \qquad c = -3; \qquad d = -436$$

for Δh in cal/g mol. The activity coefficients at 25°C could be fitted to the three-constant Redlich-Kister equations

---

[19] See Sec. 11-2.

$$\log \gamma_W = x_A^2 [A + B(4x_W - 1) + C(x_W - x_A)(4x_W - 1)^2]$$

$$\log \gamma_A = x_W^2 [A + B(4x_W - 3) + C(x_W - x_A)(4x_W - 3)^2]$$

with

$$A = 0.3338; \quad B = 0.0283; \quad C = 0.1081$$

(2) The Dimer-Trimer, Gas-Phase Model.[20]   This model differs from the Dimer-Tetramer, Gas-Phase Model only by the replacement of the tetramerzation reaction with the trimerization reaction

$$3A = A_3 \qquad K_3 = \frac{p_3}{p_1^3}$$

with

$$\log K_3 = \frac{4970}{T} - 18.591$$

where K_3 is in $mmHg^{-2}$ and T is in Kelvin. The equations describing this model are similar to those of the Dimer-Tetramer, Gas-Phase Model as is the method of solution.

At 25°C the results obtained from this model are comparable to those obtained from the previous model. This result is not surprising because the pressure of the system is quite low at this temperature and the polymerization reactions are favored by high pressure as can be seen by the rather low value of p_4 calculated for the previous model at 760 mmHg.

(3) The Two-phase Dimer Model.[21]   The binary system of acetic acid and water is assumed to be a ternary system containing water and acetic acid monomers and dimers with deviations from ideal behavior in both phases assumed to arise from the association of acetic acid. For the dimerization reaction

$$2A = A_2$$

equilibrium constants are written for the vapor and liquid phases

$$K = \frac{p_2}{p_1^2} \qquad k = \frac{\xi_2}{\xi_1^2}$$

---

[20] Based on the work of R. S. Hansen, F. A. Miller, and S. D. Christian, *J. Phys. Chem., 59*, 391 (1955).
[21] Based on the work of J. Marek, *Colln. Czech. Chem. Commun., 20*, 1490 (1955).

where ξ is a mol fraction in the ternary liquid phase and subscripts 1 and 2 refer to monomer and dimer respectively. The apparent mol fractions of acetic acid, x_A and y_A, are related to the ternary mol fractions by

$$x_A = \frac{\xi_1 + 2\xi_2}{1+\xi_2} \qquad y_A = \frac{\eta_1 + 2\eta_2}{1+\eta_2}$$

where η is a ternary mol fraction in the vapor phase. Using the equilibrium constant expressions and rearranging yields

$$\xi_1 = \frac{\sqrt{1+4kx_A(2-x_A)}-1}{2k(2-x_A)}$$

and

$$\eta_1 = \frac{\sqrt{1+4KPy_A(2-y_A)}-1}{2KP(2-y_A)}$$

which for pure acetic acid at the saturation pressure P_A° reduce to

$$\eta_{01} = \frac{\sqrt{1+4KP_A^\circ}-1}{2KP_A^\circ}$$

and

$$\xi_{01} = \frac{\sqrt{1+4k}-1}{2k}$$

For the pure acetic acid the condition of phase equilibrium for the monomer is

$$P_A^\circ \eta_{01} = P_1^\circ \xi_{01}$$

and on substituting for ξ_{01} and η_{01} the vapor pressure of the monomer P_1° is given by

$$P_1^\circ = P_A^\circ \left[\frac{\left(\sqrt{1+4KP_A^\circ}-1\right)k}{\left(\sqrt{1+4k}-1\right)KP_A^\circ} \right]$$

For the acetic acid–water mixture the condition of phase equilibrium for the monomer is written

$$P\eta_1 = P_1^\circ \xi_1$$

and on substituting for ξ_1, η_1, and P_1° and rearranging one obtains the equilibrium statement for component A

$$Py_A Z_A = P_A^\circ x_A \Gamma_A$$

where vapor-phase terms have been collected into a vapor-phase association factor Z_A which functions as a fugacity coefficient

$$Z_A = \frac{\left(\sqrt{1+4KPy_A(2-y_A)} - 1\right)KP_A^\circ}{\left(\sqrt{1+4KP_A^\circ} - 1\right)KPy_A(2-y_A)}$$

and liquid-phase terms have been collected into a liquid-phase association factor Γ_A which functions as an activity coefficient

$$\Gamma_A = \frac{\left(\sqrt{1+4kx_A(2-x_A)} - 1\right)k}{\left(\sqrt{1+4k} - 1\right)kx_A(2-x_A)}$$

The corresponding association factors for water, component B, are

$$Z_B = \frac{1+4KP(2-y_A) - \sqrt{1+4KPy_A(2-y_A)}}{2KP(2-y_A)^2}$$

and

$$\Gamma_B = \frac{1+4k(2-x_A) - \sqrt{1+4kx_A(2-x_A)}}{2k(2-x_A)^2}$$

These are obtained from the phase equilibrium statement for component B

$$P\eta_B = P_B^\circ \xi_B$$

Where relations for ξ_B and η_B are obtained by combining the dimerization equilibrium expressions with the ternary mol fraction expressions;

$$y_B = \frac{\eta_B}{1+\eta_2} = \frac{\eta_B}{1+KP\eta_1^2}$$

and

$$x_B = \frac{\xi_B}{1+\xi_2} = \frac{\xi_B}{1+k\xi_1^2}$$

On substituting for ξ_1 and η_1 and rearranging, the resulting expressions for ξ_B and η_B are inserted into the phase equilibrium statement for component B which then takes the form

$$P y_B Z_B = P_B^\circ x_B \Gamma_B$$

Because the gas-phase dimerization equilibrium constant K is known, values of Z_A and Z_B can be computed for each VLE data point. These were then used in the phase equilibrium statements for components A and B to obtain values of the liquid association factors Γ_A and Γ_B for each isobaric VLE data point. From the specified dependence of the Γ's on x_A, a value of the parameter k was determined at each temperature. These k's show the exponential dependence on temperature expected of an equilibrium constant.

$$\log k = 3.8326 - \frac{1030}{T}$$

Thus, the *xyt* data at a pressure of one atmosphere were well correlated with a single temperature-dependent parameter k.

All three models represent the VLE data for the acetic acid–water system equally well although each is based on a different microscopic view. Because of the relatively low pressures involved, gas-phase association beyond the formation of dimers is almost negligible and from *xyt* data alone it is not possible to determine whether model 1 or model 2 is more realistic. While model 3 proved effective for the acetic acid–water system, it was found to fail when acetic acid was paired with benzene and *p*-xylene thus casting doubt on the liquid-phase dimerization mechanism.

17-3 EVALUATION OF MODELS

Although the primary purpose of a model is to provide a mathematical description of the system and the major criterion of its effectiveness is how well it quantitatively represents system behavior, there are other criteria that contribute to the evaluation of a given model. Sometimes it is possible for different models to produce essentially the same results and in these cases the selection is based on these other criteria.

Realistic mechanisms.   An acceptable model should have a realistic mechanism. It might be possible to represent our system perfectly with a general mathematical expression such as a power series with several empirically determined coefficients—the brute-force approach—but we would have no confidence in its extrapolation to new conditions nor would we have gained any understanding of our system. As a general rule chemical engineers prefer mechanistic models to a brute-force approach and prefer that the number of adjustable parameters be kept as small as possible.

Successful models of complex phenomena or processes find their way into the literature and become part of our chemical engineering practice. Therefore, experienced engineers have at their disposal a vast repertoire of model mechanisms that have been proven successful which they can employ when faced with a modeling situation. Their first choice will be a model or mechanism that has worked in a similar situation. Thus, the stance of the practicing engineer is conservative—the experiences of the past are considered valuable and are discarded only if it is obvious that they are not useful or cannot be adjusted to fit the present situation.

In the context of thermodynamics there are usually constraints or limiting conditions that apply to the system which can be used to test the validity of models or model mechanisms. In addition to material balances and the first and second laws, the various criteria of spontaneity developed in Sec.7-1 and the Gibbs-Duhem equation serve as constraints. Some limiting conditions are:

- Ideal-gas behavior should be approached as the pressure decreases or the temperature increases.

- The activity coefficient of a component approaches unity as the mol fraction of that component approaches unity.

- By virtue of their definitions, certain thermodynamic quantities such as fugacities, activity coefficients, and equilibrium constants can only assume positive values.

- Equilibrium constants are expected to show an exponential dependence on absolute temperature.

Less authoritative constraints take the form of comparisons of thermodynamic properties evaluated as model parameters with the known behavior of similar systems.

Reasonable Parameters.   Whenever possible, a model parameter should be physically significant. This usually means that it conforms to the constraints listed above or to our expectations based on experience with similar systems. We should also expect consistency among the model parameters. For example, in Ex.17-4 the empirically determined equilibrium constant for the liquid-phase dimerization of acetic acid showed an exponential dependence on absolute temperature as expected, but its temperature dependence is opposite to that found for the gas-phase dimerization. While this is not impossible,[22] it should alert us to a potential problem with this model and suggests further scrutiny. Usually, if the model mechanism is realistic we will find that the parameters have physical significance. Often, the model equation will contain a grouping or "lumping" of several parameters which for the purpose of fitting data to the model equation could be treated as a single "lumped" parameter. However, if possible, one should independently evaluate as many of the parameters contributing to the "lump" as possible so that any remaining parameters can be checked for physical significance.

Other Desirable Attributes. Simplicity and ease of computability usually go hand-in-hand and are also desirable features of a good model. Because it is sometimes possible to fit data to a poor model containing many adjustable parameters, a model should be as simple as possible with as few adjustable parameters as possible if we are to judge its efficacy. Ease of computability is especially important if the model equations are to be used in a calculation requiring many iterations; for example, a model for VLE used in a plate-to-plate distillation calculation.

The Acetic acid–Water System. Example 17-4 presents three alternative models which we can evaluate. Because only dimerization is important at the low pressure of one atmosphere, there is very little difference between the effectiveness of the two gas-phase models. Both models make use of chemical equilibrium constants reported in the literature rather than treat these constants as empirical parameters to be fitted to experimental data. Thus, for both models only the parameters in the activity coefficient equations are evaluated empirically from the VLE data. Both of these models have realistic mechanisms and possess reasonable parameters and also pass the tests of simplicity and ease of computability. As they also do a respectable job of representing the VLE data, they may be deemed effective.

On the other hand, the two-phase dimerization model can be judged less effective even though it is able to represent the VLE data just as well with only one adjustable parameter, the liquid-phase equilibrium constant. The major reason for our discomfort with this model is that liquid-phase dimerization seems to be an appropriate mechanism for the acetic acid–water system but does not appear to be operative when acetic acid is paired with other substances. As no reasonable explanation of dimerization only in aqueous solutions is available, this lack of extensibility to other systems is a serious flaw. Moreover, there are serious computational difficulties associated with this model.[23]

---

[22] See Prob. 12-9.
[23] See Prob. 17-8.

PROBLEMS

17-1.   Use the HVOS model with activity coefficients estimated from UNIFAC to calculate VLE data for the propane–benzene system at 37.78°C. Compare your results with the experimental data found in the spreadsheet FG9-24B.WQ1 in the BACKUP directory of the companion CD-ROM.

17-2.   In Ex. 17-1 the program UNIFAC.EXE was used to estimate a k_{12} value for the acetone–methanol system. Use the following VLE data for this system at 45°C to determine a k_{12} via the HVOS model and use this value to calculate VLE data at 150°C. Compare your results with experimental data found in spreadsheet EX17-1.WQ1 in the BACKUP directory of the companion CD-ROM.

SYSTEM: ACETONE (1)–METHANOL (2) AT 45°C[a]

x_1	y_1	P (mmHg)
0	0	332.6
0.093	0.209	382.8
0.193	0.347	423.9
0.293	0.445	454.1
0.401	0.528	477.5
0.501	0.597	492.3
0.587	0.653	502.3
0.687	0.721	510.2
0.756	0.774	513.0
0.840	0.841	515.3
0.896	0.890	512.8
0.952	0.946	507.5
1	1	511.1

[a] Source: reference B8 in Table 10-8.

17-3.   Use the program UNIFAC.EXE to estimate activity coefficients and calculate the equilibrium composition when equal molar quantities of ethanol and acetic acid are reacted in the liquid phase to produce ethyl acetate and water at 25°C. You may use any information found in Ex. 12-12.

17-4.   The presence of dioxin in the environment is a serious concern and therefore considerable effort has been expended in identifying the sources of this contaminant. There are 22 isomers having the general formula $C_{12}H_4O_2Cl_4$, but 2, 3, 7, 8 tetrachlorodibenzo-*p*-dioxin is the most notorious and the one commonly referred to as dioxin. It has been suggested that some dioxins are produced through natural processes (the *de novo* hypothesis) such as wood burning, and the detection of dioxin in deposits formed inside the chimneys of wood-burning stoves has been cited as support for this hypothesis. Delineate plausible models of wood burning based on ultimate chemical equilibrium and use them to test this hypothesis. The detection threshold for dioxin is 3 parts per 10^{12}.

Thermochemical data for this compound and other dioxins and chlorinated phenols have been estimated at various temperatures.[24] At 500 and 1000 K reported log K' values for dioxin are 9.708 and -8.366 respectively. Assume wood to have the chemical composition of cellulose, $(C_6H_{10}O_5)_x$, and contain 84 parts per 10^6 of chlorine.[25]

17-5. On Monday, April 3, 1978 the Associated Press gave the following account of train derailment in Brownson, Nebraska.

> A derailed tank car filled with liquid phosphorus exploded Sunday, injuring three persons and scattering debris as far as a quarter-mile away, officials said. . . . Thirty cars of the 85-car train had derailed Sunday morning and the railroad reported then that the tank car was burning. It also said that, before the car exploded, the area had been evacuated as a precaution.

Examine possible mechanisms (models) and make an educated guess as to the most likely cause of the explosion. Phosphorus is a liquid at ambient temperature and burns in the presence of air. It is therefore shipped and stored under water.

17-6. Use the following isothermal data[26] at 99.1°C for the vapor density of deuterated acetic acid (mol. wt. 63.8) to discriminate between the dimer-trimer model and the dimer-tetramer model. *Hint*: Use the fact that the amount trimers and tetramers decrease as the pressure (or density) approaches zero and model the data first as dimers only. After obtaining a value of K_2 by extrapolation to zero pressure (or density), use that value of K_2 for the final model evaluation.

Pressure (mmHg)	Density (g/l)
94.9	0.537
146.9	0.587
201.5	0.841
251.7	1.079
296.8	1.302
329.8	1.465

17-7. It is believed that water dissolved in liquid chloroform may form dimers and trimers in solution. Use the following isothermal data giving the concentration of water in chloroform (g mol/l) for various values of the water activity to discriminate between four possible models: (1) no association, (2) only dimers are formed, (3) only trimers are formed, and (4) both dimers and trimers are formed. The water activity was determined by equilibrating aqueous $CaCl_2$ solutions with liquid chloroform. The solubility of chloroform in the aqueous phase is assumed negligible and activities of water in $CaCl_2$ solutions are known from published data based on other types of measurements.

[24] W. M. Shaub, *Thermochimica Acta, 58*, 11 (1982).
[25] J. R. Long and D. J. Hanson, *C & E News*, 6 June 1983, p. 23.
[26] Source: A. E. Potter, Jr., P. Bender, and H. L. Ritter, *J. Phys. Chem., 59*, 250 (1955).

CONCENTRATION AND ACTIVITY OF WATER IN CHLOROFORM AT 25°C

C_w(g mol/l)	activity
0.0738	1.00
0.0693	0.947
0.0632	0.864
0.0544	0.754
0.0455	0.631
0.0368	0.511
0.0275	0.391

17-8. Use any information in Ex.17-4 to calculate VLE for the water–acetic acid system at 760 mmHg via

(a) the dimer-tetramer, gas-phase model.
(b) the two-phase, dimer model.

Compare your results with the following experimental data. Also, comment on the ease of calculation as a criterion for model selection.

VLE DATA AT 760 mmHg
SYSTEM: WATER (1)–ACETIC ACID (2)[a]

x_1	y_1	t(°C)
0.125	0.1825	112.8
0.250	0.365	108.9
0.395	0.515	106.2
0.520	0.635	104.2
0.641	0.750	102.8
0.755	0.817	101.7
0.865	0.907	100.8

[a] Source: E. Sebastiani and L. Lacquaniti, *Chem. Eng. Sci., 22,* 1155 (1967).

17-9. Simplify the gas-phase models for the acetic acid–water system by considering only dimerization. Compare your results at 760 mmHg with experimental data. See Prob. 17-8 for VLE data.

17-10. Use a vapor-phase dimerization model to correlate the following VLE data for the formic acid–water system. The gas-phase dimerization equilibrium constant for formic acid has been measured and reported to be

$$\log K = \frac{3083}{T} - 10.743$$

and the Antoine parameters are

Antoine Parameter	Water	Formic acid
A	7.89935	7.58178
B	1623.95	699.173
C	223.59	260.714

SYSTEM: FORMIC ACID (1)–WATER (2) AT 760 mmHg[a]

x_1	y_1	$t°C$
0.100	0.049	101.8
0.171	0.093	102.9
0.260	0.164	104.2
0.368	0.282	106.0
0.478	0.433	107.1
0.536	0.518	107.6
0.589	0.595	107.6
0.621	0.641	107.3
0.679	0.721	107.1
0.782	0.838	105.9
0.845	0.898	104.6
0.960	0.976	102.3

[a] Source: J. J. Conti., D. F. Othmer, and R. Gilmont, *J. Chem. Eng. Data, 5,* 301 (1960).

Compare your correlation to a correlation that ignores chemical effects and simply uses an activity coefficient equation.

17-11. Use a gas-phase dimerization model to correlate the following VLE data for the formic acid (1)–acetic acid (2) system. Three types of dimers are possible: A_2, F_2, and AF. Reported equilibrium constants for these reactions are

$$\log K_{A_2} = \frac{3164}{T} - 10.4184$$

$$\log K_{F_2} = \frac{3083}{T} - 10.743$$

$$\log K_{AF} = \frac{3193}{T} - 10.356$$

Antoine parameters are

Antoine parameters	Acetic acid	Formic acid
A	14.39756	7.58178
B	9399.86	1699.173
C	698.05	260.714

SYSTEM: FORMIC ACID (1)–ACETIC ACID (2) AT 760 mmHg[a]

Wt. % Formic acid		
Liquid	Vapor	$t, °C$
10.0	14.4	114.2
14.0	18.6	113.3
17.6	23.6	112.3
22.9	29.4	110.8
31.4	37.8	109.4
38.3	45.6	107.9
45.5	53.4	106.6
54.9	61.1	105.3
65.1	71.1	104.0
74.5	79.2	102.9
83.6	86.6	101.9
90.5	92.6	101.2

[a] Source: J. J. Conti, D. F. Othmer, and R. Gilmont, *J. Chem. Eng. Data,* 5, 301 (1960).

17-12. A liquid of the following composition

n-pentane	20 mol %
n-hexane	20 mol %
n-heptane	20 mol %
toluene	20 mol %
benzene	10 mol %
ethanol	10 mol %

becomes saturated with water at 298 K. Estimate the mol % water in the resulting liquid.

---

[a] Source: J.J. Conti, D.F. Othmer, and R. Gilmont, *J. Chem. Eng. Data,* 5 301 (1960).

17-13. Estimate the equilibrium mol fractions when 1 mol of i-C_4H_8 reacts with 2 mols of CH_3OH in the liquid phase at 50°C to form MTBE. What pressure would be necessary to suppress the vapor phase?

17-14. A stream of mixed xylenes is available and we wish to consider the feasibility of using it to produce pure *para*-xylene by crystallization. The mother liquor from the crystallizer would be sent to a reactor containing a catalyst that promotes the xylene isomerization and the reactor effluent would be recycled to the crystallizer.

Assume that the reactor effluent is an equilibrium mixture of the xylene isomers and that the crystallizer operates at a temperature 1°C above the freezing point of the next-highest-freezing isomer. The feed stream contains 10 mol % *para*, 30 mol % *ortho*, and 60 mol % *meta* xylene.

(a) Devise a process utilizing a reactor and crystallizer that would accomplish this production of pure *para*-xylene giving special attention to the operating conditions for these two units and determine the energy requirements. Solid-liquid equilibrium for this system is treated in Ex. 11-3 and thermochemical data can be found in Prob. 12-14.

(b) With a large recycle, a reactor operating in the gas phase at reasonably high temperatures, and a crystallizer operating under conditions of solid-liquid equilibrium, an efficient process will require efficient heat exchange between the reactor effluent and the crystallizer effluent streams. Because both streams undergo a phase change, the design of a heat exchange system will not be routine but will require application of the principles developed in Sec. 14-4. Exercise your creativity and develop what you consider to be the most energy-efficient heat exchange system. For this preliminary design the following property estimates are available:

liquid c_p: 0.40 cal/g °C

vapor c_p: 0.40 cal/g °C

heat of vaporization: 80 cal/g

APPENDIXES

APPENDIX A

TABLE A-1
CRITICAL CONSTANTS OF SELECTED SUBSTANCES[a]

Substance		Molecular weight	Critical temperature, T_c (K)	Critical pressure, P_c (atm)	Critical compressibility factor, Z_c	Acentric factor, ω
Helium	He	4.00	5.3	2.26	0.300	−0.365
Neon	Ne	20.18	44.5	26.9	0.307	−0.029
Argon	A	39.94	151	48.0	0.291	0.001
Krypton	Kr	83.7	209.4	54.3	0.291	0.005
Chlorine	Cl_2	70.91	417	76.1	0.276	0.090
Hydrogen	H_2	2.02	33.3	12.80	0.304	−0.218
Nitrogen	N_2	28.02	126.2	33.5	0.291	0.039
Oxygen	O_2	32.00	154.8	50.1	0.308	0.025
Carbon dioxide	CO_2	44.01	304.2	72.9	0.274	0.239
Carbon monoxide	CO	28.01	133	34.5	0.294	0.066
Ethylene oxide	C_2H_4O	44.05	468	71.0	0.255	0.202
Nitrous oxide	N_2O	44.02	309.7	71.7	0.272	0.165
Nitric oxide	NO	30.01	180	64	0.251	0.588
Nitrogen peroxide	NO_2	46.01	431	100	0.232	0.834
Sulfur dioxide	SO_2	64.06	430.7	77.8	0.268	0.256
Sulfur trioxide	SO_3	80.06	491.4	83.8	0.262	0.481
Water	H_2O	18.02	647.4	218.3	0.230	0.344
Ammonia	NH_3	17.03	405.5	111.3	0.242	0.250

TABLE A-1 *(Continued)*

Substance		Molecular weight	Critical temperature, T_c (K)	Critical pressure, P_c (atm)	Critical compressibility factor, Z_c	Acentric factor, ω
Hydrazine	N_2H_4	32.05	653	145		0.316
Hydrogen cyanide	HCN	27.03	456.7	53.2	0.197	0.388
Carbon disulfide	CS_2	76.13	552	78	0.293	0.109
Hydrogen sulfide	H_2S	34.08	373.6	88.9	0.283	0.081
Hydrogen chloride	HCl	36.47	324.6	81.5	0.147	0.133
Methane	CH_4	16.04	191.1	45.8	0.289	0.011
Ethane	C_2H_6	30.07	305.5	48.2	0.284	0.099
Propane	C_3H_8	44.09	370.0	42.0	0.277	0.153
n-Butane	C_4H_{10}	58.12	425.2	37.5	0.274	0.199
Isobutane	C_4H_{10}	58.12	408.1	36.0	0.283	0.183
n-Pentane	C_5H_{12}	72.15	469.8	33.3	0.269	0.251
Ethylene	C_2H_4	28.05	282.4	50.0	0.268	0.089
Propene	C_3H_6	42.08	365.0	45.6	0.276	0.144
Acetylene	C_2H_2	26.04	309	61.6	0.274	0.190
Cyclohexane	C_6H_{12}	84.16	553	40.0	0.272	0.212
Benzene	C_6H_6	78.11	562	48.6	0.274	0.212
Toluene	C_7H_8	92.13	594.0	41.6	0.273	0.263
Methyl alcohol	CH_4O	32.04	513.2	78.5	0.220	0.556
Ethyl alcohol	C_2H_6O	46.07	516	63.0	0.248	0.644
n-Propyl alcohol	C_3H_8O	60.09	537	50.2	0.251	0.623
Isopropyl alcohol	C_3H_8O	60.09	508.8	53	0.278	0.665
Dioxane	$C_4H_8O_2$	88.10	585	50.7	0.253	0.281
Acetone	C_3H_6O	58.08	508.7	46.6	0.238	0.304
Ethyl methyl ketone	C_4H_8O	72.10	533	39.5	0.262	0.320
Acetic acid	$C_2H_4O_2$	60.05	594.8	57.1	0.200	0.447
Ethyl acetate	$C_4H_8O_2$	88.10	523.3	37.8	0.252	0.362
Acetonitrile	CH_3CN	41.05	547.9	47.7	0.184	0.327
Methyl chloride	CH_3Cl	50.49	416.3	65.9	0.276	0.153
Chloroform	$CHCl_3$	119.39	536.6	54	0.294	0.218
Carbon tetrachloride	CCl_4	153.84	556.4	45.0	0.272	0.193

[a] From K. A. Kobe and R. E. Lynn, Jr., "The Critical Properties of Elements and Compounds," *Chem. Rev., 52,* 117 (1953) and R. C. Reid, J. M. Prausnitz, and B. E. Poling, *The Properties of Gases and Liquids,* 4th ed., McGraw-Hill, New York, 1987.

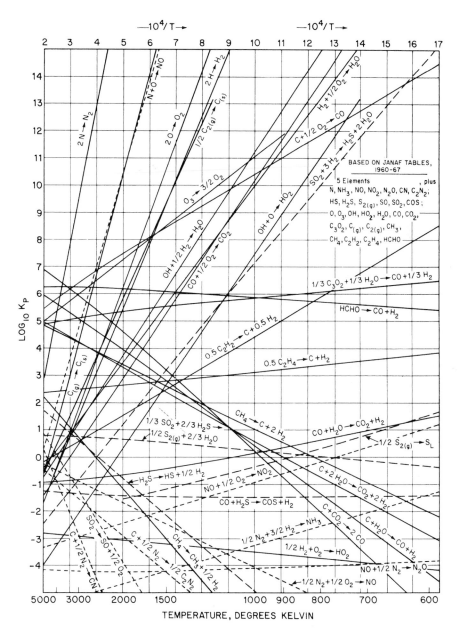

Figure A-1 Equilibrium constants over the range of 600–5000 K for selected reactions. (Michael Modell and Robert C. Reid, Thermodynamics and Its Applications, 1974, p. 396. Reprinted by permission of Prentice Hall, Inc., Englewood Cliffs, N.J.)

APPENDIX B

TABLE B-1

MOLAR HEAT CAPACITIES OF GASES AT ZERO PRESSURE,[a, b]

$$c_P^\circ = a + bT + cT^2 + dT^3$$

Substance		a	$b \times 10^2$	$c \times 10^5$	$d \times 10^9$	Temperature range (K)	Error Max. (%)	Avg. (%)
Nitrogen	N_2	6.903	−0.03753	0.1930	−0.6861	273–1800	0.59	0.34
Oxygen	O_2	6.085	0.3631	−0.1709	0.3133	273–1800	1.19	0.28
Air		6.713	0.04697	0.1147	−0.4696	273–1800	0.72	0.33
Hydrogen	H_2	6.952	−0.04576	0.09563	−0.2079	273–1800	1.01	0.26
Carbon monoxide	CO	6.726	0.04001	0.1283	−0.5307	273–1800	0.89	0.37
Carbon dioxide	CO_2	5.316	1.4285	−0.8362	1.784	273–1800	0.67	0.22
Water vapor	H_2O	7.700	0.04594	0.2521	−0.8587	273–1800	0.53	0.24
Nitric oxide	NO	7.008	−0.02244	0.2328	−1.000	273–1800	0.97	0.36
Nitrous oxide	N_2O	5.758	1.4004	−0.8508	2.526	273–1500	0.59	0.26
Nitrogen dioxide	NO_2	5.48	1.365	−0.841	1.88	273–1500	0.46	0.18
Nitrogen tetroxide	N_2O_4	7.9	4.46	−2.71	—	273–600	0.97	0.36
Ammonia	NH_3	6.5846	0.61251	0.23663	−1.5981	273–1500	0.91	0.36
Hydrazine	N_2H_4	3.890	3.554	−2.304	5.990	273–1500	1.80	0.50
Sulfur	S_2	6.499	0.5298	−0.3888	0.9520	273–1800	0.99	0.38
Sulfur dioxide	SO_2	6.157	1.384	−0.9103	2.057	273–1800	0.45	0.24
Sulfur trioxide	SO_3	3.918	3.483	−2.675	7.744	273–1300	0.29	0.13
Acetylene	C_2H_2	5.21	2.2008	−1.559	4.349	273–1500	1.46	0.59
Cyclohexane	C_6H_{12}	−15.935	16.454	−9.203	19.27	273–1500	1.57	0.37
Benzene	C_6H_6	−8.650	11.578	−7.540	18.54	273–1500	0.34	0.20
Toluene	C_7H_8	−8.213	13.357	−8.230	19.20	273–1500	0.29	0.18
Ethylbenzene	C_8H_{10}	−8.398	15.935	−10.003	23.95	273–1500	0.34	0.19
Styrene	C_8H_8	−5.968	14.354	−9.150	22.03	273–1500	0.37	0.23
Formaldehyde	CH_2O	5.447	0.9739	0.1703	−2.078	273–1500	1.41	0.62
Acetaldehyde	C_2H_4O	4.19	3.164	−0.515	−3.800	273–1000	0.40	0.17

TABLE B-1 *(Continued)*

Substance		a	$b \times 10^2$	$c \times 10^5$	$d \times 10^9$	Temperature range (K)	Error	
							Max. (%)	Avg. (%)
Methanol	CH_4O	4.55	2.186	−0.291	−1.92	273–1000	0.18	0.08
Ethanol	C_2H_6O	4.75	5.006	−2.479	4.790	273–1500	0.40	0.22
i-Propyl alcohol	C_3H_8O	0.7936	8.502	−5.016	11.56	273–1500	0.35	0.18
n-Propyl alcohol	C_3H_8O	−1.307	9.235	−5.800	14.14	273–1500	0.90	0.30
Acetone	C_3H_6O	1.625	6.661	−3.737	8.307	273–1500	0.56	0.10
Ethylene oxide	C_2H_4O	−1.12	4.925	−2.389	3.149	273–1000	0.36	0.14
Methyl chloride	CH_3Cl	3.05	2.596	−1.244	2.300	273–1500	0.75	0.16
Chloroform	$CHCl_3$	7.61	3.461	−2.668	7.344	273–1500	0.92	0.42
Carbon tetrachloride	CCl_4	12.24	3.400	−2.995	8.828	273–1500	1.21	0.57
Hydrogen cyanide	HCN	6.34	0.8375	−0.2611	—	273–1500	1.42	0.76
Hydrogen sulfide	H_2S	7.070	0.3128	0.1364	−0.7867	273–1800	0.74	0.37
Carbon disulfide	CS_2	7.390	1.489	−1.091	2.760	273–1800	0.76	0.47
Chlorine	Cl_2	6.8214	0.57095	−0.5107	1.547	273–1500	0.50	0.23
Hydrogen chloride	HCl	7.244	−0.1820	0.3170	−1.036	273–1500	0.22	0.08
Methane	CH_4	4.750	1.200	0.3030	−2.630	273–1500	1.33	0.57
Ethane	C_2H_6	1.648	4.124	−1.530	1.740	273–1500	0.83	0.28
Propane	C_3H_8	−0.966	7.279	−3.755	7.580	273–1500	0.40	0.12
n-Butane	C_4H_{10}	0.945	8.873	−4.380	8.360	273–1500	0.54	0.24
i-Butane	C_4H_{10}	−1.890	9.936	−5.495	11.92	273–1500	0.25	0.13
n-Pentane	C_5H_{12}	1.618	10.85	−5.365	10.10	273–1500	0.56	0.21
n-Hexane	C_6H_{14}	1.657	13.19	−6.844	13.78	273–1500	0.72	0.20
Ethylene	C_2H_4	0.944	3.735	−1.993	4.220	273–1500	0.54	0.13
Propylene	C_3H_6	0.753	5.691	−2.910	5.880	273–1500	0.73	0.17
1-Butene	C_4H_8	−0.240	8.650	−5.110	12.07	273–1500	0.25	0.18
i-Butene	C_4H_8	1.650	7.702	−3.981	8.020	273–1500	0.11	0.06
cis-2-Butene	C_4H_8	−1.778	8.078	−4.074	7.890	273–1500	0.78	0.14
trans-2-Butene	C_4H_8	2.340	7.220	−3.403	6.070	273–1500	0.54	0.12
Acetonitrile	CH_3CN	5.09	2.7634	−0.9111	—	273–1200	0.45	0.26

[a] Data from reference 8 in Table 12-1.
[b] c_P° (cal/g mol · K); T (K)

APPENDIX C

TABLE C-1
THERMODYNAMIC PROPERTIES OF SATURATED STEAM[a, b]

Press. kPa P	Temp. °C T	Sat. liquid v^L	Sat. vapor v^G	Sat. liquid u^L	Evap. Δu	Sat. vapor u^G	Sat. liquid h^L	Evap. Δh	Sat. vapor h^G	Sat. liquid s^L	Evap. Δs	Sat. vapor s^G
0.6113	0.01	0.001 000	206.14	.00	2375.3	2375.3	.01	2501.3	2501.4	.0000	9.1562	9.1562
1.0	6.98	0.001 000	129.21	29.30	2355.7	2385.0	29.30	2484.9	2514.2	.1059	8.8697	8.9756
1.5	13.03	0.001 001	87.98	54.71	2338.6	2393.3	54.71	2470.6	2525.3	.1957	8.6322	8.8279
2.0	17.50	0.001 001	67.00	73.48	2326.0	2399.5	73.48	2460.0	2533.5	.2607	8.4629	8.7237
2.5	21.08	0.001 002	54.25	88.48	2315.9	2404.4	88.49	2451.6	2540.0	.3120	8.3311	8.6432
3.0	24.08	0.001 003	45.67	101.04	2307.5	2408.5	101.05	2444.5	2545.5	.3545	8.2231	8.5776
4.0	28.96	0.001 004	34.80	121.45	2293.7	2415.2	121.46	2432.9	2554.4	.4226	8.0520	8.4746
5.0	32.88	0.001 005	28.19	137.81	2282.7	2420.5	137.82	2423.7	2561.5	.4764	7.9187	8.3951
7.5	40.29	0.001 008	19.24	168.78	2261.7	2430.5	168.79	2406.0	2574.8	.5764	7.6750	8.2515
10	45.81	0.001 010	14.67	191.82	2246.1	2437.9	191.83	2392.8	2584.7	.6493	7.5009	8.1502
15	53.97	0.001 014	10.02	225.92	2222.8	2448.7	225.94	2373.1	2599.1	.7549	7.2536	8.0085
20	60.06	0.001 017	7.649	251.38	2205.4	2456.7	251.40	2358.3	2609.7	.8320	7.0766	7.9085
25	64.97	0.001 020	6.204	271.90	2191.2	2463.1	271.93	2346.3	2618.2	.8931	6.9383	7.8314
30	69.10	0.001 022	5.229	289.20	2179.2	2468.4	289.23	2336.1	2625.3	.9439	6.8247	7.7686
40	75.87	0.001 027	3.993	317.53	2159.5	2477.0	317.58	2319.2	2636.8	1.0259	6.6441	7.6700
50	81.33	0.001 030	3.240	340.44	2143.4	2483.9	340.49	2305.4	2645.9	1.0910	6.5029	7.5939
75	91.78	0.001 037	2.217	384.31	2112.4	2496.7	384.39	2278.6	2663.0	1.2130	6.2434	7.4564
MPa												
0.100	99.63	0.001 043	1.6940	417.36	2088.7	2506.1	417.46	2258.0	2675.5	1.3026	6.0568	7.3594
0.125	105.99	0.001 048	1.3749	444.19	2069.3	2513.5	444.32	2241.0	2685.4	1.3740	5.9104	7.2844
0.150	111.37	0.001 053	1.1593	466.94	2052.7	2519.7	467.11	2226.5	2693.6	1.4336	5.7897	7.2233
0.175	116.06	0.001 057	1.0036	486.80	2038.1	2524.9	486.99	2213.6	2700.6	1.4849	5.6868	7.1717
0.200	120.23	0.001 061	0.8857	504.49	2025.0	2529.5	504.70	2201.9	2706.7	1.5301	5.5970	7.1271
0.225	124.00	0.001 064	0.7933	520.47	2013.1	2533.6	520.72	2191.3	2712.1	1.5706	5.5173	7.0878

TABLE C-1 *(Continued)*

Press. MPa P	Temp. °C T	Specific volume Sat. liquid v^L	Specific volume Sat. vapor v^G	Internal energy Sat. liquid u^L	Internal energy Evap. Δu	Internal energy Sat. vapor u^G	Enthalpy Sat. liquid h^L	Enthalpy Evap. Δh	Enthalpy Sat. vapor h^G	Entropy Sat. liquid s^L	Entropy Evap. Δs	Entropy Sat. vapor s^G
0.250	127.44	0.001 067	0.7187	535.10	2002.1	2537.2	535.37	2181.5	2716.9	1.6072	5.4455	7.0527
0.275	130.60	0.001 070	0.6573	548.59	1991.9	2540.5	548.89	2172.4	2721.3	1.6408	5.3801	7.0209
0.300	133.55	0.001 073	0.6058	561.15	1982.4	2543.6	561.47	2163.8	2725.3	1.6718	5.3201	6.9919
0.325	136.30	0.001 076	0.5620	572.90	1973.5	2546.4	573.25	2155.8	2729.0	1.7006	5.2646	6.9652
0.350	138.88	0.001 079	0.5243	583.95	1965.0	2548.9	584.33	2148.1	2732.4	1.7275	5.2130	6.9405
0.375	141.32	0.001 081	0.4914	594.40	1956.9	2551.3	594.81	2140.8	2735.6	1.7528	5.1647	6.9175
0.40	143.63	0.001 084	0.4625	604.31	1949.3	2553.6	604.74	2133.8	2738.6	1.7766	5.1193	6.8959
0.45	147.93	0.001 088	0.4140	622.77	1934.9	2557.6	623.25	2120.7	2743.9	1.8207	5.0359	6.8565
0.50	151.86	0.001 093	0.3749	639.68	1921.6	2561.2	640.23	2108.5	2748.7	1.8607	4.9606	6.8213
0.55	155.48	0.001 097	0.3427	655.32	1909.2	2564.5	655.93	2097.0	2753.0	1.8973	4.8920	6.7893
0.60	158.85	0.001 101	0.3157	669.90	1897.5	2567.4	670.56	2086.3	2756.8	1.9312	4.8288	6.7600
0.65	162.01	0.001 104	0.2927	683.56	1886.5	2570.1	684.28	2076.0	2760.3	1.9627	4.7703	6.7331
0.70	164.97	0.001 108	0.2729	696.44	1876.1	2572.5	697.22	2066.3	2763.5	1.9922	4.7158	6.7080
0.75	167.78	0.001 112	0.2556	708.64	1866.1	2574.7	709.47	2057.0	2766.4	2.0200	4.6647	6.6847
0.80	170.43	0.001 115	0.2404	720.22	1856.6	2576.8	721.11	2048.0	2769.1	2.0462	4.6166	6.6628
0.85	172.96	0.001 118	0.2270	731.27	1847.4	2578.7	732.22	2039.4	2771.6	2.0710	4.5711	6.6421
0.90	175.38	0.001 121	0.2150	741.83	1838.6	2580.5	742.83	2031.1	2773.9	2.0946	4.5280	6.6226
0.95	177.69	0.001 124	0.2042	751.95	1830.2	2582.1	753.02	2023.1	2776.1	2.1172	4.4869	6.6041
1.00	179.91	0.001 127	0.194 44	761.68	1822.0	2583.6	762.81	2015.3	2778.1	2.1387	4.4478	6.5865
1.10	184.09	0.001 133	0.177 53	780.09	1806.3	2586.4	781.34	2000.4	2781.7	2.1792	4.3744	6.5536
1.20	187.99	0.001 139	0.163 33	797.29	1791.5	2588.8	798.65	1986.2	2784.8	2.2166	4.3067	6.5233
1.30	191.64	0.001 144	0.151 25	813.44	1777.5	2591.0	814.93	1972.7	2787.6	2.2515	4.2438	6.4953
1.40	195.07	0.001 149	0.140 84	828.70	1764.1	2592.8	830.30	1959.7	2790.0	2.2842	4.1850	6.4693
1.50	198.32	0.001 154	0.131 77	843.16	1751.3	2594.5	844.89	1947.3	2792.2	2.3150	4.1298	6.4448
1.75	205.76	0.001 166	0.113 49	876.46	1721.4	2597.8	878.50	1917.9	2796.4	2.3851	4.0044	6.3896
2.00	212.42	0.001 177	0.099 63	906.44	1693.8	2600.3	908.79	1890.7	2799.5	2.4474	3.8935	6.3409

TABLE C-1 *(Continued)*

Press. MPa P	Temp. °C T	Specific volume		Internal energy			Enthalpy			Entropy		
		Sat. liquid v^L	Sat. vapor v^G	Sat. liquid u^L	Evap. Δu	Sat. vapor u^G	Sat. liquid h^L	Evap. Δh	Sat. vapor h^G	Sat. liquid s^L	Evap. Δs	Sat. vapor s^G
2.25	218.45	0.001 187	0.088 75	933.83	1668.2	2602.0	936.49	1865.2	2801.7	2.5035	3.7937	6.2972
2.5	223.99	0.001 197	0.079 98	959.11	1644.0	2603.1	962.11	1841.0	2803.1	2.5547	3.7028	6.2575
3.0	233.90	0.001 217	0.066 68	1004.78	1599.3	2604.1	1008.42	1795.7	2804.2	2.6457	3.5412	6.1869
3.5	242.60	0.001 235	0.057 07	1045.43	1558.3	2603.7	1049.75	1753.7	2803.4	2.7253	3.4000	6.1253
4	250.40	0.001 252	0.049 78	1082.31	1520.0	2602.3	1087.31	1714.1	2801.4	2.7964	3.2737	6.0701
5	263.99	0.001 286	0.039 44	1147.81	1449.3	2597.1	1154.23	1640.1	2794.3	2.9202	3.0532	5.9734
6	275.64	0.001 319	0.032 44	1205.44	1384.3	2589.7	1213.35	1571.0	2784.3	3.0267	2.8625	5.8892
7	285.88	0.001 351	0.027 37	1257.55	1323.0	2580.5	1267.00	1505.1	2772.1	3.1211	2.6922	5.8133
8	295.06	0.001 384	0.023 52	1305.57	1264.2	2569.8	1316.64	1441.3	2758.0	3.2068	2.5364	5.7432
9	303.40	0.001 418	0.020 48	1350.51	1207.3	2557.8	1363.26	1378.9	2742.1	3.2858	2.3915	5.6772
10	311.06	0.001 452	0.018 026	1393.04	1151.4	2544.4	1407.56	1317.1	2724.7	3.3596	2.2544	5.6141
11	318.15	0.001 489	0.015 987	1433.7	1096.0	2529.8	1450.1	1255.5	2705.6	3.4295	2.1233	5.5527
12	324.75	0.001 527	0.014 263	1473.0	1040.7	2513.7	1491.3	1193.6	2684.9	3.4962	1.9962	5.4924
13	330.93	0.001 567	0.012 780	1511.1	985.0	2496.1	1531.5	1130.7	2662.2	3.5606	1.8718	5.4323
14	336.75	0.001 611	0.011 485	1548.6	928.2	2476.8	1571.1	1066.5	2637.6	3.6232	1.7485	5.3717
15	342.24	0.001 658	0.010 337	1585.6	869.8	2455.5	1610.5	1000.0	2610.5	3.6848	1.6249	5.3098
16	347.44	0.001 711	0.009 306	1622.7	809.0	2431.7	1650.1	930.6	2580.6	3.7461	1.4994	5.2455
17	352.37	0.001 770	0.008 364	1660.2	744.8	2405.0	1690.3	856.9	2547.2	3.8079	1.3698	5.1777
18	357.06	0.001 840	0.007 489	1698.9	675.4	2374.3	1732.0	777.1	2509.1	3.8715	1.2329	5.1044
19	361.54	0.001 924	0.006 657	1739.9	598.1	2338.1	1776.5	688.0	2464.5	3.9388	1.0839	5.0228
20	365.81	0.002 036	0.005 834	1785.6	507.5	2293.0	1826.3	583.4	2409.7	4.0139	.9130	4.9269
21	369.89	0.002 207	0.004 952	1842.1	388.5	2230.6	1888.4	446.2	2334.6	4.1075	.6938	4.8013
22	373.80	0.002 742	0.003 568	1961.9	125.2	2087.1	2022.2	143.4	2165.6	4.3110	.2216	4.5327
22.09	374.14	0.003 155	0.003 155	2029.6	0	2029.6	2099.3	0	2099.3	4.4298	0	4.4298

[a] From J. H. Keenan, F. G. Keys, P. G. Hill and J. G. Moore, *Steam Tables*, © 1969, John Wiley & Sons, Inc. as used by G. J. VanWylen and R. E. Sonntag, *Fundamentals of Classical Thermodynamics*, 2nd ed., S. I. Version, John Wiley & Sons, New York, 1976. Reprinted by permission of John Wiley & Sons, Inc.

[b] v in m^3/kg, u and h in kJ/kg, s in kJ/kg K.

TABLE C-2
THERMODYNAMIC PROPERTIES OF SUPERHEATED STEAM[a,b]

T	P = .010 MPa (45.81) v	u	h	s	P = .050 MPa (81.33) v	u	h	s	P = .10 MPa (99.63) v	u	h	s
Sat.	14.674	2437.9	2584.7	8.1502	3.240	2483.9	2645.9	7.5939	1.6940	2506.1	2675.5	7.3594
50	14.869	2443.9	2592.6	8.1749								
100	17.196	2515.5	2687.5	8.4479	3.418	2511.6	2682.5	7.6947	1.6958	2506.7	2676.2	7.3614
150	19.512	2587.9	2783.0	8.6882	3.889	2585.6	2780.1	7.9401	1.9364	2582.8	2776.4	7.6134
200	21.825	2661.3	2879.5	8.9038	4.356	2659.9	2877.7	8.1580	2.172	2658.1	2875.3	7.8343
250	24.136	2736.0	2977.3	9.1002	4.820	2735.0	2976.0	8.3556	2.406	2733.7	2974.3	8.0333
300	26.445	2812.1	3076.5	9.2813	5.284	2811.3	3075.5	8.5373	2.639	2810.4	3074.3	8.2158
400	31.063	2968.9	3279.6	9.6077	6.209	2968.5	3278.9	8.8642	3.103	2967.9	3278.2	8.5435
500	35.679	3132.3	3489.1	9.8978	7.134	3132.0	3488.7	9.1546	3.565	3131.6	3488.1	8.8342
600	40.295	3302.5	3705.4	10.1608	8.057	3302.2	3705.1	9.4178	4.028	3301.9	3704.7	9.0976
700	44.911	3479.6	3928.7	10.4028	8.981	3479.4	3928.5	9.6599	4.490	3479.2	3928.2	9.3398
800	49.526	3663.8	4159.0	10.6281	9.904	3663.6	4158.9	9.8852	4.952	3663.5	4158.6	9.5652
900	54.141	3855.0	4396.4	10.8396	10.828	3854.9	4396.3	10.0967	5.414	3854.8	4396.1	9.7767
1000	58.757	4053.0	4640.6	11.0393	11.751	4052.9	4640.5	10.2964	5.875	4052.8	4640.3	9.9764
1100	63.372	4257.5	4891.2	11.2287	12.674	4257.4	4891.1	10.4859	6.337	4257.3	4891.0	10.1659
1200	67.987	4467.9	5147.8	11.4091	13.597	4467.8	5147.7	10.6662	6.799	4467.7	5147.6	10.3463
1300	72.602	4683.7	5409.7	11.5811	14.521	4683.6	5409.6	10.8382	7.260	4683.5	5409.5	10.5183

T	P = .20 MPa (120.23) v	u	h	s	P = .30 MPa (133.55) v	u	h	s	P = .40 MPa (143.63) v	u	h	s
Sat.	.8857	2529.5	2706.7	7.1272	.6058	2543.6	2725.3	6.9919	.4625	2553.6	2738.6	6.8959
150	.9596	2576.9	2768.8	7.2795	.6339	2570.8	2761.0	7.0778	.4708	2564.5	2752.8	6.9299
200	1.0803	2654.4	2870.5	7.5066	.7163	2650.7	2865.6	7.3115	.5342	2646.8	2860.5	7.1706
250	1.1988	2731.2	2971.0	7.7086	.7964	2728.7	2967.6	7.5166	.5951	2726.1	2964.2	7.3789
300	1.3162	2808.6	3071.8	7.8926	.8753	2806.7	3069.3	7.7022	.6548	2804.8	3066.8	7.5662
400	1.5493	2966.7	3276.6	8.2218	1.0315	2965.6	3275.0	8.0330	.7726	2964.4	3273.4	7.8985

TABLE C-2 *(Continued)*

T	v	u	h	s	v	u	h	s	v	u	h	s
500	1.7814	3130.8	3487.1	8.5133	1.1867	3130.0	3486.0	8.3251	.8893	3129.2	3484.9	8.1913
600	2.013	3301.4	3704.0	8.7770	1.3414	3300.8	3703.2	8.5892	1.0055	3300.2	3702.4	8.4558
700	2.244	3478.8	3927.6	9.0194	1.4957	3478.4	3927.1	8.8319	1.1215	3477.9	3926.5	8.6987
800	2.475	3663.1	4158.2	9.2449	1.6499	3662.9	4157.8	9.0576	1.2372	3662.4	4157.3	8.9244
900	2.706	3854.5	4395.8	9.4566	1.8041	3854.2	4395.4	9.2692	1.3529	3853.9	4395.1	9.1362
1000	2.937	4052.5	4640.0	9.6563	1.9581	4052.3	4639.7	9.4690	1.4685	4052.0	4639.4	9.3360
1100	3.168	4257.0	4890.7	9.8458	2.1121	4256.8	4890.4	9.6585	1.5840	4256.5	4890.2	9.5256
1200	3.399	4467.5	5147.3	10.0262	2.2661	4467.2	5147.1	9.8389	1.6996	4467.0	5146.8	9.7060
1300	3.630	4683.2	5409.3	10.1982	2.4201	4683.0	5409.0	10.0110	1.8151	4682.8	5408.8	9.8780

T	P = .50 MPa (151.86)				P = .60 MPa (158.85)				P = .80 MPa (170.43)			
	v	u	h	s	v	u	h	s	v	u	h	s
Sat.	.3749	2561.2	2748.7	6.8213	.3157	2567.4	2756.8	6.7600	.2404	2576.8	2769.1	6.6628
200	.4249	2642.9	2855.4	7.0592	.3520	2638.9	2850.1	6.9665	.2608	2630.6	2839.3	6.8158
250	.4744	2723.5	2960.7	7.2709	.3938	2720.9	2957.2	7.1816	.2931	2715.5	2950.0	7.0384
300	.5226	2802.9	3064.2	7.4599	.4344	2801.0	3061.6	7.3724	.3241	2797.2	3056.5	7.2328
350	.5701	2882.6	3167.7	7.6329	.4742	2881.2	3165.7	7.5464	.3544	2878.2	3161.7	7.4089
400	.6173	2963.2	3271.9	7.7938	.5137	2962.1	3270.3	7.7079	.3843	2959.7	3267.1	7.5716
500	.7109	3128.4	3483.9	8.0873	.5920	3127.6	3482.8	8.0021	.4433	3126.0	3480.6	7.8673
600	.8041	3299.6	3701.7	8.3522	.6697	3299.1	3700.9	8.2674	.5018	3297.9	3699.4	8.1333
700	.8969	3477.5	3925.9	8.5952	.7472	3477.0	3925.3	8.5107	.5601	3476.2	3924.2	8.3770
800	.9896	3662.1	4156.9	8.8211	.8245	3661.8	4156.5	8.7367	.6181	3661.1	4155.6	8.6033
900	1.0822	3853.6	4394.7	9.0329	.9017	3853.4	4394.4	8.9486	.6761	3852.8	4393.7	8.8153
1000	1.1747	4051.8	4639.1	9.2328	.9788	4051.5	4638.8	9.1485	.7340	4051.0	4638.2	9.0153
1100	1.2672	4256.3	4889.9	9.4224	1.0559	4256.1	4889.6	9.3381	.7919	4255.6	4889.1	9.2050
1200	1.3596	4466.8	5146.6	9.6029	1.1330	4466.5	5146.3	9.5185	.8497	4466.1	5145.9	9.3855
1300	1.4521	4682.5	5408.6	9.7749	1.2101	4682.3	5408.3	9.6906	.9076	4681.8	5407.9	9.5575

TABLE C-2 *(Continued)*

P = 1.00 MPa (179.91)

T	v	u	h	s
Sat.	.194 44	2583.6	2778.1	6.5865
200	.2060	2621.9	2827.9	6.6940
250	.2327	2709.9	2942.6	6.9247
300	.2579	2793.2	3051.2	7.1229
350	.2825	2875.2	3157.7	7.3011
400	.3066	2957.3	3263.9	7.4651
500	.3541	3124.4	3478.5	7.7622
600	.4011	3296.8	3697.9	8.0290
700	.4478	3475.3	3923.1	8.2731
800	.4943	3660.4	4154.7	8.4996
900	.5407	3852.2	4392.9	8.7118
1000	.5871	4050.5	4637.6	8.9119
1100	.6335	4255.1	4888.6	9.1017
1200	.6798	4465.6	5145.4	9.2822
1300	.7261	4681.3	5407.4	9.4543

P = 1.20 MPa (187.99)

T	v	u	h	s
Sat.	.163 33	2588.8	2784.8	6.5233
200	.169 30	2612.8	2815.9	6.5898
250	.192 34	2704.2	2935.0	6.8294
300	.2138	2789.2	3045.8	7.0317
350	.2345	2872.2	3153.6	7.2121
400	.2548	2954.9	3260.7	7.3774
500	.2946	3122.8	3476.3	7.6759
600	.3339	3295.6	3696.3	7.9435
700	.3729	3474.4	3922.0	8.1881
800	.4118	3659.7	4153.8	8.4148
900	.4505	3851.6	4392.2	8.6272
1000	.4892	4050.0	4637.0	8.8274
1100	.5278	4254.6	4888.0	9.0172
1200	.5665	4465.1	5144.9	9.1977
1300	.6051	4680.9	5407.0	9.3698

P = 1.40 MPa (195.07)

T	v	u	h	s
Sat.	.140 84	2592.8	2790.0	6.4693
200	.143 02	2603.1	2803.3	6.4975
250	.163 50	2698.3	2927.2	6.7467
300	.182 28	2785.2	3040.4	6.9534
350	.2003	2869.2	3149.5	7.1360
400	.2178	2952.5	3257.5	7.3026
500	.2521	3121.1	3474.1	7.6027
600	.2860	3294.4	3694.8	7.8710
700	.3195	3473.6	3920.8	8.1160
800	.3528	3659.0	4153.0	8.3431
900	.3861	3851.1	4391.5	8.5556
1000	.4192	4049.5	4636.4	8.7559
1100	.4524	4254.1	4887.5	8.9457
1200	.4855	4464.7	5144.4	9.1262
1300	.5186	4680.4	5406.5	9.2984

P = 1.60 MPa (201.41)

T	v	u	h	s
Sat.	.123 80	2596.0	2794.0	6.4218
225	.132 87	2644.7	2857.3	6.5518
250	.141 84	2692.3	2919.2	6.6732
300	.158 62	2781.1	3034.8	6.8844
350	.174 56	2866.1	3145.4	7.0694
400	.190 05	2950.1	3254.2	7.2374
500	.2203	3119.5	3472.0	7.5390
600	.2500	3293.3	3693.2	7.8080
700	.2794	3472.7	3919.7	8.0535

P = 1.80 MPa (207.15)

T	v	u	h	s
Sat.	.110 42	2598.4	2797.1	6.3794
225	.116 73	2636.6	2846.7	6.4808
250	.124 97	2686.0	2911.0	6.6066
300	.140 21	2776.9	3029.2	6.8226
350	.154 57	2863.0	3141.2	7.0100
400	.168 47	2947.7	3250.9	7.1794
500	.195 50	3117.9	3469.8	7.4825
600	.2220	3292.1	3691.7	7.7523
700	.2482	3471.8	3918.5	7.9983

P = 2.00 MPa (212.42)

T	v	u	h	s
Sat.	.099 63	2600.3	2799.5	6.3409
225	.103 77	2628.3	2835.8	6.4147
250	.111 44	2679.6	2902.5	6.5453
300	.125 47	2772.6	3023.5	6.7664
350	.138 57	2859.8	3137.0	6.9563
400	.151 20	2945.2	3247.6	7.1271
500	.175 68	3116.2	3467.6	7.4317
600	.199 60	3290.9	3690.1	7.7024
700	.2232	3470.9	3917.4	7.9487

TABLE C-2   (*Continued*)

(continuation of preceding pressures, T = 800–1300 °C)

T	v	u	h	s	v	u	h	s	v	u	h	s
800	.3086	3658.3	4152.1	8.2808	.2742	3657.6	4151.2	8.2258	.2467	3657.0	4150.3	8.1765
900	.3377	3850.5	4390.8	8.4935	.3001	3849.9	4390.1	8.4386	.2700	3849.3	4389.4	8.3895
1000	.3668	4049.0	4635.8	8.6938	.3260	4048.5	4635.2	8.6391	.2933	4048.0	4634.6	8.5901
1100	.3958	4253.7	4887.0	8.8837	.3518	4253.2	4886.4	8.8290	.3166	4252.7	4885.9	8.7800
1200	.4248	4464.2	5143.9	9.0643	.3776	4463.7	5143.4	9.0096	.3398	4463.3	5142.9	8.9607
1300	.4538	4679.9	5406.0	9.2364	.4034	4679.5	5405.6	9.1818	.3631	4679.0	5405.1	9.1329

T	\(P = 2.50\) MPa (223.99)				\(P = 3.00\) MPa (233.90)				\(P = 3.50\) MPa (242.60)			
	v	u	h	s	v	u	h	s	v	u	h	s
Sat.	.079 98	2603.1	2803.1	6.2575	.066 68	2604.1	2804.2	6.1869	.057 07	2603.7	2803.4	6.1253
225	.080 27	2605.6	2806.3	6.2639								
250	.087 00	2662.6	2880.1	6.4085	.070 58	2644.0	2855.8	6.2872	.058 72	2623.7	2829.2	6.1749
300	.098 90	2761.6	3008.8	6.6438	.081 14	2750.1	2993.5	6.5390	.068 42	2738.0	2977.5	6.4461
350	.109 76	2851.9	3126.3	6.8403	.090 53	2843.7	3115.3	6.7428	.076 78	2835.3	3104.0	6.6579
400	.120 10	2939.1	3239.3	7.0148	.099 36	2932.8	3230.9	6.9212	.084 53	2926.4	3222.3	6.8405
450	.130 14	3025.5	3350.8	7.1746	.107 87	3020.4	3344.0	7.0834	.091 96	3015.3	3337.2	7.0052
500	.139 98	3112.1	3462.1	7.3234	.116 19	3108.0	3456.5	7.2338	.099 18	3103.0	3450.9	7.1572
600	.159 30	3288.0	3686.3	7.5960	.132 43	3285.0	3682.3	7.5085	.113 24	3282.1	3678.4	7.4339
700	.178 32	3468.7	3914.5	7.8435	.148 38	3466.5	3911.7	7.7571	.126 99	3464.3	3908.8	7.6837
800	.197 16	3655.3	4148.2	8.0720	.164 14	3653.5	4145.9	7.9862	.140 56	3651.8	4143.7	7.9134
900	.215 90	3847.9	4387.6	8.2853	.179 80	3846.5	4385.9	8.1999	.154 02	3845.0	4384.1	8.1276
1000	.2346	4046.7	4633.1	8.4861	.195 41	4045.4	4631.6	8.4009	.167 43	4044.1	4630.1	8.3288
1100	.2532	4251.5	4884.6	8.6762	.210 98	4250.3	4883.3	8.5912	.180 80	4249.2	4881.9	8.5192
1200	.2718	4462.1	5141.7	8.8569	.226 52	4460.9	5140.5	8.7720	.194 15	4459.8	5139.3	8.7000
1300	.2905	4677.8	5404.0	9.0291	.242 06	4676.6	5402.8	8.9442	.207 49	4675.5	5401.7	8.8723

T	\(P = 4.0\) MPa (250.40)				\(P = 4.5\) MPa (257.49)				\(P = 5.0\) MPa (263.99)			
	v	u	h	s	v	u	h	s	v	u	h	s
Sat.	.049 78	2602.3	2801.4	6.0701	.044 06	2600.1	2798.3	6.0198	.039 44	2597.1	2794.3	5.9734
275	.054 57	2667.9	2886.2	6.2285	.047 30	2650.3	2863.2	6.1401	.041 41	2631.3	2838.3	6.0544
300	.058 84	2725.3	2960.7	6.3615	.051 35	2712.0	2943.1	6.2828	.045 32	2698.0	2924.5	6.2084
350	.066 45	2826.7	3092.5	6.5821	.058 40	2817.8	3080.6	6.5131	.051 94	2808.7	3068.4	6.4493
400	.073 41	2919.9	3213.6	6.7690	.064 75	2913.3	3204.7	6.7047	.057 81	2906.6	3195.7	6.6459
450	.080 02	3010.2	3330.3	6.9363	.070 74	3005.0	3323.3	6.8746	.063 30	2999.7	3316.2	6.8186
500	.086 43	3099.5	3445.3	7.0901	.076 51	3095.3	3439.6	7.0301	.068 57	3091.0	3433.8	6.9759

TABLE C-2 (Continued)

T	P = 4.0 MPa (250.40) v	u	h	s	P = 4.5 MPa (257.49) v	u	h	s	P = 5.0 MPa (263.99) v	u	h	s
600	.098 85	3279.1	3674.4	7.3688	.087 65	3276.0	3670.5	7.3110	.078 69	3273.0	3666.5	7.2589
700	.110 95	3462.1	3905.9	7.6198	.098 47	3459.9	3903.0	7.5631	.088 49	3457.6	3900.1	7.5122
800	.122 87	3650.0	4141.5	7.8502	.109 11	3648.3	4139.3	7.7942	.098 11	3646.6	4137.1	7.7440
900	.134 69	3843.6	4382.3	8.0647	.119 65	3842.2	4380.6	8.0091	.107 62	3840.7	4378.8	7.9593
1000	.146 45	4042.9	4628.7	8.2662	.130 13	4041.6	4627.2	8.2108	.117 07	4040.4	4625.7	8.1612
1100	.158 17	4248.0	4880.6	8.4567	.140 56	4246.8	4879.3	8.4015	.126 48	4245.6	4878.0	8.3520
1200	.169 87	4458.6	5138.1	8.6376	.150 98	4457.5	5136.9	8.5825	.135 87	4456.3	5135.7	8.5331
1300	.181 56	4674.3	5400.5	8.8100	.161 39	4673.1	5399.4	8.7549	.145 26	4672.0	5398.2	8.7055

T	P = 6.0 MPa (275.64) v	u	h	s	P = 7.0 MPa (285.88) v	u	h	s	P = 8.0 MPa (295.06) v	u	h	s
Sat.	.032 44	2589.7	2784.3	5.8892	.027 37	2580.5	2772.1	5.8133	.023 52	2569.8	2758.0	5.7432
300	.036 16	2667.2	2884.2	6.0674	.029 47	2632.2	2838.4	5.9305	.024 26	2590.9	2785.0	5.7906
350	.042 23	2789.6	3043.0	6.3335	.035 24	2769.4	3016.0	6.2283	.029 95	2747.7	2987.3	6.1301
400	.047 39	2892.9	3177.2	6.5408	.039 93	2878.6	3158.1	6.4478	.034 32	2863.8	3138.3	6.3634
450	.052 14	2988.9	3301.8	6.7193	.044 16	2978.0	3287.1	6.6327	.038 17	2966.7	3272.0	6.5551
500	.056 65	3082.2	3422.2	6.8803	.048 14	3073.4	3410.3	6.7975	.041 75	3064.3	3398.3	6.7240
550	.061 01	3174.6	3540.6	7.0288	.051 95	3167.2	3530.9	6.9486	.045 16	3159.8	3521.0	6.8778
600	.065 25	3266.9	3658.4	7.1677	.055 65	3260.7	3650.3	7.0894	.048 45	3254.4	3642.0	7.0206
700	.073 52	3453.1	3894.2	7.4234	.062 83	3448.5	3888.3	7.3476	.054 81	3443.9	3882.4	7.2812
800	.081 60	3643.1	4132.7	7.6566	.069 81	3639.5	4128.2	7.5822	.060 97	3636.0	4123.8	7.5173
900	.089 58	3837.8	4375.3	7.8727	.076 69	3835.0	4371.8	7.7991	.067 02	3832.1	4368.3	7.7351
1000	.097 49	4037.8	4622.7	8.0751	.083 50	4035.3	4619.8	8.0020	.073 01	4032.8	4616.9	7.9384
1100	.105 36	4243.3	4875.4	8.2661	.090 27	4240.9	4872.8	8.1933	.078 96	4238.6	4870.3	8.1300

a From J. H. Keenan, F. G. Keys, P. G. Hill, and J. G. Moore, *Steam Tables*, © 1969, John Wiley & Sons, Inc. as used by G. J. VanWylen and R. E. Sonntag, *Fundamentals of Classical Thermodynamics*, 2nd ed., S. I. Version, John Wiley & Sons, New York, 1976. Reprinted by permission of John Wiley & Sons, Inc.

b v in m^3/kg, u and h in kJ/kg, s in kJ/kg K.

APPENDIX D

Substance	Formula	Molecular weight	State	ΔH^f	ΔG^f
Methane	CH_4	16.04	g	−74.90	−50.83
Ethane	C_2H_6	30.07	g	−84.72	−32.9
Propane	C_3H_8	44.09	g	−103.9	−23.5
n-Butane	C_4H_{10}	58.12	g	−126.2	−17.2
i-Butane	C_4H_{10}	58.12	g	−134.6	−20.9
n-Pentane	C_5H_{12}	72.15	g	−146.5	−8.37
n-Hexane	C_6H_{14}	86.17	g	−167.3	−0.29
n-Heptane	C_7H_{16}	100.20	g	−187.9	8.12
n-Octane	C_8H_{18}	114.22	g	−208.6	16.5
Ethylene	C_2H_4	28.05	g	52.32	68.17
Propylene	C_3H_6	42.08	g	20.4	62.76
1-Butene	C_4H_8	56.10	g	−0.13	71.55
cis-2-Butene	C_4H_8	56.10	g	−6.99	65.90
trans-2-Butene	C_4H_8	56.10	g	−11.2	63.01
1,2-Butadiene	C_4H_6	54.09	g	162.3	198.6
1,3-Butadiene	C_4H_6	54.09	g	110.2	150.8
Acetylene	C_2H_2	26.04	g	226.9	209.3
Cyclohexane	C_6H_{12}	84.16	g	−123.2	31.8
			l	−156.3	26.7
Benzene	C_6H_6	78.11	g	82.98	129.7
			l	49.07	124.4
Toluene	C_7H_8	92.13	g	50.03	122.1
			l	12.02	113.8
Ethylbenzene	C_8H_{10}	106.16	g	29.8	130.7
			l	−12.5	119.8
Styrene	C_8H_8	104.14	g	147.5	213.9
			l	104.0	202.5
Ethylene oxide	C_2H_4O	44.05	g	−52.67	−13.1
Formaldehyde	CH_2O	30.03	g	−116.0	−110.0
Acetaldehyde	C_2H_4O	44.05	g	−116.5	−133.4
			l	−192.2	—
Methanol	CH_4O	32.04	g	−201.3	−162.6
			l	−238.7	−166.3
Ethanol	C_2H_6O	46.07	g	−235.0	−168.4
			l	−277.2	−174.3
Isopropanol	C_3H_8O	60.09	g	−272.8	−173.7
			l	−318.2	−180.5

TABLE D-1 *(Continued)*

Substance	Formula	Molecular weight	State	ΔH^f	ΔG^f
Acetone	C_3H_6O	58.08	g	−217.7	−153.2
			l	−248.3	−155.5
Acetic acid	$C_2H_4O_2$	60.05	g	−435.13	−376.9
			l	−484.41	−389.6
Ethyl acetate	$C_4H_8O_2$	88.10	g	−443.22	−327.6
			l	−479.35	−332.9
Ethylene glycol	$C_2H_6O_2$	62.07	g	−389.6	−304.7
			l	−455.23	−323.6
Phenol	C_6H_6O	94.11	g	−96.42	−32.9
			c	−165.1	−50.5
Glucose, D	$C_6H_{12}O_6$	180.16	c	−1275	−911.13

[a] Based on data from references 1-C and 3 of Table 12-1. To convert to kcal/g mol, multiply by 0.23885.

TABLE D-2
ENTHALPIES AND FREE ENERGIES OF FORMATION
OF SELECTED INORGANIC COMPOUNDS AT 298.15 K[a]
(IN UNITS OF MJ/kmol)

Substance	Formula	Molecular weight	State	ΔH^f	ΔG^f
Ammonia	NH_3	17.03	g	−46.22	−16.6
			ao	−80.89	−26.7
Calcium carbide	CaC_2	64.10	c	−62.8	−67.8
Calcium carbonate	$CaCO_3$	100.09	c	−1208	−1130
Calcium chloride	$CaCl_2$	110.99	c	−795.5	−750.7
			ai	−878.47	−815.92
Calcium hydroxide	$Ca(OH)_2$	74.10	c	−987.25	−897.4
			ai	−1003	−868.2
Calcium oxide	CaO	56.08	c	−636.0	−604.6
Calcium sulfate	$CaSO_4$	136.14	c	−1433.6	−1321
Gypsum	$CaSO_4 \cdot 2H_2O$	172.17	c	−2022.5	−1796.9
Calcium sulfide	CaS	72.14	c	−482.7	−477.7
Carbon monoxide	CO	28.01	g	−110.6	−137.4
Carbon dioxide	CO_2	44.01	g	−393.8	−394.6
			ao	−413.2	−386.5
Carbonic acid	H_2CO_3	62.03	ao	−699.2	−623.8
Carbon disulfide	CS_2	76.13	g	115.3	65.11
Hydrogen chloride	HCl	36.47	g	−92.37	−95.33
			ai	−167.6	−131.3
Hydrogen sulfide	H_2S	34.08	g	−20.2	−33.1
			ao	−39.4	−27.4
Iron oxide	Fe_2O_3	159.70	c	−822.7	−741.4
Iron oxide	Fe_3O_4	231.55	c	−1118	−1015
Iron sulfide	FeS	87.91	c	−95.12	97.64
Lead chloride	$PbCl_2$	278.12	c	−359.4	−314.2
Lead iodide	PbI_2	461.05	c	−175.2	−173.9
Nitric acid	HNO_3	63.02	l	−173.4	−79.97
			ai	−206.7	−110.6
Nitric oxide	NO	30.01	g	90.44	86.75
Nitrogen dioxide	NO_2	46.01	g	33.9	51.87
Nitrous oxide	N_2O	44.02	g	81.60	103.7
Sulfur	S_2	64.14	g	129.1	80.07
Sulfur dioxide	SO_2	64.07	g	−297.1	−300.6
			ao	−323.0	−300.7
Sulfur trioxide	SO_3	80.07	g	−395.4	−370.6
Sulfuric acid	H_2SO_4	98.08	ai	−908.12	−742.49
			l	−814.12	−690.11
Sulfurous acid	H_2SO_3	82.09	ao	−608.8	−537.8
Water	H_2O	18.02	g	−242.0	−228.7
			l	−286.0	−237.4
Hydrogen peroxide	H_2O_2	34.02	g	−136.2	−105.5
Ozone	O_3	48.00	g	142.4	162.9

[a] Based on data from references 1-A, 1-C, and 3 in Table 12-1. To convert to kcal/g mol, multiply by 0.23885. Data for all sulfur-containing species are based on elemental sulfur in the rhombic crystalline state.

TABLE D-3

FREE ENERGIES OF FORMATION OF SELECTED IONS
IN THE AQUEOUS STATE AT 298.15 K[a]
(IN UNITS OF MJ/kmol)

Ion	ΔG^f	Ion	ΔG^f
H^+	0	OH^-	-157.40
Na^+	-262.04	Cl^-	-131.25
K^+	-282.46	Br^-	-102.88
Li^+	-294.0	I^-	-51.71
NH_4^+	-79.55	NO_3^-	-110.6
Cu^+	$+50.2$	ClO_4^-	-10.8
Cu^{++}	$+65.02$	HCO_3^-	-587.44
Mg^{++}	-456.31	$CO_3^=$	-528.45
Ca^{++}	-553.40	HSO_4^-	-753.36
Zn^{++}	-147.31	$SO_4^=$	-742.47
Pb^{++}	-24.3	HSO_3^-	-527.73
Fe^{++}	-84.99	$SO_3^=$	-497.4
Fe^{+++}	-10.6	$S^=$	$+83.7$
Al^{+++}	-481.5	HS^-	$+12.6$

[a] Based on data from reference 1-A in Table 12-1. To convert to kcal/g mol, multiply by 0.23885.

APPENDIX E INVENTORY OF CD-ROM

There are six directories on the companion CD-ROM which hold an Adobe Acrobat Reader and the computing resources and auxiliary materials accompanying this textbook. If your computer doesn't have a CD-ROM drive, find a computer that has one and copy the contents of each directory (except for the ENTROPY directory) to a separate floppy disk. POLY-MATH should be installed in a directory on your hard drive. This can be done from the CD-ROM or from a floppy disk containing all the files in the CD-ROM's POLYMATH directory. Most of the other programs can be run from either the CD-ROM, a floppy disk, or a directory on your hard drive. All programs run in DOS; most can be run in Windows by accessing the MS DOS prompt or by installing icons (Windows 3.1) or shortcuts (Windows 95).

ACROBAT

This directory houses two Adobe Acrobat Reader files: ar16e301.exe and ar32e301.exe. The former installs the 16-bit reader in Windows 3.1 while the latter installs a 32-bit reader in Windows 95 or Windows NT. The readers can be used to read or print pdf files of the auxiliary material housed in the ENTROPY directory of the CD-ROM or the POLYMATH Instruction Manual.

POLYMATH

POLYMATH 4.01 is a set of numerical analysis programs for

1. solving simultaneous differential equations
2. solving simultaneous algebraic equations
3. solving simultaneous linear equations
4. performing polynomial, multiple linear, and nonlinear regression

This unabridged, commercial software is included here by special arrangement and is to be used for educational purposes only.

POLYMATH should not be run from the CD-ROM but should be installed on your hard drive because it needs to assess the available memory and identify your printer. Your system files will not be altered during installation. Instructions for installation from the CD-ROM or from a floppy disk are in the INSTALL.TXT file in this directory. Installation and program execution are possible in either DOS, Windows 3.1, or Windows 95.

Because POLYMATH is user-friendly and its operation is menu-driven, there is little need for a detailed set of instructions, however, an Operating Manual is available on the file

MANUAL.PDF. This file is housed in the subdirectory MANUAL of the POLYMATH directory and can be read or printed using the Adobe Acrobat Reader. Also, a detailed HELP section is accessible from many points in the program by pressing the F6 key. To run POLYMATH in DOS enter the directory in which it is housed, type POLYMATH, [ENTER], then make the appropriate menu choices. To run in Windows either access the DOS prompt and proceed as above or install an icon (Windows 3.1) or a shortcut (Windows 95).

The example problems in the textbook can be loaded into the POLYMATH library by entering the subdirectory LIBRARY of the POLYMATH directory on the CD-ROM and copying all files into the directory in which POLYMATH has been installed. This entails overwriting some existing empty library files. In the event that you have previously installed POLYMATH and have a set of library problems you do not wish to overwrite, you can create a subdirectory (say LIBRARY) in the directory housing POLYMATH and copy all the files from POLYMATH\LIBRARY on the CD-ROM into it. The example problems can then be accessed by using the **Change directory** command under **LIBRARY OPTIONS** to select the created subdirectory. A listing will be displayed.

For more information, visit the Website www.polymath-software.com.

VDW

The VDW directory houses a special version of the graphics program EQUATIONS OF STATE designed for use with this textbook for the running of the tutorials (macros) cited in the text. These tutorials make use of two- and three-dimensional *PVT* plots to illustrate, reinforce, and often extend principles covered in the text. The more-flexible, commercially available, parent program, EQUATIONS OF STATE, is capable of executing numerical calculations and displaying three-dimensional representations for a variety of thermodynamic processes.

This software is designed to run in DOS and runs poorly or not at all in Windows. It cannot be run from the CD-ROM but can be run from a floppy disk or a directory on the hard drive. Just copy all of the files in the VDW directory of the CD-ROM to one of these locations, preferably a directory on the hard drive. With Windows 95 the program can be run by selecting the MS DOS prompt from the Programs menu or by selecting the sequence **Start/Shut Down/Restart in MS DOS Mode**. The READ1ST.DOC file in this directory gives the simple commands needed to run the tutorials.

The following tutorials (macros) are available:

PVT3D	The *PVT* surface is explored and explained
EQSTATE	The concept and properties of equations of state are examined
1STLAW	A *PV* plot for an ideal gas is used to illustrate applications of the 1st law
CARNOT	The Carnot cycle is illustrated and worked through for an ideal gas and a van der Waals gas

2PHASE	The procedure for determining saturation conditions from the van der Waals isotherm is illustrated
JT	*PVT* and *PT* plots are used to examine the behavior of gases undergoing a Joule-Thomson expansion
RANKINE	The Rankine cycle is traced on the *PVT* surface and accompanied by calculations

The first five of these tutorials were written by me with some heavy-duty editing by Professor Kenneth Jolls. Professor Jolls wrote JT and Dr. Daniel Coy wrote RANKINE.

For more information contact

Professor Kenneth R. Jolls
Chemical Engineering Department
Sweeney Hall
Iowa State University
Ames, Iowa 50011-2230
e-mail: jolls@iastate.edu

COMPUTE

Executable Programs. There are nine programs compiled in Power Basic and bearing the extension EXE that are executable in DOS. These are housed in the COMPUTE directory on the CD-ROM and can be run from the CD-ROM by entering the COMPUTE directory and typing the program name (e.g., PREOS). The programs can also be run using Window's **MS DOS prompt** or can be installed as icons in Windows 3.1 or shortcuts in Windows 95. These programs can be copied to a floppy or a directory on your hard drive and run from there.

For each program there is a document file (in ASCII) bearing the program name and the extension DOC which provides details for using the program. It is recommended that you read these files before attempting to run the programs. These files are found in the COMPUTE directory and can be read from the screen or printed out. The following programs are available:

PREOS.EXE	Calculates thermodynamic properties from (P-R) equation of state
PRVLE.EXE	Calculates VLE from P-R equation of state
WEQ-ISOT.EXE	Fits Wilson parameters to isothermal VLE data
WEQ-ISOP.EXE	Fits Wilson parameters to non-isothermal VLE data
WILSONEQ.EXE	Calculates activity coefficients using multicomponent Wilson equation
PRKFIT.EXE	Determines k_{12} in the P-R equation from isothermal VLE data

PRSGE.EXE	Calculates solid-gas equilibria from P-R equation of state
KFIT-SGE.EXE	Determines k_{12} from isothermal SGE data
UNIFAC.EXE	Estimates activity coefficients via UNIFAC

All nine programs have the option of printing the results from the screen or creating an output file. When run in Windows the program results cannot be printed directly from the screen.

Spreadsheets. The four spreadsheets listed below are used throughout the text to execute routine calculations.

ANTOINE	Calculates vapor pressure from the Antoine equation with stored parameters for many substances
MARGULES	Fits Margules parameters to isothermal VLE data
VANLAAR	Fits van Laar parameters to isothermal VLE data
LOGKF(T)	Calculates thermochemical properties at various temperatures for gaseous substances listed in Appendix D

Figures 10-9 and 10-18 of the text show the effects of changing the parameter in the symmetric van Laar equation. The spreadsheets FG10-9 and FG10-18 are interactive versions of these figures.

There are two versions of each spreadsheet, those with the extension WQ1 run in QUATTRO PRO and those with the extension XLS run in EXCEL. Each spreadsheet has a set of detailed instructions.

BACKUP

This directory contains spreadsheets showing the detailed calculations for several Examples in the text. These were written in QUATTRO PRO 4.0 and have the extension WQ1 but are readable in EXCEL. Each spreadsheet is cited in the Example where it is used and bears that Example number as its label.

ENTROPY

This directory houses auxiliary text material related to entropy. This work is a multi-chaptered essay on entropy and its paradoxes — *ENTROPY: Reflections of a Classical Thermodynamicist*. Taken in its entirety the work presents the various views of entropy and the resulting paradoxes, and in the final chapter attempts conceptual closure. Because each of the intermediate chapters (Chaps. 3–8) deals with a single topic and is somewhat independent, selective reading should be possible. However, if the reader is unfamiliar with quantum statistical mechanics, it is recommended that Chap. 2 be read before proceeding to later chapters. Several of the chapters are based on articles previously published in *Chemical*

Engineering Education. The material is organized for easy reading with mathematical details consigned to appendices.

There are three sets of files in this directory: pdf files which are read with the Adobe Acrobat reader, doc files which are MS Word for Windows files, and cdr files which are WordPerfect 5.1 for DOS files. Each of these three sets of files could be copied to a floppy disk. The following topics are covered:

Chapter 1. *An Overview of Thermodynamics*: An attempt is made to capture the essence of thermodynamics by examining its anatomy and its methods and recognizing its strangeness.

Chapter 2. *The Microscopic Perspective*: The methods of quantum statistical mechanics are used to provide insight into the nature of temperature, energy, and entropy.

Chapter 3. *The Third Law:* The third law is examined from a historical and a microscopic perspective and absolute entropy is discussed.

Chapter 4. *Entropy and Information*: Information theory is described and the putative connection between it and entropy is discussed.

Chapter 5. *Maxwell's Demon*: The history of Maxwell's demon is surveyed and the association of the demon with the concepts of information and computing "entropy" are discussed.

Chapter 6. *Boltzmann's H-Theorem*: This historic work and the paradoxes it spawned are outlined and discussed.

Chapter 7. *The Gibbs Mixing Paradox*: This famous paradox is examined from several perspectives and a resolution is suggested.

Chapter 8. *The Mystique of Entropy*: The fascination with entropy exhibited by thinkers in fields other than thermodynamics is discussed.

Chapter 9. *A Search for Conceptual Closure*: An attempt is made to resolve the entropy enigma.

WASP

An excellent, user-friendly computerized steam table, WASP, is available as Shareware through Katmar Software and can be downloaded free of charge from their website: http://users.lia.net/katmar/

APPENDIX F PROPERTIES OF DETERMINANTS

This is an abbreviated list of determinant properties presented without proof. They are illustrated by, but not restricted to, 3rd order determinants. These operations will be useful in determining the rank of matrices found in Chap. 13.

1. The value of a 2nd order determinant can be found by

$$\begin{vmatrix} a_1 & b_1 \\ a_2 & b_2 \end{vmatrix} = a_1 b_2 - a_2 b_1$$

2. The value of a determinant is not changed by changing the columns into rows or rows into columns

$$\begin{vmatrix} a_1 & b_1 & c_1 \\ a_2 & b_2 & c_2 \\ a_3 & b_3 & c_3 \end{vmatrix} = \begin{vmatrix} a_1 & a_2 & a_3 \\ b_1 & b_2 & b_3 \\ c_1 & c_2 & c_3 \end{vmatrix}$$

3. The interchange of any two columns or any two rows changes only the sign and not the value of the determinant

$$\begin{vmatrix} a_1 & b_1 & c_1 \\ a_2 & b_2 & c_2 \\ a_3 & b_3 & c_3 \end{vmatrix} = -\begin{vmatrix} c_1 & a_1 & b_1 \\ c_2 & a_2 & b_2 \\ c_3 & a_3 & b_3 \end{vmatrix} = -\begin{vmatrix} a_2 & b_2 & c_2 \\ a_1 & b_1 & c_1 \\ a_3 & b_3 & c_3 \end{vmatrix}$$

4. If two rows or two columns of a determinant are identical, the determinant has the value of zero.

5. To multiply a determinant by any factor, k, multiply each element in one row or one column by the factor

$$k\begin{vmatrix} a_1 & b_1 & c_1 \\ a_2 & b_2 & c_2 \\ a_3 & b_3 & c_3 \end{vmatrix} = \begin{vmatrix} ka_1 & b_1 & c_1 \\ ka_2 & b_2 & c_2 \\ ka_3 & b_3 & c_3 \end{vmatrix} = \begin{vmatrix} ka_1 & kb_1 & kc_1 \\ a_2 & b_2 & c_2 \\ a_3 & b_3 & c_3 \end{vmatrix}$$

Reading this statement from right to left shows how a determinant can be factored.

6. If each element of a row or column can be expressed as the sum or difference of two terms, the determinant can be expressed as the sum or difference of two determinants.

$$\begin{vmatrix} a_{1\pm} & p; & b_1 & c_1 \\ a_{2\pm} & q; & b_2 & c_2 \\ a_{3\pm} & r; & b_3 & c_3 \end{vmatrix} = \begin{vmatrix} a_1 & b_1 & c_1 \\ a_2 & b_2 & c_2 \\ a_3 & b_3 & c_3 \end{vmatrix} \pm \begin{vmatrix} p & b_1 & c_1 \\ q & b_2 & c_2 \\ r & b_3 & c_3 \end{vmatrix}$$

7. If all but one of the elements of a row or column are zeros, the determinant can be reduced to the product of the one non-zero element and a determinant of order one less than the original determinant.

$$\begin{vmatrix} m & b_1 & c_1 \\ 0 & b_2 & c_2 \\ 0 & b_3 & c_3 \end{vmatrix} = m \begin{vmatrix} b_2 & c_2 \\ b_3 & c_3 \end{vmatrix}$$

8. If all the elements of a determinant on the left side of the diagonal are zeros, the determinant reduces to the product along the diagonal.

$$\begin{vmatrix} a_1 & b_1 & c_1 \\ 0 & b_2 & c_2 \\ 0 & 0 & c_3 \end{vmatrix} = a_1 b_2 c_3$$

INDEX

A

Absolute temperature, 9, 74-76, 97

Absolute temperature scale, 75-76

Absorption refrigeration, 657-61

Acentric factor, 55-56, 58, 123, 236-37

Acetic acid-water system, 716-19, 721

ACROBAT directory, CD-ROM, 746

Activity coefficients, 270-74, 276, 299-301, 337, 343, 382, 403, 419

    based on Henry's law, 386-90

    equations, 279-90

    experimental determination of, 274-78

    Henry's law, 278-79

    Margules equation, 279-84

    van Laar equation, 279-81, 284-86

    Wilson equation, 279-80, 286-90

Ad hoc models, 703-19

Adiabatic compressibility, 129

Adiabatic mixing, 155-57, 163-64

Adiabatic process, 4, 49-54

Adiabatic system, defined, 4-5

Ammonia:

    producing nitric acid from, 615

    temperature-entropy diagram for, 231

Antoine equation, 226-27, 237, 316

ANTOINE.WQ1 (XLS), 226, 749

Applied phase equilibrium, 309-74

    azeotropes, 327-32

    constant-pressure VLE data, 319-24

    consummate thermodynamic correlation of vapor-liquid equilibrium, 313-19

    estimates from fragmentary data, 352-59

    liquid-liquid equilibrium, 346-49

    multicomponent vapor-liquid equilibrium, 37-40

    phase behavior in partially miscible systems, 340-46

    recapitulation, 359-60

    ternary liquid-liquid equilibrium, 350-52

    thermodynamic consistency tests, 332-37

    total pressure data, 325-27

    See also Phase equilibrium

Applied thermochemistry, 176-89

ASTM distillation, 703

ATP/ADP energetics, 505

Azeotropes, 327-32

    activity coefficients from, 329-31

    effect of pressure on, 331-32

LICENSE AGREEMENT AND LIMITED WARRANTY

READ THE FOLLOWING TERMS AND CONDITIONS CAREFULLY BEFORE OPENING THIS SOFTWARE MEDIA PACKAGE. THIS LEGAL DOCUMENT IS AN AGREEMENT BETWEEN YOU AND PRENTICE-HALL, INC. (THE "COMPANY"). BY OPENING THIS SEALED SOFTWARE MEDIA PACKAGE, YOU ARE AGREEING TO BE BOUND BY THESE TERMS AND CONDITIONS. IF YOU DO NOT AGREE WITH THESE TERMS AND CONDITIONS, DO NOT OPEN THE SOFTWARE MEDIA PACKAGE. PROMPTLY RETURN THE UNOPENED SOFTWARE MEDIA PACKAGE AND ALL ACCOMPANYING ITEMS TO THE PLACE YOU OBTAINED THEM FOR A FULL REFUND OF ANY SUMS YOU HAVE PAID.

1.   **GRANT OF LICENSE:** In consideration of your payment of the license fee, which is part of the price you paid for this product, and your agreement to abide by the terms and conditions of this Agreement, the Company grants to you a nonexclusive right to use and display the copy of the enclosed software program (hereinafter the "SOFTWARE") on a single computer (i.e., with a single CPU) at a single location so long as you comply with the terms of this Agreement. The Company reserves all rights not expressly granted to you under this Agreement.

2.   **OWNERSHIP OF SOFTWARE:** You own only the magnetic or physical media (the enclosed SOFTWARE) on which the SOFTWARE is recorded or fixed, but the Company retains all the rights, title, and ownership to the SOFTWARE recorded on the original SOFTWARE copy(ies) and all subsequent copies of the SOFTWARE, regardless of the form or media on which the original or other copies may exist. This license is not a sale of the original SOFTWARE or any copy to you.

3.   **COPY RESTRICTIONS:** This SOFTWARE and the accompanying printed materials and user manual (the "Documentation") are the subject of copyright. You may not copy the Documentation or the SOFTWARE, except that you may make a single copy of the SOFTWARE for backup or archival purposes only. You may be held legally responsible for any copying or copyright infringement which is caused or encouraged by your failure to abide by the terms of this restriction.

4.   **USE RESTRICTIONS:** You may not network the SOFTWARE or otherwise use it on more than one computer or computer terminal at the same time. You may physically transfer the SOFTWARE from one computer to another provided that the SOFTWARE is used on only one computer at a time. You may not distribute copies of the SOFTWARE or Documentation to others. You may not reverse engineer, disassemble, decompile, modify, adapt, translate, or create derivative works based on the SOFTWARE or the Documentation without the prior written consent of the Company.

5.   **TRANSFER RESTRICTIONS:** The enclosed SOFTWARE is licensed only to you and may not be transferred to any one else without the prior written consent of the Company. Any unauthorized transfer of the SOFTWARE shall result in the immediate termination of this Agreement.

6.   **TERMINATION:** This license is effective until terminated. This license will terminate automatically without notice from the Company and become null and void if you fail to comply with any provisions or limitations of this license. Upon termination, you shall destroy the Documentation and all copies of the SOFTWARE. All provisions of this Agreement as to warranties, limitation of liability, remedies or damages, and our ownership rights shall survive termination.

7.   **MISCELLANEOUS:** This Agreement shall be construed in accordance with the laws of the United States of America and the State of New York and shall benefit the Company, its affiliates, and assignees.

8.   **LIMITED WARRANTY AND DISCLAIMER OF WARRANTY:** The Company warrants that the SOFTWARE, when properly used in accordance with the Documentation, will operate in substantial conformity with the description of the SOFTWARE set forth in the Documentation. The Company does not warrant that the SOFTWARE will meet your requirements or that the operation of the SOFTWARE will be uninterrupted or error-free. The Company warrants that the

media on which the SOFTWARE is delivered shall be free from defects in materials and workmanship under normal use for a period of thirty (30) days from the date of your purchase. Your only remedy and the Company's only obligation under these limited warranties is, at the Company's option, return of the warranted item for a refund of any amounts paid by you or replacement of the item. Any replacement of SOFTWARE or media under the warranties shall not extend the original warranty period. The limited warranty set forth above shall not apply to any SOFTWARE which the Company determines in good faith has been subject to misuse, neglect, improper installation, repair, alteration, or damage by you. EXCEPT FOR THE EXPRESSED WARRANTIES SET FORTH ABOVE, THE COMPANY DISCLAIMS ALL WARRANTIES, EXPRESS OR IMPLIED, INCLUDING WITHOUT LIMITATION, THE IMPLIED WARRANTIES OF MERCHANTABILITY AND FITNESS FOR A PARTICULAR PURPOSE. EXCEPT FOR THE EXPRESS WARRANTY SET FORTH ABOVE, THE COMPANY DOES NOT WARRANT, GUARANTEE, OR MAKE ANY REPRESENTATION REGARDING THE USE OR THE RESULTS OF THE USE OF THE SOFTWARE IN TERMS OF ITS CORRECTNESS, ACCURACY, RELIABILITY, CURRENTNESS, OR OTHERWISE.

IN NO EVENT, SHALL THE COMPANY OR ITS EMPLOYEES, AGENTS, SUPPLIERS, OR CONTRACTORS BE LIABLE FOR ANY INCIDENTAL, INDIRECT, SPECIAL, OR CONSEQUENTIAL DAMAGES ARISING OUT OF OR IN CONNECTION WITH THE LICENSE GRANTED UNDER THIS AGREEMENT, OR FOR LOSS OF USE, LOSS OF DATA, LOSS OF INCOME OR PROFIT, OR OTHER LOSSES, SUSTAINED AS A RESULT OF INJURY TO ANY PERSON, OR LOSS OF OR DAMAGE TO PROPERTY, OR CLAIMS OF THIRD PARTIES, EVEN IF THE COMPANY OR AN AUTHORIZED REPRESENTATIVE OF THE COMPANY HAS BEEN ADVISED OF THE POSSIBILITY OF SUCH DAMAGES. IN NO EVENT SHALL LIABILITY OF THE COMPANY FOR DAMAGES WITH RESPECT TO THE SOFTWARE EXCEED THE AMOUNTS ACTUALLY PAID BY YOU, IF ANY, FOR THE SOFTWARE.

SOME JURISDICTIONS DO NOT ALLOW THE LIMITATION OF IMPLIED WARRANTIES OR LIABILITY FOR INCIDENTAL, INDIRECT, SPECIAL, OR CONSEQUENTIAL DAMAGES, SO THE ABOVE LIMITATIONS MAY NOT ALWAYS APPLY. THE WARRANTIES IN THIS AGREEMENT GIVE YOU SPECIFIC LEGAL RIGHTS AND YOU MAY ALSO HAVE OTHER RIGHTS WHICH VARY IN ACCORDANCE WITH LOCAL LAW.

ACKNOWLEDGMENT

YOU ACKNOWLEDGE THAT YOU HAVE READ THIS AGREEMENT, UNDERSTAND IT, AND AGREE TO BE BOUND BY ITS TERMS AND CONDITIONS. YOU ALSO AGREE THAT THIS AGREEMENT IS THE COMPLETE AND EXCLUSIVE STATEMENT OF THE AGREEMENT BETWEEN YOU AND THE COMPANY AND SUPERSEDES ALL PROPOSALS OR PRIOR AGREEMENTS, ORAL, OR WRITTEN, AND ANY OTHER COMMUNICATIONS BETWEEN YOU AND THE COMPANY OR ANY REPRESENTATIVE OF THE COMPANY RELATING TO THE SUBJECT MATTER OF THIS AGREEMENT.

Should you have any questions concerning this Agreement or if you wish to contact the Company for any reason, please contact in writing at the address below.

Robin Short
Prentice Hall PTR
One Lake Street
Upper Saddle River, New Jersey 07458

ABOUT THE CD-ROM

Please see Appendix E, "Inventory of CD-ROM," for the contents of the CD and any system requirements for running the programs on it.

Technical Support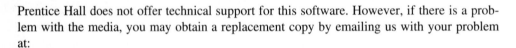

Prentice Hall does not offer technical support for this software. However, if there is a problem with the media, you may obtain a replacement copy by emailing us with your problem at:

discexchange@phptr.com